International Automobile Scientific Forum (IASF-2019)

Technologies and Components of Land Intelligent Transport Systems

IOP Conference Series: Materials Science and Engineering
Volume 819

Moscow, Russia
16 - 18 October 2019

ISBN: 978-1-7138-3160-0
ISSN: 1757-8981

Printed from e-media with permission by:

Curran Associates, Inc.
57 Morehouse Lane
Red Hook, NY 12571

Some format issues inherent in the e-media version may also appear in this print version.

Institute of Physics
Dirac House, Temple Back
Bristol BS1 6BE UK

Phone: 44 1 17 929 7481
Fax: 44 1 17 920 0979

techtracking@iop.org

Additional copies of this publication are available from:

Curran Associates, Inc.
57 Morehouse Lane
Red Hook, NY 12571 USA
Phone: 845-758-0400
Fax: 845-758-2633
Email: curran@proceedings.com
Web: www.proceedings.com

TABLE OF CONTENTS

IOP Conf. Series: Materials Science and Engineering **819** (2020) 011001 doi:10.1088/1757-899X/819/1/011001

IASF-2019 "Technologies and Components of Land Intelligent Transport Systems" was held on October 16-18, 2019. It was dedicated to the centenary of the State Research Center of the Russian Federation FSUE "NAMI".

Among the guests and participants of the Forum were representatives of the Ministry of Industry and Trade of the Russian Federation, the Ministry of Transport of the Russian Federation, representatives of automotive plants. The forum was also attended by representatives of leading universities, the Academy of Sciences of the Republic of Belarus, foreign universities and Russian and foreign companies. The total number of participants exceeded four hundred.

The forum was attended by regular partners: OICA, AVL List GMbH, Association of Russian Automakers, Association of Traffic Engineers (ATE), Publishing House AAE Press, National Technology Platform of the Russian Federation Environmentally Friendly Vehicles Green Car et al.

IOP Conf. Series: Materials Science and Engineering **819** (2020) 011002 doi:10.1088/1757-899X/819/1/011002

Peer review statement

All papers published in this volume of *IOP Conference Series: Materials Science and Engineering* have been peer reviewed through processes administered by the proceedings Editors. Reviews were conducted by expert referees to the professional and scientific standards expected of a proceedings journal published by IOP Publishing.

IOP Conf. Series: Materials Science and Engineering **819** (2020) 012001 doi:10.1088/1757-899X/819/1/012001

Mathematical model of curvilinear motion of an active road train with electromechanical transmission

O I Chudakov[1], V A Gorelov[1], V A Gartfelder[2], L S Sekletina[2]

[1] Bauman Moscow State Technical University (BMSTU), 105005, Moscow, Russian Federation, 2nd Baumanskaya St., Bldg. 5, Block 1
[2] I.N. Ulianov Chuvash State University (Chuv SU), 428015, Cheboksary, Russian Federation, Moskovskiy pr.,15

Abstract. The paper presents a mathematical model of curvilinear motion of a two-unit road train consisting of a four-axle tractor with independent wheel drive and a three-axle semitrailer on non-yielding bearing surface. The model considers variation in the normal load on fifth-wheel coupling (FWC) of the tractor during motion and cornering resistance moment caused by friction between the semitrailer and FWC bolster. Using the developed software, we can examine the effectiveness of algorithms of steering systems of the tractor and trailer unit allowing for various traction combinations of road train units.

1. Introduction

Trucking plays an important role in the overall transportation system in most countries worldwide. The efficiency of trucking can be improved by increasing performance and reducing the cost of transportation [1,2]. The performance of vehicles is determined by the average speed and load capacity. Road trains have limited opportunities to increase the average speed, so the most promising way to enhance performance is to increase the load capacity that can be achieved, particularly, by using road trains [3,4].

The use of road trains offers great opportunities in industrial and agricultural areas with poor road network, i.e. driving off-road, which is especially relevant for our country. However, the use of road trains is often limited due to inadequate haulage capacity, when it comes to the transportation of heavy machinery, equipment and weapons in tough country road conditions [5]–[9]. Development of road trains with a sufficiently high level of traction dynamics is thus a pressing scientific and practical issue, which may be effectively solved by introducing an active drive of link element wheels [10]–[14].

2. Mathematical Model of Road Train Motion

The paper presents a mathematical model of planar motion of a two-unit road train. The subject of research is a road train with 120,000 kg gross vehicle weight consisting of a four-axle tractor with electromechanical transmission (EMT) and a three-axle semitrailer. Figure 1 shows the static distribution of vertical loads.

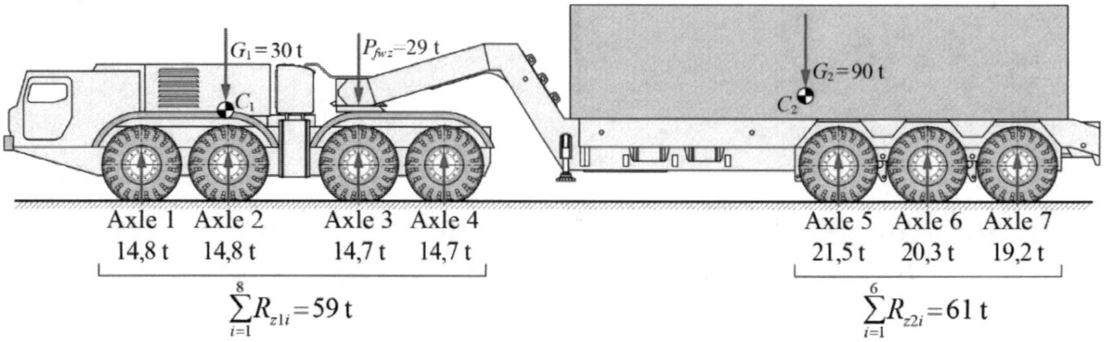

Figure 1. Static distribution of vertical loads along axes of fully loaded road train.

In the base case (road train with a passive semitrailer), each wheel drive is implemented using 60 kW traction electric motor (TEM). So, the total power of all TEMs of the tractor is 480 kW. This is the maximum power transmitted from the powertrain to TEMs. The power of active road train's TEMs has been selected so as to have equal power-to-weight ratio in all cases.

Four power distribution cases are considered: in the first one, all power is transmitted to tractor wheels (base case), in the last one, 75% of the total power is transmitted to semitrailer wheels. Figure 2 shows power distribution among road train wheels.

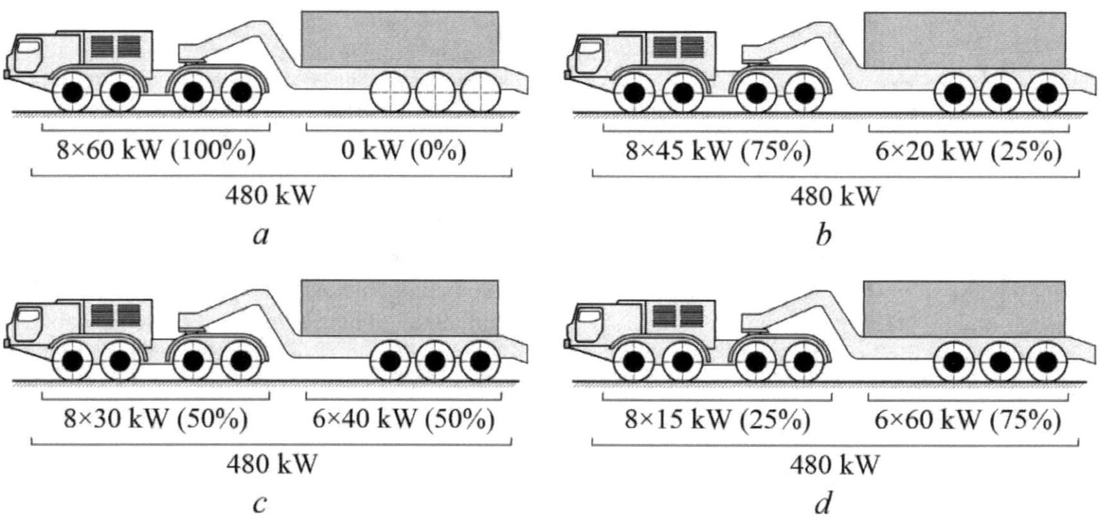

Figure 2. Power distribution (%) among road train wheels: *a* – 100/0; *b* – 75/25; *c* – 50/50; *d* – 25/75.

The mathematical model of road train planar motion described in detail in [15] may be used to study the curvilinear motion of multi-unit vehicles with active trailer units. Longitudinal and transverse angles of relative motion of moving road train units along the horizontal base are not large. Therefore, the movement of each unit as a solid body is considered in the horizontal plane taking into account the lead angle in the direction of movement on flat non-yielding surface and consists of translational and rotational motion of the center of mass (Figure 3).

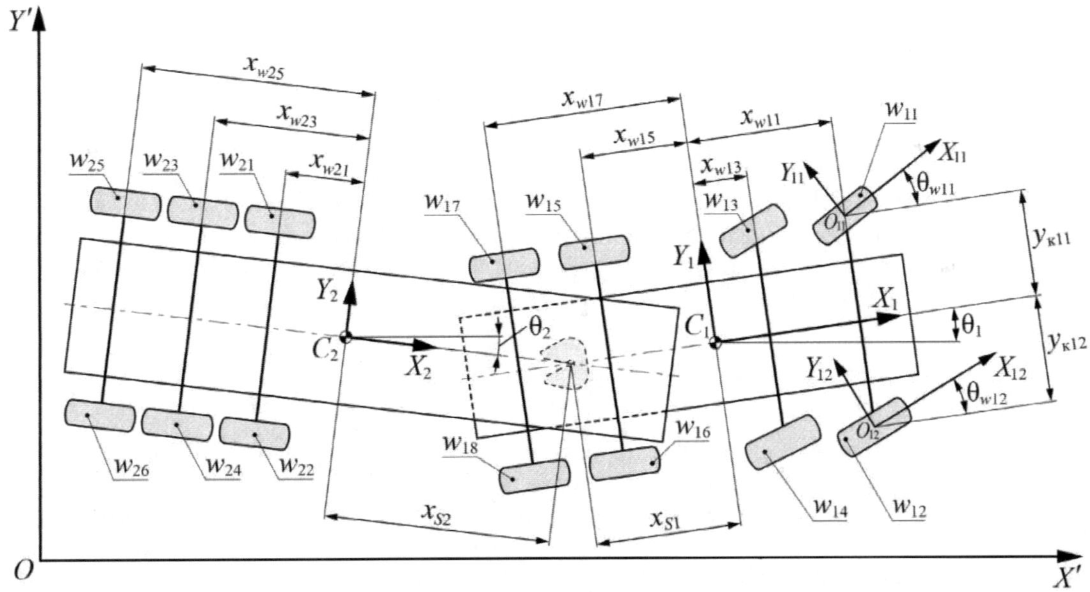

Figure 3. Calculation model of a two-unit road train.

The following coordinate systems are introduced to describe this road train motion case: fixed coordinates $X'OY'$, the origin (point O) coincides with the start point of simulated route; moving coordinates $X_1C_1Y_1$ and $X_2C_2Y_2$ referenced to centers of mass (CM) of the tractor and semitrailer, respectively; coordinates $X_{ji}O_{ji}Y_{ji}$ referenced to the i-th wheel of the j-th unit. Point O_{ji} coincides with the projection of the i-th wheel center on the bearing surface, axis $O_{ji}X_{ji}$ is directed along the projection of the wheel longitudinal plane on the bearing surface, $O_{ji}Y_{ji}$ – along the projection of wheel spindle. The dynamics of a road train unit is described by the following set of equations:

$$\begin{cases} a_{xj} = \dfrac{\mathrm{d}V_{xj}}{\mathrm{d}t} - \omega_{zj} \cdot V_{yj} = \dfrac{1}{m_j} \cdot \left(\sum\limits_{i=1} R_{xji} - m_j \cdot g \cdot \sin\alpha - P_{wx} + F_{xj} \right); \\ a_{yj} = \dfrac{\mathrm{d}V_{yj}}{\mathrm{d}t} + \omega_{zj} \cdot V_{xj} = \dfrac{1}{m_j} \cdot \left(\sum\limits_{i=1} R_{yji} - P_{wy} + F_{yj} \right); \\ J_{zj} \cdot \dfrac{\mathrm{d}\omega_{zj}}{\mathrm{d}t} = \sum\limits_{i=1} M_{pji} - \sum\limits_{i=1} R_{xji} \cdot y_{wji} + \sum\limits_{i=1} R_{yji} \cdot x_{wji} + M_j; \\ V_{x'j} = \dfrac{\mathrm{d}x'_j}{\mathrm{d}t} = V_{xj} \cdot \cos\theta_j - V_{yj} \cdot \sin\theta_j; \\ V_{y'j} = \dfrac{\mathrm{d}y'_j}{\mathrm{d}t} = V_{xj} \cdot \sin\theta_j + V_{yj} \cdot \cos\theta_j; \\ \omega_{zj} = \dfrac{\mathrm{d}\theta_j}{\mathrm{d}t}, \end{cases} \tag{1}$$

where j is the road train unit number; m_j is the mass of the j-th unit (kg); J_{zj} is the inertia moment of the j-th unit relative to C_jZ axis (kg/m^2); V_x and V_y are longitudinal and transverse components of the velocity vector of the unit CM; a_x and a_y are longitudinal and transverse components of the acceleration vector of unit CM (absolute derivative of the unit CM velocity vector); ω_{zj} is the angular velocity vector of the j-th unit (s^{-1}); θ_j is the steering angle of the j-th unit relative to OX' axis (rad); x', y' are fixed coordinates of the unit center of mass (m); R_{xji} and R_{yji} are longitudinal and transverse components of soil interaction force acting on the i-th wheel of the j-th unit (N); α is the bearing surface inclination angle (rad); P_{wx} and P_{wy} are longitudinal and transverse components of air resistance force (N); M_{pji} is the cornering resistance moment of the i-th wheel of the j-th unit; F_{xj} is the hitch force acting along C_jX_j axis (N); F_{yj} is the hitch force acting along C_jY_j axis (N); M_j is the moment applied to the j-th unit from forces acting in the hitch (Nm).

The last three equations in the set (1) establish the connection between moving and fixed coordinates. An individual set of equations is created for each unit according to (1). In this case, the traction combination of units may vary, i.e. trailer unit may be either active (with torque applied to wheels) and passive. To implement various traction combinations of units, the mathematical model of road train motion is complemented by equations describing the dynamics of a particular type of transmission.

In general, the dynamics of EMT with individually driven wheels can be described by the following set of equations:

$$\begin{cases} (J_{en} + J_G) \cdot \dot{\omega}_{en} = M_{en} - M_{CE}; \\ (J_{wji} + J_{Tji} \cdot U_{wg}{}^2) \cdot \dot{\omega}_{wji} = M_{wji} - M_{Cji}, \end{cases} \tag{2}$$

wherein J_{en}, J_G, J_T are the inertia moments of ICE crankshaft, generator shaft, shaft of the i-th TEM (kg/m^2); $\dot{\omega}_{en}$ is the angular acceleration of ICE shaft (s^{-2}); M_{en} is ICE torque (Nm); M_{CE} is the resistance moment applied to ICE shaft (Nm). U_{wg} is the wheel gear ratio; $\dot{\omega}_{wji}$ is the angular acceleration of the i-th wheel of the j-th unit (s^{-2}); M_{wji} is the torque applied to the i-th wheel of the j-th unit (Nm); M_{Cji} is the motion resistance moment applied to the i-th wheel of the j-th unit (Nm).

The mathematical model considers variation in the normal load on fifth-wheel coupling (FWC). Variation in FWC normal load significantly affects traction and dynamic properties of the road train, because the weight on its driving axles depends on it. The calculation is based on the following principle: semitrailer is treated as a seven-bearing system, i.e. the seventh response from FWC (R_{fwz}) is factored in addition to six normal responses under semitrailer wheels (R_{z2i}):

$$\begin{cases} R_{fwz} - R_{z21} + (x_{S2} - x_{w21}) \cdot k \cdot \mathrm{tg}\varphi + (y_{S2} - y_{w21}) \cdot k \cdot \mathrm{tg}\psi = 0; \\ R_{z21} - R_{z22} + (x_{w21} - x_{w22}) \cdot k \cdot \mathrm{tg}\varphi + (y_{w21} - y_{w22}) \cdot k \cdot \mathrm{tg}\psi = 0; \\ \ldots \\ R_{z25} - R_{z26} + (x_{w25} - x_{w26}) \cdot k \cdot \mathrm{tg}\varphi + (y_{w25} - y_{w26}) \cdot k \cdot \mathrm{tg}\psi = 0; \\ \sum_{i=1}^{6} R_{z2i} + R_{fwz} = m_2 \cdot g \cdot \cos\alpha; \\ \sum_{i=1}^{6} R_{z2i} \cdot x_{w2i} + R_{fwz} \cdot x_{S2} = -m_2 \cdot h_{z2} \cdot (a_{x2} + g \cdot \sin\alpha) - \sum_{i=1}^{6} M_{f2i} - P_{fwx} \cdot h_{fw}; \\ \sum_{i=1}^{6} R_{z2i} \cdot y_{w2i} + R_{fwz} \cdot y_{S2} = -m_2 \cdot h_{z2} \cdot a_{y2} - P_{fwy} \cdot h_{fw}, \end{cases} \tag{3}$$

where $i = 1 \ldots 6$ is the semitrailer wheel number; x_{S2}, y_{S2} are FWC coordinates relative to semitrailer CM (m); x_{w2i} , y_{w2i} are coordinates of the i-th wheel relative to semitrailer CM (m); k is the suspension stiffness coefficient (Nm); φ, ψ are trim and roll angles (deg); m_2 is the semitrailer mass (kg); h_{z2} is the height of semitrailer CM; a_{x2}, a_{y2} is the acceleration of semitrailer CM (m/s^2); M_{f2i} is the rolling resistance moment of the i-th wheel (Nm); P_{fwx}, P_{fwy} are longitudinal components of FWC force (N); h_{fw} is the height of FWC (m).

The normal load on FWC affects the cornering resistance moment caused by friction between the semitrailer and FWC bolster. The friction model described in [16] is used to calculate this resistance moment.

3. Research results

The mathematical model of the active road train with EMT is implemented in Matlab/Simulink dynamic mathematical simulation environment. The effect of trailer wheel activation on road train dynamics was assessed using the results of simulation of going into corner and subsequent turning with a fixed position of the steering wheel and constant speed. Figure 4 shows simulation results for two power distribution cases (100/0 and 25/75).

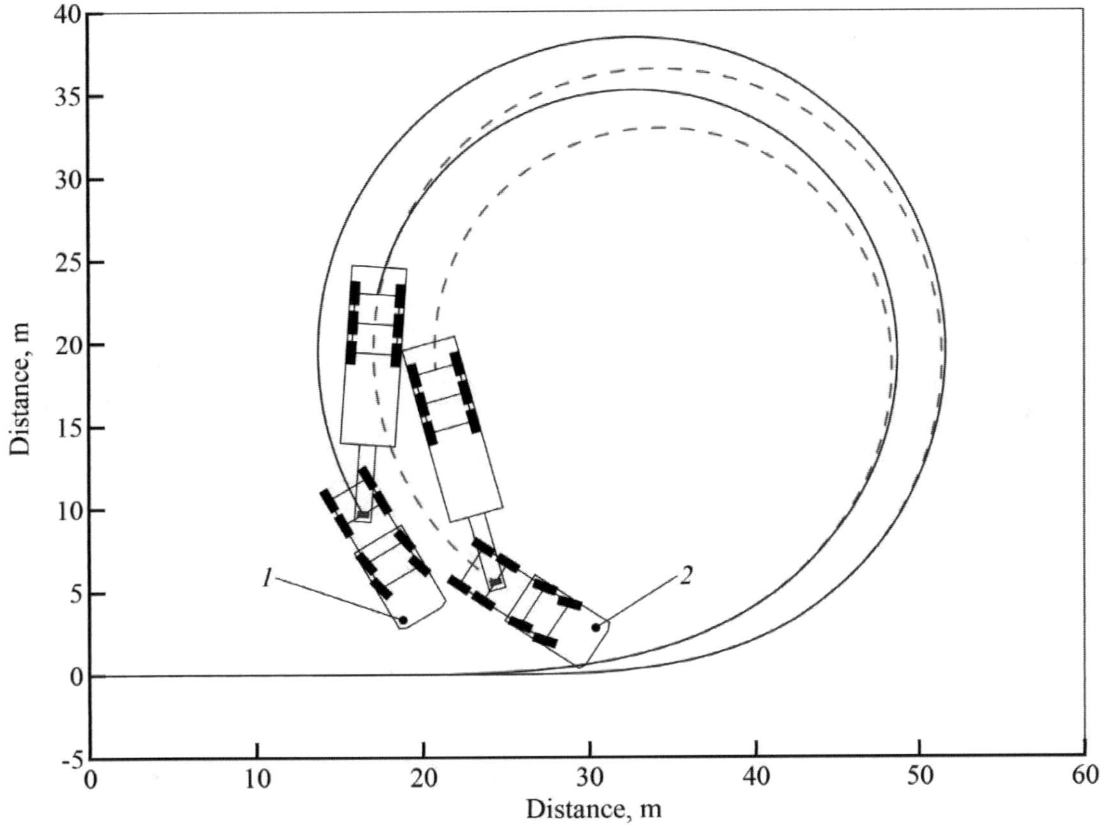

Figure 4. Road train trajectories for two cases of power distribution between the units: *1* – 100/0; *2* – 25/75.

As can be seen from Figure 4, road train trajectories differ significantly. The turning radius of the tractor with passive semitrailer is 19.18 m, but it is 17.05 m with active semitrailer. The main reason for differences is the variation of wheel slip angles when torque is applied to semitrailer wheels. Figure 5 shows the comparison of wheel slip angles of active and passive semitrailers (for wheels 1, 3 and 6). Figure 6 shows the variation of tractor trajectory curvature for four cases of power distribution between the tractor and semitrailer.

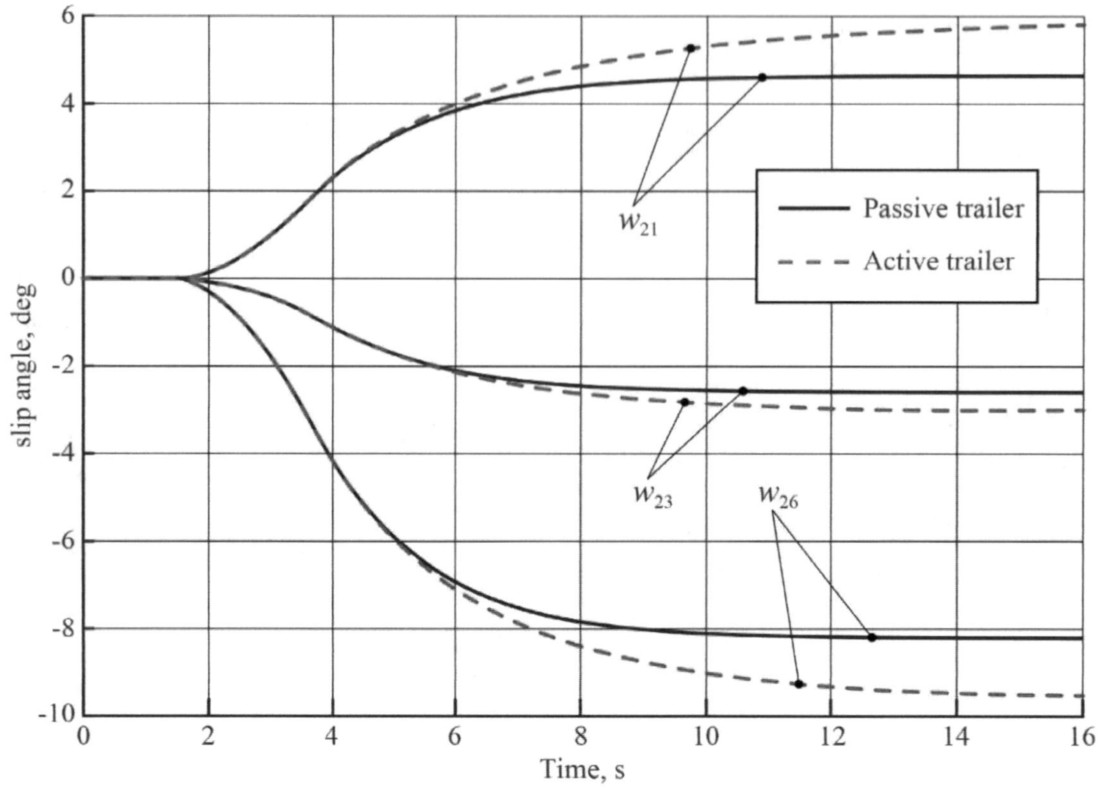

Figure 5. Variation of wheel slip angles of active and passive semitrailers.

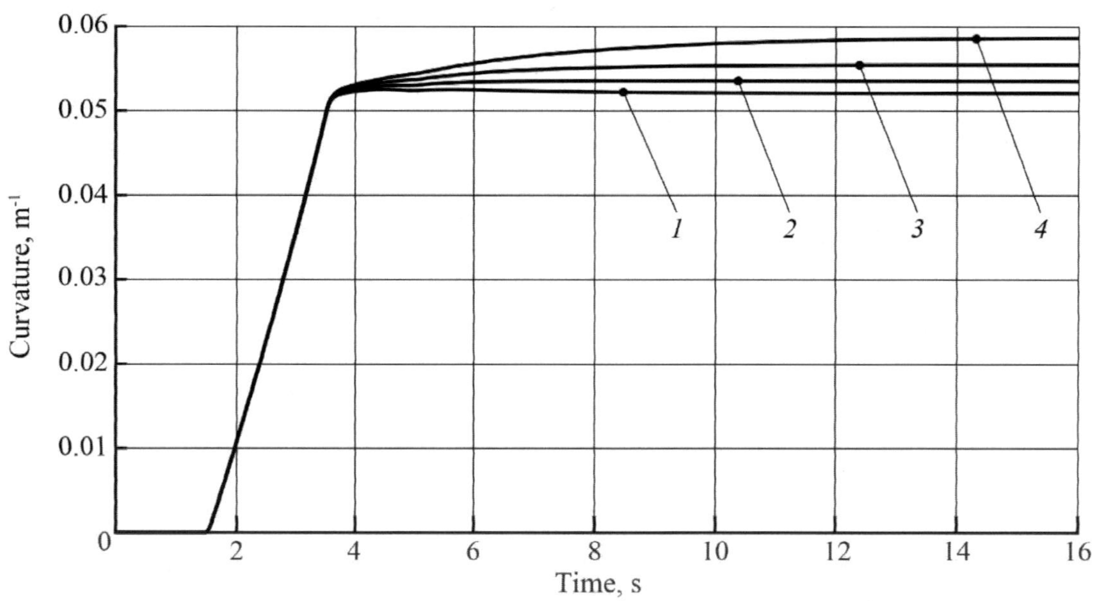

Figure 6. Variation of tractor trajectory curvature for various power distribution cases: 1 – 100/0; 2 – 75/25; 3 – 50/50; 4 – 25/75.

4. Conclusions

Based on computational experiments, we can conclude that the semitrailer wheel activation has a significant effect on the road train trajectory. The mathematical model can be used in designing automatic active drive control systems and solving a scientific and real-world problem of power distribution optimization using various electronic control systems, including those based on the hitch force factor analysis.

References

[1] Skotnikov G I, Jileykin M M and Komissarov A I 2018 *IOP Conference Series: Materials Science and Engineering* (IOP Publishing) vol **315** 012027
[2] Keller A and Aliukov S 2015 Analysis of possible ways of power distribution in an all-wheel drive vehicle *Lecture Notes in Engineering and Computer Science* **2218** pp 1154-1158.
[3] Diakov A S and Kotiev G O 2015 *MATEC Web of Conferences* (EDP Sciences) vol **224** 02096
[4] Keller A and Aliukov S 2015 Efficient Power Distribution in an All-wheel Drive Truck *Lecture Notes in Engineering and Computer Science: Proceedings of The World Congress on Engineering* pp 1201-1206
[5] Kotiev G O, Padalkin B V, Kartashov A B and Diakov A S 2017 *Journal of Engineering and Applied Sciences* (ARPN) vol **12** pp 1064–1071
[6] Keller A and Aliukov S 2016 Efficient Power Distribution in an All-Wheel Ground Vehicles *SAE Technical Paper* 2016-01-1105
[7] Bazhenov E E, Buynachev S K and Chernyshev D O 2019 *IOP Conference Series: Journal of Physics* (IOP Publishing) vol **1177** 012001
[8] Keller A , Shelepov A and Istomin D 2016 Estimation of Screw-ball Differential on Vehicle Dynamics *Procedia Engineering* **150** pp 1131-1136.
[9] Belousov B, Ksenevich T, Vantsevich V and Naumov S 2014 *SAE Technical Paper* (SAE International) vol **1** 2014-01-2288
[10] Keller A. and Aliukov S 2015 Effectiveness of Methods of Power Distribution in Transmissions of All-Wheel-Drive Trucks *SAE Technical Paper* 2015-01-2732
[11] Vol'skaya N S, Zhileykin M M and Zakharov A Y 2018 *IOP Conference Series: Materials Science and Engineering* (IOP Publishing) vol **315** 012028
[12] Boonstra S and Rieck F 2018 *World Electric Vehicle Journal* (WEVA) vol **9**
[13] Gorelov V A, Komissarov A I 2016 *Procedia Engineering* (Elsevier) vol **150** pp 1322–1328
[14] Usinin U, Gladyshev S, Grigoryev M, Shishkov A, Belousov E, Zhuravlev A M, Bychkov A and Sychev D 2015 *SAE Technical Paper* (SAE International) 2015-01-2823
[15] Gorelov V A and Tropin S L 2011 Matematicheskaya model' krivolineynogo dvizheniya avtopoyezda po nedeformiruyemomu opornomu osnovaniyu *AAI Journal* vol **2** pp 18–22
[16] Chichinadze A V 1995 *Osnovy tribologii (trenie, iznos, smazka)* (Nauka i tekhnika) p. 778

IOP Conf. Series: Materials Science and Engineering **819** (2020) 012002 doi:10.1088/1757-899X/819/1/012002

Creation of autonomous groups of combine harvesters and tractors for agriculture based on the Mivar decision-making systems "ROBO!RAZUM"

O O Varlamov[1,2], D V Aladin[1,2], L E Adamova[2,3], D A Chuvikov[2] and D V Saraev[1]

[1] Bauman Moscow State Technical University, 2-ya Baumanskaya Street 5/1, Moscow 105005, Russia
[2] MIVAR, 72 Oktyabrskaya str., Moscow 127521, Russia
[3] Russian New University, Radio str. 22, Moscow 105005, Russia

E-mail: ovar@yandex.ru

Abstract. The Mivar decision-making system (MDMS) "ROBO!RAZUM" can be used to control autonomous combine harvesters and tractors, which can additionally perform a variety of actions in the process of moving through the field: working in group interaction mode, performing tasks in different weather conditions and even with a faulty or disabled technical vision using the internal map of the area and information from internal inertial sensors or from the GLONASS system. The theoretical basis for the application of MDMS "ROBO!RAZUM" is the result of the practical implementation of MIPRA program, which dramatically reduced the requirements for equipment and time to solve the problems of STRIPS-planning robot actions.

1. Introduction

In the area of artificial intelligence, there is currently a critical mass of technologies enabling the creation of a decision-making system (DMS) for autonomous groups of combine harvesters and tractors for agriculture and other complex applications, from homeland security to mining. It is important to note that autonomous robots require logical intelligence, and the intelligence module of the DMS can be implemented with a Mivar Constructor of Expert Systems Wi!Mi "Razumator" [1-2]. This system allows constructing algorithms for planning the actions of individual robots and even heterogenous multi-level robotic systems [3]. The theoretical basis of mivar-based DMS was described in many works, for example, for autonomous vehicles [4-5], special superfast summers [6], automatic control systems [7], traffic accident analysis [8], and decision support systems for medicine [9]. It is noteworthy that the DMS for agriculture are hybrid intelligent information systems from the standpoint of system architecture [10]. These systems, aside from mivar technology-based modules, may include solutions based on meta graphs [11-12], cognitive computer graphics [13-14], neural network-based algorithms [15], and intelligent analysis methods [16].

At the conference IASF-2018, the architecture of a mivar system for enforcing traffic regulations, as well as other mivar-based products for transport [5, 8], was presented in [17]. In [18], the possibility of using mivar expert systems for planning the actions of robotic system [18] was conducted. In 2019, the results of successfully solving this problem were published in the form of a software system MIPRA [19]. We note that MIPRA was developed to solve conventional STRIPS-planning problems in a modified conventional domain called “Blocks World” [20].

IOP Conf. Series: Materials Science and Engineering **819** (2020) 012002 doi:10.1088/1757-899X/819/1/012002

Based on mivar DMS, a software framework "ROBO!RAZUM" was developed for the area of robotics. This framework can be used for controlling autonomous combine harvesters and tractors, which can perform various actions in the process of moving on the field. They also can work in group interaction mode [21]. They can operate in different weather conditions and even with a faulty or switched off machine vision system using only the internal map of the area and the information from the onboard inertial sensors or the GLONASS system. In 2016, the mivar decision-making system (MDMS) "ROBO!RAZUM" [22] was the first in the world to control a group of autonomous robots. In January 2017, we conducted field tests of the automobile control system by creating special knowledge bases and integrating "ROBO!RAZUM" in the control loop of the automobile in collaboration with the research team of Shadrin S. S. [4].

2. Analysis of the specifics of operating autonomous groups of combined harvesters and tractors

It is crucial to determine the scope and tasks for the DMS for automobile groups of combine harvesters and tractors based on the information from various customers and the analysis of the subject domain. Robotic systems based on them are often used for solving tasks of damage control and recovery after accidents and natural disasters. Such situations impose greater requirements on the diversity of situations and the autonomy of the robotic system without a communication channel with an operator. From the standpoint of mivar technologies and the decision-making speed of more than 5 million rules/sec, such toughening of requirement is not detrimental. Based on our estimates, the number of production rules in the driving regulations for driverless vehicles will not be greater than 10000. Thus, the capabilities of the MDMS "ROBO!RAZUM" guarantees the solutions of the most complex problems for automobiles and the extension of these problems for combined harvesters for work in different conditions, including accident damage control and recovery.

Thus, the specifics of the subject domain of a "tractor" include the following:

1) Tractors (robotic systems) have to negotiate complex terrain, where possible routes ("roads"), as well as obstacles and the profile of the terrain, may change dynamically.

2) robotic systems are complex multi-level systems, where a part of the robotic system or an individual mini-robotic system can perform some tasks autonomously. For example, a UAV can take off the robotic system to observe the territory for creating maps and searching possible "roads" on it. Another example of this would be special sensors determining the soil composition for "ROBO!RAZUM" to give instructions to a combined harvester to dynamically change the fertilizer composition and control the movement of the robotic system at the same time.

3) The machine vision system [23] can be external subsystems of the robotic system, which in some cases will have to move on the "field" in poor visibility conditions or with the onboard machine vision system faulty. In this case, the robot can procure the information for the DMS from external systems or using a map previously loaded in the DMS (the map is dynamically updating based on the data from the external machine vision systems) and the data from the onboard inertial sensors about the robot's movement on this "virtual map".

Besides, "combines and tractors" can work together and help each other solve complex tasks, i.e., group control of the robotic system is possible. For example, two tractors move together across rough terrain. If the first tractor gets in a "mire" and starts drowning, the second tractor can pull it out, and they continue performing their task. Such capabilities of the MDMS give birth to new requirements to robotic systems ("tractors"), which should be capable of physical and information interaction.

We note that if there are no radio communication channels, the robotic systems can exchange information via another physical channel, for example, via the optical channel. For example, one robotic system displays signals with special devices (similar to marine semaphores of the pre-radio era: flags and projectors). Besides, robotic systems can approach each other and exchange information via a special "cable." If a robotic system goes out of order, the other robotic system can approach to out of order robot, connect to it via the cable, diagnose the faulty robot and, potentially, repair it. Other options for interaction are possible.

3. Interaction between a robotic system and various obstacles

It is evident that MDMS "ROBO!RAZUM" can be given the tasks on the level of full-fledged "Mars rovers," when only the autonomous control of the group of robotic systems is possible. We note that in general case, the knowledge bases in the DMS can be complemented from the new experience of the robotic systems, similar to the way animals and even humans learn. Such an "experience-based" self-training system can be implemented as a separate module or a subsystem because mivar technologies generally allow the robots to make new rules and logical relations. After that, new rules can be added to the common knowledge base and disseminated among other robots of the groups via a special procedure.

As an example, we consider ways the robotic systems may interact with different obstacles. We have solved the task of separating the obstacles into two types: passable and impassable. For example, bushes and high grass are passable obstacles for a tractor, while trees, posts, and houses are impassable obstacles which have to be driven around. We stress that we will consider special tractors, which can move and raise special objects during accident damage control and recovery. This is a special case of regular tractors having special equipment to perform tasks on cleaning up the accident zone or delivering cargo through it.

We introduce the third type of obstacle, which we shall refer to as a "movable obstacle." We should note right away that, in some cases, the options of "moving the obstacle" may vary. The obstacle can be pushed "up" the road, hooked to the robot and pulled "back" or just pushed aside. In some cases, in shafts and urban debris, special equipment can be used - tractors with buckets, excavators, and loaders. In this case, another type of obstacle will emerge. This type of obstacle can be leveled to the ground. Besides, there may be an obstacle that can be lifted with a mobile crane and taken elsewhere. As we see, there is a large and "open" set of different types of obstacles. However, all these obstacles can be taken into account in the knowledge base of the MDMS, and the information about then will be taken into account in the operation of robotic systems and their multi-level and heterogeneous groups.

Mivar DMS allows "special tractors" to close in on the obstacle and try pushing it off the road. After this attempt, there are different options. The options strongly depend on the capabilities of the machine vision system. However, we assume that the information about the territory had been collected, and an electronic map of it had been created before the accident even in the worst-case scenario. This map has many "layers" containing information for different profiles. This map is loaded in the memory of the DMS of the robotic system, and it is used to determine the places where the robotic system can go. This can be done by explicit definition of "roads," while it is also possible to make a sort of "chessboard," where passable "square" and obstacles on them will be indicated. In 2016, this approach was implemented for the task of controlling mini-tractors and children's robot toys. The issues of creating such a map require further research.

Let us consider a group of robotic systems having a common map, and they perform a task on moving a certain cargo through an accident zone. The terrain of the zone had changed due to the accident, and new obstacles, which had not been shown on the map, had emerged. If it is possible, the group of robotic systems uses their resources and merges the data from the machine vision systems to update the map. This update can indicate free roads or detect various obstacles. The obstacles, for example, can belong to three types: 1) impassable, 2) passable, 3) movable. Generally, obstacles are of the third type, which can be removed by the robotic systems themselves, therefore, we may call these obstacles "removable." Based on the information collected, a mivar bipartite graph of possible moves is created. This graph explicitly stores "nodes," where robotic systems may be located, and "edges" the robotic systems can use to move from one node to another. Such a graph corresponds to a map received from the machine vision system, and it can be constantly updated upon getting new information in real time. RAZUMATOR can search for a route on this map with a speed of more than 5 million rules/sec [1]. In other words, mivar rule inference is used for finding the shortest path on the graph of possible moves of the robotic systems.

Then the group of robotic systems finds the root using the mivar DMS, identifies obstacles and performs actions to move cargo using the rules from its knowledge base if the route is free. If there are obstacles belonging to the second (passable) and third (impassable) type, the MDMS makes a plan for overcoming these obstacles, and the robotic systems move with the cargo. At any moment of time, the plan can be altered with the speed of more than 5 million rules/sec to take into account new obstacles or capabilities of the robotic system. However, this largely depends on the capabilities of the machine vision system.

4. Robotic system operation with a faulty machine vision system

Let us consider the possibilities of operating a robotic system in conditions when the machine vision system is out of order or cannot function because of weather conditions. So, initially, each robotic system has a map of the area of operations. We assume that the robotic system has a system of inertial sensors that can accurately determine all movements of the robotic system and its position in space (slopes, turns, etc.). The easiest way to operate is based on constructing a route based on the pre-loaded map without any machine vision, and the robotic system moves along this route based on the information from the inertial sensors. This option is used by driverless vehicles for moving along a preset trajectory. If using the first robotic system uses a method of trial and error and is stoped by an obstacle on an empty road, the obstacle bypass procedure is launched. For example, the robotic system may first try to drive around the obstacle or move it off the road. For example, an automobile may give its first spot to a tractor, which can try moving the obstacle aside based on the previously loaded map and free the road for other robotic systems. Depending on the composition of the robotic system, different options for removing or overcoming such obstacles are possible. If the obstacle cannot be removed, the next stage is initiated. But, depending on the situation, a simultaneous search of different ways of overcoming the obstacle and finding the path for the group of robotic systems is possible.

Another stage is the search for the alternative route, for example, by finding the other route on the map of the area. If the alternative routes exist, robotic systems can be sent there. A simultaneous check of all or some of the routes is also possible. The check can be done using the channel of communication between robotic systems or without it when a robotic system returns to the initial point and reports where the route is passable or not. We should stress that the capabilities of mivar technologies allow implementing this with current technology right now. Research has been carried out, and the task of planning the operation of robots (MIPRA) has been solved, so now we can pass over to creating knowledge bases and training of the mivar DMS (this will be research work followed by prototyping and field testing).

5. Conclusions

Artificial intelligence has been created, and mivar decision-making systems "ROBO!RAZUM" can control autonomous combined harvesters and tractors. These devices are complex heterogeneous and multi-level robotic systems. "ROBO!RAZUM" can control them even with external machine vision systems using just an onboard map of possible moves with dynamically changing obstacles. Besides, robotic systems can be controlled to perform complex tasks on the move [24].

6. References

[1] Varlamov O O 2018 Wi!Mi Expert System Shell as the Novel Tool for Building Knowledge-Based Systems with Linear Computational Complexity *International Review of Automatic Control* **11**(6) 314-325

[2] Chuvikov D A and Nazarov K V 2016 Designing algorithms for solving physics problems on the basis of mivar approach *International Journal of Advanced Studies* **6**(3) 31-50

[3] Panferov A A, Zhdanovich E A, Yufimychev K A and Chuvikov D A 2016 Designing algorithms for service robots on the basis of mivar approach *International Journal of Advanced Studies* **6**(3) 72-86

[4] Shadrin S S, Varlamov O O and Ivanov A M 2017 Experimental autonomous road vehicle with logical artificial intelligence *Journal of Advanced Transportation* 2492765

[5] Varlamov O O, Chuvikov D A, Aladin D V, Adamova L E and Osipov V G 2019 Logical artificial intelligence Mivar technologies for autonomous road vehicles *IOP Conference Series: Materials Science and Engineering* **534**(1) 012015
[6] Varlamov O O 2003 Exhaustive elementary-incremental summing up of numbers with linear computational complexity *Avtomatizatsiya i Sovremennye Tekhnologii* **1** 34-41
[7] Ostroukh A, Surkova N, Varlamov O, Chernenky V and Baldin A 2018 Automated process control system of mobile crushing and screening plant *Journal of Applied Engineering Science* **16**(3) 343-348
[8] Chuvikov D A, Varlamov O O, Aladin D V, Chernenkiy V M and Baldin A V 2019 Mivar models of reconstruction and expertise of emergency events of road accidents *IOP Conference Series: Materials Science and Engineering* **534**(1) 012007
[9] Varlamov O O, Chuvikov D A, Adamova L E, Petrov M A, Zabolotskaya I K and Zhilina T N 2019 Logical, Philosophical and Ethical Aspects of AI in Medicine *International Journal of Machine Learning and Computing (IJMLC)* **9**(6) 868-873
[10] Chernenkiy V, Gapanyuk Y, Terekhov V, Revunkov G and Kaganov Y 2018 The hybrid intelligent information system approach as the basis for cognitive architecture *Procedia Computer Science* **145** 143-152
[11] Chernenkiy V M, Gapanyuk Y E, Revunkov G I, Kaganov Y T, Fedorenko Y S and Minakova S V 2017 Using metagraph approach for complex domains description *CEUR Workshop Proceedings* **2022** 342-349
[12] Chernenkiy V, Gapanyuk Y, Revunkov G, Kaganov Y and Fedorenko Y 2019 Metagraph Approach as a Data Model for Cognitive Architecture *Advances in Intelligent Systems and Computing* **848** 50-55
[13] Terekhov V I, Chernenky I M, Buklin S V and Yakubov A R 2019 Cognitive Visualization in Management Decision Support Problems *Optical Memory and Neural Networks (Information Optics)* **28**(1) 27-35
[14] Chuvikov D A, Kazakova N A, Varlamov O O and Golovizin A V 2014 3D modeling and 3D objects creation technology analysis for various intelligent systems *International Journal of Advanced Studies* **4**(4) 16-22
[15] Burdakov A V, Ukharov A O, Myalkin M P and Terekhov V I 2019 Forecasting of influenza-like illness incidence in amur region with neural networks *Studies in Computational Intelligence* **799** 307-314
[16] Gapanyuk Y, Latkin I, Chernobrovkin S, Leontiev A, Ozhegov G, Opryshko A and Myalkin M 2017 Architecture and implementation of an intelligent news analysis system *CEUR Workshop Proceedings* **1975** 41-55
[17] Aladin D V, Varlamov O O, Chuvikov D A, Chernenkiy V M, Smelkova E A and Baldin A V 2019 Logic-based artificial intelligence in systems for monitoring the enforcing traffic regulations *IOP Conference Series: Materials Science and Engineering* **534**(1) 012025
[18] Varlamov O O and Aladin D V 2018 On the application of mivar networks for intellectual planning of the behavior of robots in the space of conditions *News of Kabardino-Balkarian Scientific Center of the Russian Academy of Sciences* **6-2**(86) 75-82
[19] Varlamov O O and Aladin D V 2019 Successful application of mivar expert systems for MIPRA –solving action planning problems for robotic systems in real time *Radio industry (Russia)* **29**(3) 15-25
[20] Fikes R and Nilsson N 1971 STRIPS: A new approach to the application of theorem proving to problem solving *Artificial Intelligence* **2** 189-208
[21] Saikin A M, Buznikov S E, Elkin D S and Strukov V O 2019 Integrated navigation system for unmanned vehicle convoy *International Journal of Innovative Technology and Exploring Engineering* **8**(11) 4021-4025
[22] Varlamov O O, Lazarev V M, Chuvikov D A and Punam J 2016 On prospects for design of standalone smart robots based on mivar technologies *Radio industry (Russia)* **26**(4) 96-105

[23] Endachev D V, Vasin P A and Shadrin S S 2019 Applicability of computer vision architectures and their influence on traffic safety of autonomous vehicles *International Journal of Engineering and Advanced Technology* **8**(6) 5295-5301
[24] Saikin A M and Buznikov S E 2019 Formation of new approaches to the development of perspective intelligent transportation systems *International Journal of Civil Engineering and Technology* **9**(9) 2002-2011

IOP Conf. Series: Materials Science and Engineering **819** (2020) 012003 doi:10.1088/1757-899X/819/1/012003

The employment of multi-criteria assessment methods in the study of road safety

A I Markovnina, V S Makarov and V V Belyakov

Nizhny Novgorod State Technical University n.a. R.E. Alekseev, 24 Minin St., Nizhny Novgorod, 603950, Russia

E-mail: alinomalino@yandex.ru

Abstract. Intelligent driver assistance systems and traffic assistants are becoming more technological and widespread. Due to the great diversity, there is a need to compare their effectiveness in different conditions and choose a set of systems that is the most efficient and inexpensive. The article considers possible comparison criteria. Some features that are worth considering when conducting a study are identified. The analysis of statistics on road traffic accidents.

The problem of evaluating the feasibility of installing a particular set of active safety systems requires an integrated approach. When considering a set of active safety systems for a car, it is necessary to take into account the fact that most of the systems are software extensions or additions to those already installed on the car.

Antilock Brake System ABS acts as the basis for most other systems, such as ASR Automatic Slip Regulation and BAS Brake Assist System [1]. Electronic Stability Program ESP is a complex that includes ABS, BAS, ASR, EDB (Electronic Brake Distribution), EDS (Elektronische Differenzialsperre, or Electronic differential lock) and is the basis for the Emergency Steer Assist, Drive-Steering Recommandation DSR, Downhill Assist Control and Hill-Start Assist Control. Since ASR is developed on the basis of ABS, it also acts on the brake system, but in addition it also controls the electronic differential lock and some parameters of the power plant. The Electronic Stability Program (ESP) combines the functionality of several systems, and together with the Antilock Brake System, it is the basis for the Emergency Steer Assist, Drive-Steering Recommandation, and Downhill and Hill-Start Assist Control. In its work, the ESP control unit interacts with the engine control system and the automatic transmission (through the corresponding units). Tipping over is achieved by reducing lateral acceleration by braking the front wheels and reducing engine torque. Additional pressure in the brake system is created using an active brake booster. The road train stabilization system prevents the yaw of the trailer when the car is moving, which is achieved by braking the wheels or reducing torque.

The most famous and popular active safety systems are:

- Antilock Brake System;
- Automatic Slip Regulation;
- Electronic Stability Programme;
- Electronic Brake Force Distribution;
- Brake Assist System;
- Pedestrian Detection System;

- Electronic differential lock.

For clarity, a diagram showing the software extensions and systems of active safety systems of the car is shown (figure 1).

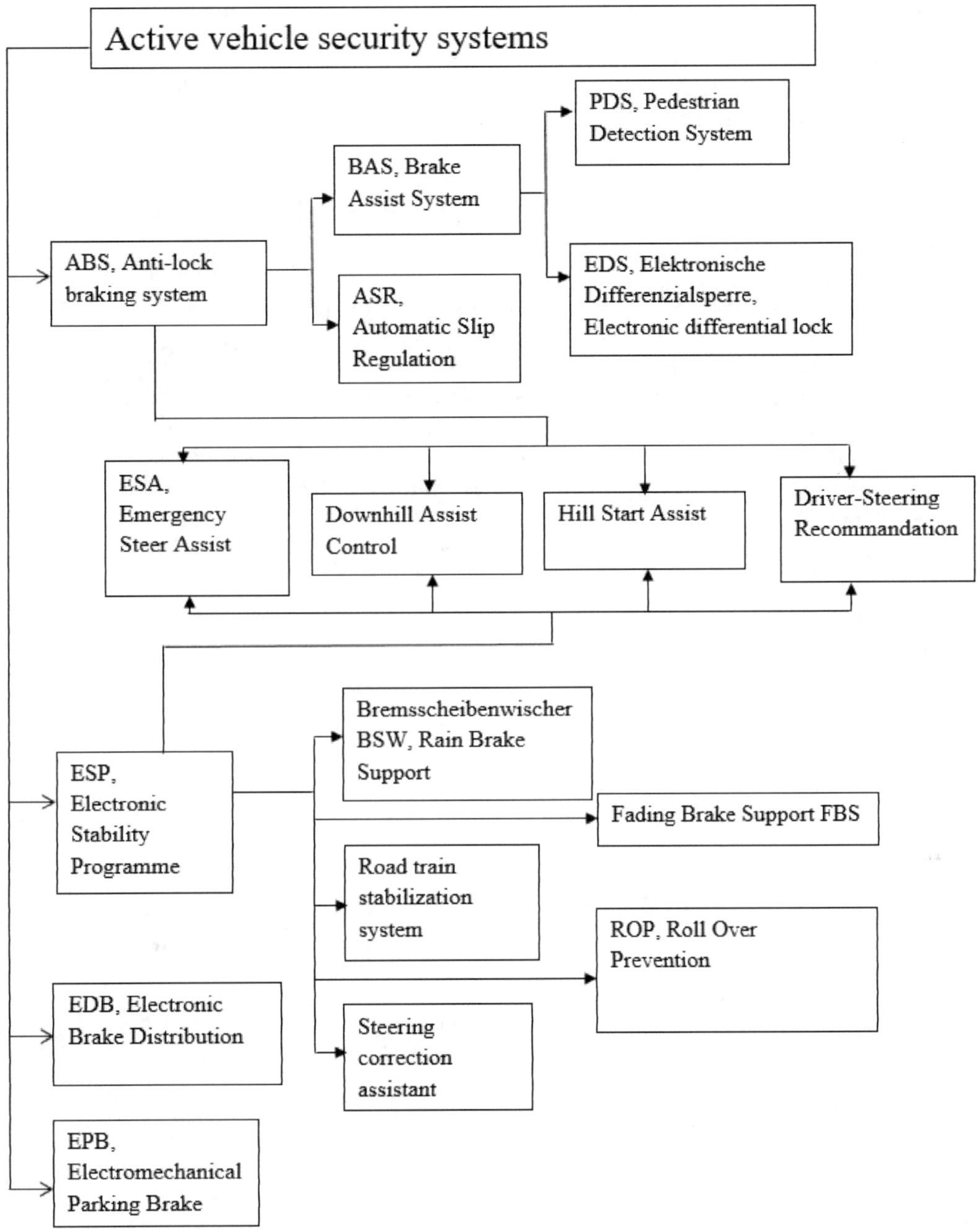

Figure 1. Scheme of active safety systems of the car.

Auxiliary active safety systems (assistants) have also been developed to help the driver in difficult driving situations. In addition to timely warning the driver of a possible danger, the systems also actively intervene in driving, using the brake system and steering system. Assistance systems are especially useful for novice drivers and older drivers.

Each assistant performs its highly specialized functions. Some systems are aimed at a smooth start to movement, others help with movement, others provide safe parking. At the same time, they are also subdivided into information systems, systems with partial intervention in driving and automatic systems that independently drive a car. Several systems perform several functions or participate in different driving modes. The Night View Assist and the Drive-Steering Recommandation perform an informative function and help with driving and parking.. Stop-and-Go traffic assistant, or traffic jam assistant, being an extension of adaptive cruise control, when turned on, it automatically starts moving, moves, brakes and stops the vehicle based on environmental data received from sensors, video cameras or radars. Adaptive cruise control performs the following tasks:

• accelerates the car to a predetermined value;

• slows it down to a complete stop;

• holds the set speed, providing not only comfort, but also by 10% or more reducing fuel consumption.

The Lane Assist, or Lane Departure Warning System, has two functions within the driving process - it informs the driver and partially affects the steering. The executive devices of the lane departure assistance system are a control lamp, a sound signal, a vibration motor on the steering wheel, a windshield heating element, an electric motor of an electromechanical power steering. The driver is warned by vibration of the steering wheel, as well as the supply of visual sound and light signals. Vibration is generated by a vibration motor integrated in the steering wheel. The heating element is located on the windshield, if necessary, automatically turns on, eliminates fogging and icing of the camera window. Correction of the trajectory is carried out by forced steering of the steering system using an electromechanical power steering or braking the wheels on one side of the car. The functional distribution of systems and assistants is shown in figure 2.

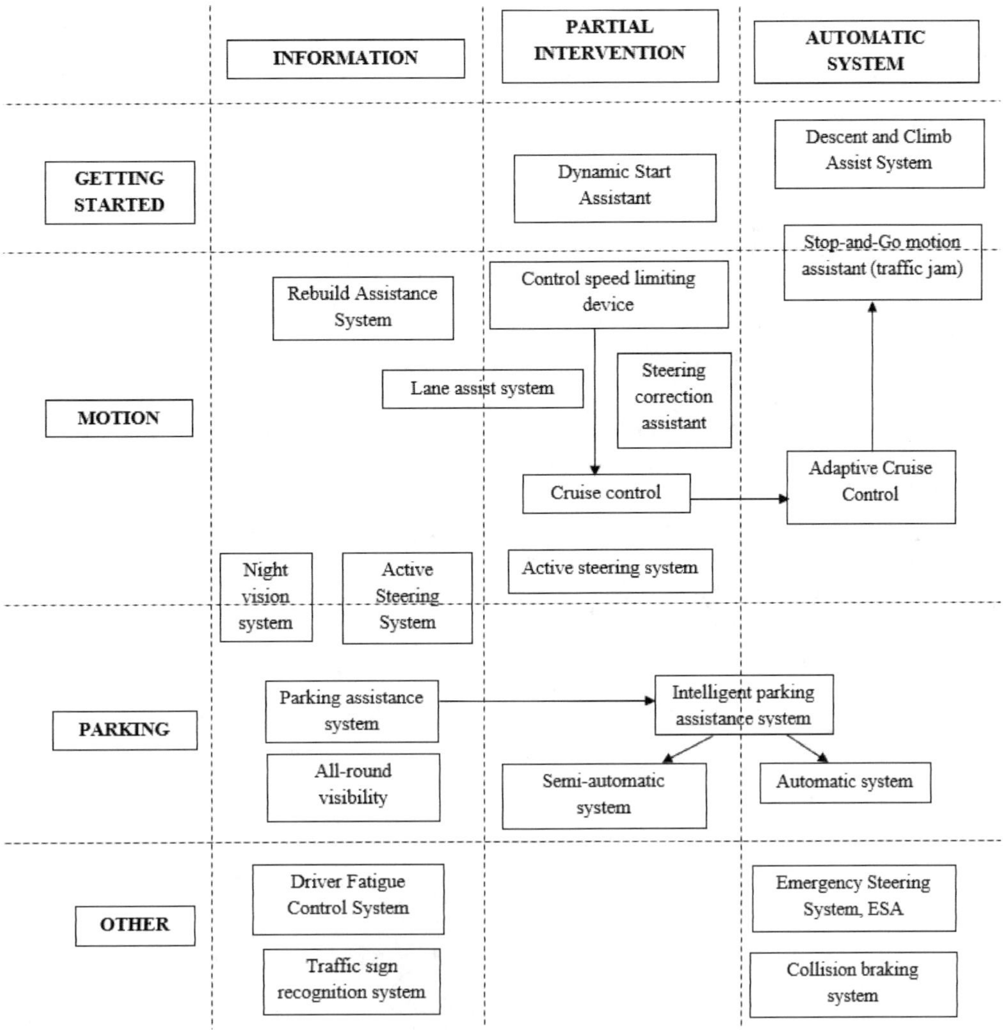

Figure 2. Functional distribution of intelligent driver assistance systems.

The question has been asked which set of systems to install on a car so that it is not expensive and most effective in the field of security. When choosing systems, it is necessary to consider that some of them are improved previous versions, so their joint installation is impossible. The optimal solution is sought by calculating the utility function. To solve the problem, comparison criteria, their values, the weight of each criterion are needed. Since the criteria will be considered as qualitative parameters, they must be expressed in points on a 5-point, 10-point or other scale [2].

Comparison criteria may be as follows:

- the likelihood of accident prevention in general;
- degree of mortality reduction in road accidents;
- the degree of reduction in injuries in road accidents;
- the life of the systems and / or their components;
- coverage of road conditions preceding or leading to accidents;
- the cost of installing systems or packages;
- the cost of maintaining and repairing systems.

It must be taken into account that at different speeds the systems react and act differently. For example, how well the emergency braking system operates at speeds of 20, 40 and 60 km / h in the city and at 80 and 100 km / h in the country. The division into urban and suburban driving can also be due

IOP Conf. Series: Materials Science and Engineering **819** (2020) 012003 doi:10.1088/1757-899X/819/1/012003

to the nature of the work or lifestyle of the driver. Intelligent systems from different manufacturers can be installed on cars of different brands, so you need to take into account some technical features. For example, the range of radars, the quality of the video cameras in conditions of pollution or lack of visibility (fog or dark).

The service life of the system as a whole can be longer than the service life of individual components, so you should consider how many times during the entire service life which spare parts will require, how much the spare part will cost and how much the replacement will cost when it is serviced. In this criterion, it is worth considering different manufacturers and different price categories.

Intelligent systems and assistants can be installed on a new car during the assembly process, but there may be a situation when it is necessary to equip a car whose design initially did not require additional devices. Accordingly, in this case, the installation cost will be different.

The degree of coverage of traffic situations can be assumed from the analysis of accident statistics from the website of the State Traffic Safety Inspectorate. Based on the number of accidents and their separation by type, as well as the conditions preceding or leading to accidents, we can assume a list of systems whose action would probably help to avoid an emergency or at least reduce the speed before an accident [3].

Example. When analyzing accident statistics for the Nizhny Novgorod region, accident with passenger cars was revealed by type: 138 collisions, 103 collisions (including collisions with pedestrians, cyclists, obstacles and standing vehicles), 13 rollovers and 3 others (such as a cargo fall). The diagram is shown in figure 3.

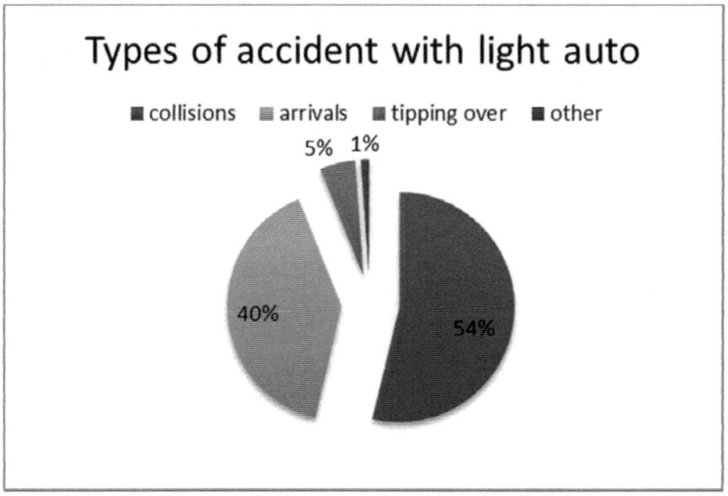

Figure 3. Types of road accident.

According to statistics, an analysis is made of the road conditions that preceded the collisions. The most common causes or previous events for the study period were: lost control, went into the oncoming lane; did not choose a safe distance; when turning left, did not provide an advantage in movement; when leaving the secondary road, did not provide an advantage in traffic (figure 4).

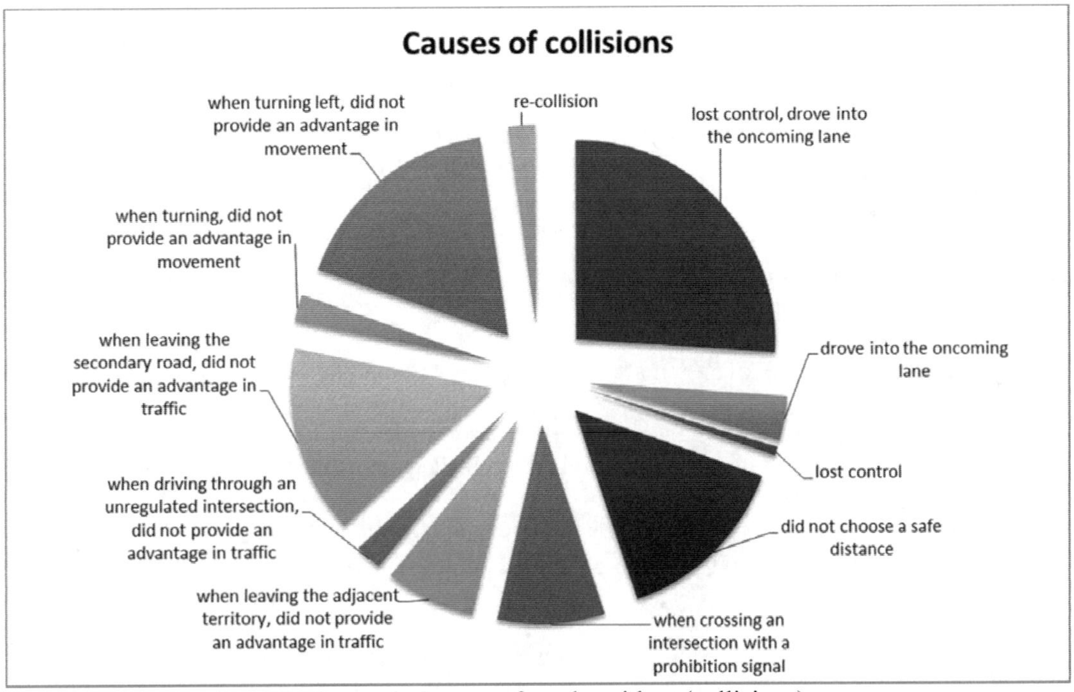

Figure 4. Causes of road accident (collisions).

Proceeding from the reasons, systems are proposed whose action could prevent them. Next, it is analyzed what percentage of all events this or that system could prevent. For example, did not choose a safe distance:

- Adaptive Cruise Control;
- Traffic jam assistant;
- Downhill Assist Control and Hill-Start Assist Control;
- Area View;
- Pedestrian Detection System, Obstacle Detection System.

Re-collision:

- Emergency Steer Assist
- Post Crash Braking.

Further elements of a multicriteria assessment are applied. Edgeworth-Pareto method and the construction of scales of criteria are applicable for graphic representation of alternatives. The Pareto method is usually considered as preliminary, and if it is possible to evaluate alternatives only by 2 criteria, then the number of scales is not limited, although it will be overloaded with a large amount of data.

In decision making, it is customary to distinguish between continuous and discrete ratings scales, quantitative and qualitative rating scales. So, for the criterion of "cost" can be used a continuous quantitative scale of assessments (in monetary units). For the criterion “system availability”, a qualitative binary scale can be used: whether or not.

The following types of scales can be used.

- Scale order — scores are ordered by increasing or decreasing expert preferences.
- The scale of equal intervals is the interval scale. For this scale there are equal distances in quality change between grades.
- The proportional scale is an ideal scale. An example is the scale of assessments according to the cost criterion, the countdown in which begins with a set value (for example, from zero cost).

IOP Conf. Series: Materials Science and Engineering **819** (2020) 012003 doi:10.1088/1757-899X/819/1/012003

Analytically, the choice of the best solution can be made by calculating the utility function.

For this, the criteria were previously expressed in points, and by the method of expert evaluations, the criteria are assigned the weight and the utility function is calculated. An example of a formula that takes into account 2 criteria is given.

$$F = S * W_S + Y * W_Y,$$

where F is the utility function;
S - evaluation of the cost criterion;
Y - assessment of the criterion for improving security;
Ws is the weight of the cost criterion;
Wy is the weight of the safety improvement criterion.

From the calculations, the set of systems with the highest value of the utility function is revealed, we assign it the best option.

The selection of comparison criteria is a difficult task, on which the accuracy and objectivity of the solution can depend, therefore, in the process of research, it is possible to attract experts, as well as consider the intervals of values.

The multicriteria optimization method helps to choose one of the likely optimal solutions, but does not guarantee that this solution will be uniquely optimal for all criteria.

References

[1] 2012 *Automatic and intelligent vehicle systems. Cars and tractors, multi-purpose wheeled and tracked vehicles, ground transportation and technological complexes, mobile robots and planet rovers* (Nizhny Novgorod: Nizhny Novgorod State Technical University n.a. R.E. Alekseeva) p 475

[2] Larichev O I 2000 *Theory and decision-making methods, as well as the Chronicle of events in the Magic Countries* (Moscow: Logos) p 296

[3] Lomakin V V et al 2011 *Vehicle safety* (Moscow: MSTU "MAMI") p 299

[4] Chernorutsky I G 2005 *Methods of decision making (*St. Petersburg: BHV Petersburg) p 416

IOP Conf. Series: Materials Science and Engineering **819** (2020) 012004 doi:10.1088/1757-899X/819/1/012004

Prospects for motorization and energy markets in the context of fully autonomous vehicles spread

M Yu Ksenofontov and S R Milyakin

Institute of Economic Forecasting, 47, Nakhimovsky prospect, Moscow, 117418, Russia

Abstract. The report is devoted to two interrelated issues: the impact of penetration of fully autonomous cars and car sharing on the car fleets, and in combination with an increase of the share of electric vehicles on the volume and structure of energy demand. For this, forecasting tools and motorization scenarios have been developed. Forecasts for car fleet and energy demand were obtained for Russia and the largest economies in the world (China, USA, EU, India, Japan).

1. Introduction

In the future, motorization and passenger cars energy demand will most probably be determined by three key factors.

1.1. The penetration of fully autonomous vehicles

Over the past decade, a significant breakthrough has been made in the development of automatic control systems: fully autonomous versions of cars have been developed, and their mass production and market entry are planned in the coming 3 - 5 years. Fully autonomous vehicles can become the basis for the "explosive" development of car sharing both in the commercial segment of passenger automobile transport (taxi, car sharing services) and in the household sector [1]. A more intensive operation will satisfy the need for transportation with a relatively smaller number of vehicles [2].

1.2. The increase in the share of electric vehicles in total sales

In many countries around the world populations who are concerned about the emissions produced by internal combustion engines express their intentions to ban the sale of cars with internal combustion engines and to pursue a policy in support of electric vehicles purchase. The penetration of electric vehicles in the future will be determined by their market price (first of all, the cost of batteries) and overcoming their limitations such as a limited range, long charge time, reduced battery capacity at low temperatures, etc.

1.3. Improving energy efficiency

The average fuel consumption rate of new cars decreased from 18 liters per 100 km in 1975 to 7.6 liters by 2015 due to improvements in the engine, reductions in the weight of cars, improvements in their aerodynamics, etc. According to EIA estimates, the average energy efficiency in the car fleet can significantly increase: from 10.5 liters per 100 km (2016) to 6.2 liters (2050). Fully autonomous shared cars can be even more efficient: by optimizing usage modes and routes, including them in territorial traffic flow coordination systems, reducing the need in finding intermediate parking, which in cities accounts for a significant proportion (30% on average) of the time spent during a trip.

2. Description of forecasting tools

We developed forecasting and analytical tools to illustrate the importance of the penetration of fully autonomous shared cars to motorization.

The calculations are carried out in four stages. At the first stage, the potential demand for passenger car services is assessed, depending on the hypothesis of economic development and population growth with the assumption that the patterns of motorization prevailing in the past are preserved. To do so an aggregate time series model was applied [3]. We use a logistic curve to approximate car ownership (the number of cars per 1000 people):

$$O_1(t) = O_{\max} \cdot \left(1+\exp\left(a(b-W(t))\right)\right)^{-1}, \tag{1}$$

Where: $O_1(t)$ is car ownership at 1st stage of calculations, $O_{\max}$ is saturation level, $W(t)$ is an independent variable, a and b are estimated coefficients.

Following Tanner's work [4], as an independent variable we consider the sum of weighted current and past values of GDP per capita:

$$W(t) = c \cdot W(t-1) + GDP(t), \tag{2}$$

Where $GDP(t)$ is GDP per capita in constant prices; c are estimated coefficients, $0 \le c \le 1$.

Using a hypothesis on the prospects for population growth we calculate car fleet.

$$Fleet_1(t) = O_1(t) \cdot Pop(t) \cdot 10^{-3}, \tag{3}$$

Where $Fleet_1(t)$ is car fleet at 1st stage of calculations, $Pop(t)$ is population.

Using a hypothesis on car retirement rate we also calculate sales of cars at 1st stage of calculations:

$$Sales_1(t) = Fleet_1(t) - \left(1 - RtrRate(t)\right) \cdot Fleet_1(t-1), \tag{4}$$

Where $Sales_1(t)$ is car sales at 1st stage of calculations, $RtrRate(t)$ is car retirement rate.

At the second stage, we consider car fleet as a combination of two functionally different types: fully autonomous shared vehicles and traditional cars (cars used personally, partly autonomous cars are also included as they don't contribute in possibilities of sharing). The estimated car fleet at the first stage is adjusted taking into account the gradual penetration of fully autonomous vehicles (which, as part of the calculations, define speed of the penetration of car sharing and the substitution coefficient of traditional cars). The scenario parameters at the second stage of the calculation are: the share of fully autonomous vehicles in total passenger car sales at the starting point (2025); growth rate of fully autonomous vehicle sales; the number of traditional cars replaced by one fully autonomous shared car in sales (in other words, a substitution coefficient); retirement rate of traditional and fully autonomous shared cars. Car sales on the second stage can be calculated as follows.

$$Sales_2(t) = Sales_1(t) - \left(SubCoeff(t) - 1\right) \cdot SalesFAC(t), \tag{5}$$

Where $Sales_2(t)$ is car sales at 2nd stage of calculations, $SubCoeff(t)$ is substitution coefficient, $SalesFAC(t)$ is sales of fully autonomous shared cars.

Then the car fleets for both types (traditional and fully autonomous) can be estimated.

At the third stage, the penetration of electric vehicles is modeled, depending on the hypotheses of their share in sales (provided by OPEC prospects). At the fourth stage, petroleum and electricity consumption by cars is calculated, depending on a hypothesis of average mileage and energy efficiency for each car type (depending on control system and energy used).

$$ConEnergy(t) = Fleet(t) \cdot Mil(t) \cdot RateCon(t), \tag{6}$$

Where $ConEnergy(t)$ is energy consumption (petroleum or electricity), $Mil(t)$ is average mileage of one car, $RateCon(t)$ is consumption rate.

For fully autonomous shared cars their additional efficiency is taken into account.

3. Scenarios of motorization

Four scenarios of motorization are considered, differing in penetration conditions and usage mode of fully autonomous shared vehicles (Table 1). A hypothesis regarding energy efficiency and the share of electric cars in total sales do not differ between scenarios.

Scenario 1 ("business-as-usual") is characterized by the inertial development of commercial sharing services, while fully autonomous vehicles do not have a significant impact on motorization. Scenario 2 ("autonomous cars in households") allows us to analyze the possibilities of reducing the personal fleet due to the more intensive use of fully autonomous vehicles compared to traditional ones within households. Fully autonomous vehicles in this scenario are mainly consumed by households. While the traditional car around 95% of time is parked, fully autonomous vehicle might serve different members of household. So it is used more intensively, as a result fully autonomous cars replace traditional cars in a 1: 2 ratio. In scenario 3 ("mobility as a service") the potential scale of "mobility as a service" systems development based on fully autonomous vehicles in large cities is considered. For many large cities it is a basic method of dealing with traffic problems and pollution. "Mobility as a service" means creating such a system that stimulates citizens to use public transport and sharing services and reduce personal cars usage. Fully autonomous shared cars are considered to be a key element of such a system. That means the penetration of fully autonomous shared vehicles in this scenario is significantly higher than in scenario 1. The substitution coefficient is higher than in Scenario 2, since autonomous cars serve not one, but several households. Scenario 4 ("distributed mobility") allows us to evaluate the effect of the distribution of traffic peaks (one in the morning and one in the evening) during the day, which increases the intensity of fully autonomous shared vehicles usage.

Table 1. The main hypotheses in different scenarios (in the case of Russia).

	2017 year	2045 year			
		Scenario 1	Scenario 2	Scenario 3	Scenario 4
Fuel consumption rate, litres per 100 km	10,3	6,2			
Electricity consumption rate, kWh per 100 km	20-	17			
The share of electric vehicles in sales,%	~0%	15%			
Sales of fully autonomous shared cars, thousand units	~0	250	1300	1300	1000
Share of fully autonomous shared cars in total sales, %	-	8%	50%	50%	50%
Substitution coefficient	-	4	2	4	8

4. Results and Discussion

The main calculation results are shown in table 2. They are given for Russia, China and for five largest economies (China, India, US, EU, Japan) aggregated.

Table 2. Calculations results for the car fleet and its energy consumption in Russia, China and five largest economies by 2045.

	2017 year	2045 year			
		Scenario 1	Scenario 2	Scenario 3	Scenario 4
Russia					
Passenger car fleet, mln. units.	44	56	45	40	33
Share of fully autonomous shared cars in the fleet,%	-	2%	10%	12%	16%
Share of electric vehicles in the fleet,%	0%	7%	7%	7%	8%
Share of electric vehicles in VMT,%	0%	7%	8%	9%	11%
Consumption of petroleum products, million tons.	54	40	39	38	36
Electricity consumption, billion kWh	0	13	14	15	16
China					
Passenger car fleet, mln. units.	185	470	390	347	238
Share of fully autonomous shared cars in the fleet,%	-	2%	21%	15%	16%
Share of electric vehicles in the fleet,%	1%	26%	26%	25%	32%
Share of electric vehicles in VMT,%	1%	26%	27%	29%	34%
Consumption of petroleum products, million tons.	178	283	269	264	221
Electricity consumption, billion kWh	4	405	417	430	477
Five countries (China, US, EU, India, Japan)					
Passenger car fleet, mln. units.	784	1 273	1 028	927	684
Share of fully autonomous shared cars in the fleet,%	-	2%	22%	16%	17%
Share of electric vehicles in the fleet,%	~0%	25%	24%	23%	27%
Share of electric vehicles in VMT,%	~0%	27%	29%	30%	36%
Consumption of petroleum products, million tons.	758	771	738	726	618
Electricity consumption, billion kWh	10	1 106	1 602	1 100	1 214

Calculation results for different scenarios give us information on how different conditions of motorization influence the car fleet and energy demand. Scenarios 3 – 4, in which fully autonomous cars are more widespread than in scenario 2, the car fleet is estimated to be significantly lower. Scenarios 2 and 3 differ primarily in usage mode (within household or within "mobility as a service" system). The more intensive usage in scenario 3 leads to lower values of car fleet (by 10-12% comparing to scenario 2) and petroleum demand from cars (by 2% comparing to scenario 2). Distribution of traffic peaks during the day in scenario 4 leads to a decrease in our estimations by an additional 30% (compared to scenario 3) for car fleet and 15 – 17% for petroleum demand. Scenario 4 is significantly different from scenario 1: the car fleet is expected to be 50% lower than in scenario 1 and petroleum demand to be 20% lower. The difference in energy demand between the scenarios is much lower than the difference in car fleet values. This is connected with the fact that despite car fleet decreases usage intensity will grow. The decrease in energy demand is a consequence of two factors: the higher electrification rate of shared cars and the relative efficiency of fully autonomous cars. Intensive usage means higher retirement rate and accelerated renewal. As a result the same hypothesis on electric cars share in total sales gives a different electric cars share in total fleet (Table 1 and 2). The higher share of electric cars in the shared car fleet combined with more intensive usage gives higher electric cars share in total VMT (Vehicle Miles Travelled) and as a result lower petroleum consumption by passenger cars. Electricity consumption grows in all the scenarios due to electric car fleet increase but we consider such a growth by 3 – 4% of total electricity consumption of considered countries in 2045.

5. Conclusion

The penetration of fully autonomous cars has a significant impact on the car fleet and its petroleum demand. In scenario 4 (in which the transport systems of cities are based on "mobility as a service" and traffic peaks are distributed during the day) the total car fleet is estimated to be 50% lower than in the "business as usual" scenario, petroleum demand will be 20% lower and electricity demand 10% higher. For energy suppliers that might mean structural changes in demand for their products: lower demand for oil and higher demand for gas or renewables.

6. References

[1] Thomopoulos N and Givoni M 2015 *Eur J Futures Res* **3(14)** 1 – 14
[2] Ksenofontov M Yu and Milyakin S R 2018 *Studies on Russian Economic Development* **29(4)** 406–14
[3] Jong G, Fox J, Daly A, Pieters M and Smit R 2004 *Transport Reviews: A Transnational Transdisciplinary Journal* **24(4)** 379-408
[4] Tanner J C 1983 *TRRL Laboratory Report* **1072** p 21

On-board driver assistance systems classification

A A Khorychev

Intellectual System Centre, FSUE "NAMI", Moscow, Russian Federation

E-mail: artem.horychev@nami.ru

Abstract. This article contains a brief overview of the known classifications of on-board driver assistance systems, the fundamental principles of a possible classification (based on a systematic approach and flow theory) and their application to systems of on-board driver assistance systems. The new classification criteria for the classification requirements is proposed for on-board driver assistance systems that take into account the peculiarities of the Russian Federation road network.

1. Introduction

In the study guide [1] the following definition of on-board driver assistance systems (hereinafter - ADAS) is introduced:
ADAS - Advanced Driver Assistance Systems are special electronic devices that allow the driver to receive in advance preventive information about possible dangers and avoid emergencies situations.
The source [2] indicates the automation levels of vehicle control tasks given in table 1.

Table 1. Vehicle control automation levels.

Automation level	Automation level description
0	No Driving Automation
1	Driver Assistance
2	Partial Driving Automation
3	Conditional Driving Automation
4	High Driving Automation
5	Full Driving Automation

On-board driver assistance systems with automation levels 1-3 are yet not classified.
The authors in [3] divide the ADAS into car control systems, driver information systems and systems for information collection and transmission.
The authors consider that car control systems are include:

- subsystems for obtain information from sensors and external sources;
- information processing subsystem;
- automatic vehicle control subsystem.

The authors consider that driver information systems are include:

- subsystems of information obtaining from sensors and external sources;
- information processing subsystem;
- information display subsystem to driver.

The authors consider that information collection and transmission systems are include:

- subsystems information from sensors and external sources obtaining;
- information storage subsystem;
- information transfer subsystem to external consumers (*V2I, V2V*).

Summing up the results of this classification, the following should be note.

The subsystems of information obtaining from sensors and external sources refers simultaneously to three types of systems: car control systems, driver information systems and information collection and transmission systems (sources, directions of transmission and consumers of information are not indicate).

The information processing subsystem refers simultaneously to three types of systems: car control systems, driver information systems and information collection and transmission systems.

The automatic vehicle control subsystem refers to Automated Control Systems (ACS) only, but it is not clear which WV's systems the ADAS belongs to.

The information display subsystem to driver refers to driver information systems only.

The storing information function is not included in the name of the category of information collection and transmission systems, although it is mention.

One of the fundamental principles of the object's classification implies that the classified object must uniquely relate to one of the many classes. Points 1–2 above contradict this principle.

The main drawback of the classification proposed by the authors is non-use of the theory of flows and actions over them. Using the flow theory makes the ADAS classification quite concrete, logical and clearly determined.

The author Elizabeth Rendon-Veles conducted an analytical review of the essential stem ADAS [4] regarding the processing of information performed by the driver while driving. She categorized ADAS into three different categories based on the sub-process of driving that they help (perception, analysis - decision and action). The review showed that these systems, in addition to being a problem when considering human adaptation issues.

2. Problem Statement

The mentioned classification about 9 years ago was proposed. Using the terminology of the source [1], WVs of the late 2000s – early 2010s with 1-level ADAS were equipped. Nowadays, level 5 self-propelled ACS is created and tested, the corresponding regulatory frameworks and requirements for the modern ADAS classification are created and improved, based on a System approach and System terminology.

3. Methodology for problem solving

The use of a System approach in automobile theory known from R.V. Rotenberg works. He looked at the system "Driver-Car-Road-Environment" [3].

The generalized functioning scheme of a technical system is known from the technical systems theory [4], and its adapted version in Figure 1 is present. Any technical system three following actions (transactions) performs: Transformation (Tf), Transportation (Tp) and Transchronation (storage) (Tc) with three flows (material objects, energy and information), in environment.

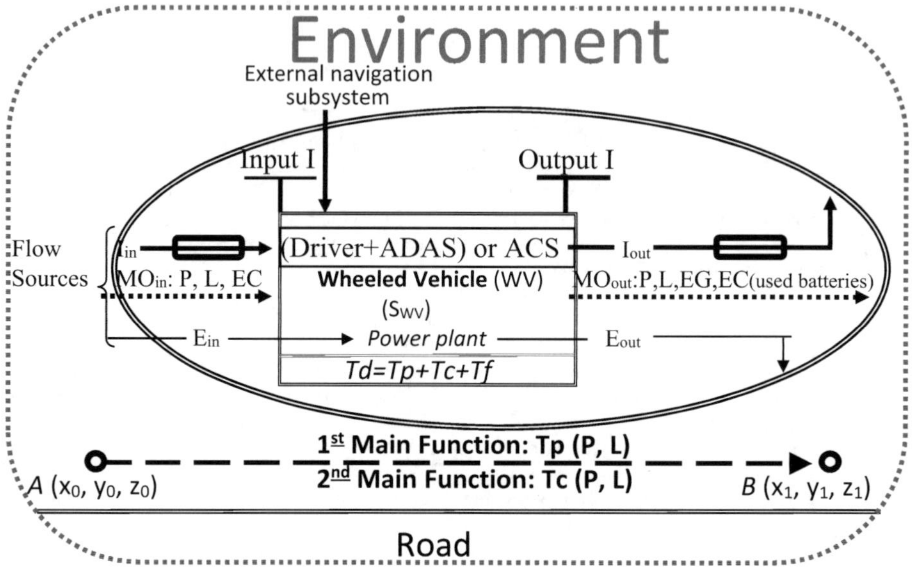

Figure 1. Technical System Generalized Functioning Scheme (for 1-3 level ADAS-equipped WV).

On figure 1: I – Information Flow; E – Energy flow; P – Passengers; L – Luggage; Flow indexes: in – Input; out – Output; EC – Energy Carrier (fuel, batteries); MO – Material Objects Flow; EG – Exhaust Gases Flow (for WV with external combustion engine only); S_{WV} – WV structure; Ta – Transaction – set of following actions in vehicle occurring: Tp (Transportation), Tc (Transchronation, or Storage), Tf (Transforming); *A* – Start point of trip with x_0, y_0, z_0 coordinates; *B* – Endpoint of trip with x_1, y_1, z_1 coordinates.

- Input & Output Information Flows;
- Input & Output Information Flow Filters (cybersecurity);
- WV's main function: Passengers & Luggage transportation and storage;
- Input & Output Energy Flows;
- Input & Output Material Objects Flows.

WVs information systems, as systems of transaction over information (Td (I)), into the following functional categories can be divide:
- Information-transforming system – Tf(I); is the traffic conditions and WV-conditions empirical information collecting and processing system; Tf(I) is divided to viewing (trivial) transformation (from one to other information kind) and parametric (quantitive) transformation;
- Storage information systems (local and «clouded»);
- Information transportation systems: input/output flows for driver/ADAS.

Separately, electronic information systems for drivers and passengers transchronation (preserving), safety systems, and first of all, accident prevention systems, should be mentioned.
The driver's functions are performing by ADAS partially, or completely replace by ACS.
The information flows (input and output) coming to and from the vehicle, when implementing by ADAS and ACS, are multiply. For this paper, we will not dwell on the energy and safety systems of the vehicle, confining ourselves to the systems responsible for driving the vehicle.
The Russian regulatory (normative) framework in the field of ADAS is currently at the initial stage of formation. An analysis of the foreign regulatory framework shows that the dynamic driving tasks (DDT) are separate into:

- lane related activities;
- velocity related activities;
- road infrastructure objects (traffic lights, road signs and markings, etc.) related activities;

IOP Conf. Series: Materials Science and Engineering **819** (2020) 012005 doi:10.1088/1757-899X/819/1/012005

- activities related by dynamic objects in environment (animals, pedestrians, pedal cyclists, motorcyclists, cars, trucks, buses, rail transport), - their recognition, moving detecting, as well as warning to the driver of the subject vehicle, pedestrians, cyclists and drivers of the surrounding vehicles.

All these actions shall be separate into active (impact on the subject vehicle control devices) and passive (warnings of the subject vehicle driver and other traffic participants).

It would be logical to assume that ACS perform active actions chiefly, and ADAS – passive action only. However, the common point of view is that ADAS perform certain control tasks in certain conditions (for example, automatic parking, highway traffic, traffic jam, etc.), while 4 or 5 automation level-ACS solve the whole DDT with passengers and/or cargo from point "A" to point "B" taking into account the traffic situation and other factors taken into account by the ordinary driver.

For the operation of ADAS and ACS, it is necessary to create classifiers of objects in the environment that can affect on the movement of the subject vehicle.

To systematize the tasks solved by the ADAS and ACS, it is necessary to identify and analyse the tasks performed by the driver.

In particular, the driver solves the following tasks.

Before driving: choosing the end point of the trip, taking into account the availability of a parking space for the subject vehicle and choosing the optimal route (by time/road conditions) taking into account the navigation information received by the navigation system).

While driving:

2a) analysis of the traffic situation on the subject vehicle's path (from 1 m to hundreds of kilometers, taking into account navigation information received from the environment), including a constant analysis of the prerequisites for the dangerous situations formation;

2b) analysis of requirements, regulations and information provided by road signs, road markings and traffic signals;

2c) analysis of road and weather conditions (road surface condition, visibility, etc.);

2d) impact on subject vehicle controls to change the trajectory, speed of the subject vehicle and on means of warning to other traffic participants in order to reach point «B» while compliance with the safety (transchronation) requirements of passengers and luggage.

Actions 1, 2a) –2c) are intellectual, actions 2d) are physical. Therefore, the tasks of subject vehicle's operating are predominantly intellectual. All intellectual actions carried out in the process of movement every second are necessary for the implementation of physical actions that ensure traffic safety. That is why the ACS also called "intelligent transport systems", and their development became possible only after the automotive electronics revolutionary progress.

Tasks performed by the driver or ACS/ADAS are divided into:

1a) "not digitizing", which solving by electronics isn't possible: the choice of the starting and ending points of the route;

1b) task solved in modern vehicles and smartphones (route selection).

Tasks 2) are divided into tactical (2a) and strategic (2b). The border between them is conventionally set at a distance of 250 m - the detection distance by modern technical control means: video cameras, radars and LIDAR's. To analyze the traffic situation at a distance up to 250 m (tactical tasks), on-board monitoring tools are enough, but for a complete analysis of the conditions in front of the subject vehicle at distance of more than 250 m – there is a need of information provided by external components of navigation systems, to solve the strategic driving tasks (choosing route selection for a detour of accident place or field of traffic jam).

The main issue in solving tactical problems 2a) is a recognition of dynamic objects. The most urgent task is the differentiation of young children from small animals, and (to a lesser extent) differentiation of people from large animals, as well as the task of differentiation of pedal cyclists from motorcyclists. For example, in case of an imminent collision with a small animal, a change in direction can lead to much more serious consequences for the driver and passengers of the subject vehicle and for the surrounding vehicles, therefore it is preferable not to change it, while a collision with a child, especially at a high speed, will surely leads to his death, therefore, an analysis of the entire environment of the subject vehicle is necessary to change the direction of movement of the subject vehicle. For timely braking in front of a child, his timely differentiation from animals at a great distance is especially relevant. For the present, the sound signal for dynamic objects located on the path of the subject vehicle is not automated in case of a collision threat.

Patents can protect the classification of road conditions in the environment of the subject vehicle. For example, Scania Sw AB obtained and maintained in Russia an invention patent "Device and system for classifying of the environment of a vehicle" [5].

The classification of vehicles in the textbook of the same name for university students is consider [6]. In particular, in the intelligent transport systems, the proposed vehicle size classification can be used in length with a single step of 0.3 m, a minimum value of 2 m and a maximum of 12 m. Modern, "passive" systems informing drivers about already formed traffic jams in the future would give way to systems which prevent them, using information about the planned route and dimensions of each of the potentially capable of participating in traffic jams, as well as the known throughput of a particular road. The same applies to public parking lots. This is an integral part of the theory of cars set [7] that is missing today, but will be formed in the near future.

The driver must control the vehicle in a way that will allow avoiding any accidents, in accordance with the Russian Traffic Rules. ADAS & ACS must make it also. Abnormal traffic situations are divided into dangerous and emergency. Modern high-quality ACS or ADAS should not allow the creation of any contingency situation, taking the necessary actions in advance (changing the speed or trajectory of movement). Currently, the operation of emergency braking systems is standardized. Emergency braking may be required when the driver loses control of the subject vehicle, or when other participants in the movement change their speed or direction unpredictably, creating a threat to the safety of the subject vehicle.

For emergency situations when an accident cannot be eliminated or the severity of its consequences cannot be eliminated by the method available under the normal conditions, vehicle classification can be used in ACS or ADAS to select a vehicle with which it is preferable to collide in a dense stream, and select the type of collision. It is known that a frontal accident is more dangerous than a glancing collision, and the severity of the consequences of collision with vehicle moving on the same direction is lower than a collision of a vehicle with an oncoming vehicle. The ADAS should, if it is possible, change over frontal collision to a glancing collision, or change the direction of movement for a collision with a vehicle moving on a same direction instead of the oncoming one. If there is a dividing fence, it is advisable to hit it, than to allow a collision with the vehicle, or hit a pedestrian or a pedal cyclist.

The need to solve problems 2b) is currently standardized abroad, and from 2021 it will be standardized in the Russian Federation. Recognizing of the LED traffic light signals remains an urgent task in solving these issues, and now its solution may consists in equipping traffic lights with systems for transmitting information to vehicles ("I2V").

2c) The tasks of weather and road conditions analyzing. This analysis is necessary to select the speed and braking options of the subject vehicle. For example, under icy conditions, the maximum speed should be limited, and the use of a standard brake system should be minimize or replaced by engine braking, as when driving on long descents. It is necessary to choose detour path or significantly reduce the speed when the road section is flooded. If the water level reaches the position of the some vehicle elements (lower body doors border or engine air intake heights (for ICE only)), need to take a stop and, if possible, drive away from the dangerous section (reach of minimal risk condition).

It is also necessary to give the driver recommendations on driving around long sections of roads with damaged pavement, first of all, for drivers of cars with low ground clearance, as well as not offering them alternative ways to avoid traffic jams if the road pavement on these ways is seriously damaged.

Another important aspect is braking of the left and right side wheels in conditions of different traction with the road, for example, when the wheels of one of the sides of the car are on snow drift or in an asphalt groove filled with water. These conditions are often found in the Russian Federation, therefore, this problem must be solved in locally-produced ACS and ADAS, in order to avoid accidents specific to the Russian Federation. For example, by introducing an algorithm in the ACS that allows to make heading corrections before braking, excluding the conditions of different traction of wheels of car different sides, leveling them.

2d) The exposures of a driver, ACS or ADAS applied on the subject vehicle control devices are well amenable to automation today. With a high quality of actions 1) - 2c) braking or changing the trajectory of the subject vehicle will be carried out in advance, for a time sufficient to prevent not only emergency situations, but dangerous situations as well.

The systematization of the ADAS functions, as a result of the completed analysis, is shown on Figure 2 (Fragment related to the main function of the vehicle).

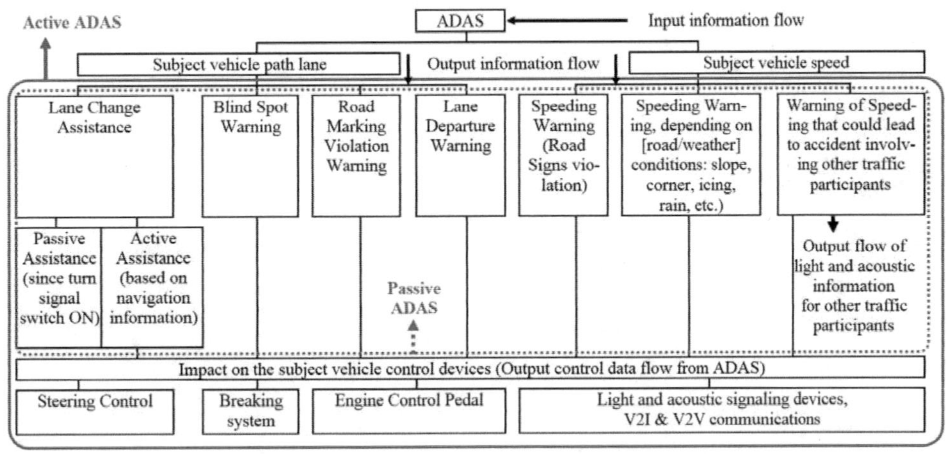

Figure 2. ADAS functions taxonomy (Fragment related to the main vehicle function).

Abbreviations on fig. 2 with corresponding ISO standards (if present):
LKAS – Lane Keeping Assist (ISO 11270), LDWS – Lane Departure Warning Systems (ISO 17361), LCDAS – Lane Change Assist System (ISO 17387), PADS – Partially Automated In-Line Driving System (ISO 21717), BSM – Blind Spot Monitoring, WDW – Wrong Direction Warning, TLR – Traffic Light Recognition, TSR – Traffic Sign Recognition, ISA – Intellectual Speed Adaptation, ISL – Intellectual Speed Limit, TA – Turning Assist, ABS – Anti-Lock Braking System (ISO 21994), AEBS – Advanced Emergency Braking System (ISO 19377), CSWS – Curve Speed Warning System (ISO 11067) CS – Cross-wind Stabilization (Mercedes-Benz), ACC – Adaptive Cruise Control (ISO 15622), PDCMS – Pedestrian Detection and Collision Mitigation System (ISO 19237), FVCWS – Front Vehicle Collision Warning System (ISO 15623), IA – Intersection Assist, RCTA – Rear Cross Traffic Alert, CAS – Collision Avoidance System, HBA – High Beam Assist, APS – Assisted Parking System (ISO 16787), PCS – Pre-Crash Safety System, BVS – Bird View System, SVS – Surround View System, DMS – Driver Monitoring System, DWM – Driver Wellness Monitoring, AA – Attention Assist (Mercedes-Benz), MALSO – Maneuvering Aids For Low Speed Operation (ISO 17386), TIWS – Traffic Impediment Warning System (ISO/TS 15624), NVS – Night-Vision System, ERBA – Extended Range Backing Aid System (ISO 22840), S&G – Stop & Go System, V2I – Vehicle-to-Infrastructure; V2V – Vehicle-to-Vehicle.

References

[1]. Zhankaziev S V Intellectual transport systems. (Study Guide) – Moscow MADI, 2016
[2]. *SAE J* 3016:2018 Taxonomy and Definitions for Terms Related to Driving Automation Systems for On-Road Motor Vehicles
[3]. Ivanov A M, Solncev A N Perspective of the development of intelligent on-board vehicle systems in the Russian Federation *AAE magazine* № 6'2010
[4]. Rendon-Velez Elizabeth Classification and Overview of Advanced Driver Assistance Systems According to the Driving Process. ASME 2010 International Design Engineering Technical Conferences and Computers and Information in Engineering Conference January 2010
[5]. Rotenberg R V Fundamentals of the reliability of the "Car-Driver-Road-Environment» system M Mashinostroenie 1986 p 214
[6]. Kravetc V N System Approach to designing competitive cars Kravetc V N, Khorychev A A Collection of the scientific articles « Problems of design, testing, operation and marketing of automotive vehicles, internal combustion engines, road construction machines, transport and technological complexes and all-terrain vehicles», International scientific-technical conference (27-29.05 2000 N. Novgorod) – N. Novgorod: N.I. Lobachevsky State University interuniversity printing-office 2000 p 560
[7]. RU № 2423665 «Device and System to classify Vehicle Environment» patent Scania Sw AB 24.05.2007

[8]. Kravetc V N Vehicles Classification. Study Guide for students 190201 (150100) specialty «Cars and tractors manufacturing» Kravets V N, Khorychev A A N. Novgorod R.E. Alexeyev NSTU 2008 p 96

[9]. Khorychev A A Car's Theory Structure and it's connection with intellectual property objects Khorychev A A, Kravets V N 5 Regional Youth scientific-technical conference «The Future of technical Science of the N. Novgorod Region" Abstract's Collection N. Novgorod 2006

IOP Conf. Series: Materials Science and Engineering **819** (2020) 012006 doi:10.1088/1757-899X/819/1/012006

About the project developing "MIPRA" – the intelligent planner in the state space for vehicles, tractors, and robots based on the architectural solutions of the Mivar systems for traffic enforcement

D V Aladin[1,2], **O O Varlamov**[1,2], **L E Adamova**[2,3], **D A Chuvikov**[2] **and D A Fedoseev**[1]

[1] Bauman Moscow State Technical University, 2-ya Baumanskaya Street 5/1, Moscow 105005, Russia
[2] MIVAR, 72 Oktyabrskaya str., Moscow 127521, Russia
[3] Russian New University, Radio str. 22, Moscow 105005, Russia

E-mail: aladin.dv@yandex.ru

Abstract. The task of creating a logical "strong" artificial intelligence (AI) to be used as new decision-making systems for autonomous vehicles, tractors, and robotic systems draws increasing attention of scientists around the world. Essentially, it is the creation of "brains" for vehicles and any other transport systems, including cyber-physical systems. The mivar technologies of logical AI allowed solving many problems at a qualitatively new level and reducing the decision-making time from billions of years to hundredths of seconds, which enables real-time vehicle control. This paper shows how the application of mivar products (mivar expert systems) allows solving another class of problems – state-space planning (STRIPS planning). In turn, the solutions of these problems enable the development of driver assistance systems at a new level and creation of a fully autonomous vehicle.

1. Introduction

Currently, there is a technological foundation for implementing a logically intelligent system helping a driver to control an automobile. Mivar traffic regulation enforcement system (MTRES) is an example of such a system. The system aims at solving the task of making recommendations for vehicle control and estimating the actions of the driver in real time. Solving this task requires logical intelligence; therefore, a Mivar Constructor of Expert Systems Wi!Mi "Razumator" was used [1]. "Razumator" allows the system to build algorithms for solving problems [2] in the given traffic situation. The theoretical basis of this system includes research in mivar technologies [3-4], including effective solutions in developing driverless automobiles with logical artificial intelligence [5-6], systems for intelligent image recognition [7], analysis of traffic accidents [8], and decision support systems for medicine [9]. It should be noted that the department of information processing systems of Bauman Moscow State Technical University conducts research on creating hybrid intelligent information systems [10] using meta graphs [11-12], cognitive computer graphics [13], neural network-based algorithms [14] and intelligent analysis methods [15]. This comprehensive approach [1-15] allows solving various practical tasks in the area of artificial intelligence.

In the IASF-2018 conference report [16], the architecture of MTRES was presented, and the prospects of using the approaches implemented in the systems were justified. Subsequent research yielded results [17] proving that the principles of developing MTRES are not limited to the tasks of assisting drivers of vehicles; the principles can be applied to implementing a mivar system for planning the actions of robots and robotic systems – MIPRA (Mivar-based intelligent planner of robot actions). The software product MIPRA was developed to solve STRIPS-planning problems in the modified conventional Blocks World domain [18].

Over the last decade, there has been a growing interest in the tasks of intelligent planning of robot actions due to the emergence of new methods and tools for training convolutional and recurrent neural networks [19-20]. There is also research looking into planning algorithms that use the sign world model as a basis for acquiring and preserving knowledge for future use in the planning of actions [21-23]. Special attention has to be paid to the works [24-25] looking into the use of hierarchical behavior planning algorithms in the coalition of agents [26] and the hierarchical reinforcement learning approach for the task of road intersection by a driverless vehicle.

This paper describes the architecture solutions of MTRES, which were essential to the development of the mivar intelligent system for planning actions of robots and robotic transport systems (MIPRA). This paper also demonstrates the advantages of using the principles of mivar logical planning in the area of intelligent transport solutions.

2. Statement of the problem of the mivar planning of actions for a robotic system

The task domain for MIPRA was described in detail in [17]. The task of planning the actions of a robotic system is a modification of the conventional Blocks World domain [18]. This domain consists of only the blocks themselves, a table, and a robot. The blocks form towers with different height, including towers one block high. The table only fits M towers. The task of planning comes down to making a plan of actions for the robot to rearrange the positions of blocks in the space from the initial state to the desired state. The robot is fitted with a manipulator for handling the blocks and machine vision systems for determining the positions of blocks in space.

3. Architecture solutions in developing MTRES and MIPRA

In this section, we demonstrate the pass from the architectural solution of the MTRES over to the creation of the logical intelligent planner MIPRA.

Figure 1 shows the architecture of MTRES. The system input information includes data from machine vision systems [27], automobile sensors, and a navigator [28]. The information is processed to fit the format of the parameters of mivar models, which are made by experts in collaboration with knowledge engineers. Mivar models correspond to the section of the Russian Federation traffic regulations. After that, the parameters are saved in the working memory. It is a section of the memory where a set of the facts describing the subject domain, i.e., the road situation, is accumulated.

The statement of the problem of making recommendations on controlling a vehicle requires sending the current road situation parameters in the system, where the algorithm of making a maneuver is constructed. Each step of the maneuver is placed in a stack of actions. The finished stack is sent in the driver alerting system, thus terminating the processing of the task.

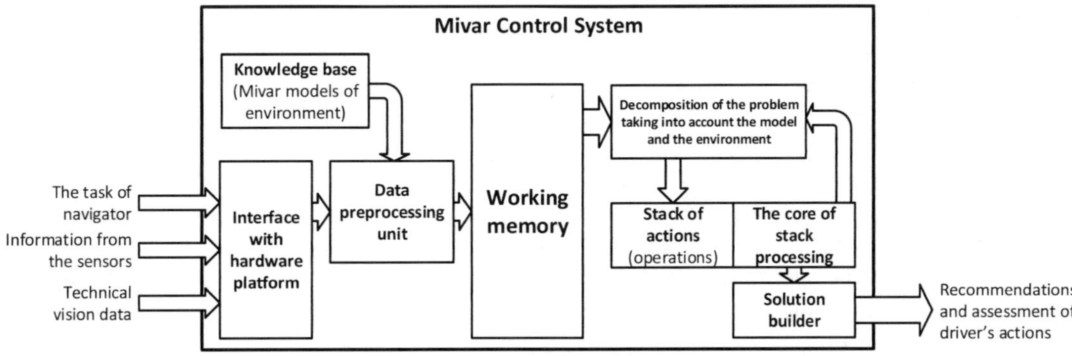

Figure 1. MTRES architecture.

If the system has to evaluate the actions of the driver, the system is given the information about the performed maneuver along with the characteristics of the road situation. "Razumator" constructs a set of actions required from the driver. This set is then compared to the actions performed by the drivers. The result of the comparison is given to the driver alerting system.

This structure of MTRES was presented at the beginning of 2018 [16]. Further research revealed that the proposed system architecture approaches could be successfully used to develop MIPRA [17]. Figure 2 shows the architecture of MIPRA.

The structural differences between the architectures of MIPRA and MTRES are related to the definition of information streams given below. A planning task is submitted to the input of MIPRA. A mivar model is created for this task. The models' generation algorithm aims at decomposing the planning task into sub-goals, the completion of which yields the desired result. The mivar model is loaded in the working memory, where it will organize the planning process. The user of the system signals the start of work, and MIPRA gives the robot a plan of performing the first step. The plan of a step is a set of actions. Subsequent performing of these actions allows getting closer to achieving an intermediate goal. After the step actions are performed, information about the current state of the subject domain is loaded into MIPRA. Based on the current situation, a new step plan is constructed and submitted to the robot for execution. The iteration "loading current situation → generating step execution plan" is repeated until all the conditions of achieving intermediate goals are met.

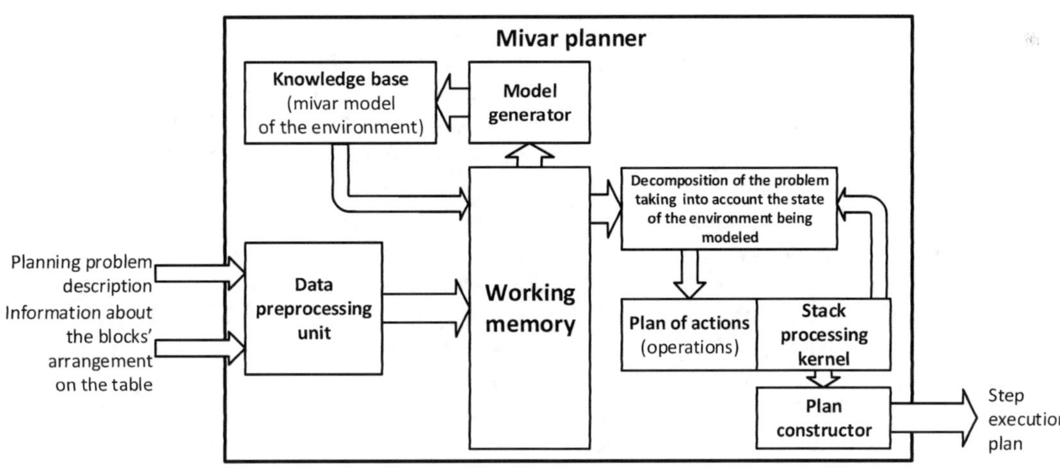

Figure 2. MIPRA architecture.

4. Results of applying the architecture solutions of MTRES in MIPRA

Practical experiments [17] revealed that the mivar planner does not need many hours of computing on a powerful multi-processor to solve this task. On a regular computer, the mivar planner solves the tasks of rearranging blocks in the following time: 10 blocks — 0.98 seconds; 50 blocks — 3.44 seconds; 100 blocks — 10.57 seconds; 200 blocks — 38.32 seconds; 300 blocks — 84.07 seconds. The mivar approach implemented in MIPRA allowed working with more than three towers on the table and with a variable number of blocks even in dynamically changing conditions. We should note that the computational complexity of planning of actions in a domain 300 blocks large (MIPRA generates individual mivar models of each planning task in the domain; the model has 3311 parameters and 1200 rules) using conventional exhaustive search methods will be 1200! (factorial) with the total computation time of more than several billion years.

5. Conclusions

The principles of developing the system of traffic regulations' enforcement MTRES [16] demonstrated their efficacy for developing the mivar intelligent planning system in the state space [17].

The mivar approach and the system architecture concepts used in developing the MTRES allowed moving to the development of a new class of system - intelligent planners with logical reasoning. The developed software product MIPRA, which is based on the mivar system "Razumator," qualitatively reduces the computation complexity and accelerates the speed of constructing the algorithm for solving the STRIPS tasks for rearranging blocks in the Blocks World domain by many orders of magnitude.

The investigation of MIPRA allowed preparing a theoretical and practical basis for proving that the use of mivar technologies allows drastically reduce the computational complexity of the task of planning the actions of robots, groups of robots, multi-level heterogeneous robotic systems [25] and cyber-physical systems of different purpose and application.

The theoretical basis, the test setup, and the practical results obtained in the development of MIPRA were used in the further development of the MTRES. The obtained theoretical results allow the real-life implementation of the project called: "smart cars on smart roads in the smart city" [29].

6. References

[1] Varlamov O O 2018 Wi!Mi Expert System Shell as the Novel Tool for Building Knowledge-Based Systems with Linear Computational Complexity *International Review of Automatic Control* **11**(6) 314-325

[2] Chuvikov D A and Nazarov K V 2016 Designing algorithms for solving physics problems on the basis of mivar approach *International Journal of Advanced Studies* **6**(3) 31-50

[3] Varlamov O O 2003 Exhaustive elementary-incremental summing up of numbers with linear computational complexity *Avtomatizatsiya i Sovremennye Tekhnologii* **1** 34-41

[4] Ostroukh A, Surkova N, Varlamov O, Chernenky V and Baldin A 2018 Automated process control system of mobile crushing and screening plant *Journal of Applied Engineering Science* **16**(3) 343-348

[5] Shadrin S S, Varlamov O O and Ivanov A M 2017 Experimental autonomous road vehicle with logical artificial intelligence *Journal of Advanced Transportation* 2492765

[6] Varlamov O O, Chuvikov D A, Aladin D V, Adamova L E and Osipov V G 2019 Logical artificial intelligence Mivar technologies for autonomous road vehicles *IOP Conference Series: Materials Science and Engineering* **534**(1) 012015

[7] Chuvikov D A, Kazakova N A, Varlamov O O and Golovizin A V 2014 3D modeling and 3D objects creation technology analysis for various intelligent systems *International Journal of Advanced Studies* **4**(4) 16-22

[8] Chuvikov D A, Varlamov O O, Aladin D V, Chernenkiy V M and Baldin A V 2019 Mivar models of reconstruction and expertise of emergency events of road accidents *IOP Conference Series: Materials Science and Engineering* **534**(1) 012007

[9] Varlamov O O, Chuvikov D A, Adamova L E, Petrov M A, Zabolotskaya I K and Zhilina T N 2019 Logical, Philosophical and Ethical Aspects of AI in Medicine *International Journal of Machine Learning and Computing (IJMLC)* **9**(6) 868-873

[10] Chernenkiy V, Gapanyuk Y, Terekhov V, Revunkov G and Kaganov Y 2018 The hybrid intelligent information system approach as the basis for cognitive architecture *Procedia Computer Science* **145** 143-152

[11] Chernenkiy V M, Gapanyuk Y E, Revunkov G I, Kaganov Y T, Fedorenko Y S and Minakova S V 2017 Using metagraph approach for complex domains description *CEUR Workshop Proceedings* **2022** 342-349

[12] Chernenkiy V, Gapanyuk Y, Revunkov G, Kaganov Y and Fedorenko Y 2019 Metagraph Approach as a Data Model for Cognitive Architecture *Advances in Intelligent Systems and Computing* **848** 50-55

[13] Terekhov V I, Chernenky I M, Buklin S V and Yakubov A R 2019 Cognitive Visualization in Management Decision Support Problems *Optical Memory and Neural Networks (Information Optics)* **28**(1) 27-35

[14] Burdakov A V, Ukharov A O, Myalkin M P and Terekhov V I 2019 Forecasting of influenza-like illness incidence in amur region with neural networks *Studies in Computational Intelligence* **799** 307-314

[15] Gapanyuk Y, Latkin I, Chernobrovkin S, Leontiev A, Ozhegov G, Opryshko A and Myalkin M 2017 Architecture and implementation of an intelligent news analysis system *CEUR Workshop Proceedings* **1975** 41-55

[16] Aladin D V, Varlamov O O, Chuvikov D A, Chernenkiy V M, Smelkova E A and Baldin A V 2019 Logic-based artificial intelligence in systems for monitoring the enforcing traffic regulations *IOP Conference Series: Materials Science and Engineering* **534**(1) 012025

[17] Varlamov O O and Aladin D V 2019 Successful application of mivar expert systems for MIPRA –solving action planning problems for robotic systems in real time *Radio industry (Russia)* **29**(3) 15-25

[18] Fikes R and Nilsson N 1971 STRIPS: A new approach to the application of theorem proving to problem solving *Artificial Intelligence* **2** 189-208

[19] Ayunts E and Panov A I 2017 Task Planning in "Block World" with Deep Reinforcement Learning *Advances in Intelligent Systems and Computing* **636** 3-9

[20] Svaco M, Jerbic B, Polancec M and Suligoj F 2019 A Reinforcement Learning Based Algorithm for Robot Action Planning *Mechanisms and Machine Science* **67** 493-503

[21] Panov A I 2017 Behavior planning of intelligent agent with sign world model *Biologically inspired cognitive architectures* **19** 21-31

[22] Panov A I and Yakovlev K 2017 Behavior and Path Planning for the Coalition of Cognitive Robots in Smart Relocation Tasks *Advances in Intelligent Systems and Computing* **447** 3-20

[23] Osipov G S, Panov A I and Chudova N V 2015 Behavior control as a function of consciousness. II. Synthesis of a behavior plan *Journal of Computer and Systems Sciences International* **54**(6) 882-896 ()

[24] Kiselev G and Panov A 2019 Hierarchical Psychologically Inspired Planning for Human-Robot Interaction Tasks *Lecture Notes in Computer Science* **11659** 150-160

[25] Shikunov M and Panov A I 2020 Hierarchical Reinforcement Learning Approach for the Road Intersection Task *Advances in Intelligent Systems and Computing* **948** 495-506

[26] Saikin A M, Buznikov S E, Elkin D S and Strukov V O 2019 Integrated navigation system for unmanned vehicle convoy *International Journal of Innovative Technology and Exploring Engineering* **8**(11) 4021-4025

[27] Endachev D V, Vasin P A and Shadrin S S 2019 Applicability of computer vision architectures and their influence on traffic safety of autonomous vehicles *International Journal of Engineering and Advanced Technology* **8**(6) 5295-5301

[28] Saikin A M and Buznikov S E 2019 Formation of new approaches to the development of perspective intelligent transportation systems *International Journal of Civil Engineering and Technology* **9**(9) 2002-2011

[29] Panferov A A, Zhdanovich E A, Yufimychev K A and Chuvikov D A 2016 Designing algorithms for service robots on the basis of mivar approach *International Journal of Advanced Studies* **6**(3) 72-86

IOP Conf. Series: Materials Science and Engineering **819** (2020) 012007 doi:10.1088/1757-899X/819/1/012007

Features and prospects of the electric vehicle in the Vietnamese automobile market

Nguyen Khac Minh

Moscow automobile and road construction state technical university (MADI), 64, Leningradsky prospect, Moscow, 125319, Russian Federation

Abstract. In this scientific paper, the author attempts exploring the notable features and the prospects for the advancement of the electric vehicle in the Vietnamese automobile market. The need for utilizing electric vehicles, the challenges associated with the development and marketing of them, and advantages for the Vietnamese market are broke down in detail. An analysis of reliable data showed that in recent years, the automotive industry in Vietnam has developed quickly. Vietnam has many advantageous conditions for development and success in the electric vehicle market. However, to rapidly effectively execute the strategy of electric vehicles, more preferential policies of the Vietnamese government are required and it is necessary to build close relationships with the world's automobile manufacturers.

1. Introduction

The electric vehicle in the Vietnamese automobile market is young and promising. In Vietnam, motorcycles and electric motorbikes are currently the most famous methods of transportation. Vietnam's economy is growing strongly, interest in new vehicles is developing in the nation. JATO Dynamics Ltd is a global supplier of automotive business intelligence, pointed out that Vietnam's automobile market has become the second fastest-growing market in the world in 2016. Accordingly, the Vietnamese market increases by 11.1 % per year in the period of 2017 – 2021 (BMO research). In 2016, Vietnam spent about US$ 500,000 on electric vehicle purchases overseas, including Japan, China, and the United States. The Ministry of Industry and Trade Vietnam expects new vehicle sales to more than double to 600,000 units per year by 2025 [1].

The number of vehicles in Vietnam is increasing rapidly in recent years. This is the main cause of traffic blockage and environmental pollution in big cities like Hanoi and Ho Chi Minh. In fact, according to the World Air Quality Report 2018, Hanoi is one of the two most polluted cities in Southeast Asia. The concentration of dust in the air in Hanoi in 2018 was 40.8 μg/m3 - four times higher than the annual limit recommended by WHO (10 g/m3) [2]. Meanwhile, Ho Chi Minh City is the main source of greenhouse gas emissions in Vietnam - 38.5 million tons of carbon dioxide emissions each year, accounting for about 16% of national emissions. In particular, the amount of greenhouse gas emissions in the urban transport sector accounts for 45%. Greenhouse gas emissions adversely affect environmental pollution. The Government of Vietnam has set a goal of developing sustainable and environmentally friendly automobile industry. In particular, the research and development of hybrid and electric vehicles are most concerned.

2. Demand for electric vehicles in Vietnam

According to the survey of JATO Dynamics Ltd, based on the number of tourist vehicles, pickup trucks and commercial vehicles among automobile markets in the world People's Republic of China and the United States of America are the leading countries about the number of vehicles sold, meanwhile owning growth rates of 27.1 % and 48.2 % respectively in 2015 - 2016, Vietnam and Singapore are the leading vehicle markets in the world in terms of growth.

Frost & Sullivan, a global growth consulting company, conducted a study in 2018 titled "The Future of Electric Cars in Southeast Asia". The study was commissioned by Nissan and is based on 1,800 customer surveys in six ASEAN countries. Of all the countries surveyed, relatively low rates of development of electric vehicles in Thailand and Malaysia (figure 1). Among the respondents interviewed, Vietnam has a high development rate 84% and 33% of the people surveyed said that they would consider electric vehicles when they decided on their next purchase (figure 2) [3].

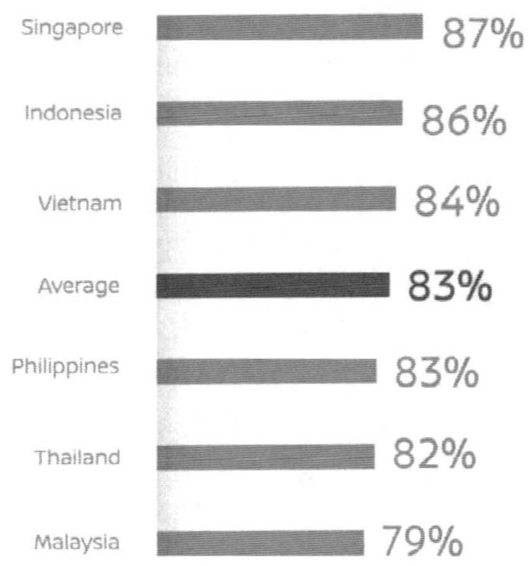

Figure 1. Association of EVs with BEVs.

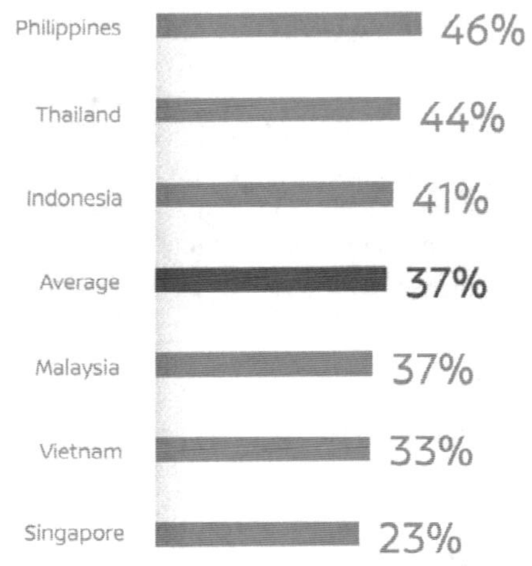

Figure 2. Intention to Buy an EV as the Next Purchase.

The report of the Vietnam Automobile Manufacturers Association (VAMA) in December 2018 shows that the entire automobile market in Vietnam with sales of 34,234 units increased by 10.8% compared with the previous month and reached a growth rate of 22.8% compared to last year [4].

The statistics above show that the number of vehicles sold in the Vietnamese automobile market is not too large, but in recent years, Vietnam's automobile market has reached a very high growth rate. Vietnam has significant potential for the development of the vehicle market.

3. Obstacles to the development of the electric vehicle market

In the future, the market for electric vehicles in Vietnam is in great demand and development prospects, but there are still barriers to the expansion of the electric vehicle market.

Firstly, the traditional vehicle market in Vietnam has been developing for a long time. But the infrastructure for servicing electric vehicles, such as electric charging stations, service stations, is not keeping up with the growing demand.

Secondly, vehicle manufacturers and builders in Vietnam have very little experience in electric vehicles. The fact is that electric vehicles located in Vietnam are imported from the USA, Japan, and China. Finally, the government does not have advantageous preferences to help development for the electric vehicle market [5]. In addition to import duties on electric vehicles shipped to Vietnam, special consumption tax rates range from 15% to 70%, which raises the prices of vehicles to 20%. According to free trade agreements in which Vietnam is a signatory to the Convention, some electric vehicles are shielded from taxes since 2018. These include electric vehicles from Korea and China. Electric vehicles from Japan are taxed at 4%, while Tesla electric vehicles from the USA are taxed at 70% under the most favoured nation status.

4. Advantageous conditions for the development of the electric vehicle market

In Vietnam, there are many advantageous conditions for the development of the electric vehicle market. The population of Vietnam is about 97 million people, which occupies the 14th place in the world. The population aged 24 years to 64 years makes up more than 69% of the total population of the country. From 2008 to 2018, the annual GDP is constantly increasing from 5 % to 7 % per year. This population and economic characteristic is the basis for the development of the automotive industry in general and electric vehicles in particular.

Besides, automobile manufacturers in the world have long been working in the Vietnamese market. In 2017, Mitsubishi Motors Vietnam completed the installation of the first charging station for electric vehicles in Hanoi, located at the headquarters of the Ministry of Industry and Trade, and also referred to the use of this agency electric car. Mitsubishi Motors Vietnam is developing new charging stations in Da Nang and Hoi An (Quang Nam). In 2019, the Fuso trademark (Germany) introduced light trucks using electric motors to the world market, including in Vietnam. This electric vehicle only uses an electric motor and does not discharge pollutant emissions into the environment. This vehicle can carry 2-3 tons of cargo and the range without recharging is about 100 km, charging time takes only one hour to charge 80% of the battery capacity, battery life is 10 years [6, 7].

In 2017, the participation of the automotive industry of VIN Group, the country's largest private conglomerate, raised new hope for the industrialization of Vietnam. VIN Group is chaired by Pham Nhat Vuong, Vietnam's first billionaire with a net worth of $8 billion as of March 21, 2019. The company aims to become a leading automaker in Southeast Asia with an annual capacity of 500,000 units and a localization rate of 60% by 2025 [8, 9, 10].

VinFast has certain advantages in the domestic market, including the potential of the Vietnamese automobile market. VinFast's task, as a new automaker, is to determine their business strategies without encountering inherited problems; good business and financial results of VIN Group and its ecosystem; strong support from the Vietnamese government; and nationalistic sentiment that will encourage Vietnamese buyers to choose their products.

During the year, VinFast managed to develop two prototypes, sedan, and SUV, which debuted with great success at the Paris Motor Show in 2018. VinFast plans to launch its first vehicles on the market in 2019.

Finally, following the ASEAN Free Trade Agreement (AFTA), import duties on manufactured vehicles in ASEAN countries were abrogated from the beginning of 2018 [11, 12]. Given the advantageous conditions outlined above, the Vietnamese government and experts believe in a bright future for their automotive industry.

5. Conclusion

In recent years, Vietnam's vehicle market has experienced significant growth. This is the main reason leading to severe environmental pollution in large cities. Electric vehicles are an excellent solution to the pressing problem of environmental pollution the depletion of fossil fuel reserves. Based on advantageous conditions for the operation of electric vehicles, the market for electric vehicles in Vietnam will grow rapidly in the near future. However, to successfully implement the electric vehicle strategy in Vietnam, the electric vehicle market needs more preferential policies from the Vietnamese government and better relations with vehicle manufacturers around the world.

6. References

[1] Dylan Pastoor 2018 *Vinfast and the Electric Vehicle Market in Vietnam*
[2] IQAir AirVisual 2018 *World Air Quality Report Region & City PM2.5 Ranking*
[3] Frost and Sullivan 2018 *Future of Electric Vehicles in South East Asia*
[4] Vietnam automobile manufacturers' association 2018 *Cover Letter Sales report Dec 2018*
[5] Nguyen X T and Nguyen Q H 2015 *Armand Peugeot Chair International Conference 3rd Electromobility Challenging Issues December* (Singapore) pp 1-9
[6] Karpukhin K, Shorin A, Terenchenko A and Kondrashov N 2017 *ARPN Journal of engineering and applied sciences* **12** pp 4189-95

[7] Shorin A, Karpukhin K, Terenchenko A and Kondrashov N 2018 *International Journal of Mechanical Engineering and Technology* **9** pp 1903-09
[8] https://www.forbes.com/profile/pham-nhat-vuong/#42f38bcc382e (date of the application 12.09.2019)
[9] Le H H 2019 *Vietnam's Industrialization Ambitions The Case Of Vingroup And The Automotive Industry* No 2 (Singapore) p 29
[10] https://www.autocar.co.uk/paris-motor-show-2018 (date of the application 12.09.2019)
[11] Karpukhin K, Terenchenko A and Kolbasov A 2018 *FISITA World Automotive Congress 2018*
[12] Kolbasov A, Karpukhin K, Terenchenko A and Girutskiy O 2019 *IOP Conf. Series: Materials Science and Engineering* **534** pp 1-13

IOP Conf. Series: Materials Science and Engineering **819** (2020) 012008 doi:10.1088/1757-899X/819/1/012008

"Forecast" - adaptive forecast diagnostic intelligent system for vehicles

A A Poddubnaya and A V Keller

FSUE "NAMI", 2 Avtomotornaya St., Moscow, 125438, Russia

E-mail: poddubnaya575@gmail.com

Abstract. The article contains general information about promising vehicle diagnostic systems. Existing diagnostic systems, including those built into modern vehicles (MV), are not able to predict the moment of failure of components and assemblies, but only state the fact of a malfunction. To diagnose the current state and predict the residual life of the vehicle in motion mode, a mathematical model, based on machine learning technologies and data from standard additional sensors and MV detectors, is suggested. Using this approach will make it possible to forecast the defect before its actual occurrence.

1. Introduction

For autonomous transport and connected vehicles, the diagnosis of a vehicle's technical condition is a basic safety standard.* The issue of determining the mechanical failure in unmanned vehicles is extremely relevant due to the lack of driver who can assess uncharacteristic noises or external vibrations. Errors received from vehicle tires are not sufficiently informative when assessing the current state of MV and do not predict breakdowns or faults. An expanded self-diagnosis system should be laid at the stage of design of autonomous vehicles. In the process of exploitation the data from sensors and reliability monitoring systems will be processed on board MV and transferred further to the ITS - intelligent transport system, as well as to the servers of owners and manufacturers. (* according to the research of the European Commission, more details further in the article).

2. Main part

Almost all modern cars are modified with a variety of full-time detecting devices and sensors, fixing faults and operation errors of some units by electrical parameters and fixing "extreme" system states in codes. Error icons appear on the vehicle dashboard when the system detects a fault. If the driver notices an error indicator or an incorrect operation of certain unites or systems and needs to find out the technical condition of the vehicle, some specialized diagnostics is carried out.

To identify the technical problem computer diagnostics of the vehicle is carried out by a certified technical specialist: a scanner with software is connected to the on-board systems through special diagnostic connectors (CAN bus data) which reads all the codes transmitted by the car about possible malfunctions on the main units. Error codes are currently vendor-specific, set by OEM and available for reading and monitoring on a limited list of codes. The received codes are decrypted by specialists using special programs and on the basis of the received data a conclusion about the presence of certain failures or malfunctions is made. On-board data consists of thousands of signals from sensors and ECUs that get transmitted through the CAN network.

They are sent repeatedly with a certain frequency and form continuous data streams that can be used both for driving a vehicle and for signaling the status of various components of the vehicle. Research on monitoring and signal analysis of standard sensors is carried out by vehicle manufactures and so far continuous registration on board has been limited to the vehicles during testing periods when they are developing new models or modernized equipment. These systems are expensive and designed for product development.

In the automotive industry there is little research in predictive prognostics using on-board data from standard sensors. Currently, such methods either require the participation of a person for an expert assessment or the technical condition of the car is assessed by monitoring signals and comparing them with perfect process model. In the problem development review, the main approaches to the solution are formulated in the following methods: Model-Based Diagnostics (MBD) method and the Condition Based Maintenance - CBM method [1]. There are effective studies based on onboard forecasting methods:

- D'Silva carried out a complex stationary system based on a method on the formation of a cross-correlation matrix, including pairwise correlations between signals, where the Mahalanobis distance is used as an assessment scale to search for deviations and malfunctions [2]. The full correlation matrix is used to determine vehicle status. Normal workspace is determined from experimental data. The system works on board with saved normal operation models, and this was demonstrated on simulated data.
- stationary signals for finding damage were used by Vachkov [3] and Kargupta et al. [4]. Their systems consist of an onboard part that continuously monitors the vehicle and loads models into the onboard analogue of OEM monitoring systems. An autonomous system includes a database in which data models, faulty and faultless systems are stored [5]. Also, for embedded on-board systems with limited resources, methods are developed in which sudden changes in the correlation matrix are signs of wear or failure [6].

Until recently car manufacturers have not been interested in developing technologies of exploitation monitoring, but in the present, in connection with the emergence of contractual relations on the principle of "full-cycle service" related to the development of rental, commercial and autonomous vehicles, the topics of reliability forecasting are foregrounded. Commercial vehicle manufacturers have not yet released any advanced forecasting solutions to the market. There are simple preventative maintenance solutions for assessment and prediction of wear that track wear and use of brake pads, clutches, and similar wear equipment. All of them are based on data streams that are aggregated on board and transmitted to a remote office. Mercedes and MAN, among other things, offer to the customers direct remote monitoring and preventative maintenance recommendation solutions. Volvo for commercial vehicles includes forecasting systems offered with maintenance contracts. Volkswagen, BMW [7] and GM [8] have methods for predicting future service needs based on telematics solutions and on-board data. VW and BMW offer preventative maintenance as a maintenance solution for the owner, and GM publishes recommended repairs through the OnStar portal.

Developed embedded on-board solutions have unlimited access to real-time data streams. This provides fast detection since the detection algorithms are located close to the data source. On-board solutions usually have limited computing and storage capabilities, since the hardware must be automotive-grade, for example, resistant to water, shock and electromagnetic interference, as well as inexpensive. Typically, automotive electronics are usually two to three generations behind the consumer market.

The conceptual model which is being developed by the authors is planned to be used in terms of the implementation of remote diagnostics services of the intelligent transport system (ITS) for the connected transport. This service is supposed to be mandatory for the purpose of ensuring exploitation of connected and autonomous vehicles. The service will guarantee feedback on the exploitation of the full product life cycle to manufacturers.

IOP Conf. Series: Materials Science and Engineering **819** (2020) 012008 doi:10.1088/1757-899X/819/1/012008

The direction of the ongoing research coincides with the direction of work within the framework of the European Commission C-ITS, according to the statement of which (Fig. 1) (Cooperative Intelligent Transport System Delegated Act), adopted and agreed by the key stakeholders from the automotive, motorcycle, agricultural, and telecommunications industries; international technical organizations are developing integrated solutions for the priority of the road safety.

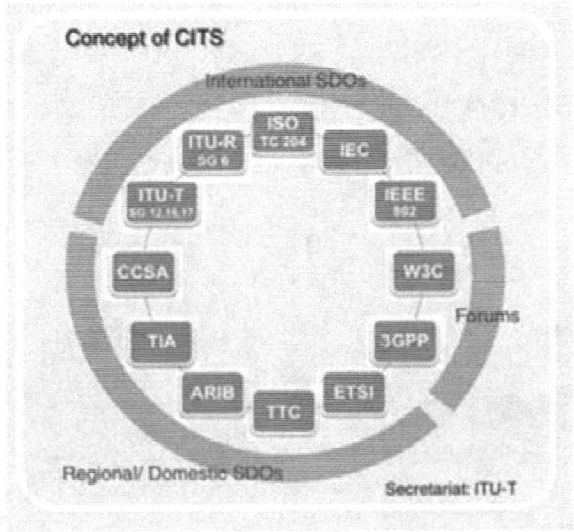

Figure 1. Scheme of the interaction of International Technical organizations, which are making standards for the development of ITS.

The 3GPP international consortium develops technical specifications and technical reports in the field of network technologies in mobile systems together with the European Telecommunications Standardization Institute (ETSI) which transfers the worked out documentation on communication standards and ITS services to the ITU (International Telecommunication Union), which is a specialized institution within the United Nations (UN). It is responsible for issues related to information and communication technology. ITU transfers data to the United Sustainable Nations Development Group - the United Nations development group, which approves international standards at the ISO site.

In international concepts of the ITS development, special attention is paid to the issues of diagnostics and monitoring of connected and autonomous vehicle technical condition. The 3GPP international consortium is developing a standard for connected vehicles * 3GPP 22.885 p. 5.27 – "Remote diagnosis and just in time repair notification", which provides for the installation on the connected vehicle devices with a hardware-software complex that supports interaction C-V2X, (Vehicle-to-everything) and collects diagnostic data from sensors inside the vehicle.

Diagnosis of the technical condition of the vehicle is solved by developing a prognostic model for monitoring electrical, mechanical and hydraulic failures of the vehicle components and assemblies.

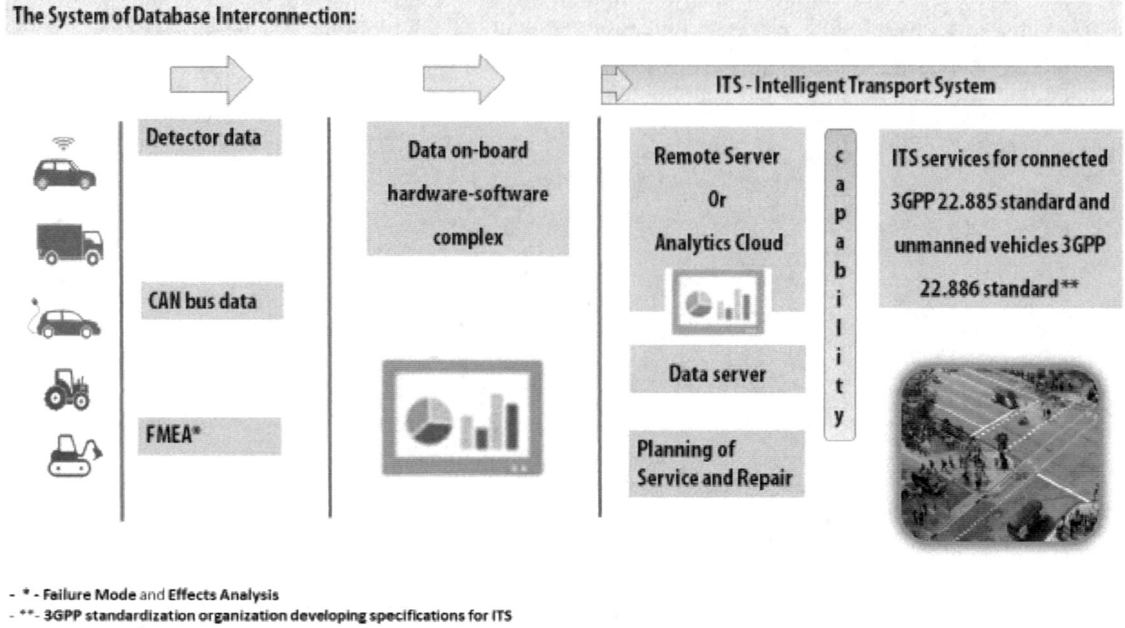

Figure 2. The relationship of databases in the system under development.

The novelty of the idea and its advantages lie in the creation of a model for processing data obtained from various monitoring and diagnostic systems, using a neural network to predict the operation time before the failure of a unit, taking into account the optimal load mode.

Figure 2 shows the relationship between the databases of the described system, which may be applicable for various vehicles. Information is collected as follows: additional sensors are installed on the vehicle, the data received from them together with the data from CAN-BUS and theoretical and statistical results are received and processed on the onboard hardware-software complex by a neural network. After this, the initial data conversion and predictive analysis send it to a remote ITS server with hardware-software complex. The mathematical models used in the analytical process rank incoming data by the integrated indicators for assessing the residual life of the diagnosed units. The analytical system of the ITS in the process of exploitation, taking into account current loads impact assessment and modeling previous and future operating parameters, creates a virtual dynamic operational model for a particular vehicle.

Information from sensors and detectors, based on vibration measurements, acoustic emission and generated heat intensity, determines the technical condition of the hydraulic or mechanical units of the vehicle and is used for the construction of a virtual mathematical model on the ITS. The used complex acoustic emission energy parameter adequately estimates the change in the friction coefficient in the kinematic pair. The complex energy parameter of acoustic emission, in Fig. 3 - parameter D, is calculated on the base of the analysis of a working friction pair acoustic emission signals at ultrasonic frequencies of 20-300 KHz, with ranges from 10 to 10,000 rpm. The sources of acoustic-emission signal formation in the ultrasonic frequency range are elastic deformation waves formed by the dissipation of fracture energy in the structure of materials [9]. Figure 3 shows the confirmed results obtained with the help of an acoustic emission analyzer in experimental modeling of a friction pair.

The transition to the evaluation of the signal by acoustic emission method makes it possible to evaluate the magnitude and nature of the change in energy dissipation that occurs when objects interact in the ultra-acoustic range, to obtain information on friction in the unit and to diagnose this process with an assessment of the condition of the friction pair, attachment quality, lubrication nature and conditions and other parameters in contact. So, when the level of the acoustic emission signal in the friction unit is changing, it is possible to evaluate lubrication quality of the controlled unit. Constant monitoring of the quality of the lubricant will reduce the wear rate and development of defects.

To assess energy losses with the acoustic emission method a diagnostic device is used: it fixes the integral indicator - complex energy parameter of acoustic emission. According to the previous studies' results, the RF patent No. 2427815 - G01M13 / 02 - transmission mechanisms test, "Method for the diagnosis of mechanical transmissions" is registered [10].

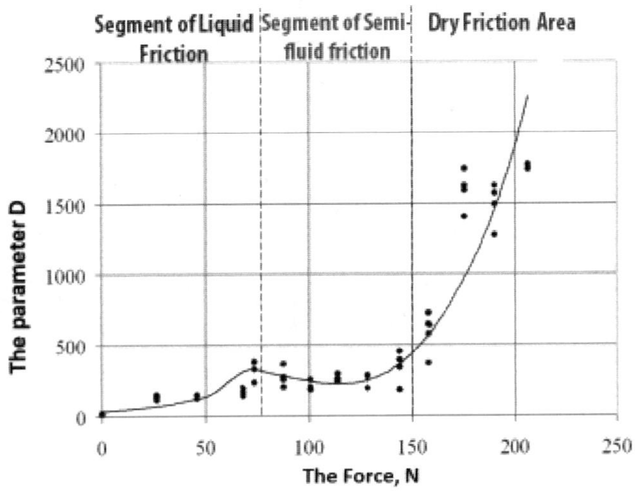

Figure 3. The ratio of the complex energy parameter **D** to the force in the simulated kinematic pair, **N** force on the contact spot.

Calculation methods have objectively established the proportionality of the integral indicator of the acoustic emission sensor signal and the friction coefficient, both for gears and bearings. It is this area that has expanded the existing non-destructive testing methods capabilities, which will allow solving practical problems of mechanism condition monitoring and forecast, rationally distributing forces and means during repair and inspection of vehicle technical condition. Traditional approach resource forecasting is based on the design and construction parameters of unit elements and failure statistics and does not give a real result on operational reliability. It is proposed to change the approach to data processing methods using machine learning. In this case, element failures are considered as some abstract random events of the multifactor process, and the diverse physical conditions of products are reduced to two states: serviceability and malfunction. Prediction problems must be considered with errors in the initial and boundary conditions even when the non-stationary process can be considered as a strictly deterministic process, that is, its outcome is completely determined by the algorithm, the values of the input variables and the initial state of the system. Based on mathematical models of vehicle components and the dependences of sensor and detector data changes, when compared with operating conditions and CAN data, creation of an analytical complex applicable to a wide range of vehicles is proposed.

The complexity of the implementation of the diagnostic complex is not only in the necessity to take into account the operating conditions of a particular vehicle, but also in the need to carry out maintenance in accordance with these conditions. A modern service system must interconnect individual vehicle technical condition continuous monitoring data, planning of their operation by a trucking company and readiness of the service department to fulfill the required vehicle maintenance and repair works.

A general problem that the operating organization is faced with when using schedules of service maintenance and planned replacements of units is the fulfillment of an inappropriate amount of asset maintenance. Since calendar-based maintenance does not take into account the asset's performance, the frequency of maintenance works can often be either too high or too low. These problems can be prevented by optimizing and improving preventative maintenance programs.

The system of preventive diagnostics based on artificial intelligence will make it possible in future to abandon the system of scheduled preventive repairs and switch to maintenance by actual condition [11], experts estimate that it will reduce maintenance costs by 75%, the number of services by more than 50% and the number of failures by 70% in the first year of operation.

3. Conclusions

Advantages of maintenance according to the proposed system:

- The system can be applied in various industries and is adaptable;
- Registration of signs of failure signs and analysis of their variations is performed. Assessment of failure probability of both the system as a whole and its individual units;
- Failures of units are transferred from the category of suddenness to the category of predictability, due to early detection and notification of personnel about a developing malfunction;
- Spare parts logistics is being optimized - timely order and delivery.

The expected economic effect from the introduction of the proposed preventive diagnostics system is possible due to reduction of downtime caused by malfunctions, maintenance and repair costs. The proposed system solves the problems of insufficient competence of employees in assessing the technical condition, and will also help maintain the working condition of aging equipment in a limited budget.

Realized analogs in the field of objective resource forecasting do not exist. The advantage is the prospect of integration into the ITS (Intelligent Transport System), and the creation of a predictive service for any type of vehicle, and most importantly this is the only way to diagnose autonomous vehicles in the field and on the road.

References

[1] Isermann and Rolf 2006 Fault-Diagnosis Systems - An Introduction from Fault Detection to Fault Tolerance ISBN 978-3-540-30368-8

[2] D'Silva S H 2008 Diagnostics based on the statistical correlation of sensors *SAE International Journal of Passenger Cars - Electronic and Electrical Systems* **1(1)** 2008-01-0129

[3] Vachkov G 2006 Intelligent data analysis for performance evaluation and fault diagnosis in complex systems *IEEE International Conference on Fuzzy Systems* pp 6322-6329

[4] Kargupta H et al. 2003 "VEDAS: A mobile and distributed data stream mining system for real-time vehicle monitoring," *Int. SIAM Data Mining Conference*

[5] Mesgarpour M, Landa-Silva D and Dickinson I 2013 Overview of Telematics-Based Prognostics and Health Management Systems for Commercial Vehicles *13th International Conference on Transport Systems Telematics TST 2013* pp 123-130

[6] Kargupta H, Gilligan M, Puttagunta V, Sarkar K, Klein M, Lenzi N and Johnson D 2010 MineFleet®: The Vehicle Data Stream Mining System for Ubiquitous Environments **6202** Lecture Notes in Computer Science (Berlin: Springer) pp 235-254

[7] Weiss V H 2012 Ingenieur.de predictive maintenance: Vorhersagemodelle krempeln die wartung um https://www.ingenieur.de/technik/forschung/predictive-maintenance-vorhersagemodelle-krempeln-wartung-um/ 2019-08-23

[8] OnStar 2014 On Star on-star services https://www.onstar.com 2018-08-23.

[9] Poddubnaya A A, Fokin A S, Ivanov S L, Kremcheev E A and Potapenko V S Assessment of the lubricity of gear oils of mining machines *Notes of the Mining Institute* **178** http://pmi.spmi.ru/index.php/ pmi / article / view / 2176

[10] RF patent No. 2427815 - G01M13 / 02 test transmission mechanisms Method for the diagnosis of mechanical transmissions http://www.freepatent.ru/patents/2427815

[11] Antonenko I N Management of equipment maintenance and repair: automation capabilities - NPP SpetsTek http://trim.ru/sites/default/files/files/pdf/maintenance_food_industry.pdf 2019-05-20

Testing and calculation of vehicle adaptive suspension air springs

D Yurlin, S Bakhmutov, V Kulagin, R Kuleshov

FSUE "NAMI", Moscow, Russian Federation

E-mail: dmitry.yurlin@nami.ru

Abstract. The paper is dedicated to describing the methods of measuring the characteristics of suspension air springs (elastic pneumatic elements) on a test bench, as well as to the methods of evaluating their effect on vehicle parameters. The results of the measurement of air spring characteristics on the test bench under various conditions are presented together with the analysis of changes in their properties. A method for setting a universal force (load bearing) characteristic of the air spring, as well as a method for simulating it in the vehicle are suggested. The results of the vehicle simulation with characteristics of the air springs of different stiffness are shown, a numerical interval of the possible change in the vehicle behaviour parameters in case of change of suspension stiffness in the process of driving is specified.

1. Current trends in the development of suspension air springs

Current leaders of the automotive industry switch to adjustable suspension springs, whereby the adjustment is performed not only by the road clearance but also by their stiffness. An example is the suspension stiffness adjustment system implemented by Porsche in such models as the Panamera and Cayenne. The new versions of these vehicles are equipped with the system of air springs with three chambers (volumes) switching between which allows achieving 3 different stiffnesses in one air spring unit. The description of the Porsche system is given in figure 1.

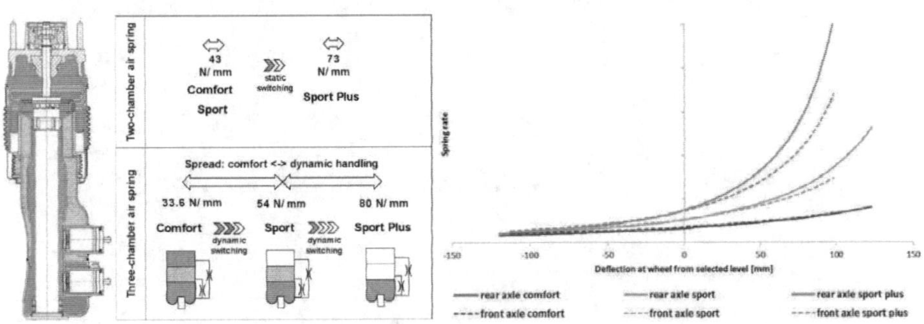

Figure 1. Design of multi-chamber (polyvolume) suspension air spring by Porsche.

Within the projects which are being implemented by FSUE "NAMI", there is a need of analysis of capability and feasibility of such systems and of the advantage of their application. As a result, works on experimental, analytical and computational analysis of the air suspension systems were carried out.

2. Measurement of the air spring stiffness on the test bench

In order to check the feasibility of considerable change in the air spring stiffness by reducing its internal volume, an experiment was carried out where the air spring stiffness was measured with various internal volumes, and also the effect of the internal system pressure and compression velocity (rate) on the stiffness were investigated. The external view of the test object and of the experimental test bench installation is shown in figure 2. The installation control system is presented in figure 3.

The test setup included special washers the removal of which leads to the increase of the internal volume and reduction of the stiffness. The test bench is equipped with a hydraulic installation for compressing the air spring with various frequencies, as well as with a sensor measuring the force generated by the sleeve-type element in the axial direction.

Figure 2. Experimental installation view.

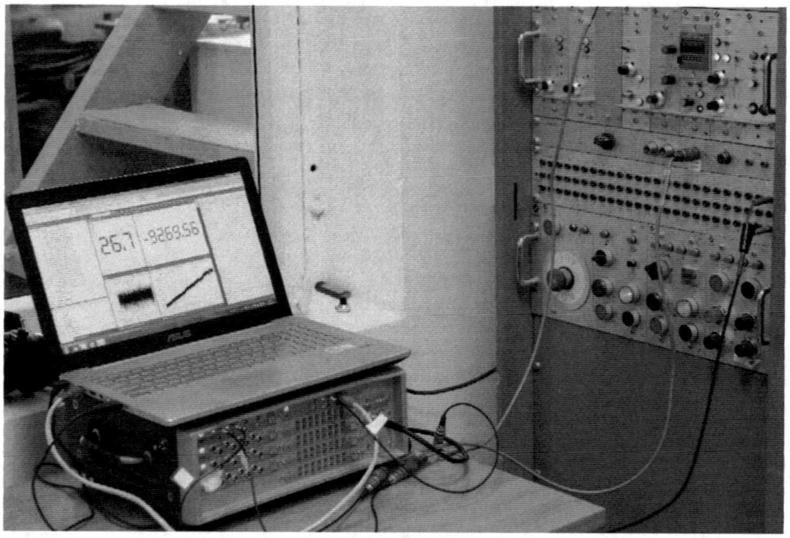

Figure 3. Experimental installation control system.

The graphs of dependence between the axial force and the piston displacement are shown in figure 4. The record is made for 3 variants of frequencies: 0.1 Hz, 1 Hz and 2.5 Hz, as well as for different internal volumes. To simplify the visualization, the results are shown only for 2 volumes of 3.68 l and

2.93 l. The resulting stiffness for different frequencies, as well as the dynamic stiffness coefficient are shown in Table 1. The measurement was carried out under the pressure providing the force of 10500 N in the design position.

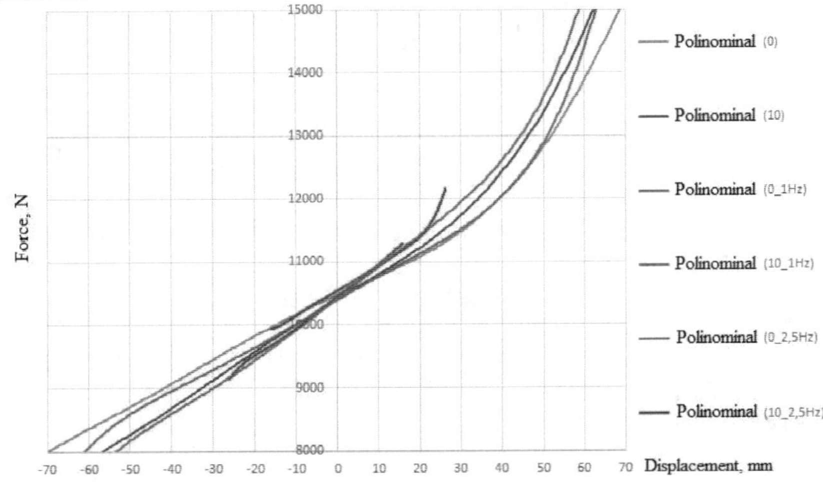

Figure 4. Record of force on sleeve-type element with different volumes and frequencies

Table 1. Stiffness of sleeve-type element with different volumes.

V_intern	V_washer	Number of washers	Stiffness at different frequencies, N/mm			Coeff_dyn_stiffness
1	1	-	0 Hz	1 Hz	2.5 Hz	-
3.68	0	0	29.3	37.8	37.1	1.27-1.29
2.93	0.75	10	39.2	50.4	50.1	1.28-1.29

Figure 5 shows the dependence between the stiffness of the air spring and the internal volume and vibration frequency. Analysis of the results shows that increase of vibration frequency above 1 Hz does not result in additional increase of stiffness. With the amplitude used, the frequency of 1 Hz corresponds to the relative velocity of 0.22 m/s. The coefficient of dynamic stiffness (ratio of dynamic stiffness to quasistatic one) is 1.28 on average.

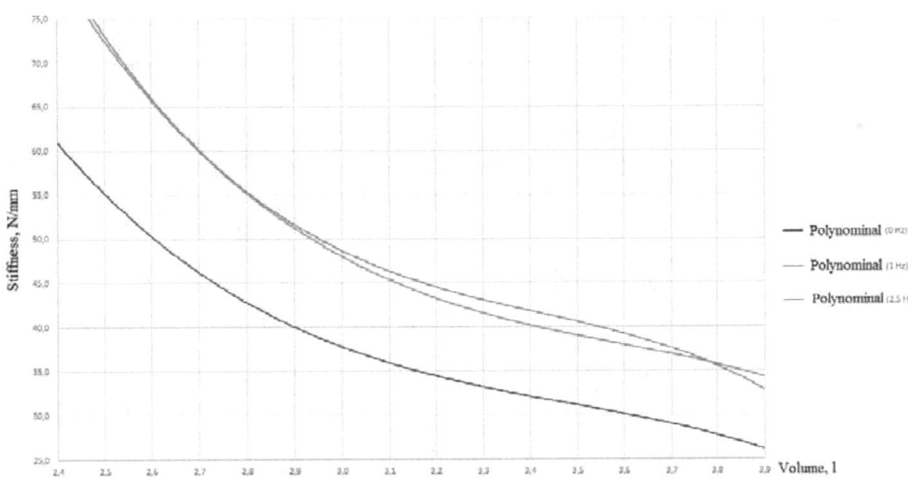

Figure 5. Effect of variable volume on the air spring stiffness.

To simplify the tests, the measured internal pressure in the initial (design) position is replaced by the "installation (mounting) force" - the axial force, developed by the air spring with the initial length at the moment of installation on the bench. Figure 6 provides the recorded force data for different installation force (internal pressure) and two variants of frequencies: 0.1 Hz and 1 Hz.

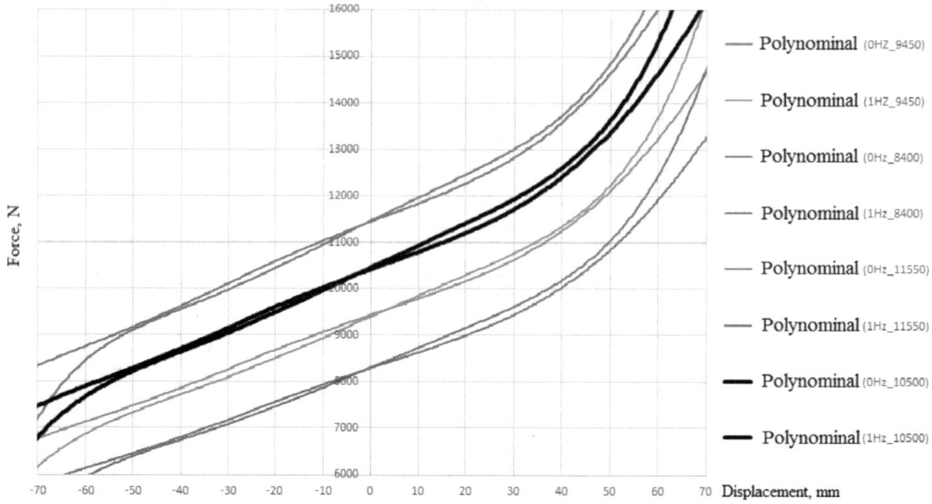

Figure 6. Record of force on sleeve-type element at different internal pressure.

Based on the data from Figure 6, the data according to Table 2 can be obtained. The stiffness of the air spring increases along with the increase of the installation force.

Table 2. Measurement results for different installation force.

Pressure, MPa	Force, N	Frequency, Hz	Stiffness, N/mm
9.68	11550	0 Hz	39.43
9.69	11550	1 Hz	51.07
8.81	10500	0 Hz	38.16
8.84	10500	1 Hz	48.95
7.98	9450	0 Hz	35.57
7.94	9450	1 Hz	45.99
7.01	8400	0 Hz	34.30
7.02	8400	1 Hz	43.34

Figure 7 provides measurement results with different installation forces. The stiffness increases along with the internal pressure increase, the dynamic stiffness is higher than the quasistatic one by 28% on average.

IOP Conf. Series: Materials Science and Engineering **819** (2020) 012009 doi:10.1088/1757-899X/819/1/012009

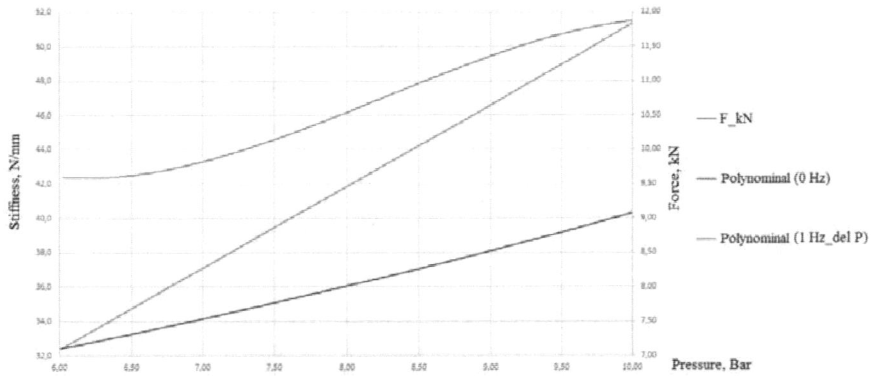

Figure 7. Effect of internal pressure on the air spring stiffness.

3. Universal description of the air spring force

Based on the performed measurements, a universal description of the sleeve-type element force depending on the internal volume, pressure and piston velocity may be suggested. A polynomial function obtained when measuring the air spring characteristics and accepted as a basic one (Fbaz(h)) may be used for that. The design position of the air spring is considered to be its axial length when installed in a vehicle and when the vehicle is motionless. The design force is considered to be the value of the load borne by the air spring under these conditions. Accordingly, on the test bench, the design position of the air spring is reproduced and the initial load considered as the design force is generated. In order to simplify the comparison of the functions obtained with different parameters, all the graphs are brought to the zero initial value – the measurement is carried out with different design forces which are then subtracted from the obtained values. The dependence is built based on the piston displacement, similar to the measurements on the test bench. In order to take into account the effect of the volume, pressure and velocity, special coefficients are introduced which also depend on the piston displacement, as the effect of these parameters changes depending on the sleeve-type element compression ratio (degree) (K_v(h), K_f(h), K_velos(h,velos)). The coefficients are built by way of identifying correlations between the basic characteristic, auxiliary characteristic and current properties of the system. Based on the test results, different characteristics which can be divided into "basic" and "auxiliary" are obtained for individual parameters. Knowing the value of the defining parameter change which led to the transition of the basic characteristic to the auxiliary one, and knowing also the parameter value for the "new" – "current" – state of the system, the value of the coefficient, by which the basic characteristic values should be multiplied in order to obtain its current condition (Figure 8), may be defined.

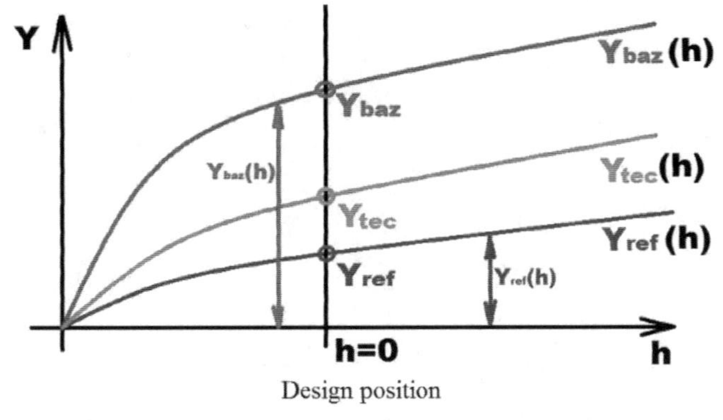

Figure 8. Characteristics for different states of the system.

Upon that, the progression of the characteristic growth for different piston positions may vary, which should be considered in case of rearrangement. The Kvsp auxiliary coefficient which is defined for each position of the piston within the analyzed range (corresponding to the possible piston displacement in real systems) is used for this. The auxiliary coefficient takes on values less than one in accordance with the ratio (correlation) of the values of the investigated characteristics (Figure 9).

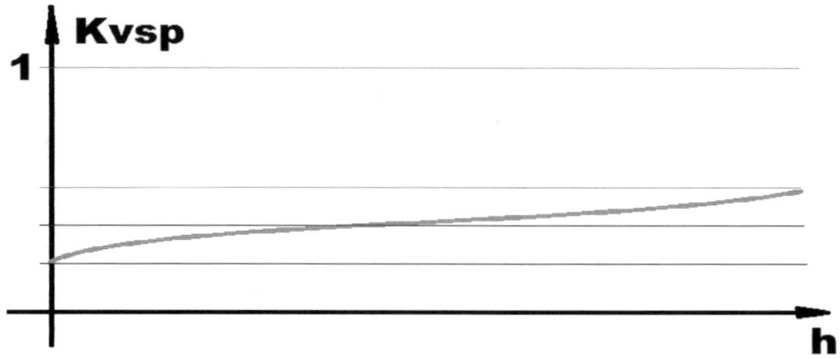

Figure 9. Auxiliary coefficient characteristic.

To carry the basic and auxiliary characteristics to the current state of the system, the Kosn main coefficient is used which corresponds to the ratio of the differences of characteristic values in the design position – the basic characteristic and the current one to the basic and auxiliary ones – Equation (3). The basic characteristic was chosen from the test results, with the parameters which were used in every test (during the tests with different volume, different pressure and velocity). The basic characteristic corresponds to the volume of 3 litre (9 auxiliary washers), the force of 10500 N and the velocity close to 0 m/s. The auxiliary characteristic corresponds to the test results with the system properties considerably differing from the basic one in terms of the investigated parameter. The characteristic for 10500 N, 0 m/s and 3.68 l is used for the volume coefficient. The characteristic of 8300 N, 0 m/s and 3 l is used for the force (pressure) coefficient. 10500 N, 0.22 m/s (1 Hz) and 3 l are used for the velocity coefficient.

The equation of each of the 3 coefficients is made by an identical pattern:

$$K = 1 +/- Kvsp * Kosn \tag{1}$$

Kvsp is the auxiliary coefficient that defines the ratio or correlation of the basic and auxiliary characteristics.

Kosn is the coefficient that defines the ratio of system properties in static position.

$$Kvsp = \frac{\mathrm{Ybaz(z) - Yref(z)}}{Ybaz(z)} \tag{2}$$

$$Kosn = \frac{\mathrm{Ybaz(z) - Ytec(z)}}{Ybaz - Yref} \tag{3}$$

According to the provided analytics and formulas (1), (2) and (3), the coefficients of pressure (force), volume and velocity will be as follows.

Volume coefficient:

$$Kv = 1 - KvspV * KosnV \tag{4}$$

$$KvspV = \frac{FbazV(z) - FrefV(z)}{FbazV(z)} \tag{5}$$

$$KosnV = \frac{Vbaz - Vtec}{Vbaz - Vref} \tag{6}$$

Force (pressure) coefficient:

$$Kf = 1 - KvspF * KosnF \tag{7}$$

$$KvspV = \frac{FbazF(z) - FrefF(z)}{FbazF(z)} \tag{8}$$

$$KosnF = \frac{Fbaz - Ftec}{Fbaz - Fref} \tag{9}$$

Velocity coefficient:

$$Kvelos = 1 + KvspVel * KosnVel(Velos) \tag{10}$$

$$KvspVel = \frac{FbazVel(z) - FrefVel(z)}{FbazVel(z)} \tag{11}$$

$$KosnVel = \begin{cases} \frac{VelosTec}{VelRef} \; if\ Velos \leq 0{,}22 \\ 1\ if\ Velos > 0{,}22 \end{cases} \tag{12}$$

Thus, the equation of force of the sleeve-type element will correspond to the function of the piston displacement (13).

$$Ffin = Ftec + \mathrm{Fbaz(h)} * Kf(z) * Kv(z) * Kvelos(z, velos) \tag{13}$$

According to the previously presented system parameters, the volume coefficient took on the values according to Figure 10, the force coefficient – according to Figure 11 and the velocity coefficient – according to Figure 12. At the same time, the main coefficient took on the value of 0.71 for the volume (Vtec = 3.48 l) and 0.3 for the force (Ftec = 9790 N). For the velocity, it amounted to 1 (as the quasistatic situation was considered). According to Equation 13, a curve characteristic of the air spring was built (Figure 13). Figure 13 A shows the curve according to the specified parameters, Figure 13B provides the curve at the velocity above 0.22 m/s.

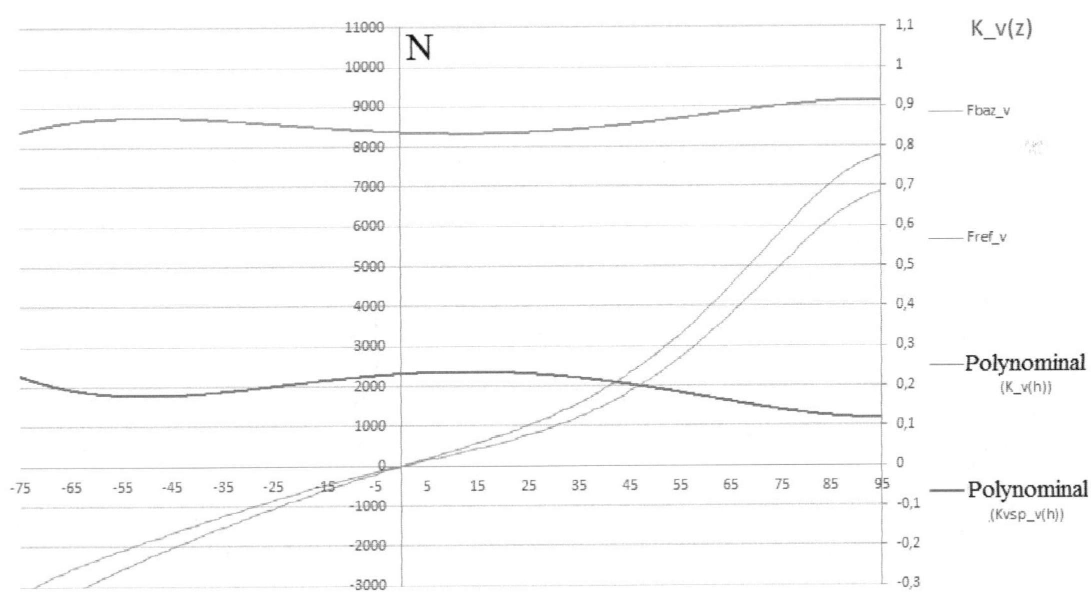

Figure 10. Resulting value of the volume coefficient.

IOP Conf. Series: Materials Science and Engineering **819** (2020) 012009 doi:10.1088/1757-899X/819/1/012009

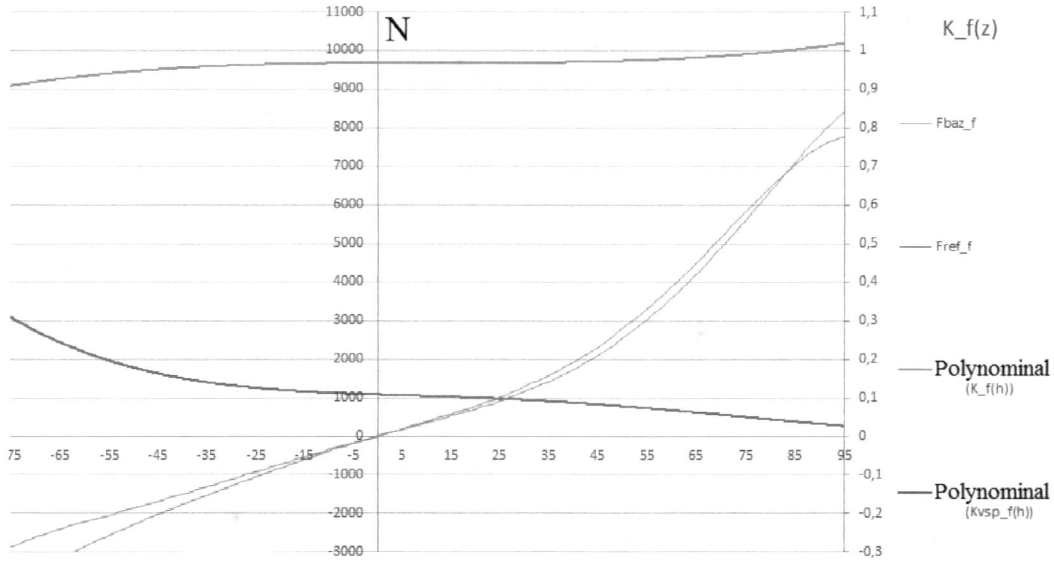

Figure 11. Resulting value of the force coefficient.

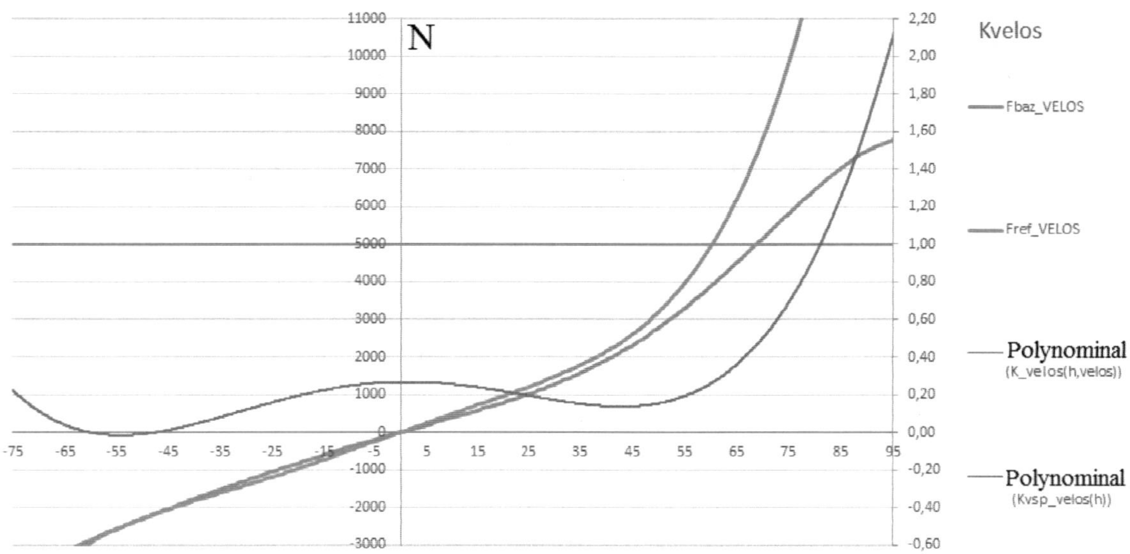

Figure 12. Resulting value of the velocity coefficient.

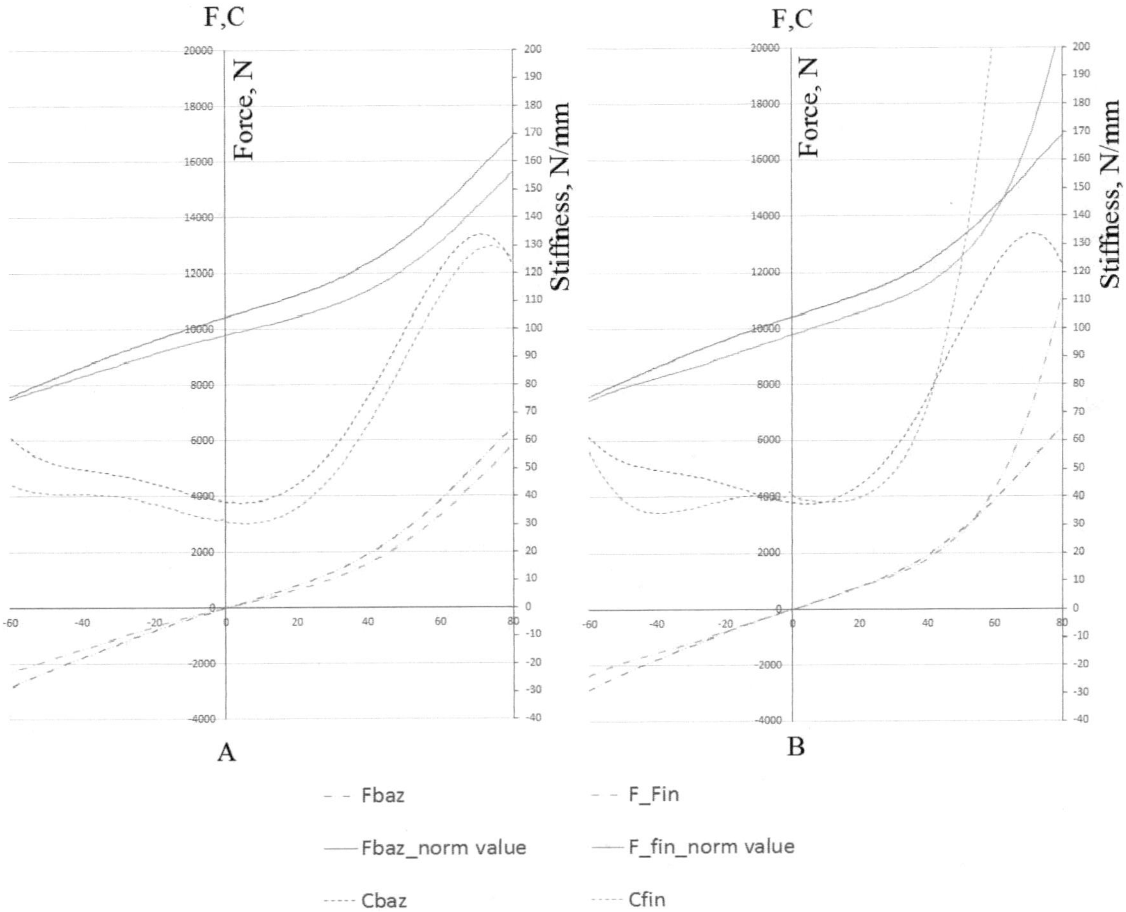

Figure 13. Resulting curve of the air spring. A – for quasistatic state, B – for dynamic state.

To check the accuracy of the calculation performed, the data obtained by calculation shall be compared with the data obtained during tests for the same system properties. Figures 14 A and B provide the comparison of simulation (computed) and experimental data for the installation force of 9450 N, volume of 3 l and two variants of velocity – quasistatic and 1 m/s.

IOP Conf. Series: Materials Science and Engineering **819** (2020) 012009 doi:10.1088/1757-899X/819/1/012009

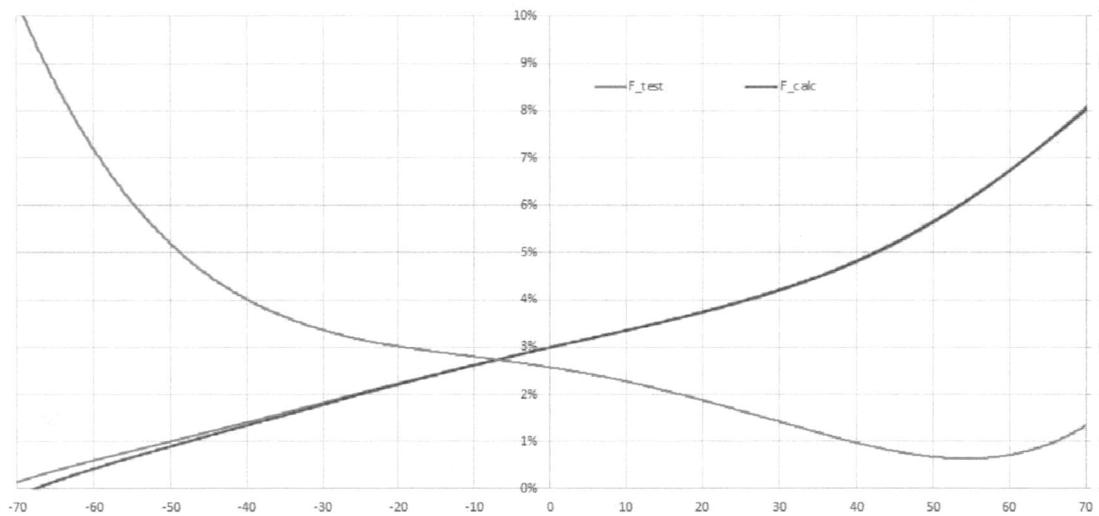

(A) Percentage difference

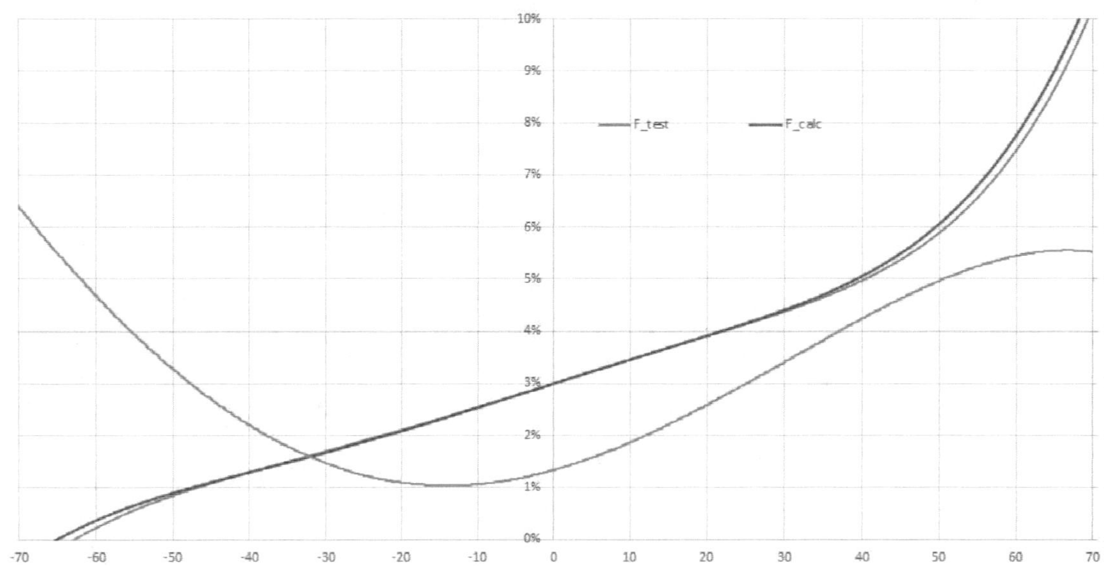

(B) Percentage difference

Figure 14. Tolerance of simulation (computed) and experimental characteristics at minimum piston speed (Simulation (computed) and experimental force characteristic on the right axis). Percentage difference on the main axis.

It is evident from the tolerance graphs analysis that the data deviation for the displacement of +/- 50 mm does not exceed 5%. With bigger displacements, the tolerance increases along with the value of the force itself, however, no high calculation accuracy is required within these ranges due to no need to assess the ride comfort. It is the progressivity value that is estimated, which corresponds to the required level. In general, it is obvious from the graphs and diagrams that the simulation (computed) and experimental characteristics almost match, which indicates high calculation accuracy.

4. Description of the prototype design of multi-chamber (polyvolume) air spring system

To test the system in a real vehicle, a prototype system with two gas chambers (volumes) in each air spring is used. Its characteristics and calculated stiffness are given in Table 3.

Table 3. Volume characteristics of the air springs.

Axle	Front		Rear	
Static force on the air spring	F=9790 N		F=8080 N	
Volume	3.48	2.87	3.32	2.35
Stiffness 0 Hz	31	39	31	42
Stiffness 1 Hz	41	50	40	54

The prototype air spring view is given in Figure 15. The curve for the front air spring with the volume of 3.48 l is used as an example in Figure 13.

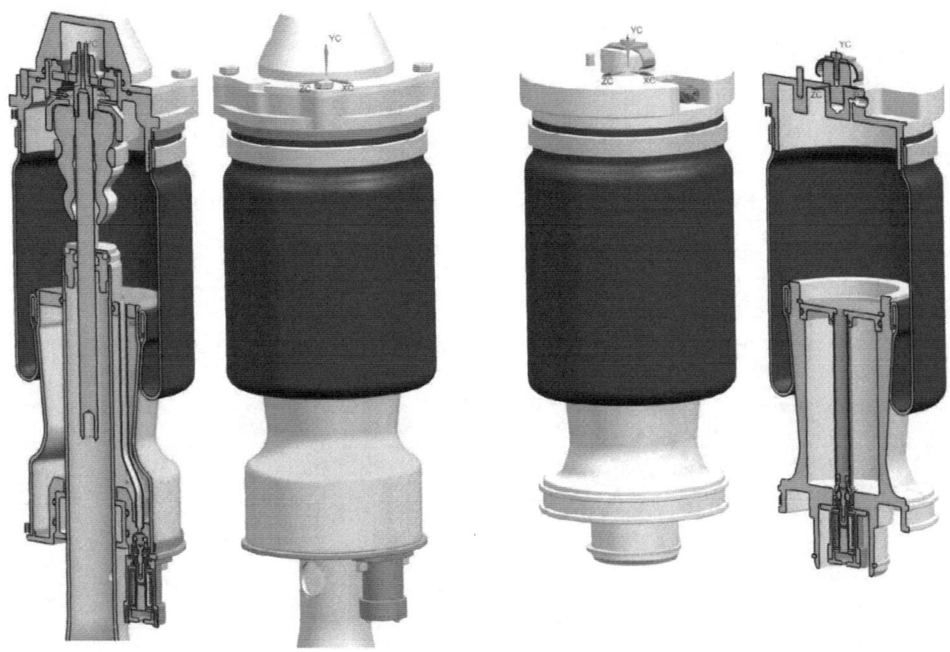

Figure 15. Prototype two-chamber air spring view.

5. Computer analysis of the air spring influence on the vehicle behaviour

To assess the vehicle behaviour with the real characteristic of the air suspension, a complex model is developed with the vehicle description in the form of a multibody model in the ADAMS environment and the air spring description by a separate program in the Matlab environment. The system parameters and the piston speed obtained from the multibody model are used at the input to the air spring program. The layout of the Matlab model is given in Figure 16.

IOP Conf. Series: Materials Science and Engineering **819** (2020) 012009 doi:10.1088/1757-899X/819/1/012009

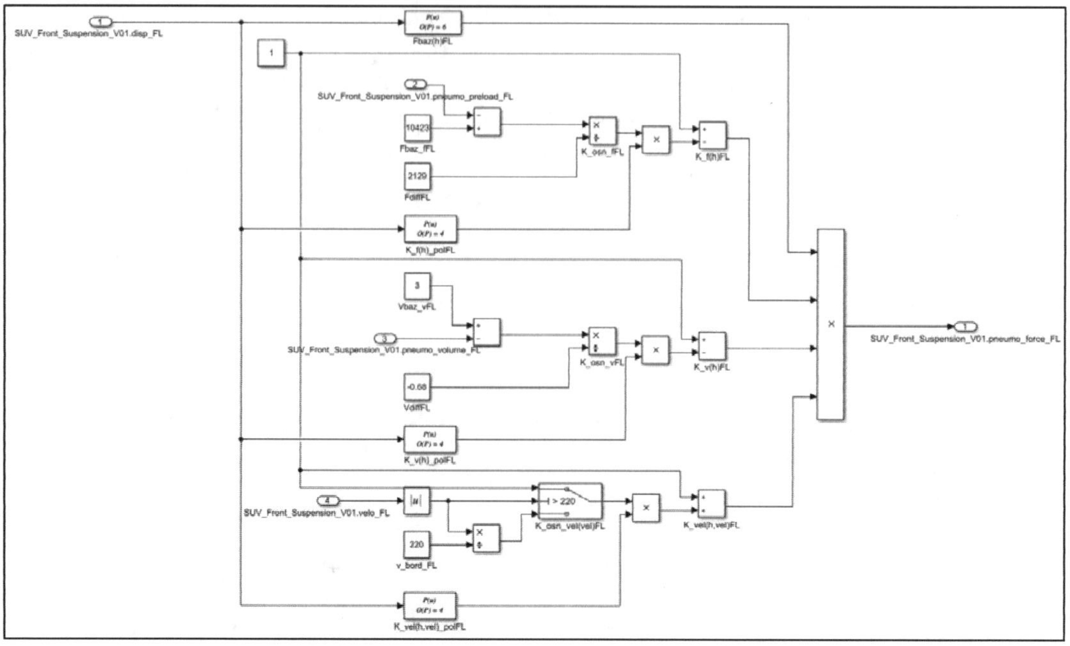

Figure 16. Layout of the air spring model in Matlab environment.

Three coefficients are calculated in this model with the subsequent calculation of the force on the air spring. The $Kvsp$ coefficients are presented in the form of polynomials in accordance with the basic and auxiliary characteristic graphs.

To assess the stiffness of the in-vehicle air springs, it is necessary to conduct a number of special maneuvers with measuring the key parameters, the value of which will give an opportunity to estimate to which extent the selected stiffness suits the vehicle.

It is customary to use the maneuvers specified below for the vehicle properties assessment.

For stability and steerability assessment:

- driving with a constant radius;
- driving with a constant radius and steering angle;
- sine driving;
- slalom;
- "fish hook";
- lane change;
- double lane change (elk test);
- acceleration;
- braking.

For vibration load assessment:

- cobblestone driving;
- single bumping test.

Upon that, not all the specified maneuvers are indicative for assessing the contribution of the air springs only. In the dynamic maneuvers, dampers (shock absorbers) are more significant, while during the maneuvers related to big lateral pitches antiroll bars contribute more. Among the stability and steerability maneuvers, the air spring contribution can be detected when changing the lane. Increase in the air spring stiffness by 30% leads to decrease in the maximum roll angle by 2.5% at the first peak and 8% at the second one or by 3% at the first one and 12.5% at the second one in case of the deactivated antiroll bar (Figure 17). The graphs where both air spring chambers are used (soft characteristic) are designated as "Air soft". For the variant with one chamber (hard characteristic), the "Air hard" designation is used. The graph with the activated antiroll bar is designated as "ARB on", the one with the deactivated antiroll bar is designated as "ARB off".

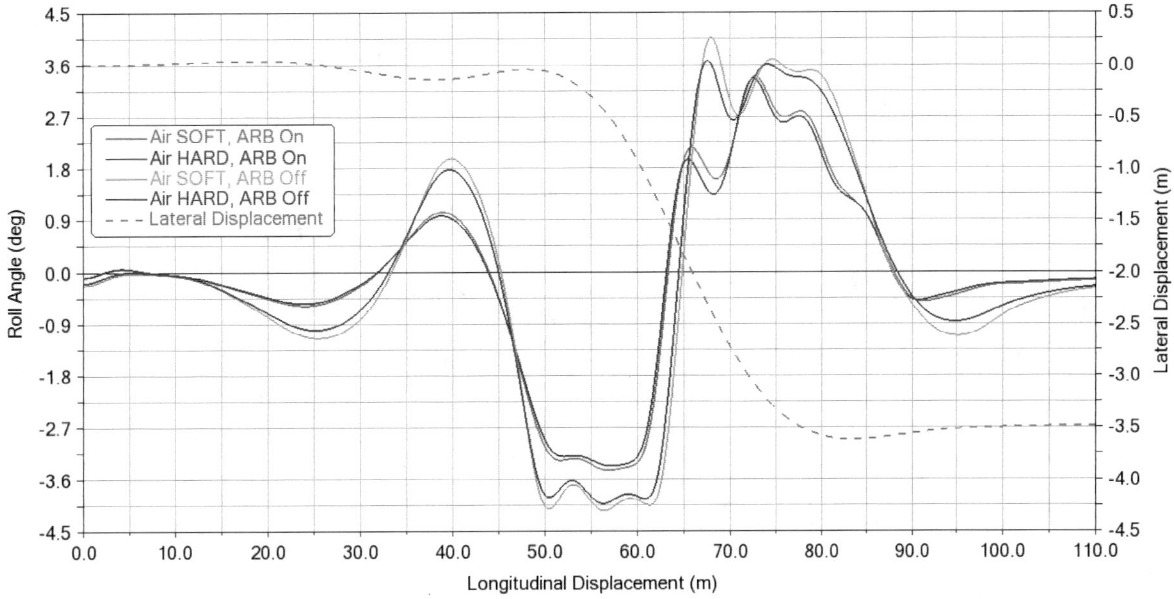

Figure 17. Dependence of roll angle and lateral displacement on longitudinal displacement for lane change at 80 km/h. Lateral vehicle displacement on the right axis. Longitudinal vehicle displacement on the x-axis.

The air springs show themselves most clearly during braking and acceleration. Without switching off the antiroll bars, the decrease in the longitudinal pitching reaches 10%, irrespective of the "soft" or "hard" characteristic of the dampers (Figure 18). The graphs with low resistance of the dampers are designated as "Damper soft", the graphs with high resistance – as "Damper hard".

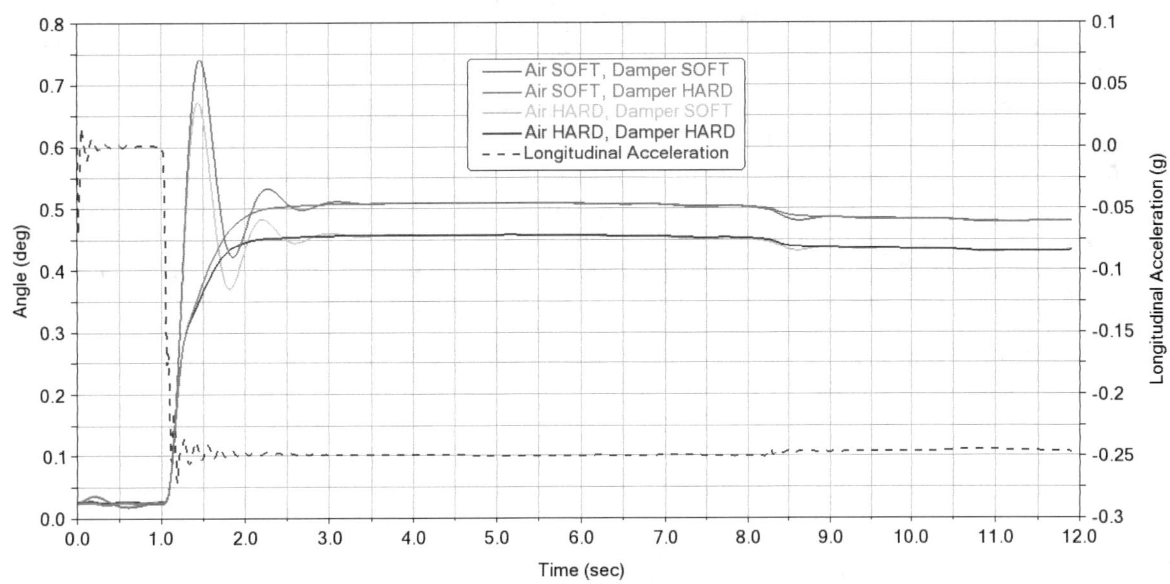

Figure 18. Dependence of longitudinal pitching angle and longitudinal acceleration on time for braking with 0.25g acceleration from 100 km/h

At the same time, when driving on cobblestone pavement, the decrease in maximum values of accelerations reaches 34% (8% for hard dampers) and that of the mean square deviation of acceleration is by 17% (5% for hard dampers) – Figure 19.

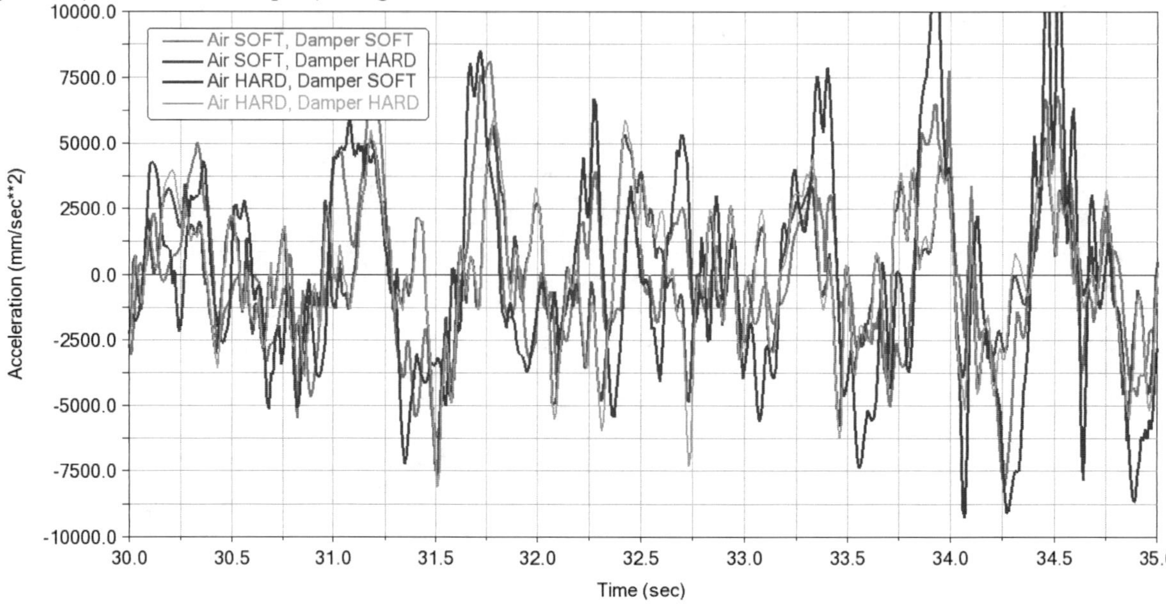

Figure 19. Graph of vertical vibration acceleration dynamics at front passenger seat rails.

6. Conclusion

The paper presents a method to measure the force (load bearing) characteristic of adaptive suspension air springs. The provided data enable making a conclusion and obtaining polynomial descriptions of properties of the air springs when changing their internal volume, pressure in the design position and piston velocity. It was discovered that the dynamic stiffness of the suspension is 28-30% higher than the quasistatic one. At the same time, the velocity increase for more than 0.22 m/s does not result in any additional increase in stiffness. A method of universal description of a sleeve-type element force by multiplying the basic characteristic polynomial by 3 coefficients defining the main properties of the system (internal volume, initial pressure, piston velocity) is presented.

A method of investigating the air springs installed in a vehicle is described, also a method of simulating the vehicle behaviour for performing such research is presented. Upon that, the real air spring characteristics and the most distinctive maneuvers are used in the model. The air spring volumes (chambers) themselves were selected according to the real prototype of such system manufactured for the vehicle behaviour study. The simulation results with the numerical parameters indicative of expediency of implementing air springs with two and more chambers in the vehicle suspension have been obtained.

References

[1] Gantikow M, Boyraz E, Kallert N, Legierski R *The new multi-chamber air spring by Porsche – future innovation in chassis mechatronics and integration* 7th International Munich Chassis Symposium 2016 Wiesbaden: Spinger Fachmedien 2017 pp 437-456

[2] GOST 31191.1-2004 (ISO 2631-1:1997) Vibration and shock. Measurement and evaluation of human exposure to whole-body vibration

[3] GOST 31507 – 2012 Handling and stability. Technical requirements. Test methods

[4] Pevzner Ya M, Gorelik A M *Pneumatic and hydro-pneumatic suspensions* Ed. Gelfgat D B Moscow Leningradskaya printing house of Gosgortechizdat Publ, 1963 p 321

[5] Chassis Handbook Fundamentals, Driving Dynamics, Components, Mechatronics, Perspectives

This book is based on the 2nd edition of the German book a hrwerkhandbuch edited by Bernd HeiЯing and Metin Ersoy 1st ed

[6] *Spring design manual* prepared under the auspices of the SAE spring committee 2nd ed 1996

[7] Peyman K E *Interconnected Air Suspensions with Independent Height and Stiffness Tuning*. A thesis presented to the University of Waterloo in fulfillment of the thesis requirement for the degree of Master of Science in Mechanical Engineering Waterloo Ontario Canada 2014

[8] Nieto A J, Morales A L, Chicharro J M, Pintado P *An adaptive pneumatic suspension system for improving ride comfort and handling* Journal of Vibration and Control 2016 Vol. 22(6) pp 1492–1503

[9] Nahvi H *Evaluation of Whole-Body Vibration and Ride Comfort in a Passenger Car*. Mechanical Eng Dept Isfahan University of Tech Isfahan 84156-83111 Iran 2009

[10] Khettou N et al, *Contribution to the modelling of a pneumatic semi-active Control of vehicle suspension* Vojnotehnički glasnik / Military Technical Courier Vol LXIII No 4 2015, pp 99–115

IOP Conf. Series: Materials Science and Engineering **819** (2020) 012010 doi:10.1088/1757-899X/819/1/012010

Simulation based feasibility confirmation of using hybrid powertrain system in unmanned dump trucks

A B Kartashov[1] **and G I Skotnikov**[2]

Wheeled Vehicles Department, Bauman Moscow State Technical University, Moscow, Russia

E-mail: [1] kartashov@bmstu.ru, [2] skotnikovg@bmstu.ru

Abstract. The use of medium-duty vehicles (up to 50 tons) with a simultaneous reduction in technological and operational downtime, which is achieved by automated traffic control and loading and unloading systems, as well as optimization of technological maintenance modes, can be considered a new trend in the development of mining transport. The solution that meets the requirements and current development trends is the use of medium-duty unmanned dump trucks (UDT) in the interests of the Russian coal industry. An assessment of the most important performance characteristics of UDT, such as energy efficiency indicators, is considered in this research. The purpose of the research is to confirm the feasibility of using hybrid power system (HPS) in the construction of UDT. A series HPS is considered, characterized by the use of two traction electric motors: one drives the wheels of the front axle, the second drives the wheels of the rear bogie axle. The feasibility confirmation is carried out by comparing the energy efficiency and mobility of the studied vehicle with similar, equipped with other types of powertrains by simulation methods. The presented simulation results show the efficiency of HPS application in comparison with a traditional powertrain and a mechanical transmission for UDT when driving in drive cycles similar to a given one. The use of HPS allows reducing energy consumption for motion by 20-25% due to the recovery of braking energy.

1. Introduction

Currently, more than 1.5 billion tons of coal, 80% of copper ore, two-thirds of gold, more than 50% of bauxite, phosphate, lead, zinc and other metals and minerals are mined in hundreds of the largest quarries in the world. At the same time, there is a steady tendency towards an increase in capacity and depth of quarries (the growth rate of which is 10-15 m per year) and the resulting demand for high-performance energy-efficient mining and transportation equipment. About 14 thousand units of transport equipment operate at 760 quarries in 63 countries of the world. The largest share (about 10 thousand units) are dump trucks with a carrying capacity of more than 90 tons. At the same time, mining dump trucks with a carrying capacity of 120 tons and above account for more than 60% of the total fleet of heavy vehicles [1].

The use of medium-duty vehicles (up to 50 tons) with a simultaneous reduction in technological and operational downtime, which is achieved by automated traffic control and loading and unloading systems, as well as optimization of technological maintenance modes, can be considered a new trend in the development of mining transport [2].

Requirements for the directions of development of mining transport are defined in the long-term program for the development of the coal industry of the Russian Federation for the period until 2030.

IOP Conf. Series: Materials Science and Engineering **819** (2020) 012010 doi:10.1088/1757-899X/819/1/012010

One of the directions of development is the need to develop and implement systems of unmanned mining of minerals, including the use of modern software and hardware systems integrated into the overall management system of a mining enterprise.

The solution that meets the requirements and current development trends is the use of medium-duty unmanned dump trucks (UDT) in the interests of the Russian coal industry.

An assessment of the most important performance characteristics of UDT, such as energy efficiency indicators, is considered in this research.

The object of research is an articulated six by six UDT, which gross weight is 75 tons. The purpose of the research is to confirm the feasibility of using hybrid power system (HPS) in the construction of these vehicles. A series HPS scheme is considered, characterized by the use of two traction electric motors (TEM): one drives the wheels of the front axle, another one drives the wheels of the rear bogie axle. In Figure 1b, solid lines show mechanical connections, dotted lines show electrical connections. The goal is achieved by solving the following tasks:

- estimation of energy efficiency of UDT with a HPS,
- estimation of mobility of UDT with a HPS.

Studies of the use of HPS for dump trucks are relevant at the present time, so many articles, such us [3 – 7], are devoted to this problem.

The energy efficiency of a UDT with a HPS is estimated by comparing it with a diesel analog with a mechanical transmission (Figure 1 a), as well as with a UDP using only an electrochemical storage (electric UDT, Figure 1 c).

The mobility of UDT during its movement in unprepared terrain is understood as a set of properties that determine the ability to move in given drive cycle with high speed and efficiency [8-11]. In this research, the mobility of UDT is defined by estimation of deviating of the actual UDP speed from the given one in the proposed drive cycle. Using of driving cycles for the same goals was applied by authors of studies [12, 13].

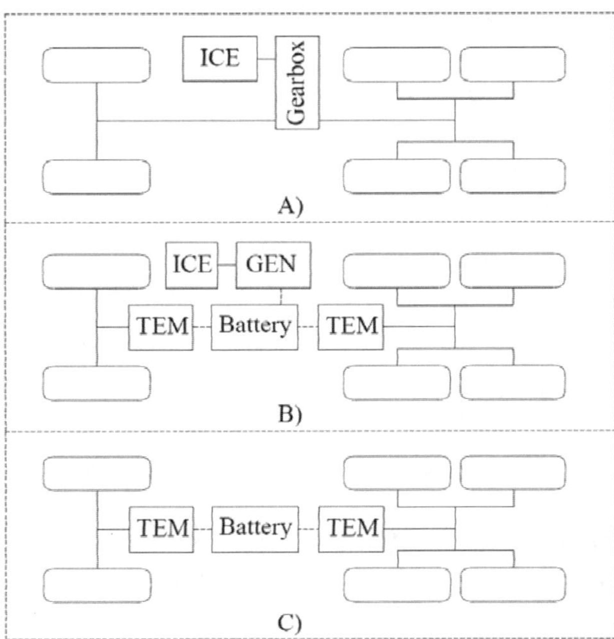

Figure 1. UPD Scheme: a) with diesel ICE and mechanical transmission; b) with HPS; c) with electrochemical storage.

2. Research methodology

Simulation mathematical models have been developed for estimation of the UDT energy efficiency and the ability to move under given conditions at maximum speed.

The developed mathematical models are implemented in the MATLAB\Simulink software and allow simulating UDT movement along a given driving cycle (coal mine) in order to obtain quantitative indicators of UDT energy efficiency and mobility with the selected type of powertrain.

The developed mathematical model for evaluating the energy efficiency and mobility of the UDT consists of several structural modules, including lower-level modules (battery module, generator module, ICE module, etc.) and upper-level modules (HPS module, UDT 6x6 chassis module). It is possible to synthesize simulation models of various powertrains of UDT to assess their mobility and energy efficiency using a combination of these modules. The structure of the UDT simulation model with a diesel engine and a mechanical transmission is shown in Figure 2 a, with HPS - in Figure 2 b, of the electric UDT - in Figure 2 c. The simulation model of the UDT chassis includes model of the interaction of the wheel with the supporting surface and a model of the linear motion of the UDT.

The UDT dynamic was simulated only for the case of the linear motion along a rigid road without normal reactions redistribution between the sides of the vehicle because of the little effect of the curvilinear motion on the values of capacity of energy storage devices, TEM, generator and ICE characteristics. For the same reasons the work of the suspension wasn't taken into account during the creation of the model. The structure of the simulation model of the 6x6 UDT chassis motion is presented in Figure 2 d. The research is using the model of interaction of the wheel with the supporting surface described in the [13], because more complicated contact models [14 - 17] slow down the speed of computational experiments.

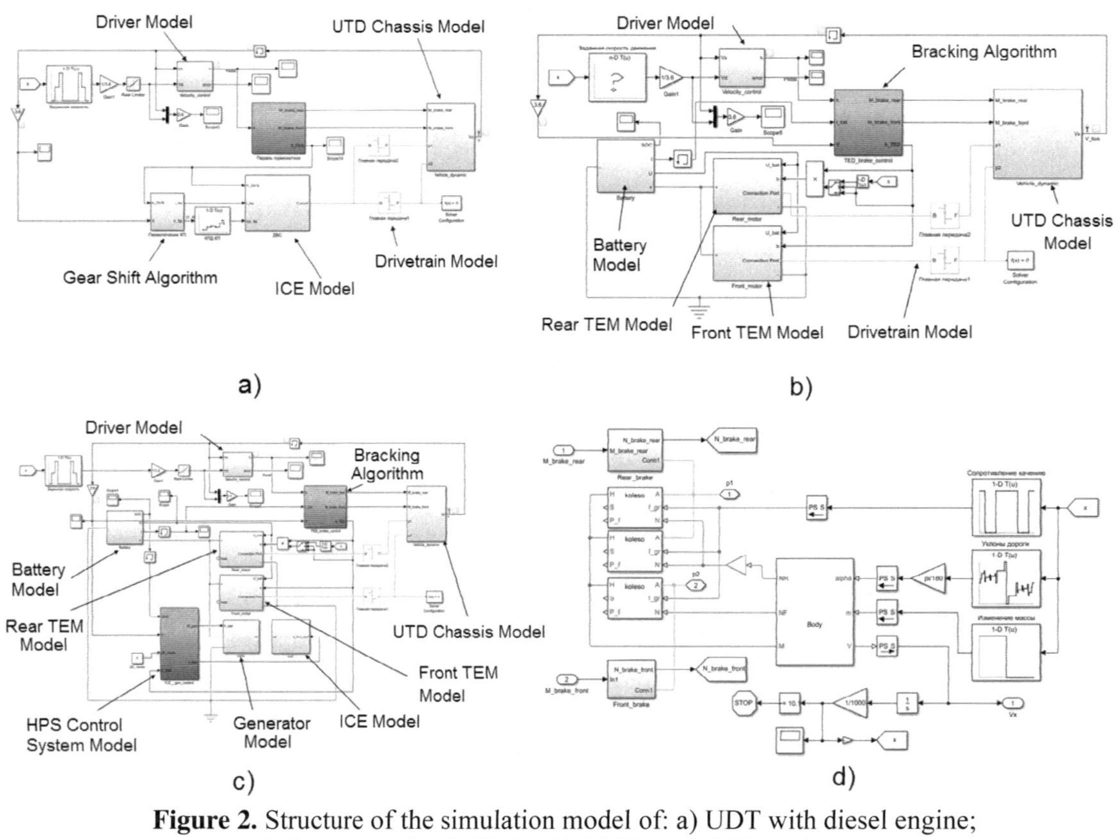

Figure 2. Structure of the simulation model of: a) UDT with diesel engine; b) electric UDT; c) UDT with HPS; d) 6x6 UDT chassis.

IOP Conf. Series: Materials Science and Engineering **819** (2020) 012010 doi:10.1088/1757-899X/819/1/012010

The operation of the internal combustion engine is described by the following differential equation:

$$J_{ICE} \cdot \dot{\omega} = h \cdot M_{ICE} - M_{gen} \quad (1)$$

where J_{ICE} – moment of inertia of the internal combustion engine elements reduced to the crankshaft , $\dot{\omega}$ – angular acceleration of the ICE output shaft, M_{ICE} – torque developed on the ICE output shaft, which is determined by the given engine external speed characteristic (ESC), h – degree of ICE power use, M_{gen} – the generator torque.

The fuel consumption is determined by the interpolation table, depending on the ICE speed and power.

The generator and TEM are modeled by the Servomotor element from Simscape/Electronics library. The Servomotor element initial data are the dependencies of the electric machine maximum moment from the speed, the power of the losses from torque and speed and the rotor inertial characteristics. The electric machines are controlled by the mechanical torque. The electric motor model is described by the following equations:

$$\begin{cases} N_{el} = I_{cur} \cdot U_{dc} \\ N_{el} = M_{mech} \cdot \omega + N_{los} \end{cases} \quad (2)$$

where I_{cur} – current consumed by the motor, U_{dc} – DC link voltage, N_{los} – power loss, ω – angular velocity of the electric machine shaft, M_{mech} – TEM shaft torque. Power loss N_{los} at each operating mode of the electric machine is determined in accordance with its passport characteristics.

As a model of the battery, the battery element Battery of the Simscape / Electronics library is used. The model takes into account the following parameters: rated voltage, capacity, internal resistance. The power loss N_{los} is estimated by the following relationship:

$$N_{los} = I_{cur}^2 R_{bat} \quad (3)$$

where I_{cur} – current on battery, R_{bat} – internal battery resistance.

The ability to reproduce both standard driving cycles and those which are obtained by recording the parameters of the vehicle on the route is provided by the driver's model. The bus control is to maintain the set speed by adjusting the torque to the TEM with the help of a P-controller. Control action varies in the interval [-1; 1]. Negative values correspond to the brake pedal pressing, positive values - the accelerator pedal pressing.

The hybrid powertrain control system model based on the input signals: the degree of accelerator and brake pedals pressing, the vehicle speed, the battery charge level, current flowed in the battery, etc., determines the modes of operation of the internal combustion engine and generator, as well as the distribution of the braking torque between the TEM and the mechanical brakes.

The initial data for simulation are shown in Table 1. The parameters of the diesel UDT gearbox are shown in Table 2.

The drive cycle is based on data of dump trucks operation in quarries and regulatory documents. It represents equal uphill and downhill driving areas (5 km). While the UPD moves uphill, it has a gross mass. While the UDT moves downhill, it has curb mass. Sections of the drive cycle differ in slope, rolling resistance and maximum speed determined from the safety requirements [18, 19, 20].

Graphs of changes in the UDT parameters and motion conditions are shown in Figure 5.

The energy efficiency evaluation of a 6x6 UDT with HPS was carried out by comparing its energy consumption in driving cycle with the energy consumption of diesel UDT when driving in the same conditions. The similar approach was used by authors of research [21].

The UDT with HPS has slightly higher dynamic characteristics than diesel UDT, so the energy consumption in the drive cycle was compared in several stages:

1) first, the simulation of the UDT motion with a diesel engine was carried out to obtain velocity time history while moving in drive cycle;

2) then, simulation of the UDT with HPS motion was carried out with maintaining the velocity obtained at the previous stage.

This approach allows to most objectively compare the energy consumption of the UDT with different powertrains, since UDT with HPS, having higher power-to-weight ratio then diesel UDT, can spend more energy by going through the drive cycle in less time with greater accelerations and speed.

Table 1. Initial Data for UDT Motion Simulation.

Parameter	Value
Gross Mass, kg	75000
Curb Mass, kg	30000
Spacing Between Centre of Mass and Front Axle, m	3.619
Spacing Between Centre of Mass and Centre of Bogie, m	1.552
Rear Axle Spacing, m	1.960
Wheel Radius, m	0.922
Wheel Moment of Inertia, $kg \cdot m^2$	425
Height of the Centre of Mass, m	2
Frontal Area, m^2	8.075
Aerodynamic Drag Coefficient	0.9
Front TEM Characteristic	See Fig. 3 a
Rear TEM Characteristic	See Fig. 3 b
Generator Characteristic	See Fig. 4
ICE Characteristic	See Fig. 4
Front Axle Ratio	21.27
Rear Axle Ratio	21.8
Efficiency Factor of Axles, %	0.96
Gearbox Ratios	See Table 2
Gearbox Efficiency Factor	See Table 2
Gear Ratio Between TEM and Rear Axle	1.5
Nominal Normal Wheel Load, kN	150

Table 2. Diesel UDT Gearbox parameters.

Gear Number	Gear Ration	Efficiency Factor, %
1	5,615	98,0
2	3,997	96,1
3	2,845	96,0
4	2,026	95,4
5	1,429	96,5
6	1,017	95,9
7	0,724	95,4
8	0,516	94,6

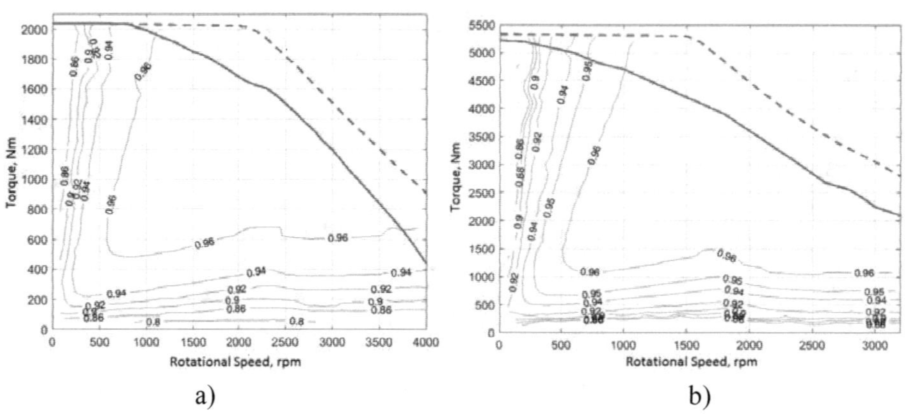

a) b)

Figure 3. Characteristics of TEM: a - Front, b – Rear.

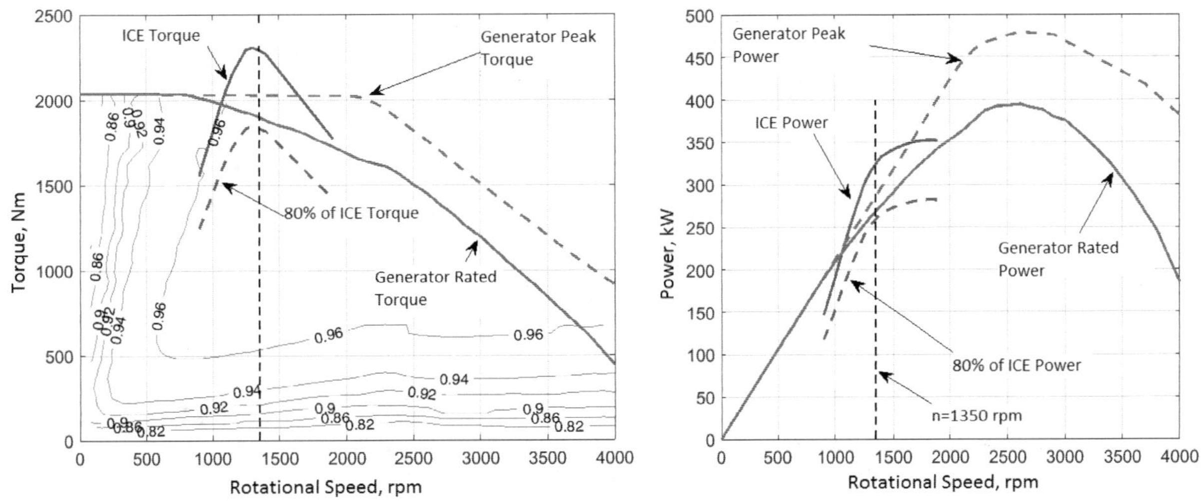

Figure 4. Characteristics of the Generator Set.

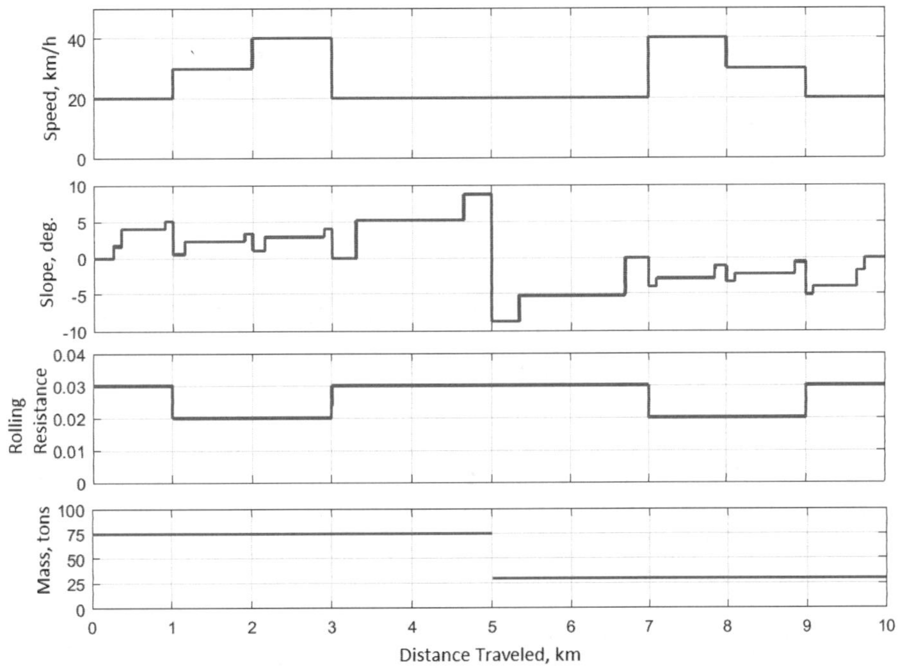

Figure 5. UDT Drive Cycle.

3. Results

Figure 6 shows a plot of the change in speed, and Figure 7 shows energy consumption when diesel UDT moves in drive cycle. Energy consumption was determined by integrating over time the power developed by the ICE. The energy consumption of diesel UDT for a drive cycle amounted to 122.2 kW·h.

IOP Conf. Series: Materials Science and Engineering **819** (2020) 012010 doi:10.1088/1757-899X/819/1/012010

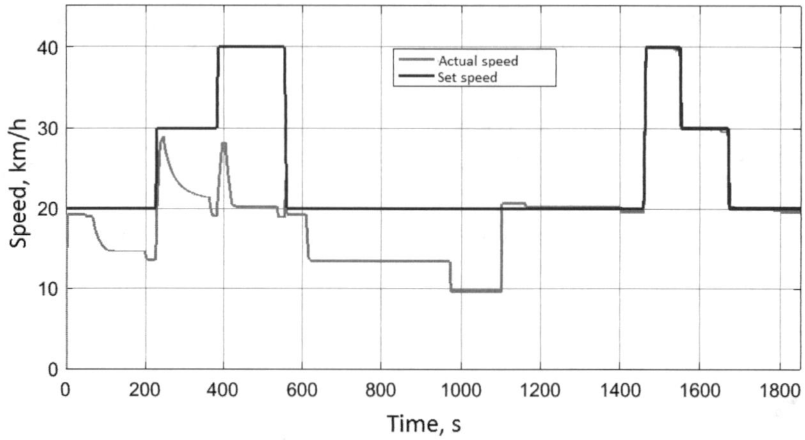

Figure 6. Diesel UDT Speed.

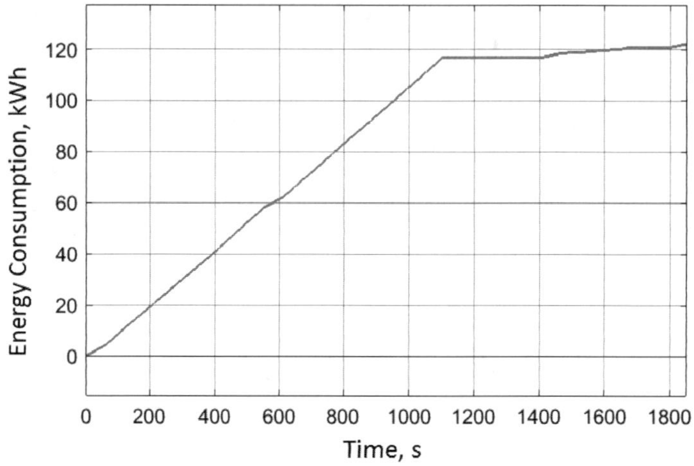

Figure 7. Diesel UDT Energy Consumption.

Figure 8 shows a plot of the change in speed, and Figure 9 shows energy consumption when UDT with HPS moves in drive cycle at the diesel UDT speed. The energy consumption of UDT with HPS for a drive cycle amounted to 94.8 kW·h.

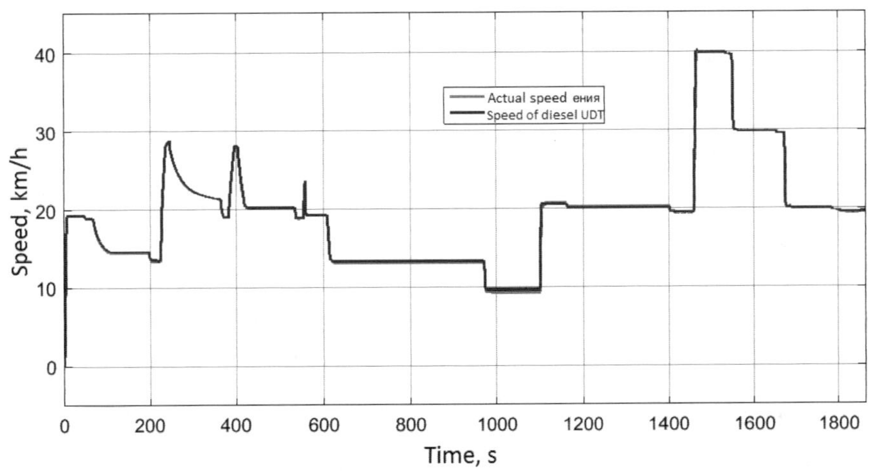

Figure 8. Speed of UDT with HPS.

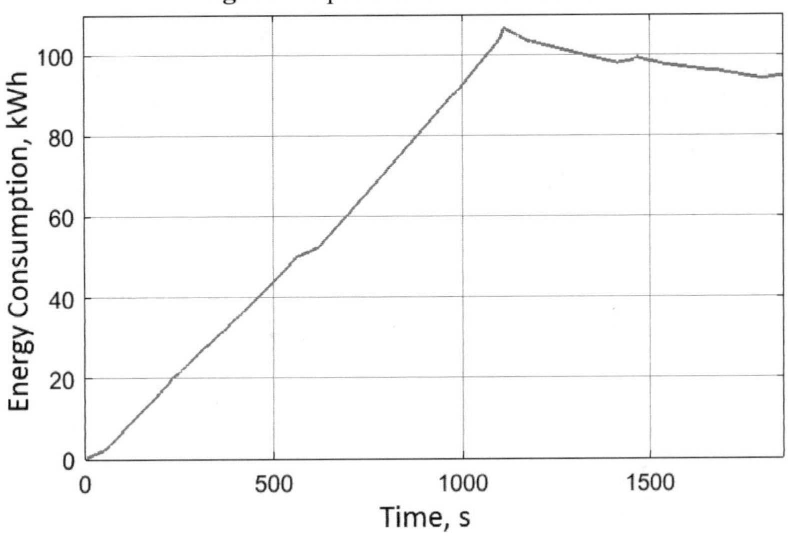

Figure 9. Energy Consumption of UDT with HPS.

Energy consumption of UDT with HPS was determined by integrating over time the power developed by the ICE and the battery. The decrease in the total energy consumption at the end of the driving cycle (when driving downhill) is due to negative battery power while it is being charged from the generator and during regenerative braking.

The maximum total energy expended for the motion of UDT with HPS is 106.6 kWh. The decrease in the total energy spent on motion towards the end of the drive cycle is explained by the recovery of braking energy during the descent, which is the last part of the drive cycle. This means that the UDT with HPS, with selected units consumes 20–22% less energy than diesel one when moving at the same conditions.

Plot of UDT with HPS energy consumption obtained when simulating motion in a drive cycle at the highest possible speed is shown in Figure 10.

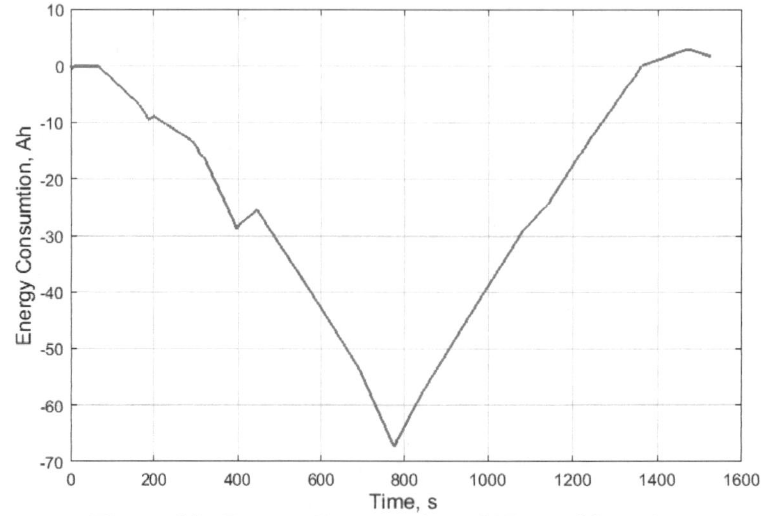

Figure 10. Energy Consumption of UDT with HPS.

According to the plot shown in Figure 10, the energy consumed during driving uphill (the falling part of the curve) is less than the total energy generated both by the generator set and recovered during driving downhill.

This means that in order to ensure the motion of UDT with HPS, the installation of a stationary charging station in the quarry is not required. The resulting "surplus" of energy (the ascending part of the curve, fig. 10) suggests that when the UDT moves in real conditions of the quarry on part of the drive cycle, the generator set may be switched off, because the energy stored in batteries will be enough.

Figures 11 and 12 show plots of changes in current and power of HPS units when the UDT moves in the drive cycle. The battery current rises to 650 A. The power of charge and discharge of the battery reaches 380 kW.

The results of actual currents and required power changes when UDT with HPS at drive cycle must be taken into account when choosing the number and type of batteries.

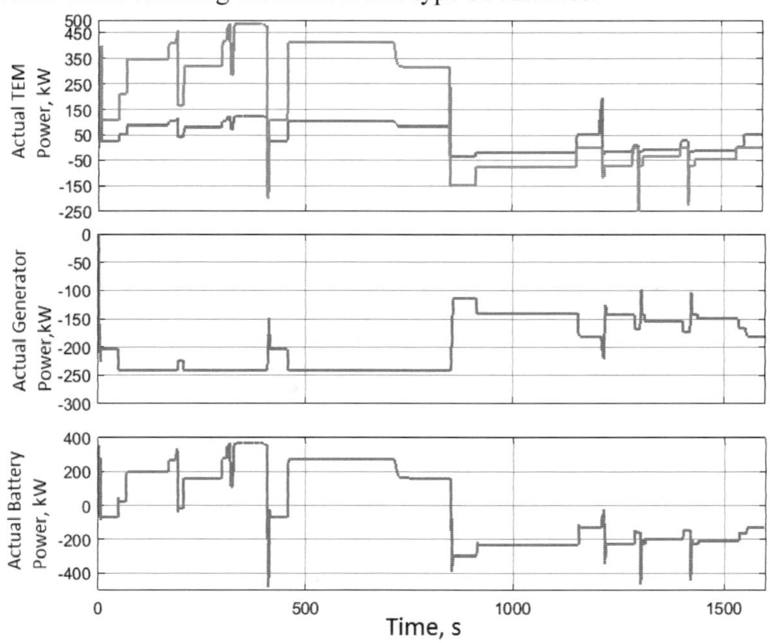

Figure 11. Actual Power of UDT HPS units.

IOP Conf. Series: Materials Science and Engineering **819** (2020) 012010 doi:10.1088/1757-899X/819/1/012010

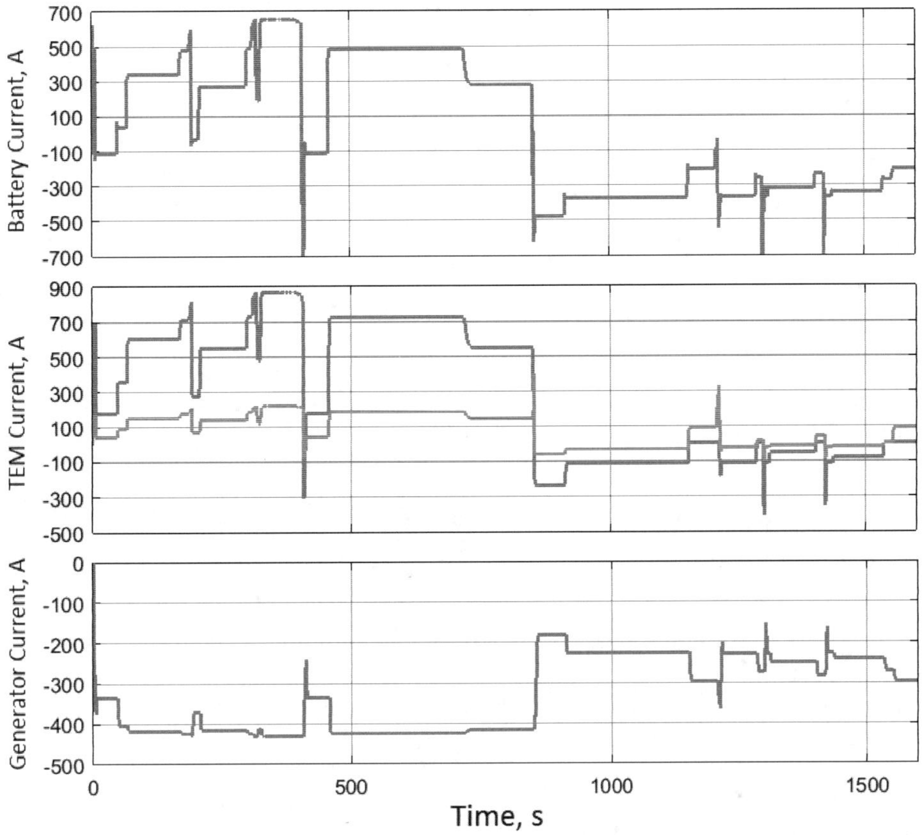

Figure 12. Actual Currents of UDT HPS units.

Figure 13 shows: when UDT with HPS is driving uphill, the slope of which is 9 degrees, with a movement resistance of 0.03 deviation from the required speed are insignificant and do not exceed 5 km/h. This confirms that the 6x6 UDT with HPS has sufficient mobility.

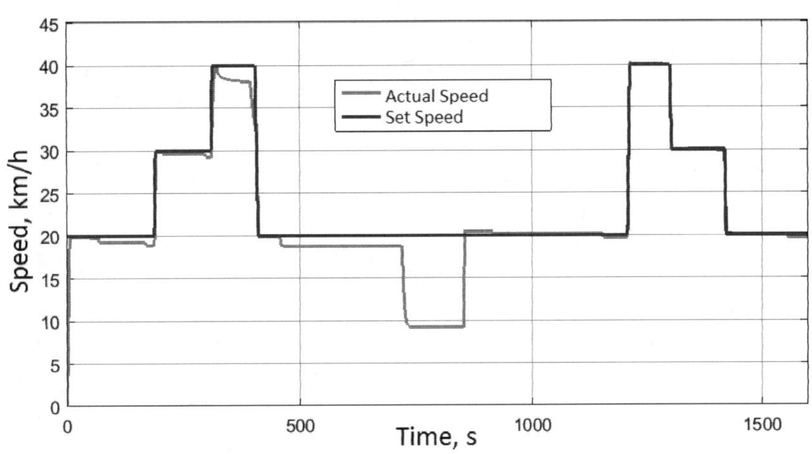

Figure 13. Speed of UDT with HPS when moving in the drive cycle.

As a result of the simulation of motion in the drive cycle of an electric UDT, a plot of energy consumption from battery was obtained (Figure 14).

IOP Conf. Series: Materials Science and Engineering **819** (2020) 012010 doi:10.1088/1757-899X/819/1/012010

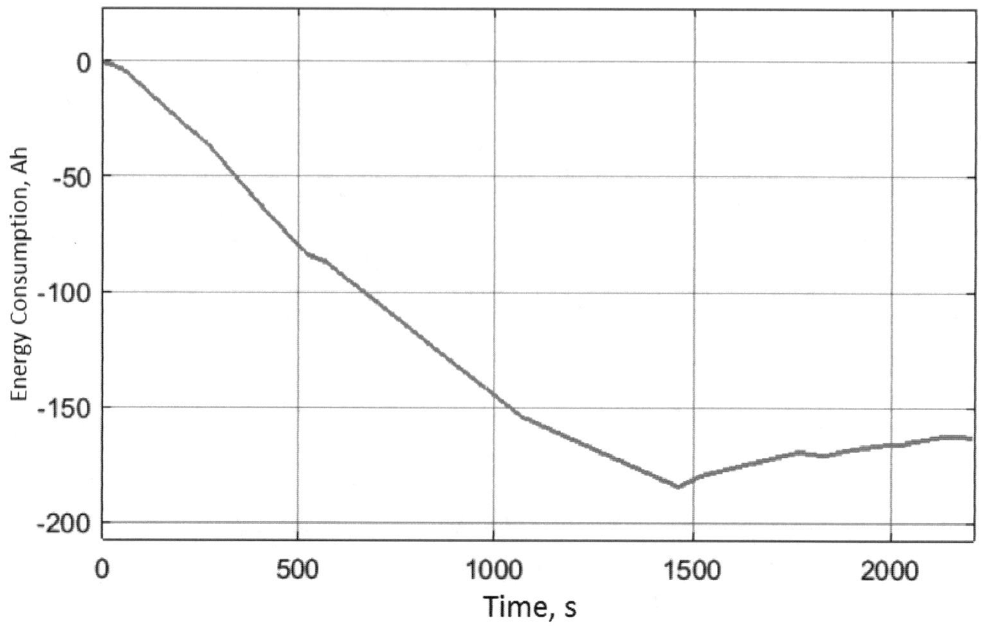

Figure 14. Energy Consumption of electrical UDT.

According to the plot shown in Figure 14, the energy consumed during driving uphill (the falling part of the curve) is greater than the total energy generated both by the generator set and recovered during driving downhill. This means, that to ensure the motion of the electric UDT, it will be necessary to install a stationary charging station in the quarry, and when calculating the time for which useful work is performed, it is necessary to take into account the time taken to recharge the batteries.

There are plots of changes in current and power of electrical UDT units when it moves in the drive cycle in Figures 15 and 16. The battery current rises to 700 A. The power of charge and discharge of the battery reaches 400 kW.

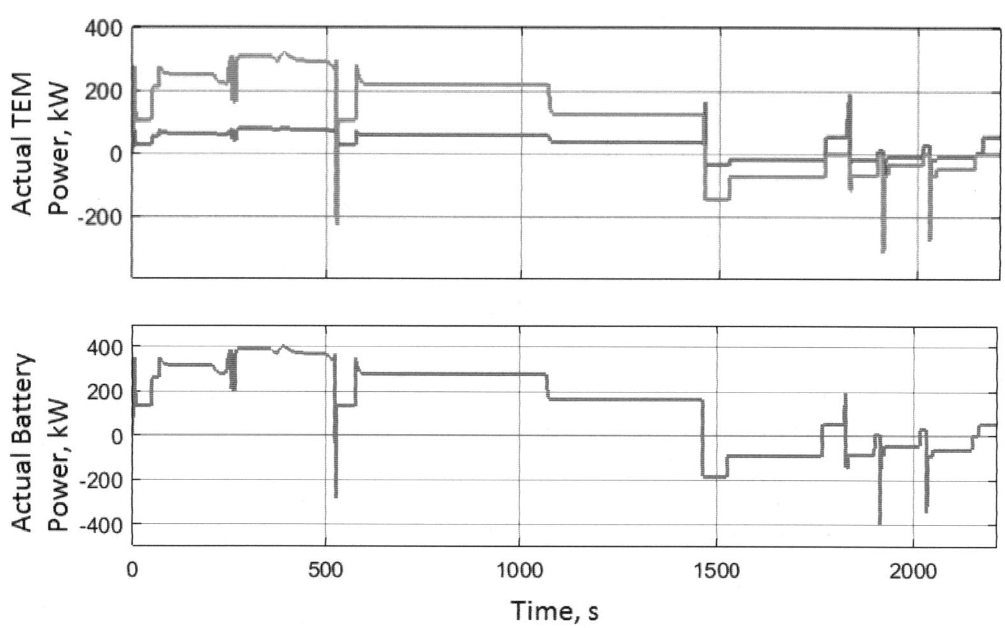

Figure 15. Actual Power of electrical UDT units.

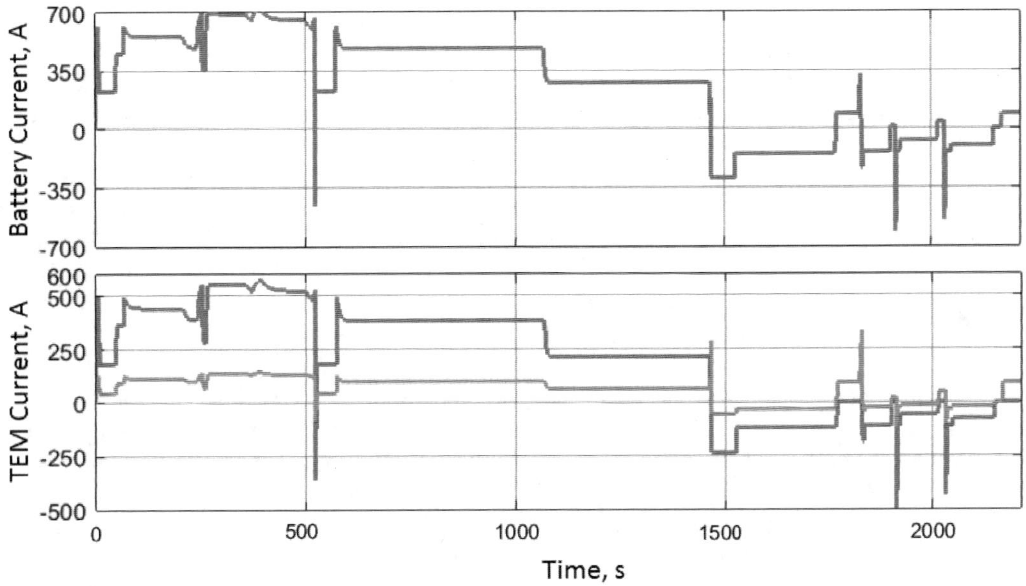

Figure 16. Actual Currents of electrical UDT units.

Based on the results of a simulation of the motion of the electric UDT, a plot of the actual speed versus time during the motion at drive cycle is obtained (Figure 17).

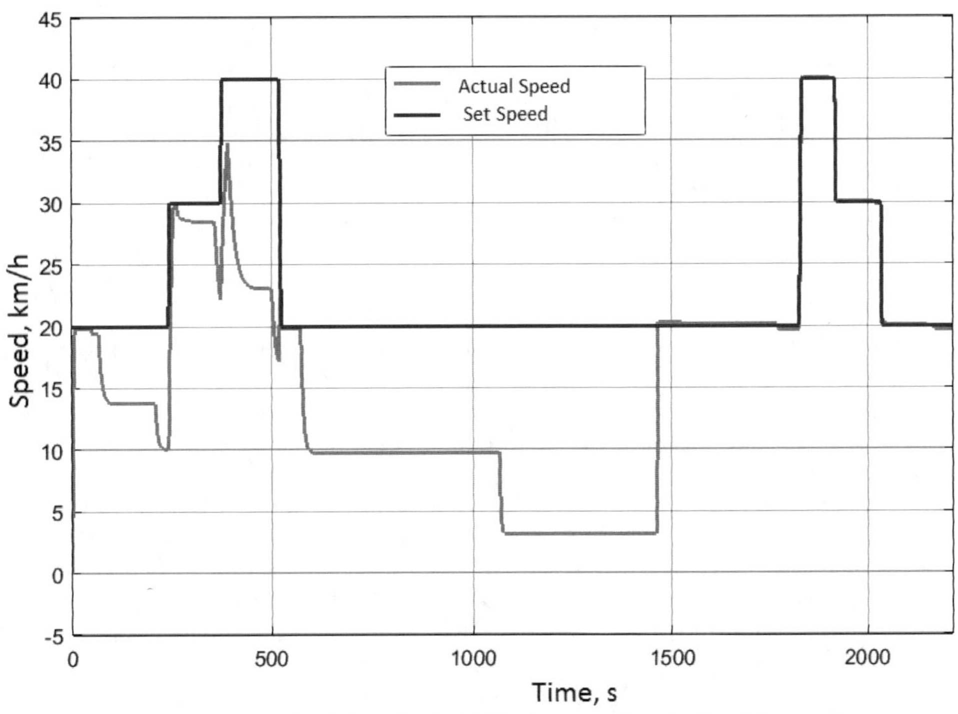

Figure 17. Speed of electrical UDT when moving in the drive cycle.

Figure 17 shows: when electrical UDT driving uphill, the slope of which is 9 degrees, with a movement resistance of 0.03 deviation from the required speed are significant and exceed 5 km/h. This confirms that electrical UDT has insufficient mobility.

4. Conclusion

As part of the research, mathematical simulation of movement in the drive cycle of the diesel UDT, UDT with HPS and electrical UDT was carried out. The simulation model of the diesel UDT with a mechanical transmission takes into account gear shifting in the gearbox and efficiency of both each gear and the rest of the mechanical transmission. The simulation model of the UDT with HPS, as well as the electric one, takes into account the limitations of the charge and discharge currents of the battery, the efficiency of the TEM, generator, and axles. Preliminary results of the simulation of UDT motion in the drive cycle showed that the UDT with HPS has higher power-to-weight ratio then diesel one, therefore, to compare the energy efficiency when modeling the motion in drive cycle of the UDT with HPS, speed versus time record of the diesel UDT was used.

According to the calculated energy consumption of the UDT in the drive cycle, it is possible to determine the required capacity of the battery. As shown in the plots, the capacity of the battery of an electric UDT should be at least one and a half times greater than that of UDT with HPS, in accordance with the energy consumption in the drive cycle and the maximum charge and discharge currents occurring in the battery. An increase in the capacity of the battery causes a proportional increase in its mass, and with it an increase in the curb mass of the UDT and a decrease in its carrying capacity.

The presented simulation results show the efficiency of HPS application in comparison with a traditional powertrain and a mechanical transmission for UDT when driving in drive cycles similar to a given one. The use of HPS allows reducing energy consumption for motion by 20-25% due to the recovery of braking energy.

5. References

[1] Mariev P L 2003 Technological open pit transport with brand of 'Belaz': Perspective trends in development *Gornyi Zhurnal* **7** p 43

[2] Koryagin M and Voronov A 2017 Improving the organization of the shovel-Truck systems in open-pit coal mines *TRANSPORT PROBLEMS* **12** pp 113-122

[3] Xia Y, Tang H, Tan Q and Zhu G 2019 Nonlinear Variable Structure Excitation Control for Electric Brake of Hybrid Motor Dump Truck *Hunan Daxue Xuebao/Journal of Hunan University Natural Sciences* **46** pp 28-35

[4] Dvornikovs I, Marinbahs M, Zarembo J, Groza E and Ketners, K 2019 Investigation of traction motor dynamic using computer simulation and method of mutual loading of two pair motors, *ELMA 2019 – Proceedings*, 16th Conference on Electrical Machines, Drives and Power Systems June 2019

[5] Sheshko O E 2018 Comparative assessment of environmental impact of dump trucks and diesel trolley cars in open pit mining *Mining Informational and Analytical Bulletin* **2018** pp 119-125

[6] Paraszczak J, Svedlund E, Fytas K and Laflamme M 2014 Electrification of loaders and trucks – A step towards more sustainable underground mining *Renewable Energy and Power Quality Journal* **11** pp 81-86

[7] Lajunen A, Sainio P, Laurila L, Pippuri-Mäkeläinen J and Tammi K 2018 Overview of powertrain electrification and future scenarios for non-road mobile machinery *Energies* **11**

[8] Keller A and Aliukov S 2015 Analysis of possible ways of power distribution in an all-wheel drive vehicle Lecture Notes in Engineering and Computer Science 2218 pp 1154-1158

[9] Keller A. and Aliukov S 2015 Effectiveness of Methods of Power Distribution in Transmissions of All-Wheel-Drive Trucks SAE Technical Paper 2015-01-2732

[10] Keller A and Aliukov S 2016 Efficient Power Distribution in an All-Wheel Ground Vehicles SAE Technical Paper 2016-01-1105

[11] Keller A , Aliukov S and Anchukov V 2017 Mathematical model of the truck for investigation of stability and control of movement Lecture Notes in Engineering and Computer Science pp 711-716

[12] Xiaogang Wu, Han Chen and Demenkov N P 2016 Comparative study on energy consumption for plug-in hybrid electric vehicles based on trip characteristics *Int. J. Electric and Hybrid Vehicles* **8**
[13] Gorelov V A, Butarovich D O, Staduhin A A and Skotnikov G I 2019 Simulation-based identification of the parameters of a minibus hybrid powertrain *IOP Conf. Series: Materials Science and Engineering* **534**
[14] Gorelov V A, Ankinovich G G and Chudakov O I 2017 Mathematical model of rectilinear dynamics on non-deformable support base of saddle road-train with active semi-trailer *VESTNIK MASHINOSTROENIYA* **3** pp 37-42
[15] Volskaia V N, Zhileykin M M and Zakharov A Y 2018 Mathematical model of rolling an elastic wheel over deformable support base *IOP Conf. Series: Materials Science and Engineering (Electronic Materials)* **315**
[17] Gorelov V A and Komissarov A I 2016 Mathematical Model of the Straight-line Rolling Tire – Rigid Terrain Irregularities Interaction *Procedia Engineering* **150**
[18] Sarkisov P, Prokop G, Gladov G and Popov S 2015 Physical tire model based on understanding of separate components of tire deformation *The Dynamics of Vehicles on Roads and Tracks*
[14] Voronov A Y 2015 Optimizaciya pokazatelej ekspluatacionnoj proizvoditel'nosti ekskavatorno-avtomobil'nyh kompleksov razrezov (Ph.D. Dissertation — Kemerovo: KuzGTU)
[19] Zaharov A Y 2012 *Opredelenie parametrov trassy transportirovaniya po otdel'nomu marshrutu* (Kemerovo: KuzGTU) 17 p
[20] Anistratov Y I *2010 Spravochnik po otkrytym gornym rabotam* (Moscow: NTC «GORNOE DELO») 700 p
[21] Mariev P L, Voitov V T and Shishko S A 2015 Design procedure and results of analysis of efficiency coefficient of hydromechanical and electromechanical transmission of mining dump trucks are given *Gornyi Zhurnal* **2015** pp 83-87

Acknowledgments
The research was carried out at BMSTU with the financial support of the Ministry of Education and Science of the Russian Federation under Agreement 14.577.21.0287 (Unique identifier: RFMEFI57718X0287).

IOP Conf. Series: Materials Science and Engineering **819** (2020) 012011 doi:10.1088/1757-899X/819/1/012011

Development of Calculation Research Method for Exhaust System Main Elements Acoustic Characteristics

R I Rakhmatov[1,2,3], **V E Krutolapov**[1], **V N Zuzov**[3]

[1]FSUE "NAMI", Moscow, Russia
[2]Moscow Polytechnic University, Moscow, Russia
[3]Bauman Moscow State Technical University, Moscow, Russia

E-mail: rakhmatjon.rakhmatov@nami.ru

Abstract: This paper presents a calculation method for researching acoustic characteristics of the exhaust system main elements in modern motor vehicles. The acoustic characteristics of the absorbing materials used in the muffler have been defined by method of calculation and experiment.

1. Introduction

In recent years, the requirements for the vibroacoustic (NVH) characteristics of motor vehicles have become more stringent. The output noise remains the main source of noise of the running engine, whereby the acoustic power of a not muffled exhaust noise reaches 100 W (up to 140 dBA) and is tens and hundreds of times superior to those of other elements and systems of the vehicle. The vibration frequency of an acoustic wave generated as a result of operation of an 8-cylinder 4-stroke internal combustion engine (ICE) with 650 rpm at idle will be 43.3 Hz, according to formula 1:

$$f = \frac{n \cdot i}{60 \cdot \tau} = \frac{650 \cdot 8}{60 \cdot 2} = 43.3\ Hz \tag{1}$$

Where: n – engine crankshaft speed per minute; i – number of cylinders; $\tau = 2$ for four-stroke engines,
$\tau = 1$ for two-stroke engines.
The wavelength at the frequency of 43.3 Hz at the exhaust gas temperature of 20°C is:

$$\lambda = \frac{\vartheta}{f} = \frac{340\ \frac{M}{s}}{43{,}3\ Hz} = 7.85\ m \tag{2}$$

The length of the whole exhaust system is 5.8 meters. The exhaust pipe cut falls on 3/4 of the wavelength at the frequency of 43.3 Hz. Respectively, it falls within the maximum sound pressure amplitude.

2. Main part

To solve this problem, a classical Helmholtz resonator, which is the most efficient within the narrow-band frequency spectrum, has been developed. First of all, the resonator quality factor depends on the chamber (cavity) volume. Figure 1 shows the classical Helmholtz resonator scheme:

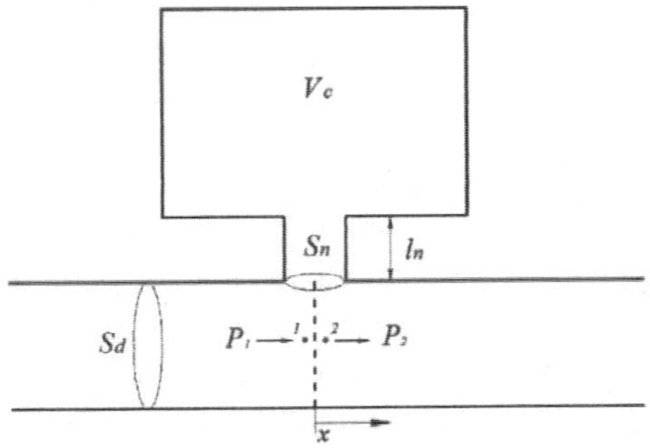

Figure 1. Helmholtz resonator scheme.

In order to find the required value of transmission loss (TL) of the Helmholtz resonator the following formula was used:

$$TL = 20log_{10}(\frac{1}{2}\left|2 + \frac{\rho_0 c_0}{S_d}\frac{1}{Z_r}\right|) \tag{3}$$

where

$$Z_r = j(\omega\frac{\rho_0 l_n}{S_n} - \frac{1}{\omega}\frac{\rho_0 c_0^2}{V_c}) \tag{4}$$

where $\rho_0 = 1.2\ kg/m^3$ is the air density at 20°C, $c_0 = 343\ m/s$ - acoustic velocity at 20°C, S_d – longitudinal pipe area, Z_r – acoustic impedance, l_n – resonator's tube length, S_n – resonator's tube area, V_c – volume of Helmholtz resonator chamber, ω – circular frequency

Parametric optimization was carried out for frequency adjustment and adjustment of the quality factor of the Helmholtz resonator. According to the results of the parametric optimization, the following geometric parameters of the resonator have been obtained: the tube length is 75 mm, the tube diameter is 22 mm, the tube thickness is 2 mm and the volume of the Helmholtz resonator chamber is 7452330.5 mm^3. The results of the analytical calculation based on the results of the parametric optimization are shown in Figure 2.

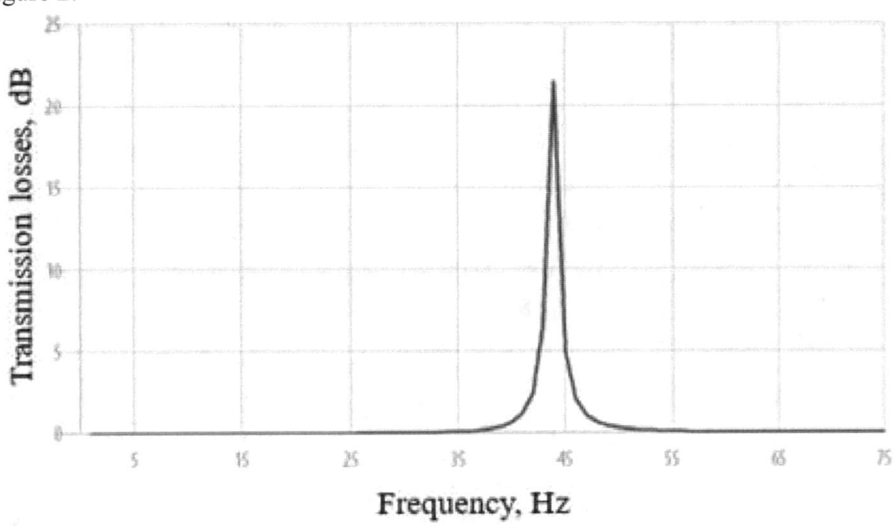

Figure 2. Analytical calculation results for Helmholtz resonator used for the muffler.

The resonator is tuned to the frequency of 43,3 Hz (Figure 2). The transmission loss at this frequency is 22 dB.

For the exact mathematical calculation of the acoustic behavior of the main elements of the exhaust system, it is necessary to solve a diffractive problem for the wave equation describing the acoustic propagation in a gaseous medium filling this element under complex boundary conditions.

The whole exhaust system design can be divided acoustically into a limited set of elements having the same operating principle. These are reactive elements made in the form of chamber systems connected to each other and to the volume of the waveguide using pipes, gaps and holes. And also the dissipative elements containing different sound-absorbing materials. The whole system can be divided into a number of separate known acoustic elements, interconnected in a certain way. As a result of theoretical research for each element of this limited number of elements, a mathematical formula has been derived that describes its acoustic characteristics. The interaction of each of these elements with the neighboring one in the exhaust system is set through the input and output impedances Z (acoustic impedance), which is determined by the sound pressure P and the vibrational velocity of the particles of the medium V. The connection of these parameters for each element is determined by the table (transfer matrix) T, which fully characterizes the acoustic behavior of the element. In a matrix form this ratio can be written as follows:

$$Z_{in} = T\,Z_{out}\,,\ \text{where}\ Z_{in} = ;\ Z_{out} = \begin{bmatrix} P_{out} \\ V_{out} \end{bmatrix};\ T = \begin{bmatrix} T_{11}T_{12} \\ T_{21}T_{22} \end{bmatrix} \tag{5}$$

or in a form of an equation system:

$$P_{in} = T_{11}\,P_{out} + T_{12}\,V_{out} \tag{6}$$

$$V_{in} = T_{21}\,P_{out} + T_{22}\,V_{out} \tag{7}$$

The quadripole (four-pole) method applied in the electroacoustics equivalent circuit theory is used in calculation of the entire system. Acoustic calculation is replaced with calculation of equivalent circuits, the characteristics of which are defined by well-known transfer matrices [1, 2, 3, 4].

This method is successfully used in calculation of simple elements of the exhaust system. As the design becomes more complicated, the calculations' accuracy falls dramatically, because it is difficult to build an equivalent circuit fully describing the operation of the elements and the exhaust system on the whole, it is difficult to derive an accurate analytical formula for a separate specific element, it is difficult to gather sufficient experimental statistics for deriving an empirical formula. In case of calculation using this method, it is difficult to consider the effect of temperature and gas flow on the acoustic parameters.

The finite-element modelling method has none of these disadvantages. It allows to accurately simulate acoustic behaviour of an exhaust system element of any complexity and to investigate the effect of the exhaust system geometric parameters and of such properties of the gas medium as density, temperature and flow rate on its acoustic parameters.

The finite-element modelling method is based on representation of the exhaust system element being investigated in the form of a gas volume filling this element and divided into a number of discrete volumes – finite elements approximating the geometric form of the whole element.

The calculation of transmission losses is performed based on the finite-element solution to the Helmholtz wave equation:

$$\frac{\partial^2 p}{\partial x^2} + \frac{\partial^2 p}{\partial y^2} + \frac{\partial^2 p}{\partial z^2} + \frac{\omega}{c} p = 0 \tag{8}$$

with the corresponding boundary conditions at the calculated volume boundaries, where p (x,y,z) is vibration velocity potential, ω is cyclic frequency, c is acoustic velocity in the medium.

The channel modes which were located at the end face of the muffler cylindrical surface "input" were selected as a source of sound waves. Any acoustic wave which propagates inside the channel is divided into incident and reflected waves.

$$p = p^+ + p^- \tag{9}$$

The peculiarity of channel modes is as follows: only a plane wave propagates at low frequencies inside the channel, and at higher frequencies, the acoustic field may be more complex, but in any case, the pressure may be considered a mathematical superposition of channel modes, the formulas of which are shown below:

$$p^+ = \sum_{propagating\ mode} \alpha_i^+ \psi_i^+ \quad (10)$$

$$p^- = \sum_{propagating\ mode} \alpha_i^- \psi_i^- \quad (11)$$

where α is a contribution factor, ψ is a mode form.

The boundary conditions for cylindrical surface reflecting walls are given as an example (see Fig. 3 and Formulas 4, 5).

$$\left[\frac{\partial p}{\partial x}\right]_{x=0} = \left[\frac{\partial p}{\partial x}\right]_{x=r} = 0 \quad (12)$$

$$\left[\frac{\partial p}{\partial y}\right]_{y=0} = \left[\frac{\partial p}{\partial y}\right]_{y=r} = 0 \quad (13)$$

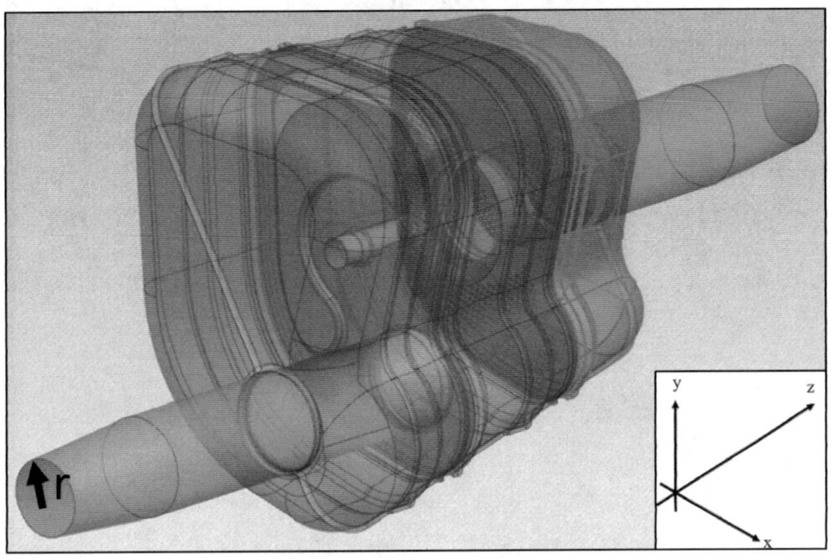

Figure 3. Boundary conditions: reflecting wall of muffler model cylindrical surface.

Solution to the Helmholtz equation for round channel modes

$$p(x, y, z) = A\cos(k_x x)\cos(k_y y)e^{ik_z z} \quad (14)$$

true as long as the following ratio is maintained:

$$k_x^2 + k_y^2 + k_z^2 = k^2 \quad (15)$$

The following conditions shall be met for defining the boundary conditions of the channel modes with reflecting walls:

$$k_x = m\frac{\pi}{r},\ k_y = n\frac{\pi}{r} \quad (16)$$

From equation 7 taking into account equation 8, we define

$$k_z = \sqrt{k^2 - \frac{\pi^2}{r^2}m^2 - \frac{\pi^2}{r^2}n^2} \quad (17)$$

In case of excitation by a channel mode, the pressure profile in the cross-section is defined by Bessel functions and represents a set of waves. M and n are wave numbers showing how many half-waves fit along the radius of the cross-section.

IOP Conf. Series: Materials Science and Engineering **819** (2020) 012011 doi:10.1088/1757-899X/819/1/012011

Also, the red color shows (Figure 3) the position of the dissipative element (basalt fiber) in the muffler, the acoustic characteristics of which were determined by the calculation and experiment method.

The determination of the acoustic characteristics of the dissipative element (basalt fiber) used in the muffler was performed using the method of four microphones. A material sample of the d thickness is installed in the pipe between two pairs of microphones. On one end of the pipe, S speaker is placed, which generates sound waves in the pipes, and on the other end, there is L load.

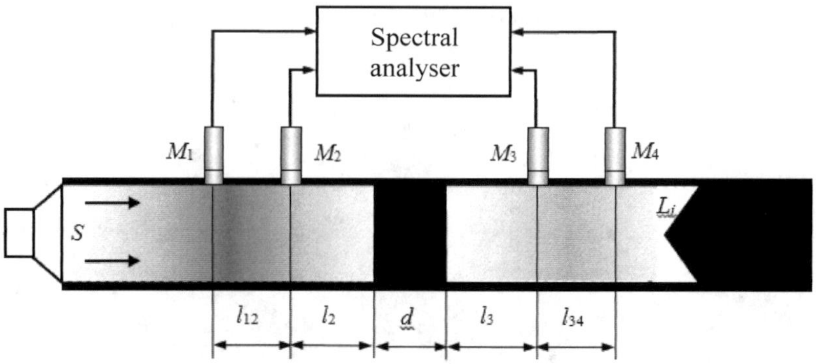

Figure 4. Test bench layout for determining acoustic parameters of dissipative elements.

The measurements and determination of coefficients of the dissipative element sample transfer matrix were carried out using the technique described in TL Transmission Losses Determination section. The difference is in the fact that for such symmetrical system, "sound-absorbing material" dissipative element, the following ratios occur between the transfer matrix coefficients:

$$T_{11} = T_{22} \tag{18}$$

$$T_{11}\,T_{22} - T_{12}T_{21} = 1 \tag{19}$$

Taking into account these ratios, the coefficients may be determined according to the measurement results with only one load according to the following formulas:

$$T_{11} = T_{22} = \frac{P_d V_d + P_u V_u}{P_d V_u + P_u V_d} \tag{20}$$

$$T_{12} = \frac{P_u^2 - P_d^2}{P_d V_u + P_u V_d} \tag{21}$$

$$T_{21} = \frac{V_u^2 - V_d^2}{P_d V_u + P_u V_d} \tag{22}$$

On the other hand, for the sound-absorbing material (SAM) sample, the transfer matrix coefficients are given by:

$$\begin{bmatrix} T_{11} & T_{12} \\ T_{21} & T_{22} \end{bmatrix} = \begin{bmatrix} \cos kd & (j/W)\sin kd \\ jW\sin kd & \cos kd \end{bmatrix} \tag{23}$$

where k, W are the constant of distribution and wave drag of the dissipative element respectively.

The obtained coefficients of the SAM transfer matrices were further used in the final procedure on determining acoustic parameters in accordance with the developed calculation program [5]. The results of the SAM wave parameters calculations are given in Figures 5 and 6.

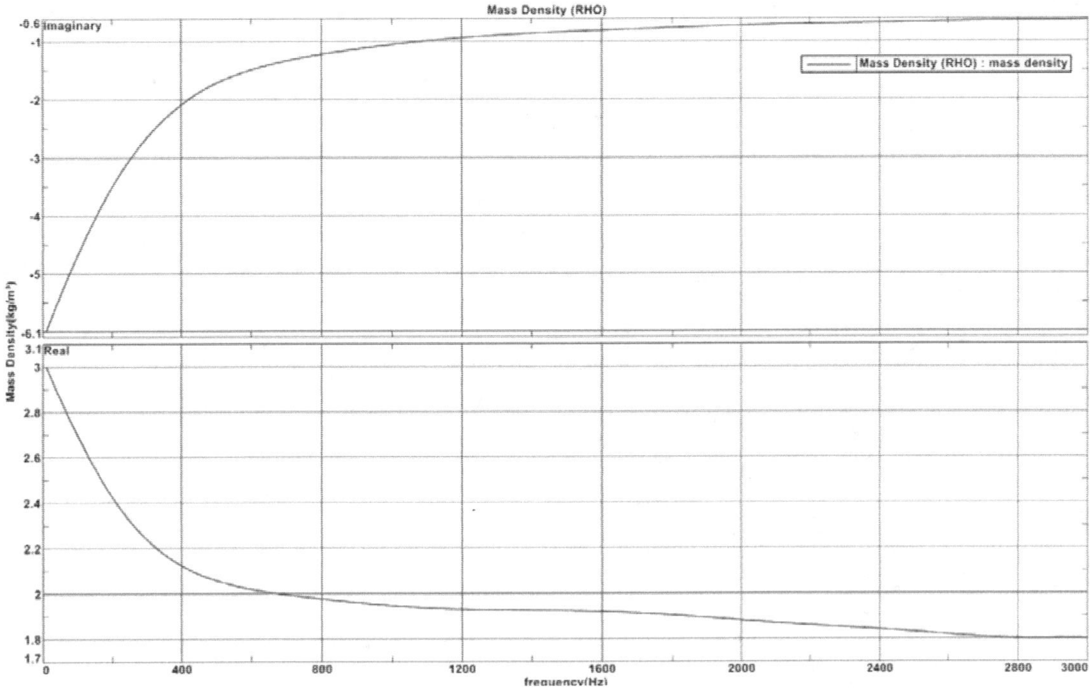

Figure 5. Comprehensive density averaged by the SAM volume.

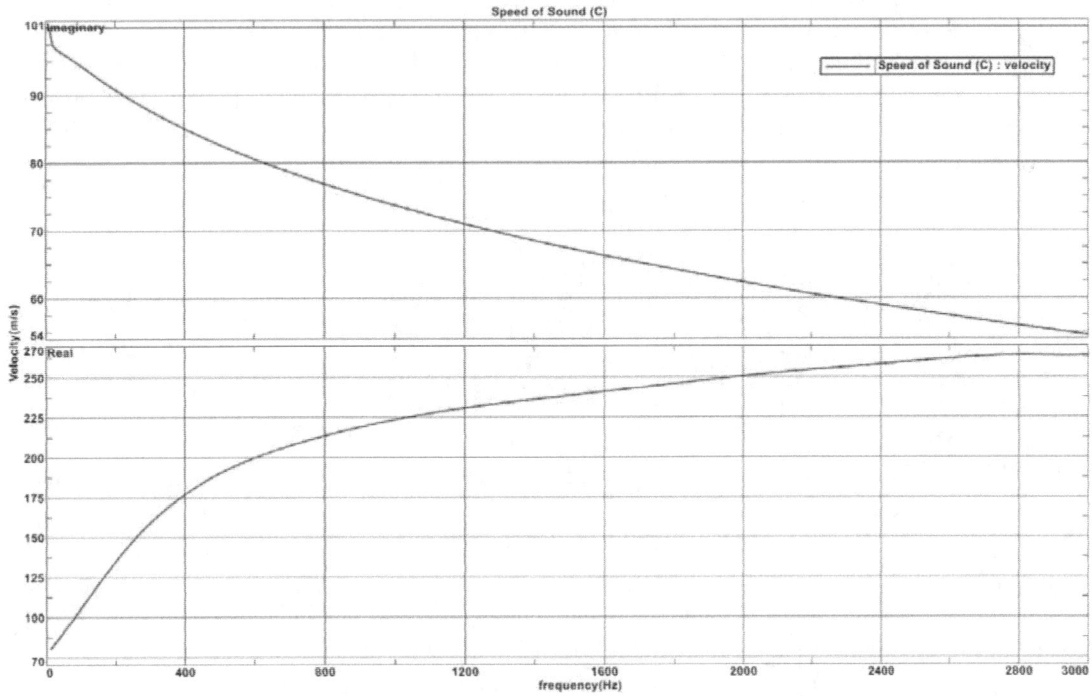

Figure 6. Comprehensive acoustic velocity averaged by the SAM volume.

Thus, we obtained the SAM parameters, which were used in the muffler calculation. The calculation results for the muffler transmission loss are shown in Figure 7.

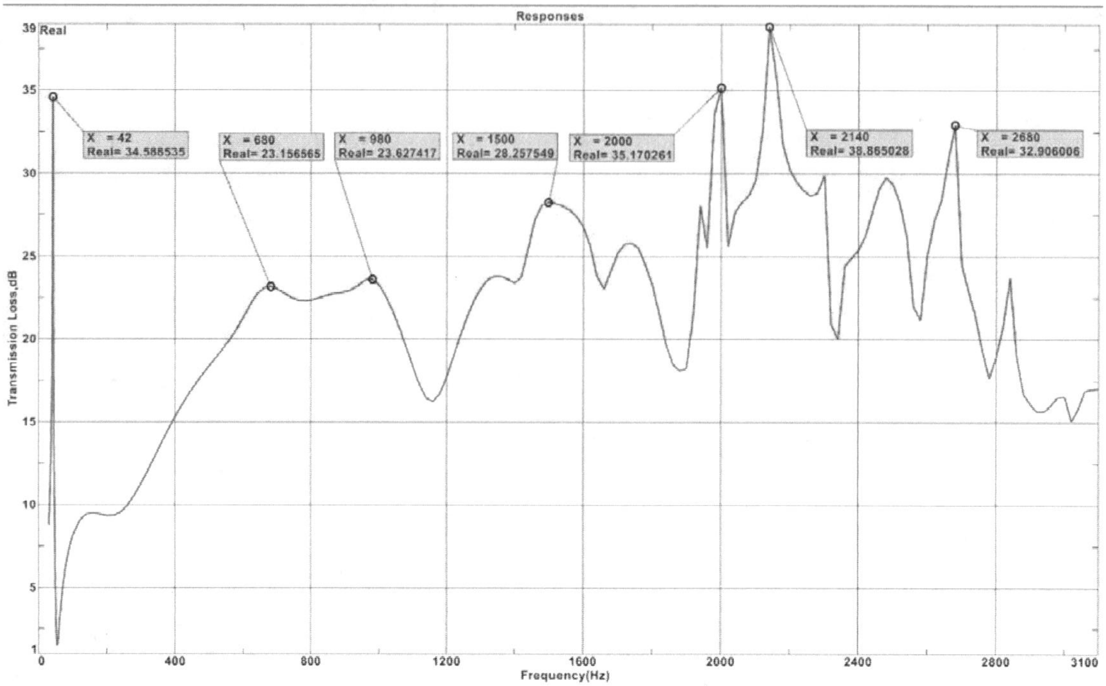

Figure 7. Calculation results for the muffler transmission loss.

To check the accuracy of the calculation using the finite-element model, the muffler bench tests were conducted. Under the bench conditions, within the operating frequency range, the TL transmission loss parameter is determined and then compared with the calculated values.

The experimental method of the TL determination is based on the sound pressure measurement in several points at the input and output of the muffler under investigation, calculation division into the incident and reflected wave, subsequent calculation of sound wave power at the input Win and output Wout of the muffler (Figure 8) and the TL determination using the formula:

$$TL = 10 \lg (Win / Wout) \tag{24}$$

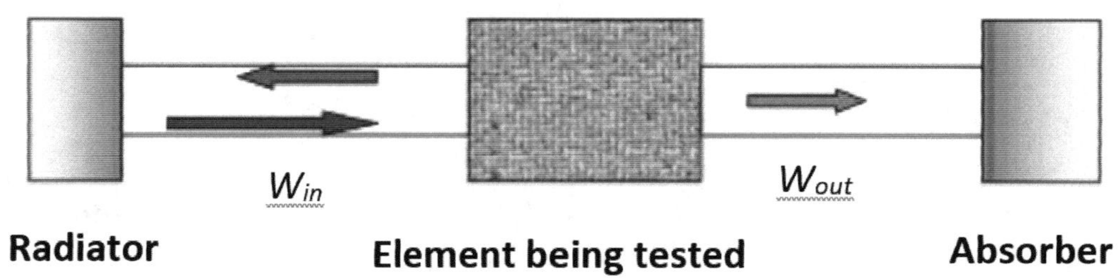

Figure 8. Exhaust system test setup.

The exhaust system element being tested (muffler) is located between the source of the sound waves on one end of the pipe and the sound-absorbing structure on the other end of the pipe. The input measurement microphones unit is installed between the acoustic radiator and the sample being tested, while the output measurement microphones unit is installed between the sample being tested and the sound absorber. During the testing, both amplitude and phase relations between the measured signals are measured in the form of a complex transmission (transfer) function [5].

Having determined the sound-pressure spectra at the positions of microphones P_1 and P_2 at the input of the exhaust system element, P_3 and P_4 at the output of the exhaust system element as well as transmission functions $H_{21}=P_2/P_1$, $H_{31}=P_3/P_1$, $H_{41}=P_4/P_1$, one can find the sound-pressure and oscillation (particle) velocity spectra before the exhaust system element P_u and V_u and after the exhaust system element P_d and V_d:

$$P_u = P_1 \frac{H_{21}\sin[k_0(l_{12}+l_2)]-\sin(k_0 l_2)}{\sin(k_0 l_{12})} = P_1\tilde{P}_u \tag{25}$$

$$P_d = P_1 \frac{H_{31}\sin[k_0(l_{34}+l_3)]-H_{41}\sin(k_0 l_3)}{\sin(k_0 l_{34})} = P_1\tilde{P}_u \tag{26}$$

$$V_u = jP_1 \frac{H_{21}\cos[k_0(l_{12}+l_2)]-\cos(k_0 l_2)}{W_0\sin(k_0 l_{12})} = jP_1\tilde{V}_u \tag{27}$$

$$V_d = jP_1 \frac{H_{41}\cos(k_0 l_3)-H_{31}\cos[k_0(l_{34}+l_3)]}{W_0\sin(k_0 l_{34})} = jP_1\tilde{V}_u \tag{28}$$

where $k_0 = \omega/c_0$ is the wave number; ω is the cyclic frequency; c_0 is the air acoustic velocity; $W_0 = \rho_0 c_0$ is the air wave drag; ρ_0 is the air density, l_{12} is the distance between the microphones at the element input; l_2 is the distance between the input microphones and the element; l_{34} is the distance between the microphones at the element output; l_3 is the distance from the element to the output microphones.

The sound-pressure and oscillation (particle) velocity spectra on both sides of the exhaust system element are related according to the following matrix equation:

$$\begin{bmatrix} P_u \\ V_u \end{bmatrix}_{x=0} = \begin{bmatrix} T_{11} & T_{12} \\ T_{21} & T_{22} \end{bmatrix} \begin{bmatrix} P_d \\ \mathrm{V}_d \end{bmatrix}_{x=d} \tag{29}$$

where T is the transmission matrix of the exhaust system element under study.

This matrix equation represents a system of two equations with four unknown coefficients of the transmission matrix of the exhaust system element T_{11}, T_{12}, T_{21}, T_{22}. Two more equations, which can be obtained by additional similar measurements with the second load on the measurement pipe output, thus implementing the two-load method, are necessary to find these coefficients.

Thus, we have:

$$\begin{bmatrix} P_{ui} \\ V_{ui} \end{bmatrix}_{x=0} = \begin{bmatrix} T_{11} & T_{12} \\ T_{21} & T_{22} \end{bmatrix} \begin{bmatrix} P_{di} \\ \mathrm{V}_{di} \end{bmatrix}_{x=d}, \quad i=1,2 \tag{30}$$

Here, the value of index i corresponds to the load number at the measurement pipe output. The solution of this equation system gives the following expressions for the transmission matrix coefficients:

$$T_{11} = \frac{P_{u1}V_{d2} - P_{u2}V_{d1}}{P_{d1}V_{d2} - P_{d2}V_{d1}}; \quad T_{12} = \frac{P_{d1}P_{u2} - P_{d2}P_{u1}}{P_{d1}V_{d2} - P_{d2}V_{d1}} \tag{31}$$

IOP Conf. Series: Materials Science and Engineering **819** (2020) 012011 doi:10.1088/1757-899X/819/1/012011

$$T_{21} = \frac{V_{u1}V_{d2} - V_{u2}V_{d1}}{P_{d1}V_{d2} - P_{d2}V_{d1}}; \quad T_{22} = \frac{P_{d1}V_{u2} - P_{d2}V_{u1}}{P_{d1}V_{d2} - P_{d2}V_{d1}} \tag{32}$$

The TL transmission loss parameter is defined by the following expression:

$$TL = 20 \cdot \log\left(\frac{1}{2} \cdot \left| T_{11} + \frac{T_{12}}{W_0} + W_0 \cdot T_{21} + T_{22} \right|\right) \tag{33}$$

The bench basic diagram and structural setup are shown in Figures 9 and 10.

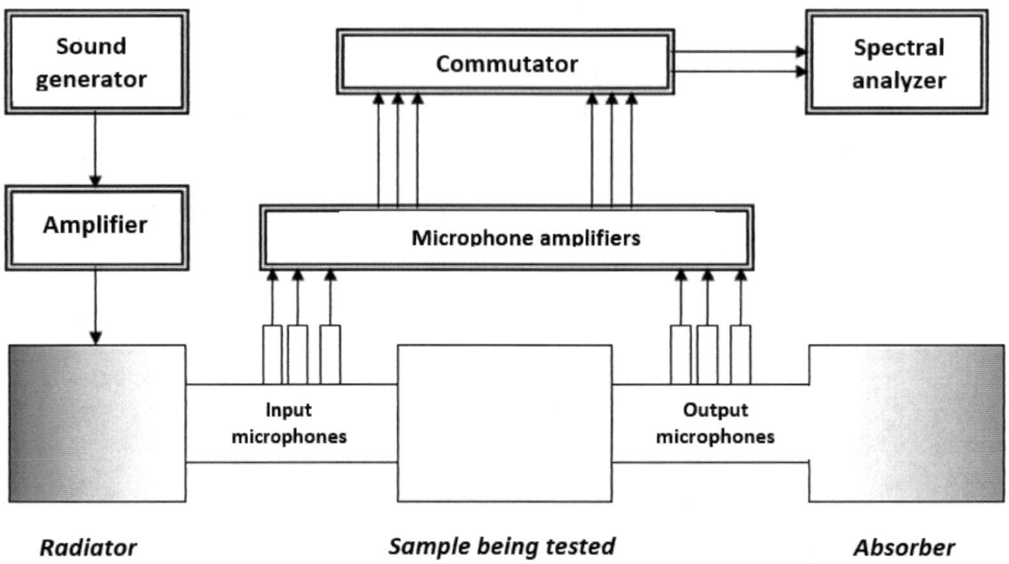

Figure 9. Test Bench Basic Diagram.

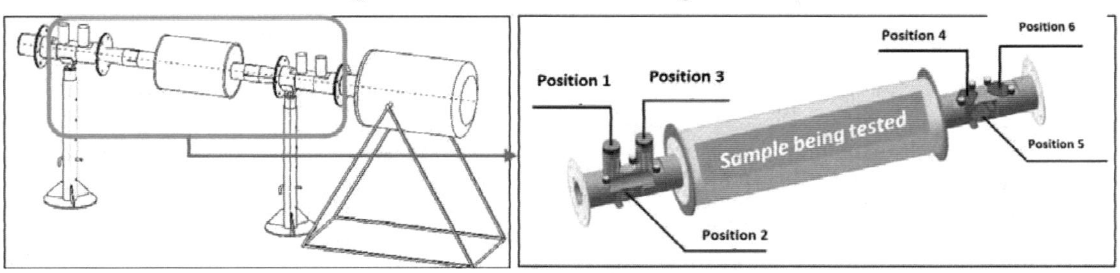

Figure 10. Test Bench Structural Setup.

The bench consists of five series-connected units, the middle one of which is the exhaust system element being tested. The input and output microphone measuring units, which are also connected through the adapter flanges with gaskets to the acoustic radiator at the input and the acoustic absorber at the output, are attached to the input and output of the test sample through adapter flanges with gaskets.

The microphone measuring units are hollow metal cylinders, each of which has 3 holes for measuring microphones with the diameter of 1/2 inch. The positions of the measuring microphones are indicated sequentially as No. 1, 2 and 3 for the input measuring unit and No. 4, 5 and 6 for the output measuring unit. The microphones are installed in the operational position on the measuring units using holders made in the form of glasses. The holders set the position of the microphone measuring membrane at the level of the inner surface of the measuring unit, and also provide reliable and stable fixation, electrical, vibration and acoustic isolation of the microphone housing from the bench.

As an acoustic radiator, a dynamic head (speaker) with a passband of 50 ... 10 000 Hz is used, mounted in a soundproof column, connected with its radiating hole to the inlet flange of the input measuring microphones unit. Basalt fiber placed in a box is used as an acoustic absorber. The inlet hole of the absorber is connected to the output flange of the output measuring microphones unit. According to the test procedure, it is necessary to carry out measurements with two versions of absorbers having different absorption coefficients. As a second acoustic absorber, an additional disk in the form of acoustic foam is used. The absorption coefficients of the two loads are shown in Figure 2.6.

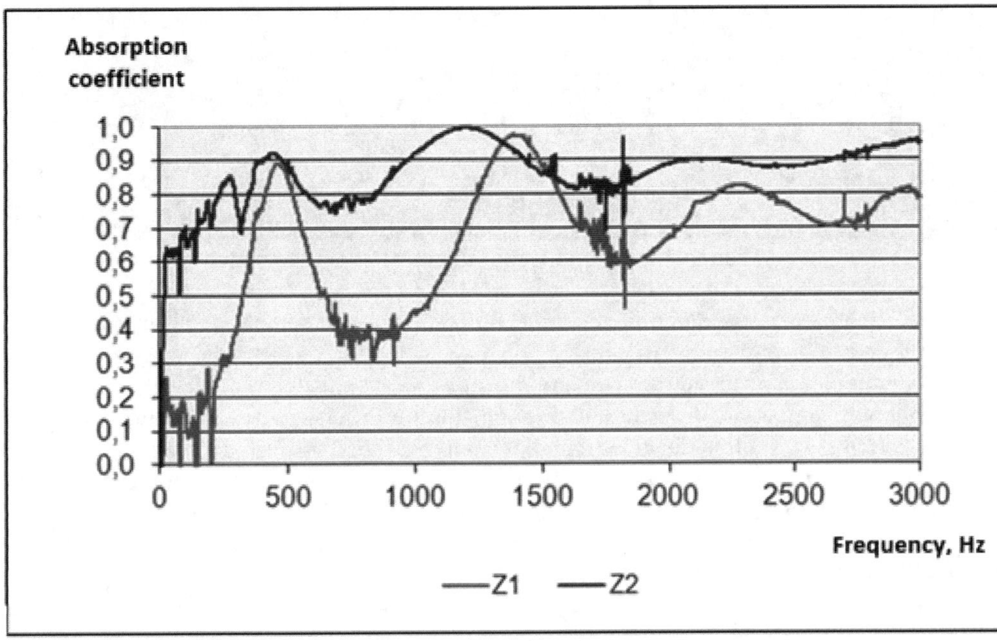

Figure 11. Experimental research results for absorbers. Absorption coefficients of the two loads: Z1- basalt fiber, Z2- acoustic foam.

Used equipment: Brüel & Kjær spectrum analyzer 3560C (with 3109 module) consisting of two signal generators capable of generating both a sinusoidal signal in the range from 2 Hz to 25.6 kHz, and white or pink noise, a measuring amplifier and a spectrum analyzer; Brüel & Kjær integrated microphones 4191C-001 with the frequency range of 3,15 Hz - 40 kHz (measured level 20 - 162 dB). Figure 12 shows the installation of the muffler on the bench.

Figure 12. Installation of the muffler on the bench.

The target TL parameter is determined based on measurement results using the calculations that include the air acoustic velocity and air density. Since these parameters are sensitive to temperature and atmospheric pressure, it is necessary to measure the specified values.

The microphone signals are transmitted through the amplifiers and commutator to the spectrum analyzer that determines the transfer functions between the signals. If all six microphones are used at the same time, a hardware multiplexer shall be used as a commutator, providing simultaneous measurement of all transfer functions in a complex form. However, to determine the target TL parameter, transfer functions are used, which are measured pairwise, and all the measurements shall be performed with only one pair of microphones. In this case, the microphones are one-by-one installed in all the necessary pairs of measuring positions, and the remaining holes during each measurement are closed with plugs imitating the measuring microphones.

The upper limit of the frequency range of the bench Fu is selected from the condition of ensuring the propagation of plane waves in the pipe and preventing the occurrence of transverse oscillations of higher modes. For cylindrical pipes with an inner channel diameter d, this condition is formalized by ratio $d < 0.58*\lambda u$, where $\lambda u = c / Fu$ is the wave length on the upper boundary frequency Fu, c is the acoustic velocity in the medium. For a pipe with the diameter of 60 mm, we obtain upper frequency limit Fu = 3286 Hz. On the other hand, the s distance between the microphones along the longitudinal axis of the pipe has an effect on λu and is chosen from $s < 0.45*\lambda u$ ratio, which also imposes a constraint on the frequency upper limit: $Fu < 0.45\ c / s$.

The lower limit of the bench operating frequency also depends on the distance between the microphones s: to obtain the measurement error of no more than 1%, the distance between the microphones s shall not exceed 5% of the sound wave length λd, i.e. $Fd = 0.05\ c / s$.

To extend the operating frequency range of the bench, the measurement microphones units are designed & built to allow measurement with three variants of the distance between the microphones: 30 mm, 70 mm and 100 mm. The tests for each sample are performed twice: for the low-frequency range of 170-1000 Hz, when the measurement microphones are installed at the distance of 100 mm, and for the high-frequency range of 1000-3300 Hz with the microphones installed at the distance of 30 mm. The measurement result processing and all calculations shall be performed for each frequency range, and the final result shall be formed by consolidation of the obtained results for the common frequency range of 170 - 3300 Hz.

As, when tested, only transmission functions between measuring channels 1 and 2 are measured, there is no need for absolute calibration of each measuring path. For amplitude-phase calibration and

IOP Conf. Series: Materials Science and Engineering **819** (2020) 012011 doi:10.1088/1757-899X/819/1/012011

definition of the complex amplitude-phase calibration factor, double measurement of the transmission function with mutual change of microphone positions in mounting seats of the test bench shall be used. The calibration factor defined this way will take into account differences in transmission of signal amplitude and phases for all measuring paths of channels 1 and 2. All values being measured shall be defined as a complex value and the calibration factor is as follows:

$$H_c = |H_c| \times e^{i\varphi} \tag{34}$$

To obtain reliable values of the transmission function for the measured acoustic signals, when measuring, it is necessary to obtain the maximum value of the coherence function by selection of the optimum ratio between the acoustic signal level and amplification coefficient of the measuring paths. The following condition may be considered as acceptable:

$$0.95 \leq \text{Coherence Function} \leq 1.0 \tag{35}$$

The exhaust system main element (muffler) acoustic characteristic calculation and experimental research result is given in figure 13:

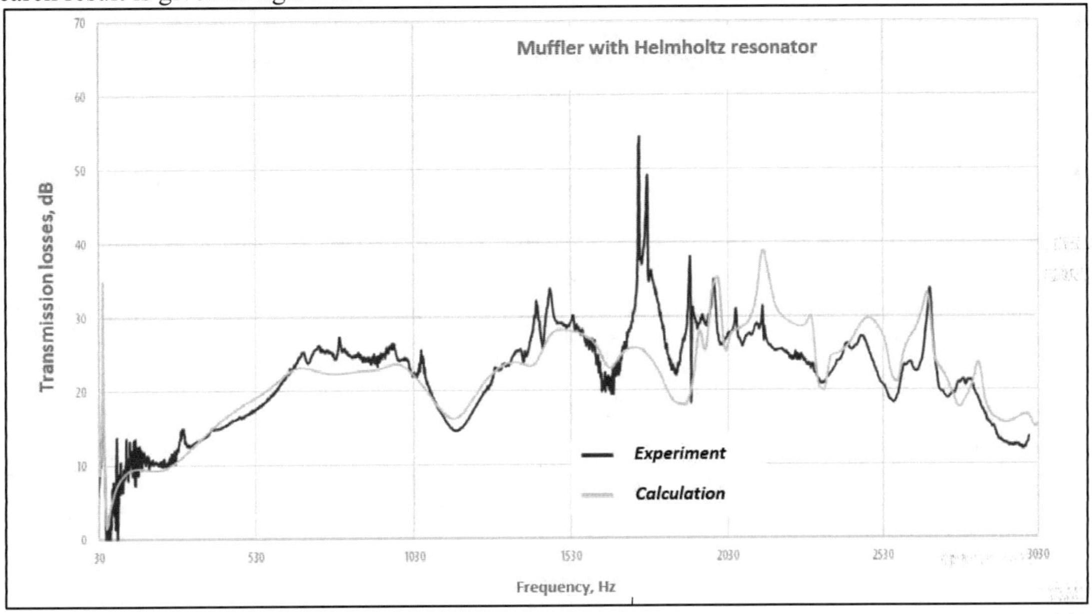

Figure 13. Results of calculation and experimental research of the muffler by transmission loss parameter.

At frequencies up to 200 Hz, there is a "diffusion" of experimental results connected with "cutting" (roll-off) of the unpowered installation speaker characteristics at low frequencies.

At frequencies from 1600 Hz, there is a deviation from calculated values connected with influence of natural frequencies and oscillation forms of housing parts and partitions of the muffler.

3. Conclusions

1. The developed calculation method allows defining acoustic characteristics of the main elements of the exhaust system;
2. The boundary conditions of the developed method, in particular, characteristics of basalt fiber were defined by calculation and experimental method;
3. The developed calculation method was validated with experiment;
4. For more exact calculation, it is necessary to consider natural frequencies and oscillation forms of housing parts of the exhaust system element.

References

[1] Song B 2000 A transfer-matrix approach for estimating the characteristic impedance and wave number of limp and rigid porous materials *Journal of the Acoustical Society of America* vol 107 pp 1131-1152

[2] Muehleisen R 2005 Measurements and empirical model of the acoustic properties of reticulated vitreous carbon *Journal of the Acoustical Society of America* vol 117, pp 536–544

[3] Rakhmatov R I, Galevko V V, Nadareishvili G G, Yudin S I 2014 Tendencies of development of exhaust systems of contemporary vehicles *Natural and technical sciences* pp 92-96

[4] Manjal M L 1990 Theory of a two sourcelocation method for direct experimental evaluation of the four-pole parameters of an aeroacoustic element *Journal of Sound and Vibration* vol 141 pp 323–333

[5] Prasad M G A 1987 four load method for evaluation of acoustical source impedance in a duct *Journal of Sound and Vibration* vol 114 pp 347–356

[6] Rakhmatov R I 2018 improvement methods of research and design exhaust system vehicles on nvh parameters *FSUE NAMI* p 174

[7] Rakhmatov R I 2019 Improvement of acoustic characteristics of motor vehicle intake system based on calculation and experimental research *Lecture Notes in Mechanical Engineering* pp 2311-2322

[8] Rakhmatov R I Modern Vehicle exhaust system design dynamic analysis *Akustika* vol 32 pp 351-354

[9] Rakhmatov R I, Krutolapov V E, Zuzov V N 2019 Development of design solutions of exhaust system for gasdynamic noise reduction *Vibroengineering PROCEDIA* vol 25 pp 135-138

[10] Rakhmatov R I, Krutolapov V E, Zuzov V N 2019 Development of a design method for determining the attachment of the exhaust mounting mounts *Akustika* vol 34 pp 141-147

[11] Vasilyev A V 2019 About the approaches to mathematical description and calculation of tire road noise radiation *Akustika* vol 32 pp 97-99

Modeling the algorithm of the automatic emergency braking system with the prediction of the coefficient of tire grip

V V Petin and A V Keller

FSUE "NAMI", 2 Avtomotornaya St., Moscow, 125438, Russia

E-mail: viktor.petin@nami.ru

Abstract. The problem of improving road safety is currently one of the most significant in the automotive industry. One of the promising ways to improve road safety is the development and introducing of intelligent active safety systems for automobiles. To solve the problem of improving safety, world leaders in the automotive industry are actively engaged in the development and implementation of intelligent control systems that provide control over the movement of the car and can partially intervene in driving. The article is devoted to the effectiveness of the automatic emergency braking system in accordance with the climatic and road conditions of Russia. The report discusses the problems of determining the coupling properties of the wheels of a car with a road on various surfaces, discusses the problems of predicting braking distance. Options for improving forecasting and building algorithms for the automatic emergency braking system are considered.

1. Introduction

The main reason for the occurrence of an accident is still a violation of traffic rules by drivers of vehicles. Also, an increase of accidents with the participation of people who at the time of the accident did not have the right to drive a vehicle was recorded - 3.4%. In total, 18,214 people died on Russian roads in car accidents in 2018. More detailed accident statistics for the past year can be found on the official website of the State traffic inspectorate of the Ministry of Internal Affairs of Russia (traffic police) in the section "Indicators of the state of road safety." [1] Based on the statistics, it can be concluded that the total number of accidents is reducing, but still remains at a high level and requires new solutions in terms of increasing the level of traffic safety. Let us consider in more detail the statistics of accidents and highlight some of emergency scenarios and their number that could be prevented with the help of an automatic emergency braking system and a warning about the possibility of a collision.

Table 1. Accident statistics for 2018.

Name of indicator	Accident	
	absolute	by type
Accidents and victims - total	168099	
in vehicle collisions		71167
with collisions with a standing transport		4903
with a collision with a pedestrian		48834
hitting an obstacle		11565
with a collision with a cyclist		5294
horse-drawn carriage		28
Of them		
at pedestrian crossings		13921
in cities and towns		52106
on public roads		57191

To considered collision scenarios, one can also add those that cannot be found in standard classifications. We can refer to them accidents caused by poor road conditions, late driver reaction, poor health condition, driving in bad weather conditions, etc. Taking into account for 18 thousands of people die every year, and economic damage is about 300 billion rubles, we can conclude that it is necessary to implement this system. Based on the considered statistics, we can conclude that ADAS and particularly the automatic emergency braking system, can significantly reduce the number of collisions, but its effectiveness and implementation safety remain in question.

2. The difficulties of implementation

The automatic emergency braking algorithm has to predict the braking distance of the vehicle, clearly detect and classify an obstacle and decide to brake automatically only when the driver does not have time to avoid a collision on his own, and does not apply actions for braking or emergency maneuvering. Otherwise the activation of braking prevents the driver from avoiding the collision by himself. For such scenarios, when the driver reacts to an emergency, a system such as braking assistance should work. The braking assistance system is activated, when the automatic emergency braking algorithm detects an accident of a collision, while the driver applies braking force. In case of insufficient pressure in the brake system, the brake mechanism is automatically applied to achieve a higher deceleration value of the car in order to prevent a collision [2].

Each driver has a different driving style, from calm to aggressive. In this case, drivers with an aggressive driving style may experience early warning signals about a potential collision. In addition, automatic control of the brake system should not interfere with the driver's avoidance maneuver, since improper operation of automatic emergency braking can lead to an emergency. Consequently, either the driver will ignore warnings or forcibly shut down the system. To avoid this, it is necessary to use such algorithms that will accurately assess the environment and identify emergencies. The algorithm of the automatic emergency braking system should not interfere with vehicle control.

3. Potential obstacle detection

One of the most important factors in the operation of the system is its stable functioning and the absence of false activations. One of the reasons for false triggering may be the detection of an imaginary obstacle on the way of the car's movement by sensory equipment. In order to avoid false detection, filtering of these sensors is used. Filtration of false objects is fundamentally difficult to organize using only radar data. For this reason, cars use not one sensor, but a set of sensors. Moreover, in difficult climatic conditions, it is difficult to organize stable operation of the system using only one sensor. The most

IOP Conf. Series: Materials Science and Engineering **819** (2020) 012012 doi:10.1088/1757-899X/819/1/012012

popular and affordable solution is to use a combination of a radar sensor + front camera. This configuration is called «sensor fusion».

Processing in a separate electronic unit of the car is performed at a high level. This helps solve the problem of filtering data from the radar.It uses data about speed, angle of rotation, as well as information from other vision systems. Screening of false objects is carried out using the following algorithms:

- the dependence of the effective dispersion area (EDA) on the distance;
- low-pass filtering (by the lifetime of the object);
- tracking movement in time;
- forecasting the trajectory of the movement of objects.

Today, most automatic emergency braking systems show stable operation in a very limited range of conditions. One of the reasons for such cases is the limited list of objects that the sensor system is able to recognize with the least likelihood of false detection. One of the most effective ways to achieve high results in the field of detection is the Sensor fusion (a combination of several sensors) [3]. Sensor fusion includes a combination of several sensors. The most common solution to such recognition system is a radar sensor that works in conjunction with an object recognition camera. This option of fusion of sensors has the potential to collect additional information that allows you to detect and classify an object with high accuracy, which can reduce the likelihood of false detection of obstacles, which is an indicator of the reliability of detection of the object. In Figure 1 we can see an example of object recognition by a camera with classification and distance determination.

Figure 1. An example of recognition of an object by the front camera.

During detecting an object with a high probability of truth, control of the possibility of a collision is carried out by an algorithm for calculating the emergency braking system. It is based on the following parameters:

- position of a potentially dangerous target;
- vector of the direction of movement of the target;
- relative speed of movement;
- prediction of possible deceleration on this section of the road [4].

Next, we will examine in detail the part of the algorithm that is responsible for predicting the achievable deceleration on this section of the road.

IOP Conf. Series: Materials Science and Engineering **819** (2020) 012012 doi:10.1088/1757-899X/819/1/012012

4. System operation algorithm

Guided by existing rules, we cannot use emergency braking earlier than 3 seconds before a possible collision. Thus, the range of the system should be within the timelines shown in Figure 2.

These requirements are regulated in the following documents:

- Standard ISO 15623: 2013. This standard establishes requirements for collision avoidance systems with a vehicle in front. This system is one of the functions of the automatic emergency braking system.
- UNECE Regulation No. 131. These documents contain uniform requirements regarding automatic emergency braking systems installed on mechanical vehicles of categories M2, M3, N2, and N3. The main focus of these standards is on roads.

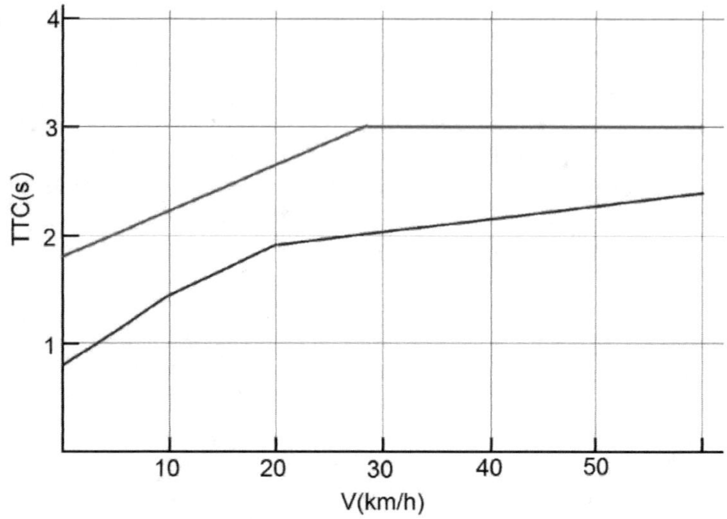

Figure 2. The response area of the automatic emergency braking system.

Thus, we see that the braking distance and the start time of the response should vary within these limits. The distance the car will go from the moment the brake system starts to compensate for speed to avoid a collision will directly depend on the deceleration that the car can develop on this section of the road.

The calculation of the time to stop the car (t_o) is made by adding the time intervals of the stages of braking system and represented by formula 1.

$$t_o = \frac{1}{2} t_n + t_r + t_p + \frac{K_e V_0}{\varphi g} \quad (1)$$

t_r – driver reaction time, s; t_p – response time of the brake drive, s; t_n – time to reach deceleration value, s; K_e – braking efficiency coefficient; V_0– vehicle speed immediately before braking, m/s; φ– coefficient of adhesion of car wheels to the road surface; g– acceleration of gravity.

Braking schedule (jst) by time can be seen in Figure 2. This figure shows the time intervals and the process of reducing speed, depending on the achieved deceleration of the vehicle at the moment. As shown in the graph, part of the schedule takes up the portion of time allotted for the driver's reaction. This section can vary from 0.5s to 1.5s depending on the health of the driver, his experience and age. Given modern high-speed driving modes, the car will cover a fairly large distance for this, as it might seem at first glance, a short period of time [5].

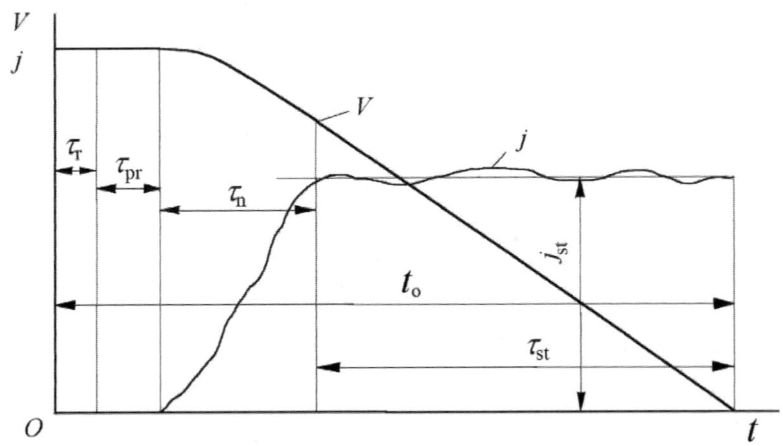

Figure 3. Calculation of stopping time with a driver.

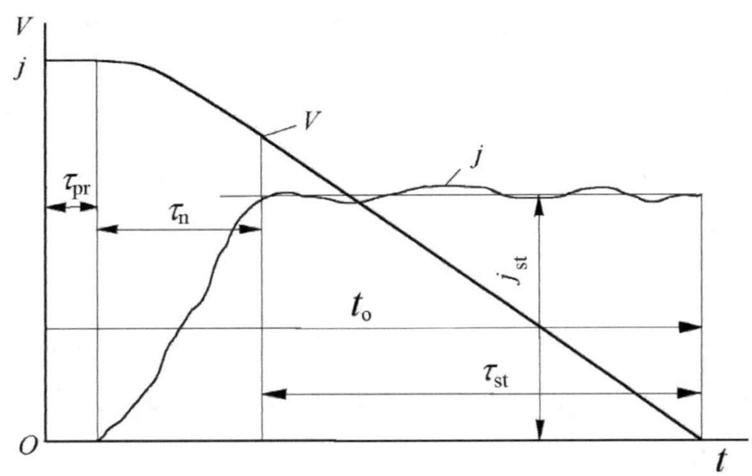

Figure 4. Calculation of the stopping time of automatic braking.

$$t_o = \frac{1}{2} t_n + t_{pr} + \frac{K_e V_0}{\varphi g} \quad (2)$$

where t_p – driver reaction time, s; t_{pr} – response time of the brake system, s; t_n – achieving the necessary deceleration, s; K_e – braking efficiency coefficient; V_0– vehicle speed immediately before braking, m / s; φ– coefficient of adhesion of car wheels to the road surface; g– acceleration of gravity.

As we can see from the graphs, the system does not require reaction time to act on the brake pedal [6].

To test the algorithm, a mathematical model of the car's movement was created taking into account changes in the coefficient of adhesion of tires to the roadway in Simulink. We can see the external view of the created model in Figure 5. On it we see several blocks that simulate sensors used in fuzzy logic algorithms to determine the current external conditions in which the car is moving. Also created a model of the movement of the vehicle in front. In the model of movement of the vehicle in front, the initial speed, distance, the moment the braking starts and the deceleration value are set. Based on these data, the algorithm should predict at what distance to a potential obstacle it is necessary to activate the brake system taking into consideration the value of the brake deceleration that can be applied under these conditions. From the block of the braking distance forecasting algorithm, a signal is sent with a request to start braking of the model of the road surface, where the road parameters are entered (dry asphalt, wet

IOP Conf. Series: Materials Science and Engineering **819** (2020) 012012 doi:10.1088/1757-899X/819/1/012012

asphalt, snowy road, etc.). Further, the vehicle's movement model calculates the braking movement, according to the results of which we can observe how accurately the algorithm calculated the necessary time for the activation of the braking request.

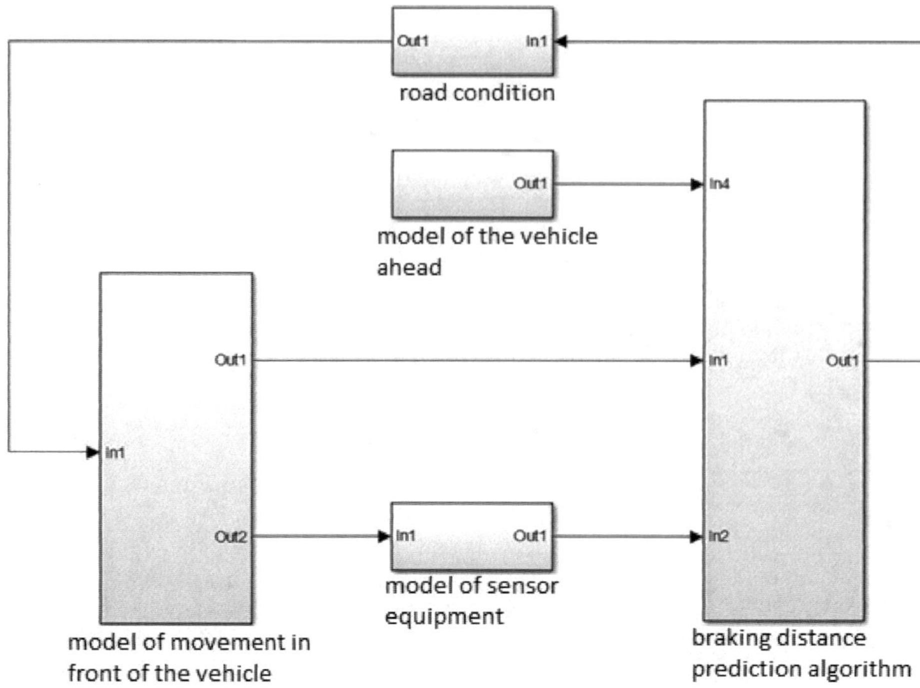

Figure 5. Appearance of the model in Simulink.

The range of the change in the braking distance of the car, depending on the condition of the roadway, presented in the figure 6.

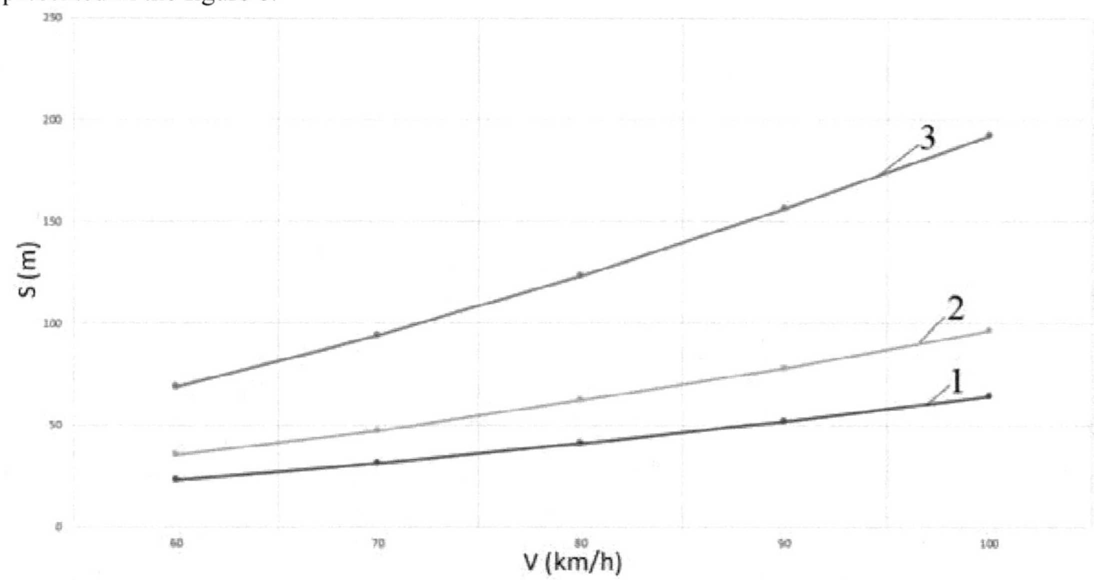

Figure 6. Braking distance in various road conditions.

The size of the braking distance varies depending on the coupling properties of the pair "tire-road". A change in this coefficient can be seen in Figure 7.

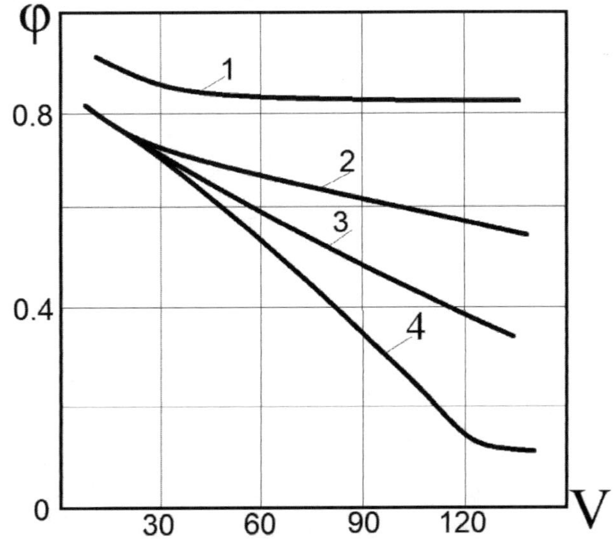

Figure 7. The dependence of the coupling properties of the pair "tire-road" on the linear speed at different thicknesses of the surface layer of water
ϕ – tire grip coefficient; V – vehicle speed.
dry asphalt, 2- 0.2 mm water layer, 3- 1.0 mm water layer, 4- 2.0 mm water layer [7].

Due to such a wide spread of the braking coefficient, depending on the size of the water layer on the road, a change in the braking distance of the car also happen in a wide range. As sources of information for predicting the coefficient of tires with the road, it is proposed to consider information:

- Ambient temperature sensor;
- Precipitation sensor;
- Current operating mode of the wiper;
- Anti-lock braking system;
- The operation of the system of ESP;
- Recognition of the current state and type of pavement using the front camera.

The values of these parameters are modeled based on the recorded values of real sensors in various conditions. A graph of the distance to the vehicle in front with automatic braking is shown in Figure 8.

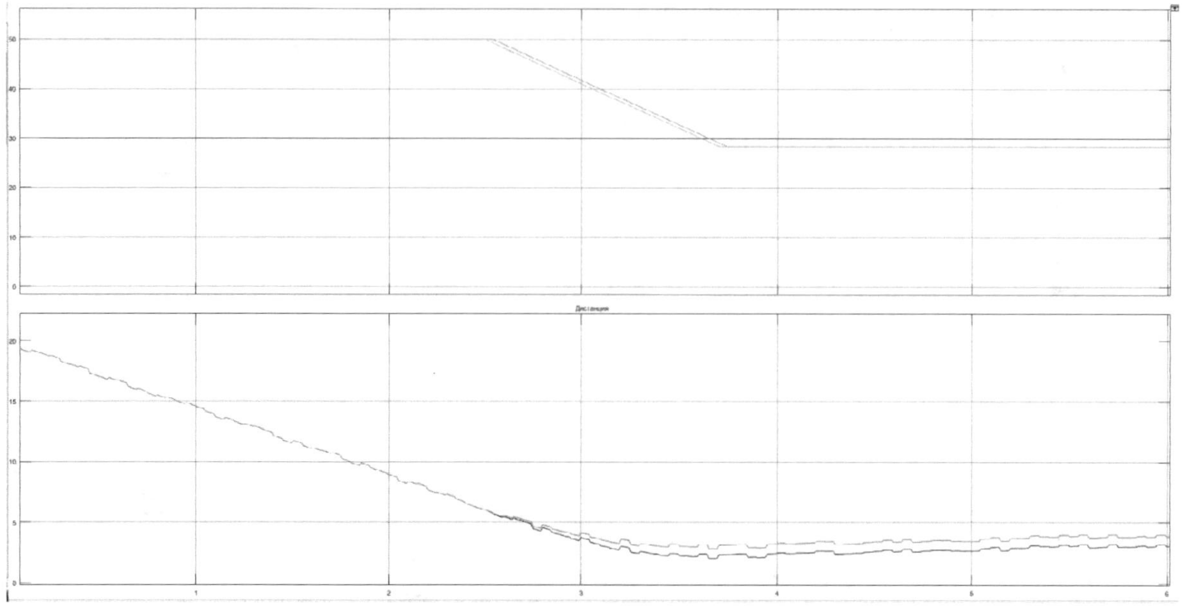

Figure 8. Graph of changes in distance and speed at the moment of activation of the system of automatic emergency braking.

In the figure above, we see two graphs that display the simulation of changes in the speed of the tested cars. In the lower part, we can see two graphs of the distance change. One taking into account additional information from sensors and systems, and the other without taking into account this information, with a statically specified coefficient of tire grip. To conduct this test, the following scenario model was created:

Temperature	+20 °C
Precipitation	Rain
Wiper Intensity	At 2 second of simulation, the intensity decreases
Driver reaction	Absent
Initial distance between cars	20 m
Speed of the car in front	30 km/h
Initial vehicle speed	50 km/h

As we can see on the top of the graph, the moment of the start of braking and speed reduction occurs a bit later, unlike the model, which does not use information about the intensity of the wipers. Based on this, the calculation of the distance to the collision is less accurate in the first case and can activate the response too early, and as described above, early system activation is one of the reasons why drivers force-shut off the automatic emergency braking system.

5. Conclusion

A modern automatic emergency braking system can reduce more than 30% of typical collisions.

The introduction of active safety systems can significantly increase the competitiveness of cars in the international market.

Considered the most common and effective combination of sensors for automatic emergency braking.

The adaptive system for calculating the coefficient of adhesion of tires to the road is able to most accurately calculate the required braking distance and the response time of the system. Mathematical modeling of the algorithm showed an improvement in braking distance calculation.

References

[1] Komzalov A M and Shilov N G 2017 Application of modern technologies in assistance systems for a driver of a car *Izv. universities. Instrument making* ***60****(11) pp 1077-1082*

[2] *2014* "Taxonomy and Definitions for Terms Related to On-Road Motor Vehicle Automated Driving Systems" *SAE International Surface Vehicle Information Report SAE Standard J3016 Issued Jan.*

[3] Murao S, Yoto A and Yokogoshi H 2013 Effect of Smelling Green Tea on Mental Status and EEG Activity *International Journal of Affective Engineering* **12**(*2) pp 37-43*

[4] NHTSA "Department of Transportation Releases Policy on Automated Vehicle Development" 14-13Thursday May 30 2013

[5] Lerner N, Singer J, Huey R, Brown T, Marshall D, Chrysler S, Schmitt R, Baldwin C, Eisert J, Lewis B, Bakkar A, and Chiang D 2014 Crash Warning Interface Metrics (CWIM): Phase 3 NHTSA (Washington, D.C.)

[6] Yoshihira M, Watanabe S, Nishira H and Kishi N 2016 Developing an Autonomous Vehicle Control System for Intersections Using Obstacle Blind Spot Detection Frames *SAE Technical Paper 2016-01-0143*

[7] Revin A A 1983 *Improving the efficiency, stability and controllability when braking vehicles* Thesis (Volgograd)

IOP Conf. Series: Materials Science and Engineering **819** (2020) 012013 doi:10.1088/1757-899X/819/1/012013

Some aspects of modern automobile transport development

V I Kotliarenko[1]

[1] FSUE "NAMI", 125438 Avtomotornaya st., 2, Moscow, Russia

Abstract. The paper treats modern automobile transport development tendencies. In this regard, the main focus is on safety and ecology. Road accidents (hereinafter referred to as RA) are one of the main causes of death in the world, and every year they claim over 1.35 million lives, lead to injuries of up to 50 million people, and are ranked eighth among the main causes of death for all age groups. The economic losses caused by RAs in some countries reach 5% of the gross domestic product, in the Russian Federation this share is around 2%. The transport sector produces 18% of all global anthropogenic emissions and 23% of greenhouse gas emissions related to the energy industry. 3 million people die annually due to air pollution. A part of the listed problems can be solved by means of increasing vehicles' autonomy and by creating driverless vehicles, the advantages of which are evident: safety - minimization of RAs which mostly occur due to human error; environmental friendliness - quantity optimization of the vehicle fleet, road traffic organization (optimal selecting route and driving modes) and decrease of fuel consumption (selecting driving modes); economic efficiency - decrease of costs for insurance and rapid-response (emergency) medicine in case of RA, decrease of cargo and people transportation costs by means of economizing driver's salary and economizing fuel.

1. Introduction

The strategic goal of the development of the transport system of any state is to satisfy the needs of the economy and society in quality transport services. In the Russian Federation, due to its geographical features and the specifics of the territorial distribution of productive forces, transport plays a key role in the economy. Therefore, the government has always paid and pays much attention to the effective, safe and high-quality development of the transport system. In this case, special attention is paid to road transport.

2. Freight and passenger transportation

In the Russian Federation, more than two thirds of cargo is transported by road (Figure 1 [1, 2]).

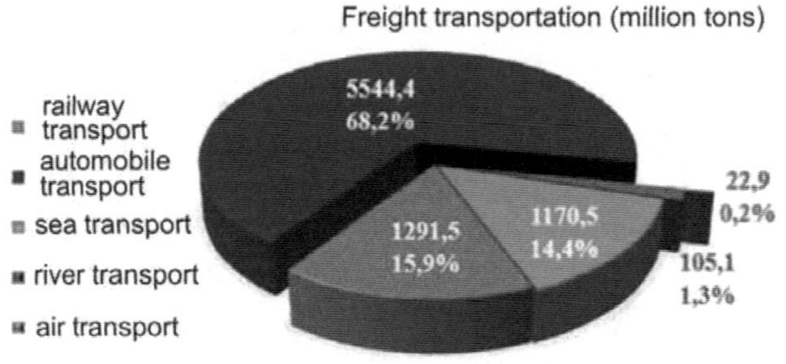

Figure 1. The structure of cargo transportation in the Russian Federation.

The share of passenger transportation by road is 66%, and with urban electric transport – 89% (Figure 2 [1, 2]).

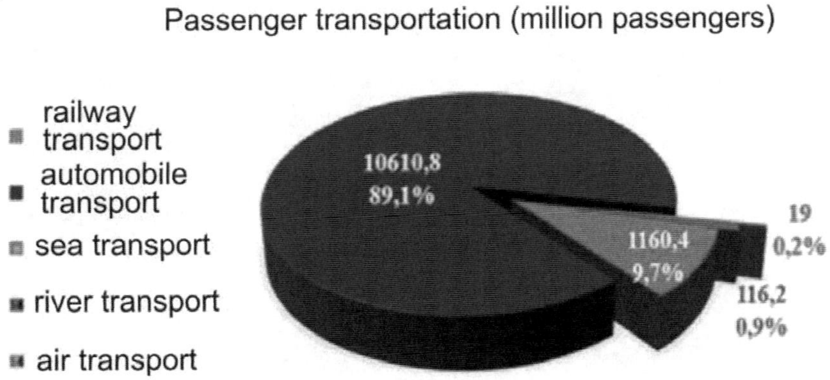

Figure 2. The structure of passenger transportation in the Russian Federation.

At the same time, transportation carried out on personal vehicles, which dominates the general fleet of motor vehicles of the country, is not taken into account.

Fleet of vehicles

At present, the domestic fleet of motor vehicles, registered with the State traffic inspectorate, reached more than 57.1 million units excluding trailers and semi-trailers (Figure 3).

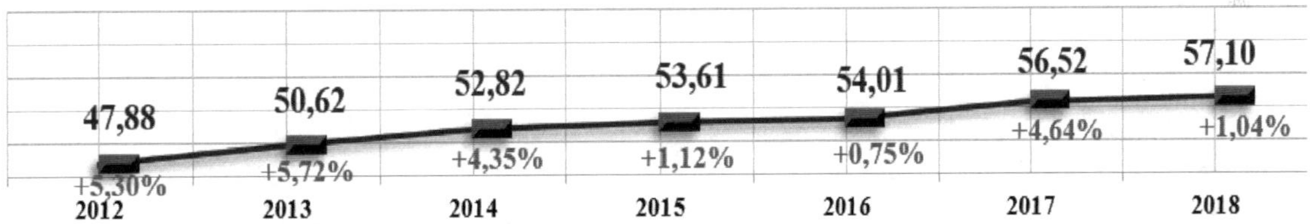

Figure 3. The dynamics of the growth of the fleet of vehicles in the Russian Federation (million units).

The number of cars reached 47.4 million units or 78.3% of the total number of vehicles, buses – 0.86 million (1.5%), trucks - 6.48 million (10.7%), motorcycles – 2.32 million (3.8%), trailers and semi-trailers – 3.4 million (5.7%). At the same time, 54.6 million units or 90% of the country's total vehicle fleet are owned by individuals [1, 7].

3. Fleet Age

A characteristic feature of the fleet is a significant proportion of outdated vehicle models with long service life. About 47.5% of cars, 61.7% of trucks, 50.6% of buses and 89.5% of motorcycles have been in operation for over 10 years. Over 35% of the fleet has been in operation for more than 15 years (Figure 4 [1]).

The consequence of the presented statistics is a decrease in the technical reliability of vehicles (as mileage increases, fatigue stresses in the structure increase by 1.4 - 2.5 times), lag behind modern requirements for passive, active and environmental safety, increasing the likelihood of sudden failure of components and assemblies in the process of traffic and, as a result, an increase in the probability of an accident due to a technical malfunction.

IOP Conf. Series: Materials Science and Engineering **819** (2020) 012013 doi:10.1088/1757-899X/819/1/012013

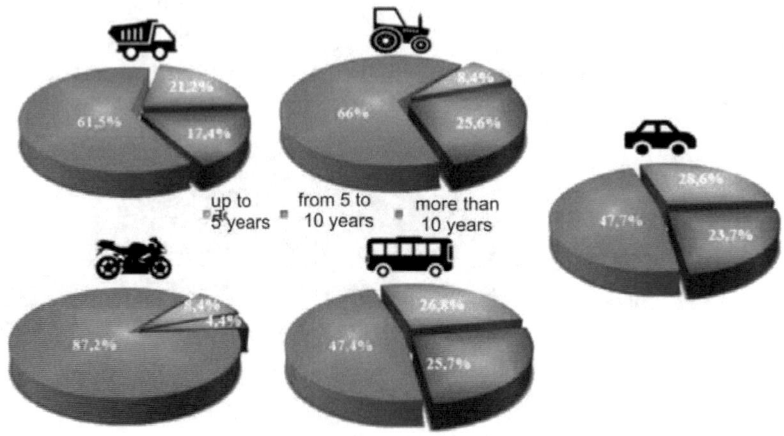

Figure 4. Distribution of the fleet of vehicles by types and times of operation.

The other side of these statistics is the low environmental friendliness of the domestic automobile fleet (Table 1 [7]).

Table 1. Ecological class of the domestic automobile fleet

Vehicles registered in the Russian Federation	Number of vehicles
Ecological class 0	2577456
Ecological class 1	753498
Ecological class 2	3231777
Ecological class 3	5932642
Ecological class 4	14885092
Ecological class 5	10195149
Ecological class 6	35637
Ecological class not established	17168375

4. Traffic accidents

Worldwide, among all modes of transport, road transport is the most dangerous. It accounts for 97% of all deaths in transport [8] or 1.35 million deaths a year. Traffic accidents cause injuries to 50 million people and take eighth place among the leading causes of death for all age groups at the world level [10].

According to the United Nations Economic Commission for Europe (hereinafter - UNECE), road traffic crashes cause significant economic damage, which in some countries reaches 5% of the gross domestic product. The annual economic losses of the Russian Federation from accidents amount to about 2% of the gross domestic product [6].

These losses are directly related to material costs in case of mechanical damage to vehicles and road infrastructure, as well as to lost income from death and disability of the able-bodied population, expenses for treating victims, and social support for people with disabilities, as a result of which road safety issues remain the most relevant in the modern world, including Russia.

Over the past several years, measures have been taken in the Russian Federation to increase the protection of road users. Due to this, the trend of decreasing the main accident indicators [1], which has been existing since 2012, continues (Figure 5).

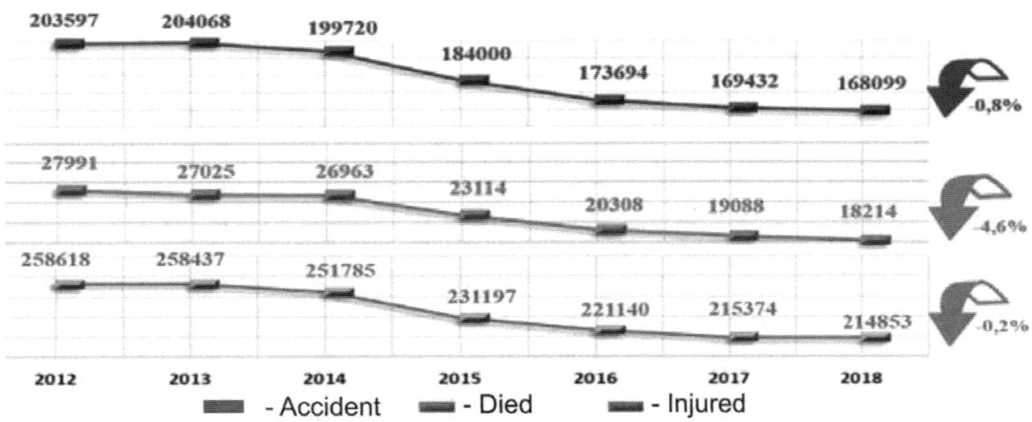

Figure 5. Dynamics of the main accident indicators in the Russian Federation.

However, despite some positive changes, the level of road traffic accidents in the country remains high. At the end of 2018, 168.1 thousand incidents were registered, in which 18.2 thousand were killed and 214.9 thousand people were injured.

The reasons for 88.1% of incidents were violations of the Traffic Laws of the Russian Federation committed by drivers of vehicles. Moreover, almost every tenth accident occurred with the participation of drivers in a state of intoxication.

The most common types of accidents are vehicle collisions (42.3%) and pedestrian collisions (29.1%). Most pedestrian accidents (67.5%) are due to violations by drivers. Almost three quarters of these incidents occurred in the dark, in which 84.6% of pedestrians died. These incidents are characterized by the most serious consequences - fatal injuries sustained by pedestrians in almost every second case [1].

International priorities in the field of road safety in developed countries are defined by the slogan "above zero", that is, targeting zero mortality on the roads. In accordance with the tasks set by the Transport Strategy until 2030, one of the most important tasks is to reduce the number of deaths in a traffic accident per year per 10 thousand cars by more than 60% [4].

This problem is planned to be solved not only through the formation of a regulatory framework, improved traffic management and control activities, but also through the use of intelligent transport systems in road transport and road infrastructure, as well as improving the design of vehicles, including the development of unmanned vehicles.

5. Ecology

According to UNECE, 3 million people die every year from environmental pollution. In one projection, energy-related CO2 emissions are expected to grow by 40% between 2013 and 2040. The transport sector accounts for 23% of global energy-related greenhouse gas emissions and 18% of all anthropogenic emissions in the global economy. Moreover, from 70% to 84% of fuel energy is lost in engines and inefficient drives [8].

The 2010 report, Energy for a Sustainable Future, set out recommendations that laid the foundation for the three goals of the Sustainable Energy for All (SE4ALL) initiative: affordability, efficiency and renewable energy.

In order to implement the Climate Doctrine of the Russian Federation, approved by Decree of the President of the Russian Federation dated December 17, 2009 No. 861-rp, the Government of the Russian Federation must ensure by 2020 the reduction of greenhouse gas emissions to no more than 75 percent of the volume of these emissions in 1990 [5].

Reducing the negative impact of the transport system on the environment is also one of the important goals of the transport strategy. It is planned that resource-saving, energy-efficient and environmentally friendly materials and technologies, alternative fuels, and environmental safety systems will be introduced in the transport complex.

Compared with 2007, carbon dioxide emissions from the transport complex should decrease by 15% by 2030 (Figure 6), the share of alternative fuels in the total fuel consumption of vehicles will increase to 30%, the share of vehicles using alternative fuels (hybrid and electric vehicles, vehicles with gas and hydrogen engines) can reach half the fleet [4].

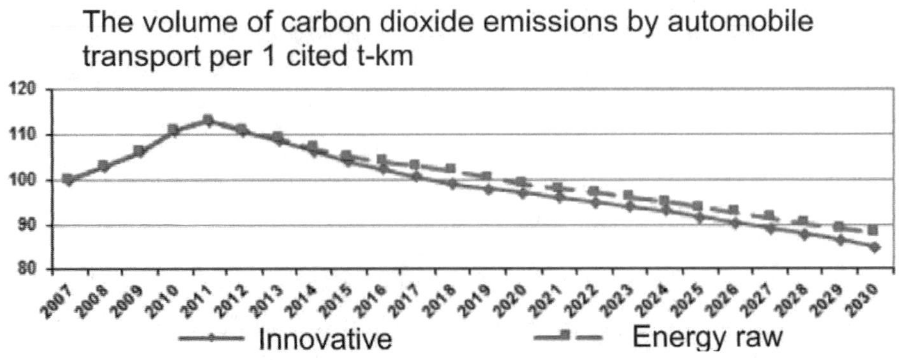

Figure 6. Forecast of carbon dioxide emissions by automobile transport until 2030.

Measures to reduce the negative impact of road transport on the environment and public health can be divided into the following areas:

- improving environmental safety and energy efficiency of operating vehicles;
- improving the quality of traditional motor fuels and expand the use of alternative energy sources;
- improving transport infrastructure, transport planning, traffic management, development of public modes of transport;
- measures in the field of managing transport behavior (mobility) of the population (switching to public transport, taxi and car sharing, including unmanned cars).

6. Trends

In accordance with the goals set by the Transport Strategy until 2030, one of the most important tasks in the field of automobile transport is to increase road safety and reduce harmful emissions into the atmosphere. These tasks should be largely solved by improving the design of vehicles.

According to the Strategy for the Development of the Automotive Industry until 2025, approved by Decree of the Government of the Russian Federation dated April 28, 2018 No. 831-r (hereinafter referred to as the Strategy), the global automotive industry is currently characterized by a set of global trends that also affect the development of the Russian automobile market and the automotive industry generally. The main ones are as follows [3].

Electrification of vehicles. According to experts, in 2020, the share of electric vehicles in the total world output will be at least 17%, of which 75% will be in hybrid cars, and the total number in the global fleet will reach 20 million or 1.7%.

Improving the autonomy of vehicles. Increasing the autonomy of vehicles will require addressing issues, including cybersecurity, regulation of property rights and data management, the distribution of responsibilities in road accidents, and the development of a new regulatory environment.

Implementation of network (telecommunication) technologies in transport systems. The introduction of telematics systems can increase transport capacity by 25%, passenger turnover by 20%, cargo turnover by 5-10% while reducing the number of accidents to 60%.

For the Russian Federation, the priority areas for the development of the automotive industry are improving the environmental performance of vehicles, passive and active safety, autonomy and robotization of vehicles, the use of intelligent security and control systems, informatization and computerization.

It is assumed that the implementation of the measures provided by the Strategy will ensure the renewal of the fleet of vehicles of all types and the emergence of a new line of products (electric vehicles and unmanned vehicles) with sales growth rates of about 40-50 percent per year [3].

7. Findings

By 2030, annual passenger traffic will increase by 50 percent, and global freight traffic by 70 percent. By 2050, approximately 1.2 billion more cars will enter the road [9]. All this leads to unsustainable mobility - deaths in transport, the intensive use of fossil fuels, large greenhouse gas emissions and increased environmental pollution by harmful emissions and noise.

Some of these problems can be solved by increasing the autonomy of vehicles and creating unmanned vehicles, the advantages of which are obvious:

- safety - minimization of accidents that occur mainly due to the human factor, transportation of goods in hazardous areas, during natural and man-made disasters or military operations;
- environmental friendliness - reducing the global environmental load both due to quantitative optimization of the car fleet, and due to the organization of road traffic (optimal choice of route and driving modes) and reduction of fuel consumption;
- cost-effectiveness - reducing the cost of insurance and quick-response medicine in case of an accident, reducing the cost of transporting goods and people by saving on wages, driver rest time and fuel economy, increasing the efficiency of road use by centralizing traffic flow and increasing their capacity;
- social orientation - the ability to independently travel in a robotic car for people without a driver's license, including minors, and people with visual impairments.

At the same time, it seems expedient, first of all, to automate urban passenger (buses, trolleybuses and taxis / car sharing) and freight (mail, school meals, etc.) transportations carried out on established routes. For these transportations, it is not necessary to have constant communication with satellite systems of global positioning and Internet connection, it is only necessary to have an appropriate processor with an operating algorithm and memory with a detailed high-resolution terrain map (HD-map).

References

[1] State report on the state of road safety in the Russian Federation, Ministry of Internal Affairs of Russia, June 2019
[2] Information and statistical bulletin "Transport of Russia January-December 2018", Ministry of Transport of Russia, April 3, 2019
[3] The order of the Government of the Russian Federation of April 28, 2018, No. 831-r "On approval of the Strategy for the development of the automotive industry until 2025"
[4] The order of the Government of the Russian Federation of November 22, 2008, No. 1734-p
[5] The order of the President of the Russian Federation dated December 17, 2009 No. 861-rp "On the Climate Doctrine of the Russian Federation"
[6] The order of the Government of the Russian Federation of 08.01.2018 No. 1-r e "On approval of the Road Safety Strategy in the Russian Federation for 2018-2024"
[7] Form No. 560 of the federal statistical monitoring for January-December 2018, (Section. 3. The number of motor vehicles, trailers and semi-trailers for them, registered)
[8] Global mobility report 2017, Tracking Sector Performance. Internet: https://openknowledge.worldbank.org/bitstream/handle/10986/28542/120500.pdf?sequence=6
[9] Sustainable Mobility for AllTM. 2017-2018. Annual report. Internet: http://www.sum4all.org
[10] Road safety for all. January 2019. Internet: http://www.unece.org/fileadmin/DAM/trans/roadsafe/publications/Road_Safety_for_All.pdf/

IOP Conf. Series: Materials Science and Engineering **819** (2020) 012014 doi:10.1088/1757-899X/819/1/012014

Mathematical model of an electric vehicle with a non-flat battery of photovoltaic converters

V N Kozlov[1], A F Kolbasov[1], K E Karpukhin[1] and N T Katanaev[2]

[1] FSUE "NAMI", 2 Avtomotornaya St., Moscow, 125438, Russian Federation
[2] Moscow Polytechnic University, 38 B. Semenovskaya St., Moscow, 107023, Russian Federation

E-mail: v.kozlov@nami.ru

Abstract. One of the major environmental problems is emissions of CO2 and other harmful substances from motor vehicles (MV). According to the tests, the most environmentally friendly MV (measurement of emissions on MVs) are electric vehicles (EV). But a low emission level can only be ensured by the use of environmentally friendly sources of electricity used to charge EV: hydroelectric power plants, wind power plants, tidal, geothermal, wave, solar power plants, etc. However, the transmission of electric energy is connected with losses. One of the ways to use them is the application of batteries of photovoltaic converters (PVC) within the EV structure as the most suitable in terms of weight and size characteristics. The use of PVC in MV is not a new idea, but due to the increase in the efficiency of PVC, their relevance becomes more obvious. Purpose of work: development of mathematical models, scientific substantiation and rationale of the calculation method of EV energy efficiency increase by enlarging the cruising range when using PVC batteries. Research methods: methods of ground MV movement theory, system design, theory of electric circuits, mathematical modeling, etc. Results: development and scientific substantiation of the calculation method for a non-flat PVC battery within the EV structure and its effect on run or mileage change. Summary: 1) the mathematical model of EV rectilinear motion on flat non-deformable horizontal ground, suitable for the study of the processes of energy exchange between the EV and PVC batteries and the effect of these processes on the mileage, has been developed; 2) the application of the calculation method for non-flat PVC batteries to be used within the EV structure has been scientifically substantiated.

1. Introduction

To solve environmental problems associated with emissions of harmful substances by motor vehicles (MV), it is necessary to develop environmentally friendly MV.

It is known that electric vehicles (EV) are environmentally friendly transport. But the environmental friendliness of an EV depends on the way it receives electric energy from the electric grid. For example, coal-fired power plants emit 1001 g of CO_2 into the atmosphere (not counting more harmful substances) when generating 1 kW·h of electricity, fuel oil-burning power plants – 840 g, gas-burning power plants – 469 g. The average emissions of CO_2 per 1 kW·h by countries are as follows: China – 745 g, Russia – 597 g, the UK – 225 g [[1]].

Low emissions of CO_2 and other harmful substances can only be ensured by the application of environmentally friendly renewable energy sources used for EV charging [[2]]. In addition, in EV and vehicles with combined power units, braking in most cases occurs not by mechanical brakes, but by electric drives or actuators in the recuperation mode, which significantly reduces the release of microparticles of dust from brake pads and discs, including cancerogenic ones [[3]]. Also widely known is the problem of an insufficient EV cruising range due to the small amount of electric energy stored on board of EV, compared to the chemical energy contained in gasoline used to refuel the tank of the vehicle [[4]].

Therefore, the task of increasing the cruising range of EV, including application of batteries of PVC, is relevant. Among the MV manufacturers that use PVC batteries in EV are Volkswagen, Toyota, Sanyo, Ford, etc. [[5]].

There are also works of Russian researchers, which are devoted to the application of PVC batteries in EV [[6]-[8], [1], [9], [10]]. But no one has yet worked particularly on the specifics of non-flat PVC batteries and connection between the MV body geometry and electric parameters of PVC batteries and cruising range.

For mathematical modeling of different variants of a battery of PVC within the MV structure, it is necessary to use, among other things, a mathematical model (hereinafter "model") of the EV itself consisting of other models, for example, those of wheels, transmission, electric machine (EM), battery, etc.

2. Vehicle movement model

The rolling resistance force according to Coulomb's law [[11]]

$$F_f = m \cdot g \cdot \cos\alpha \cdot f, \text{N},$$

where:
m – the weight of the EV, kg;
g – free fall acceleration on the Earth's surface, 9.81 m/s²;
α – angle of inclination of the support surface (road), rad;
f – coefficient of rolling resistance of the tire from the translational speed.

In the absence of data on the terrain, the α angle of inclination of the road is assumed to be 0.

Aerodynamic drag (resistance) force of the EV [[9]]

$$F_w = \frac{C_x S_M \rho_{air} V^2}{2}, \text{N},$$

where:
C_x – coefficient of aerodynamic resistance of the vehicle (according to the vehicle manufacturer);
S_M – midsection area, m^2 (according to the vehicle manufacturer);
ρ_{air} – density of air (under normal conditions 1.204 kg/m³);
V – speed of the rectilinear movement of the vehicle, m/s.

Force of inertia of the EV

$$F_a = m \cdot a, \text{N},$$

where:
a – acceleration of the EV,
$a = \frac{dV}{dt}$, m/s².

The total resistance force acting on an EV and consisting of the three forces listed above

$$F_\Sigma = F_f + F_w + F_a\ , \text{N}.$$

3. Wheel model

The coefficient of rolling resistance of the tire from the translational speed [[11], [12]]:

$$f = f_0 + k_f \cdot V^2, \text{N},$$

where:
f_0 – coefficient of rolling resistance of a wheel at a speed close to 0;
k_f – coefficient of increase of rolling resistance of a wheel from speed.

Torque on the wheel axis taking into account the moment of inertia of the wheel

$$M_w = F_\Sigma r + a \cdot \frac{I_w n_w}{r}, \text{N·m},$$

where:
r – average radius of EV wheels, m;
I_w – EV wheel moment of inertia, kg·m²;
N_w – number of EV wheels.

4. Transmission model

For the transmission model, the rectilinear movement of the vehicle is assumed without taking into account the wheels slip. In this case, all the wheels rotate at the same speed, and the angular velocity of the EM is determined as

$$\omega_{EM} = \frac{V u_{tr}}{r}, \text{rad/s},$$

where:
u_{tr} – transmission gear ratio.

The torque on the shaft of the EM when transmitted through the transmission from the wheel is calculated as

$$M_{EM} = \frac{M_w}{u_{tr}} \eta_{tr}^{sgn(M_w)}, \text{N·m},$$

where:
M_w – torque at the wheels, N·m;
H_{tr} – transmission efficiency;
$sgn\ (M_w)$ – wheel torque sign function: mechanical efficiency is determined by the value of friction forces, so when accelerating it increases the load moment on the EM (sign -1), and when recuperating-reduces it (sign +1).

The moment of inertia of the rotor of the EM, given to the torque on the axis of the EM

$$M_I = I_{EM} \frac{d\omega_{EM}}{dt} = I_{EM} \frac{dV}{dt} \frac{u_{tr}}{r} = I_{EM}\, a \frac{u_{tr}}{r}, \text{N·m},$$

where:
I_{EM} – moment of inertia of the rotor of the EM, kg·m².

5. Electric machine model

The torque on the axis of an EM is the sum of the torques M_{EM} and M_I. The efficiency of the EM is given by a table-valued function from the angular velocity and the load moment:

$$\eta_{EM} = f(\omega_{EM}, M_{EM} + M_I).$$

Power consumed by an EM taking into account its efficiency

$$P_{EM} = (M_{EM} + M_I) \cdot \omega_{EM} \cdot \eta_{EM}^{sgn(M_{EM}+M_I)}, \text{W},$$

IOP Conf. Series: Materials Science and Engineering **819** (2020) 012014 doi:10.1088/1757-899X/819/1/012014

where:
η_{EM} – EM efficiency.

6. Inverter model

The current strength at the input of the traction inverter taking into account its efficiency

$$i_{dc} = \frac{P_{EM} \cdot \eta_{dc}^{sgn(P_{EM})}}{u_{b,hv}}, \text{ A},$$

where:
η_{dc} – traction inverter efficiency.
$u_{b,hv}$ – GB1 high-voltage (traction) battery voltage, V.

The total power losses in the power leads (wires) of the electric drive and at the equivalent resistance of the battery

$$P_{lost} = i_{dc}^2 R_{bat} + i_{dc}^2 \frac{l_{wire} \rho_{wire}}{q_{wire}}, \text{ W}.$$

where:
R_{bat} – equivalent active resistance of the battery, Ohm;
R_{wire} – active resistance of wires from the battery to inverter, Ohm;
l_{wire} – wire length, m;
q_{wire} – wire cross section, mm²;
ρ_{wire} – wires material resistivity, for copper at 20 °C it would be around 0.01724-0.0180 Ohm·mm²/m.

7. Model of onboard consumers

The current strength taken by the load of onboard consumers with power P_{add} is

$$i_{add} = \frac{P_{add}}{u_{b,lv}}, \text{ A},$$

where:
P_{add} – power of onboard consumers, W;
$u_{b,lv}$ – GB2 low-voltage (12 V) battery voltage, V.

At the same time, random oscillations with an amplitude of ±25% from the average value must be added to the current strength value of onboard consumers (A1-An) to simulate random deviations.

8. Traction battery model

In the pulse mode of the DC/DC converter, the charge and discharge processes of the low-voltage buffer battery are taken into account, so additional consumers are powered by the low-voltage battery, and the charge/discharge power of the traction battery

$$P_b = P_{Ch} \cdot \eta_{dcdc} - P_{dc} - P_{lost}, \text{ W},$$

where:
P_{Ch} – electric power consumed by the DC/DC converter, W;
η_{dcdc} – DC/DC converter efficiency;
P_{dc} – electric power consumed by the traction inverter, W;
P_{lost} – the total electrical power losses, W.

9. Buffer battery model

The differential equation of energy exchange between the onboard consumers, traction battery, buffer battery and PVC battery

$$\frac{dE_{GB2}}{dt} = P_{b.PVC} \cdot \eta_{MPPT} - P_{Ch} - P_{add} - i_{GB2}^2 R_{GB2} - P_{lost.cond2}, \text{W},$$

where:
$P_{b.PVC}$ – electrical power generated by the PVC battery, W;
η_{MPPT} – MPPT converter efficiency;
P_{Ch} – electric power consumed by the DC/DC converter, W;
P_{add} – power of additional (onboard) consumers, W;
$P_{lost.cond2}$ – power losses in the power wires between GB2 buffer battery and MPPT converter, W;
i_{GB2} – charge/discharge current of the GB2 buffer battery, A;
R_{GB2} – equivalent active resistance of the GB2 battery, Ohm;
E_{GB2} – GB2 traction battery energy, J.

The power losses in the power wires of the charge/discharge circuits of the GB2 buffer battery may be neglected due to their small length and active resistance.

High-voltage traction battery voltage $u_{\mathrm{b,lw}} = f(\mathrm{E}_{GB2})$ in the first approximation can be taken as a constant.

10. DCDC converter model

The power consumed by DC/DC converter [[9]]

$$P_{Ch} = u_{b,lv} i_{dc,lv} \cdot \eta_{dc}^{sgn(i_{dc,lv})}, \text{W},$$

where:
$i_{dc,lv}$ – current on the low-voltage side, for maximum efficiency $i_{dc,\mathrm{lv}} = 125$ A.
$\eta_{dc}^{sgn(i_{dc,lv})}$ – converter efficiency taking into account the direction of current.

From the above equations by means of simple transformations it is possible to deduce the equation of rectilinear motion of the EV:

$$\frac{dV}{dt} = \frac{\left(M_{EM} u_{tr} \eta_{tr}^{\text{-sgn}(M_W)}\right)\frac{1}{r} - F_f - F_w}{m + (I_w n_w)\frac{1}{r^2}},$$

which, taking into account the absence of the moment of inertia of the transmission and the change of the sign of the torque of the wheel, is identical to the differential equation of rectilinear motion of the EV [[9]].

11. MPPT converter model

The MPPT converter efficiency is a function of the power and voltage of the PVC battery [[9]]:

$$\eta_{mppt} = f(P_{PVC}, u_{PVC}),$$

where:
P_{PVC} – power generated by PVC, W,
u_{PVC} – voltage of the PVC battery, V.

The current output of the MPPT converter

$$i_{mppt,out.} = \frac{i_{PVC} \cdot u_{PVC} \cdot \eta_{mppt}}{u_{b,lv}}, \text{A},$$

where i_{PVC} – current at the output of the PVC battery, A.

12. Model of the PVC battery

The power generated by each individual PVC is directly proportional to the flow of solar radiation taken by it, if other conditions are equal [[13]]:

IOP Conf. Series: Materials Science and Engineering **819** (2020) 012014 doi:10.1088/1757-899X/819/1/012014

$$P_{pvc} = Ipvc \cdot Upvc \frac{E}{E_{nom}}, \text{W},$$

where:
$P_{pvc,nom.}$ – PVC nominal power at nominal solar radiation density, W;
E – current density of the solar radiation, W/m²;
E_{nom} – nominal density of the solar radiation, W/m²;
I_{pvc} – current generated by PVC, A;
U_{pvc} – voltage generated by PVC, V.

Approximation of PVC CVC (Current-Voltage Characteristic) can be performed by various methods [[10], [13], [14]]. An alternative function can be proposed based on the curves of the tire adhesion to the road (figure **Ошибка! Источник ссылки не найден.**) [[9]].

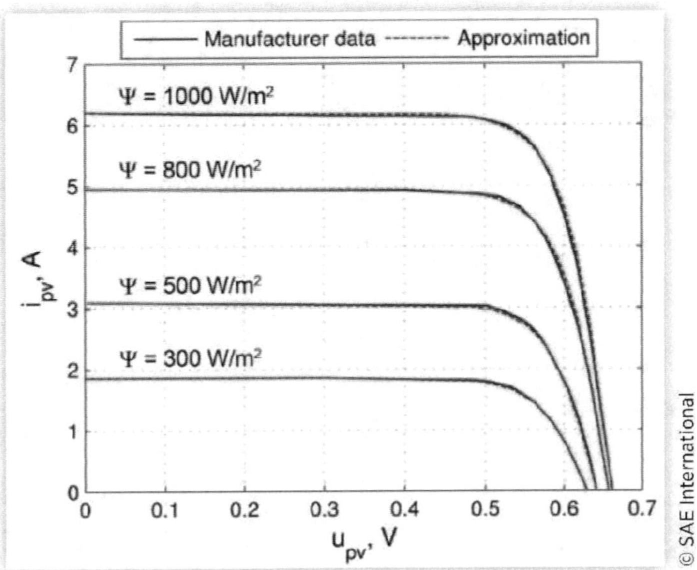

Figure 1. PVC CVC and their approximation.

The tire model called "Magic Formula" provides a good approximation of such characteristics and is used to approximate PVC current depending on the voltage [[15]]

$$i_{pv} = D \cdot sin\left(C \cdot tan^{-1}\left(B \cdot u - E(B \cdot u - tan^{-1}(B \cdot u))\right)\right), \text{A},$$

where:
u – argument in the following form: $u = u_{piv} - u_{pv}$, V;
u_{pv} – PVC voltage, V;
u_{piv} – PVC idle voltage, V;
B, C, E – dimensionless empirical coefficients;
D – maximum current of PVC depending on the intensity of solar radiation, A.

In case of series connection of PVCs into the battery in accordance with Kirchhoff's laws, the current through all the elements is the same, so the CVCs of the individual elements are added by voltage. In case of parallel connection of PVC in the battery, the voltage applied to parallel circuits is the same, therefore CVCs of separate circuits are added by current. The geometric interpretation of the calculation algorithms is shown respectively in figures **Ошибка! Источник ссылки не найден.** and **Ошибка! Источник ссылки не найден.**.

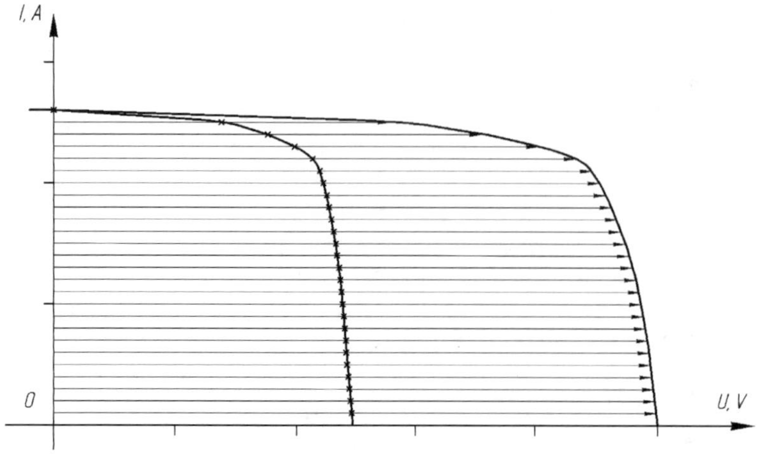

Figure 2. Addition of PVC CVC by voltage.

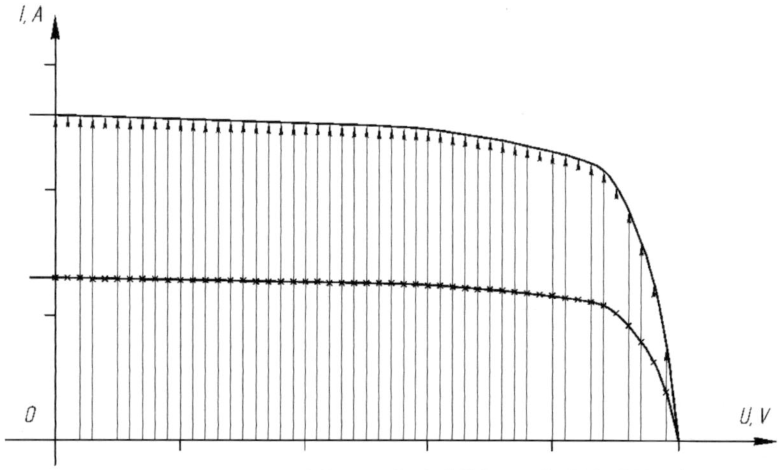

Figure 3. Addition of PVC CVC by current.

13. MPPT algorithm

The control system (CS) of the PVC battery is an MPPT and implements one of the algorithms for finding the maximum power point. In the simulation, the task of PVC CS (MPPT) comes down to the selection of the point at which the power generated by the PVC battery is maximum:

$$P_{pv,nom.} = max_{i=0}^{m}(U_i \cdot I_i), \text{ W},$$

where:
i – number of the instant count of the simulated CVC;
m – maximum number of the count of the simulated CVC;
U_i – voltage value in each of the counts of the simulated CVC;
I_i – current in each of the counts of the simulated CVC.

14. Solar illumination model

There are several methods for calculating the density of solar radiation: Atwater's, Perrin's, Yang's, Bird's, Gueymard's, REST, Winter's, Iqbal's methods (table 1). Of these, the Perrin's, Gueymard's, REST and Iqbal's methods do not take into account diffuse solar radiation, which unacceptably reduces the accuracy of the calculation of the total solar radiation flux. Of the remaining ones, the Bird's method gives the minimum root mean square error of calculating the total solar radiation among them [[16]].

Table 1. Root mean square error of calculations of solar radiation energy density according to the tested methods.

No.	Method	Root mean square error, %		
		Direct SR	Full SR	Diffuse SR
1	Atwater	16	15	51
2	Bird	15	14	43
3	Perrin	21	---	---
4	Yang	18	19	48
5	Gueymard	49	---	---
6	REST	22	---	---
7	Winter	26	15	49
8	Iqbal	28	---	---

15. Test cycles

The results of calculations performed in accordance with the models depend on the operation modes of the vehicle and PVC battery. Operation modes are determined by test cycles. Standard cycles are selected to make it possible to compare both the initial parameters and the obtained results of the calculations.

The standards for certification of vehicles and EV are defined by UNECE Regulation No. 101, according to which the experimental assessment of the characteristics of the vehicle being developed is conducted. UNECE Regulation No. 101 currently uses the New European Driving Cycle (NEDC) combined conditions (cycle), which is selected as the first basic cycle for EV testing.

The New European Driving Cycle (NEDC) combined conditions (cycle) data are taken from UNECE Regulation No. 101 [[17]].

16. Conclusion

– The mathematical model has been developed that integrates not only mathematical models of rectilinear motion of an EV, the Bird's sky model and the PVC element model, but also the PVC battery model taking into account the geometry of the MV body.

– This mathematical model makes it possible not only to calculate more accurately the amount of electricity generated by a non-flat PVC battery, but also to develop recommendations for a more rational use of PVC for recharging traction batteries of electric transport.

References

[1] Karpukhin K, Terenchenko A, Kolbasov A 2018 The creation of modern electric vehicles with additional source of energy *IOP Conference Series: Earth and Environmental Science* 012043
[2] Kolbasov A 2012 [On the future of electric vehicles in Russia] *Modern materials, equipment and technology: papers of the Intern. scientific - practical conf.* (Kursk) pp 139-141
[3] Azarov V, Koutenev V, Stepanov V 2013 [On the actual emission of particulate matter by vehicles] *Zhurnal avtomobil'nykh inzhenerov* **4 (81)** pp 45-47
[4] Kozlov V, Kolbasov A 2018 [On the problem of energy storage capacity and accumulation of electric power in electric traction driverless vehicles] *Trudy NAMI* **2 (273)** pp 35-40
[5] Kozlov V, Kolbasov A 2019 [Analytical review of photovoltaic cells application in vehicles] *Trudy NAMI* **2 (277)** pp 32-44
[6] Bakhmutov S, Karpukhin K 2012 ["Pure" cars: the directions of realization and reached results] *Zhurnal avtomobil'nykh inzhenerov* **6 (77)** pp 51-54
[7] Saikin A, Karpukhin K, Pereladov A, Endachev D, Terenchenko A 2016 [Features of creation a promising environmentally friendly driverless electric vehicle] *Driverless vehicles: problems*

and prospects: papers of the 94th Intern. Scientific and technical. conf. (Nizhny Novgorod: NGTU im. R. E. Alekseeva Publ.) pp 44-52

[8] Nagaytsev M, Eidinov A 2014 [*Vehicles with combined power units (CPU)*] (Moscow: TRP Publ.)

[9] Kulikov I, Karpukhin K 2018 Model analysis of efficiency and energy distribution in the powertrain of an electric vehicle equipped with a solar cell battery. *SAE Technical Paper* 2018-01-5026 DOI: 10.4271/2018-01-5026

[10] Bauer P, Preitl Zs, Peter T, Gaspar P et al. 2006 Control oriented modelling of a series hybrid solar vehicle *Proceedings of the International Workshop "Hybrid and Solar Vehicles"* (Italy: University of Salerno)

[11] Petrushov V 2008 [*Vehicles and road trains: New technologies for research of rolling and air resistances*] (Moscow: Torus Publ.) p 352

[12] Genta G 2006 Motor vehicle dynamics: modeling and simulation *Series on Advances in Mathematics for Applied Sciences* **43** (Singapore, New Jersey, London, Hong Kong: World Scientific Publishing Co. Pte. Ltd.) ISBN 9810229119

[13] Tiwari G, Dubey S 2010 Fundamentals of photovoltaic modules and their applications *RSC Energy Series* (Cambridge: RSC Publ.) ISBN: 978 1 84973 020 4

[14] Saloux E, Teyssedou A, Sorin M 2011 Explicit model of photovoltaic panels to determine voltages and currents at the maximum power point *Solar Energy* **85(5)** 713-722 DOI: 10.1016/j.solener.2010.12.022

[15] Pacejka H Besselink I 2012 *Tire and vehicle dynamics: third edition* (Amsterdam: Elsevier Ltd.) ISBN 978-0-08-097016-5

[16] Elistratov V, Aronova E 2012 [*Solar power plants. Assessment of solar radiation intake: Study Guide*] (Saint Petersburg: Polytechnic University Publ.)

[17] 2013 Uniform provisions concerning the approval of passenger cars powered by an internal combustion engine only, or powered by a hybrid electric power train with regard to the measurement of the emission of carbon dioxide and fuel consumption and/or the measurement of electric energy consumption and electric range, and of categories M1 and N1 vehicles powered by an electric power train only with regard to the measurement of electric energy consumption and electric range *UNECE Regulation* **101**

The choice of a performance criterion for a high-voltage battery of a vehicle

I A Lyubimov and R Kh Kurmaev

FSUE "NAMI", 2 Avtomotornaya St., Moscow, 125438, Russia

Abstract. Currently, major vehicle manufacturers are working to improve environmental friendliness by introducing alternative energy in motor vehicles: vehicles with combined power plants (hybrids) based on ICE, electric vehicles based on promising current sources like electrochemical generator based on fuel elements of "hydrogen-air" and traditional electric vehicles. All of the above vehicles use High-voltage battery running. The use of batteries in hybrid vehicles and electric vehicles puts forward certain requirements for them: large specific energy capacity; high resource indicator; fairly low cost. This article seeks a criterion for the performance of a high-voltage battery used as an energy source in electric vehicles. Performed an analysis of scientific publications and experiments describing the dependence of the parameters of the battery on the temperature and operating mode of the vehicle. The main parameters are determined and an assessment of their impact on vehicle performance is carried out.

1. Introduction

Currently, major manufacturers of automotive vehicles are conducting research to improve their fuel economy and environmental friendliness. Although today many traditional methods are known for improving the fuel efficiency of an internal combustion engine (ICE) (improving the working process, improving ICE systems and elements, using new operational materials, etc.) all of them individually or together can reduce fuel consumption only by a few percent, and a decrease in the emission of harmful substances from automobile exhaust gases, at best, by several tens of percent. Therefore, vehicle and engine developers are increasingly applying to unconventional ways to increase fuel economy and improve the ecology of cars.

Many world automobile countries connect the solution of the problem of integrated environmental safety and the problem of energy conservation by standardizing CO2 emissions, as well as the introduction of alternative energy in motor vehicles: vehicles with combined power plants (hybrids) based on ICE, electric vehicles based on promising current sources like electrochemical generator based on fuel elements of "hydrogen-air" and traditional electric vehicles. All of the above vehicles use High-voltage battery (HVB) running.

2. Battery types

There are many different types of batteries. Also in the development and preparation for mass production are new types of batteries. The use of batteries in hybrid vehicles and electric vehicles puts forward certain requirements for them: large specific energy capacity; high resource indicator; fairly low cost (Figure 1) [1].

IOP Conf. Series: Materials Science and Engineering **819** (2020) 012015 doi:10.1088/1757-899X/819/1/012015

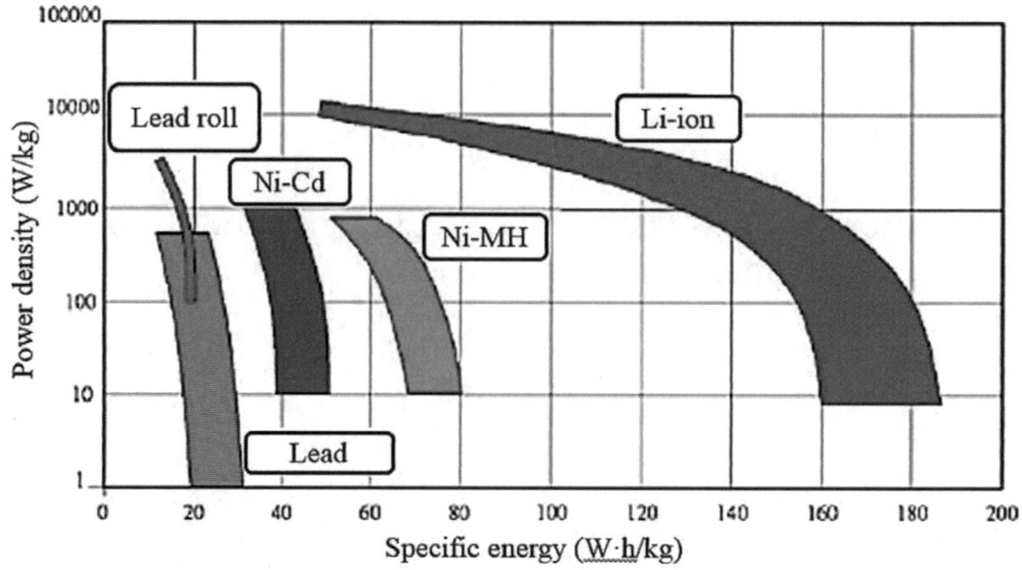

Figure 1. The power density and specific energy of different types of batteries (Ni-Cd - nickel-cadmium, Ni-MH - nickel-metal hydride, Li-ion - lithium-ion) [1].

Based on the analysis of manufactured vehicles, the following conclusions can be drawn on the trends in the production of hybrid vehicles and electric vehicles - in most cases, Li-ion HVBs with a capacity of up to 20 kW·h in hybrid cars and up to 120 kW·h in electric vehicles are used. They took their leading positions due to low self-discharge and a large number of charge-discharge cycles; Li-ion batteries are most preferred for use in vehicles [1, 2].

Lithium-ion HVBs can operate in a fairly wide temperature range, but there are limitations that should be kept in mind all the time, especially in countries with a cold climate [3].

3. Analysis of scientific publications and experiments describing the dependence of the parameters of the battery

Based on the previously described, the influence of ambient temperature on the characteristics of lithium-ion batteries will be investigated. Many scientific works and experiments show a significant decrease in battery capacity with decreasing temperature (Figure 2) [4].

In this case, the loss of battery capacity occurs mainly due to an increase in resistance, which causes a warming effect during charging and discharging.

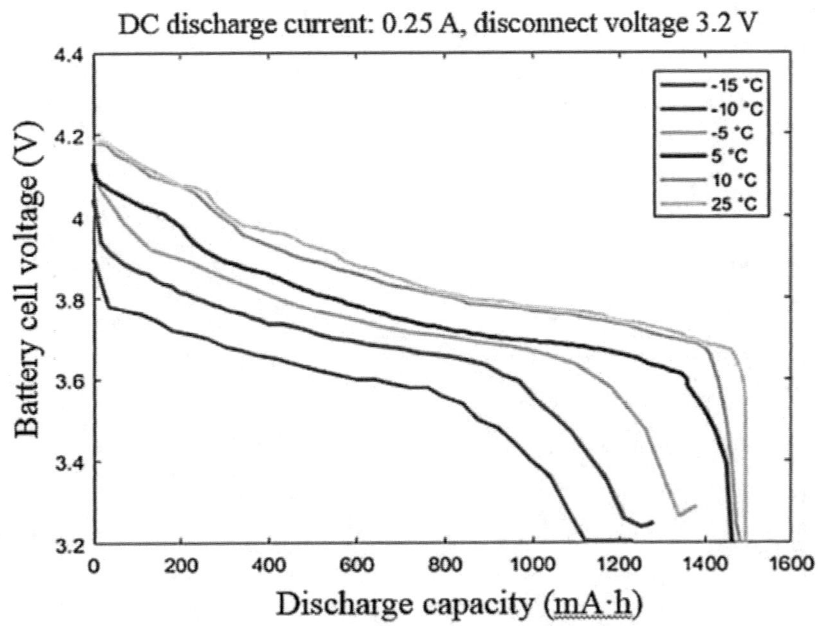

Figure 2. An example of the dependence of the cell voltage of a lithium-ion battery on capacity at different temperatures [4].

Elevated temperatures also negatively affect capacitance. Figure 3 below shows typical self-discharge rates for a lithium-ion battery [5].

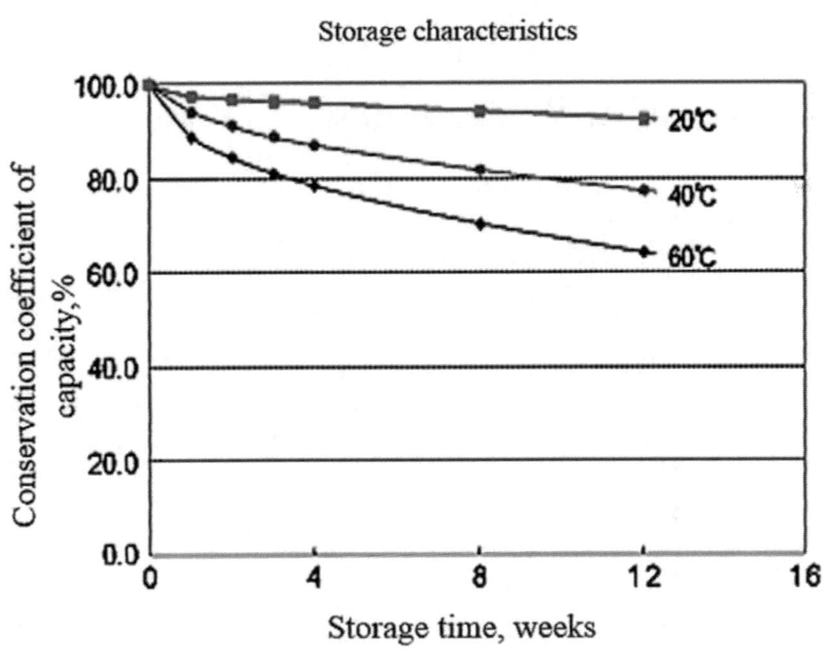

Figure 3. An example of the dependence of the conservation capacity coefficient of a lithium-ion battery on the storage time given the temperature [5].

The rate of undesirable chemical reactions that cause internal leakage of current between the positive and negative electrodes of the element, like all chemical reactions, increases with temperature, thereby increasing the self-discharge rate of the battery. In addition to self-discharge, elevated temperatures can cause a fire. All these facts show the need for a thermomanagement and safety systems for a high voltage battery.

The release of thermal energy during a charge-discharge of a battery depends on the magnitude of the currents, the initial temperature, and the initial state of charge of the battery (Figures 4 and 5) [6].

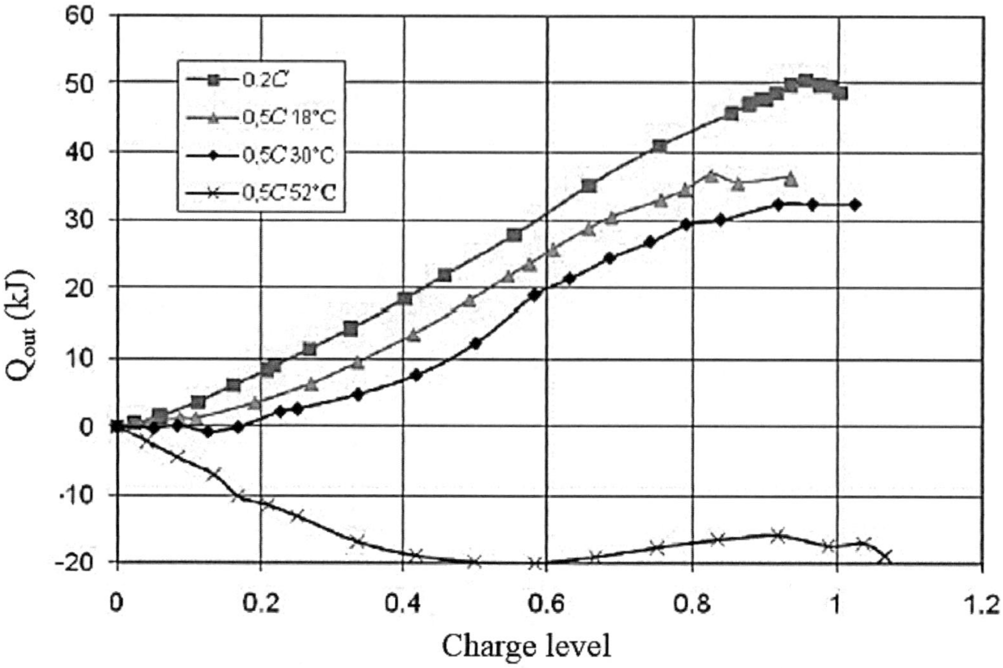

Figure 4. The total amount of heat removed from the battery during charging [6].

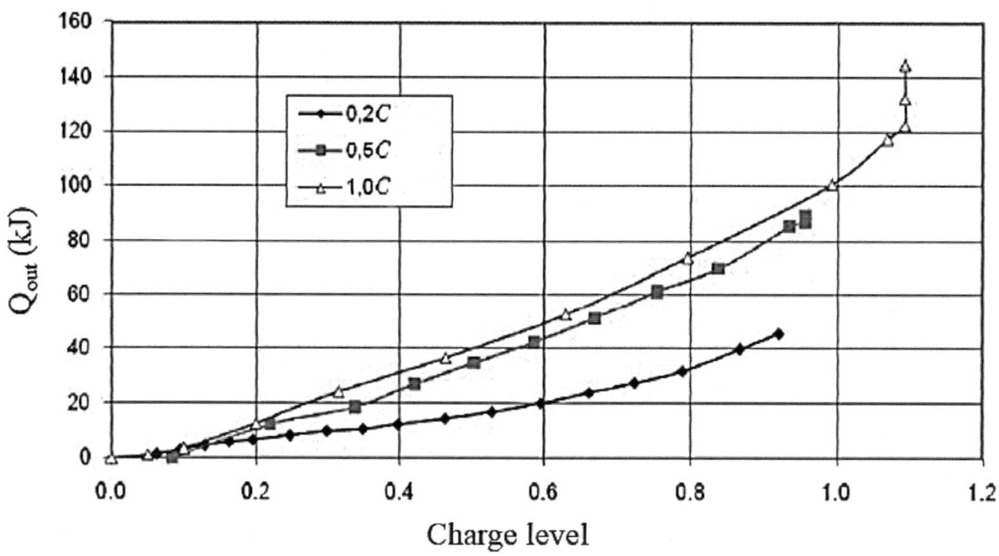

Figure 5. The total amount of heat removed from the battery during the discharge process [6].

The study shows that, ceteris paribus, the considered type of battery is better to operate in a certain temperature range, which ensures a decrease in internal resistance, a decrease in the final temperature and a reduction in heat generation in the battery [6]. Given all of the above, we can confidently talk about the need for a battery thermomanagement system.

4. Main parameters and assessment of their impact

Vehicle movement is a multifactorial process, depending on a huge number of conditions and parameters. To increase the mileage of an electric vehicle, which is an increase in the efficiency of the vehicle, which in turn directly depends on the efficiency of the battery, it is necessary to evaluate the level of influence of various conditions and characteristics on the vehicle and HVB (Figure 6).

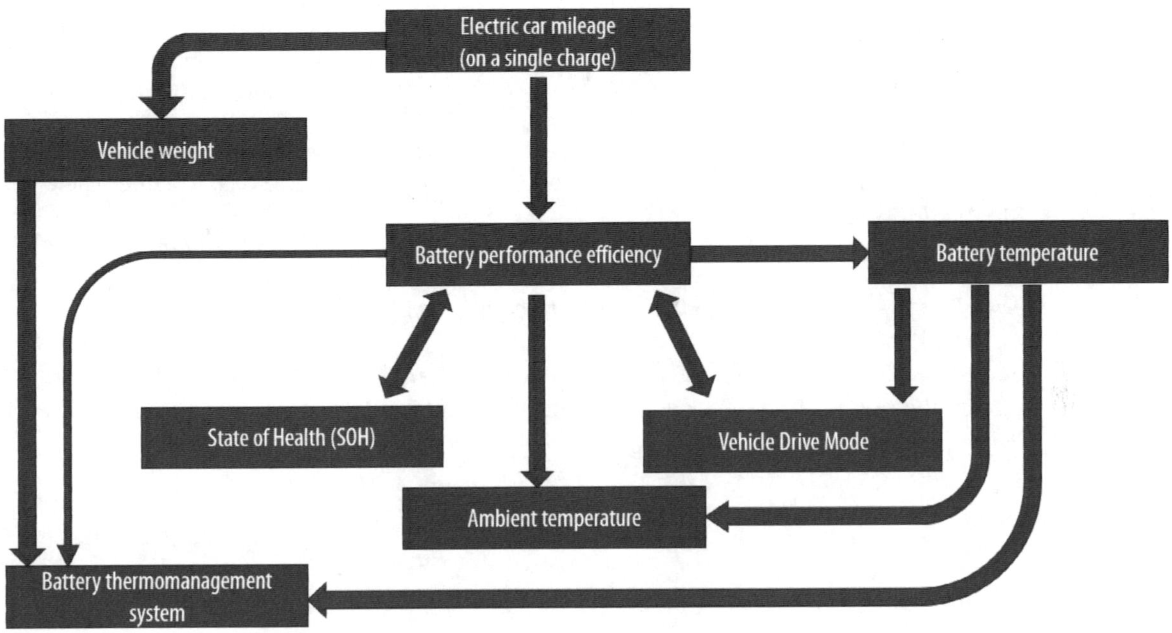

Figure 6. Scheme of the dependence of the mileage of electric vehicles on various conditions.

Battery performance efficiency - in fact, these are the best working conditions that provide the greatest battery capacity at a given vehicle driving mode. The battery state of health (SOH) is a value that reflects the current state of the battery compared to its ideal state, is selected by the manufacturer, does not correspond to a specific physical parameter. Since this parameter is a conditional quantity, we will not consider further analysis. The most important conditions and having a direct effect are the HVB temperature and the vehicle's driving mode, which sets the charge or discharge mode of the battery, which in turn affects the temperature of the HVB cells. In addition, the ambient temperature has a direct effect on the HVB temperature, which in turn affects the operation of the thermomanagement system. The thermomanagement system of the HVB, functioning according to certain algorithms, must ensure the HVB operation in a certain temperature range. These algorithms should take into account the most important conditions:

- HVB temperature;
- Vehicle drive mode;
- Ambient temperature.

5. Conclusions

Thus, the criterion for battery performance is the cell temperature, because this indicator has the greatest impact on battery capacity and safety. From this it follows that in order to increase the efficiency of a vehicle on electric traction, it is necessary to determine the characteristics of a high-voltage battery depending on the ambient temperature and to develop a specific algorithm for the functioning of the temperature control system depending on the most significant parameters.

6. References

[1] Bystrov Yu A, Kudryavtsev N A, Krasnybryzhiy A V, Loginova M M and Zhdanov V V 2007 *Development of lithium-ion batteries and batteries with high specific characteristics for use in marine facilities for various purposes* Underwater research and robotics **2**(4) pp 34-37

[2] Kurmaev R Kh, Umnitsyn A A, Struchkov V S, Karpukhin K E and Lyubimov I A 2018 *Development and research of temperature control system of a hight-voltage battery of a perspective electric vehicle* Trans and Motauto World **4** pp 151-154

[3] Struchkov V S, Kurmaev R H, Iakunov D M and Lubimov IA 2019 *Thermostating system with intelligent management for electric vehicles* IASF-2018 IOP Conf. Series: Materials Science and Engineering **534**

[4] Wang C-Y, Zhang G, Ge S, Xu T, Ji Y, Yang X-G and Leng Y 2016 *Lithium-ion battery structure that self-heats at low temperatures* Nature **529** pp 515-518

[5] *Battery Performance Characteristics* Electropedia www.mpoweruk.com/performance.htm (28.08.2019)

[6] Klimenko G K, Lyapin A A and Marakhtanov M K 2013 *Investigation of the thermal state of the battery in a duty cycle* Engineering Journal: Science and Innovation **10**

IOP Conf. Series: Materials Science and Engineering **819** (2020) 012016 doi:10.1088/1757-899X/819/1/012016

System of Requirements and Testing Procedures for Autonomous Driving Technologies

A M Ivanov and S S Shadrin

Moscow automobile and road construction state technical university (MADI), 64, Leningradsky prospect, Moscow, 125319, Russian Federation

E-mail: shadrin@madi.ru

Abstract. This article is dedicated to the subject of autonomous driving technologies' testing system development for the purposes of technology certification on local market with respect to the local operational peculiarities and road safety aspects. The overview of UNECE document 'Proposal for the Future Certification of Automated/Autonomous Driving Systems' (ECE/TRANS/WP.29/GRVA/2019/13) is presented. Proposals for critical testing scenarios localization and complex virtual tests conduction are introduced, practical results are presented.

1. Introduction

The aim of our study is to develop a testing system for autonomous vehicles with respect to the peculiarities of road and weather-climatic conditions of the Russian Federation and Customs Union countries. The results of our study can be considered in the national certification system at type approval level, as well as in any process of developing autonomous driving technologies.

Why do we think that this subject is so important and relevant? The rapid worldwide attempts to implement autonomous driving technologies with ADAS (advanced driver assistance systems) functionality at the first stage [1] and the lack of legal regulation in Russia have led to the situation when we have a lot of modern road vehicles with ADAS on public roads. These ADAS systems (autonomous emergency braking, lane keeping, automated parking, etc.) have not been tested and may not have been adapted for use in Russian conditions [2]. Of course, it is a potential threat to national road safety.

Among our objectives we emphasize the analysis of foreign experience, localization of critical testing scenarios, making amendments to the Technical Regulations of the Customs Union (TR CU 018) [3].

It is important to mention, that testing of autonomous driving technologies is very important and can be applied to the development process and to the process of technology certification. These processes are essentially different if we are not talking about self-certification regime. Car manufacturers and their suppliers are involved in the development process [4], and testing in certification is carried out by approval authorities [5]. We make focus of our study on certification process only.

2. Testing in Certification

With the introduction of automated driving systems, the complexity and thereby the number of software-based functions will continue to increase in vehicles. When we are talking about automated and autonomous road vehicles in comparison with conventional vehicles, the potentially affected safety-areas and variances of scenarios will increase and cannot fully be assessed with a limited number of tests that are performed on a test track or test bench. Figure 1 illustrates enormous number of test scenarios for vehicles with high SAE levels of automation [6].

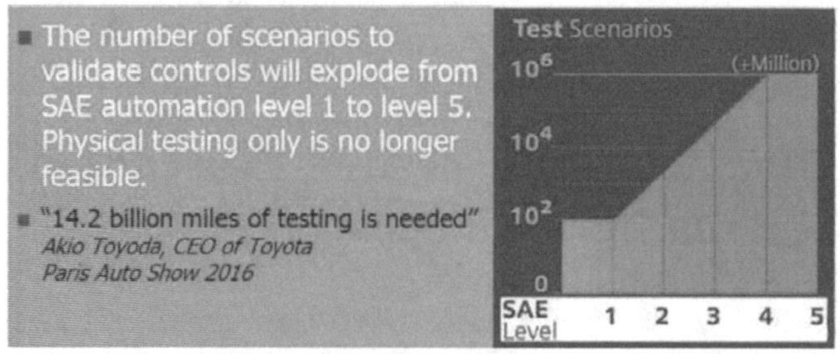

Figure 1. Number of test scenarios required for validation and verification cycles in autonomous vehicles.

Physical testing only is no longer feasible, as well as 'classical' certification approach. The most relevant document in this field was recently published by the United Nations Economic Commission for Europe (UNECE) WP.29 on automated/autonomous and connected vehicles, and it is called 'Proposal for the Future Certification of Automated/Autonomous Driving Systems' [7]. The concept for the future certification of automated and autonomous driving systems is shown in figure 2. This approach consists of three components or three pillars. The first one correlates with the audit of developmental process and involves virtual testing of exotic scenarios with low statistical probability in real life – see figure 3. The second pillar involves validation of previously used mathematical models in virtual tests, as well as conduction of critical scenarios on proving ground. The third pillar deals with typical scenarios and tests on public roads.

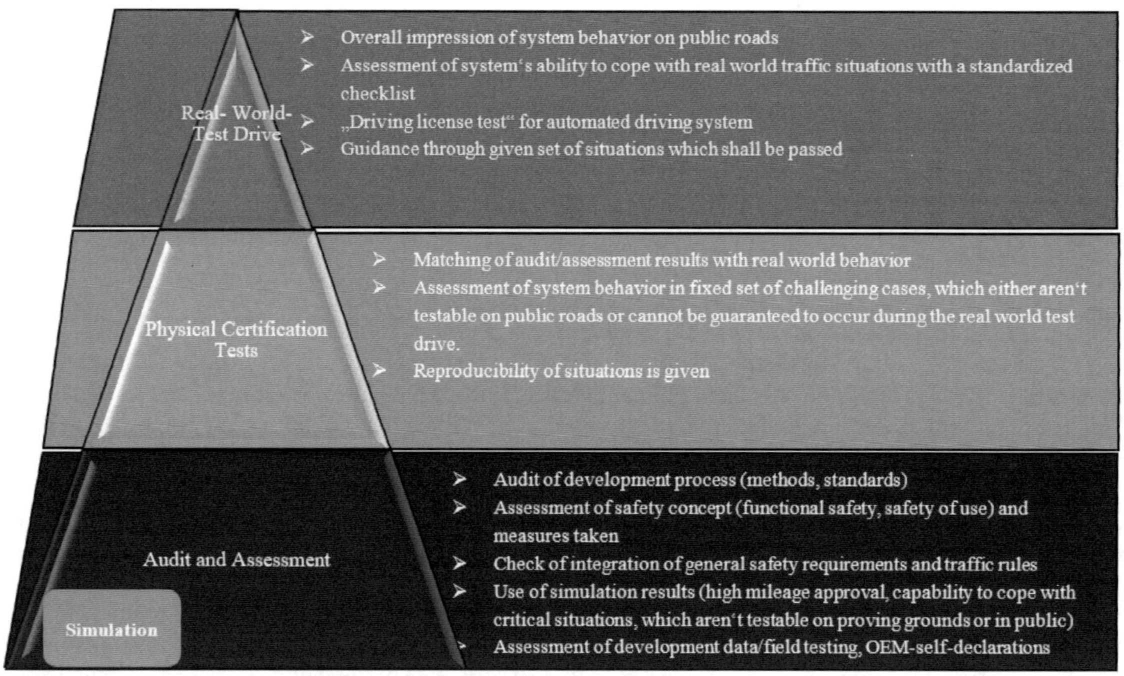

Figure 2. Concept for the future certification of autonomous driving systems – the three pillars.

It is important to mention, that the choice of testing scenarios at each component level of the above approach will be different for each market and have to consider the peculiarities of operational conditions (weather, climate, typical road pavement, cultural and psychophysiological features, etc.).

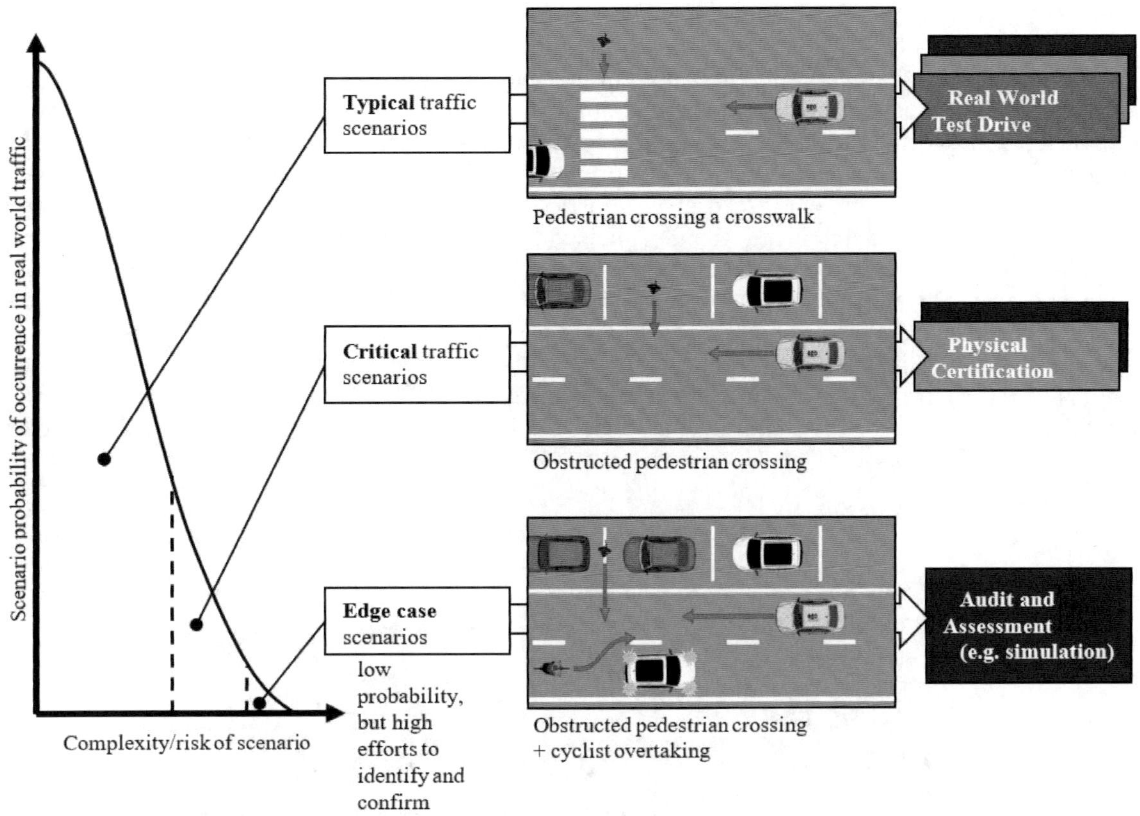

Figure 3. Example of different test cases with respect to the pillars' functions.

We've got fully-detailed statistics of road accidents in Russia from the Ministry of the Interior (MVD), and we're trying to localize relevant Russian scenarios for the purposes of automated and autonomous road vehicles testing in certification – it is the subject of future work.

Currently, we propose using the next sequence of actions for critical testing scenarios localization and initial states for each scenario detection:

1. Real road accidents statistics analysis, applicable for local market.
2. Definition of the test's purpose, which may include: the task of functional or operational safety monitoring; determination of operating conditions, reliability metrics, fault tolerance, performance variation between the same cars, as well as during the whole life cycle, etc.
3. Creation of test scenarios or selection and/or modifying from the existing ones (e.g. EuroNCAP, ADAC, ISO):
    a. Definition of ranges of changeable parameters of the system 'driver—car—road—environment' [8].
    b. The choice of evaluation criteria and their permissible values.
4. Flow simulation of all possible variations of the developed test scenarios in a virtual environment performing:
    a. Critical testing scenarios detection with respect to the test's purpose.
5. Critical scenarios' field tests performing:
    a. Virtual model's parameters validation, the above cycle repetition.

IOP Conf. Series: Materials Science and Engineering **819** (2020) 012016 doi:10.1088/1757-899X/819/1/012016

6. Making resolution or performing changes in the design or control algorithms to achieve goals of testing.

3. Virtual Tests

Moscow Automobile and Road Construction State Technical University (MADI) is the first organization in Russia which has officially purchased the license for Prescan software (Siemens). This software is a physics-based simulation platform that is used in the automotive industry for development of ADAS. We have tested several software solutions from different companies, but have chosen Prescan software, because it has most detailed models for vehicles, sensors (radars, lidars, cameras, etc.), and infrastructure elements. The next important reason is that this software has an opportunity to simulate different weather conditions (rain, fog, snow, etc.), road characteristics (roughness, holes, inclinations, speed bumps, friction coefficients, etc.), lighting (day, night, glare of the sun, etc.), custom objects, and custom control algorithms. Thus, this software solution is very flexible.

Several examples of ADAS virtual testing (Advanced Emergency Braking System – AEBS) with respect to the road and weather-climatic conditions in Russia are shown below.

We simulated straight driving of a host car with initial speed of 60 km/h with a stationary object (Euro NCAP car target) on a way. It's one of the standard Euro NCAP tests [9], but we have modified it with another environmental and road conditions [10, 11]: winter time, snow weather and wet road conditions have been used (figure 4).

Figure 4. Visualization of virtual test conditions.

The AEBS control algorithm gave a warning at 2.6s TTC (time to collision), then it applied 40% of braking at 1.6s TTC, and 100% of braking at 0.6s TTC [12]. Virtual vehicle under test was equipped with long range radar (LRR – 120m range, 9 degrees narrow beam) and short range radar (SRR – 30m range, 90 degrees wide beam). The test scenarios were as follows: 1) driving with idealized radar sensors; 2) driving with characteristics of real radars (Continental ARS510 as a LLR, and Continental SRR510 as a SRR with appropriate characteristics [13]).

Figure 5 illustrates virtual AEBS operation with idealized sensors during snowfall. Collision has occurred at driving speed of 28.05 km/h because of lower road friction coefficient only.

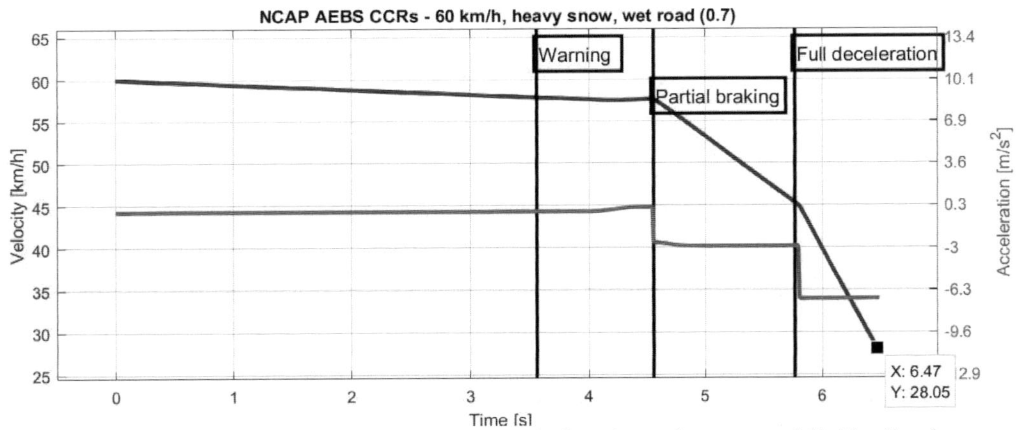

Figure 5. AEBS virtual test – stationary target, wet asphalt, winter time, snowfall, idealized sensors.

For the second test specific radar's antenna characteristics (figure 6) have been used, as well as the specific characteristics of atmospheric attenuation (figure 7).

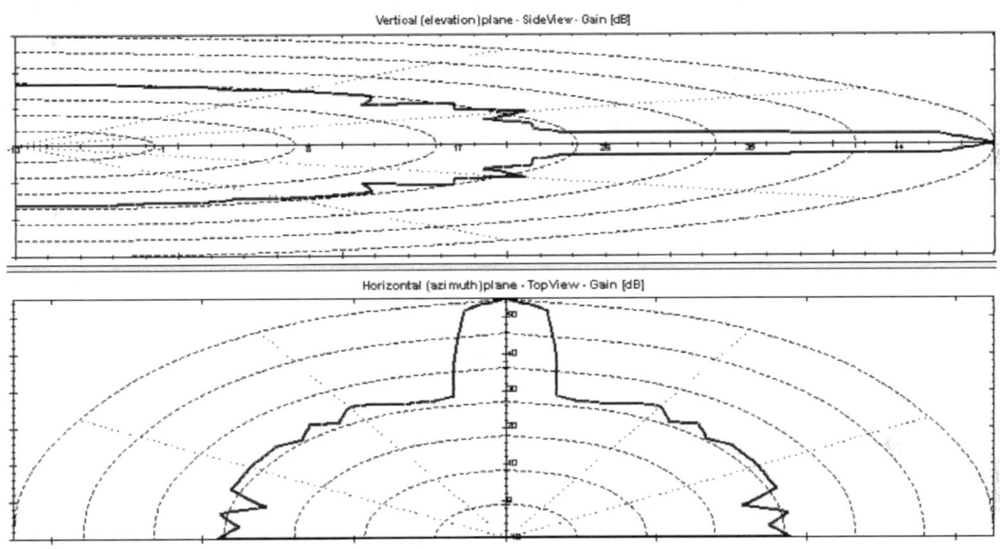

Figure 6. Radar antenna patterns.

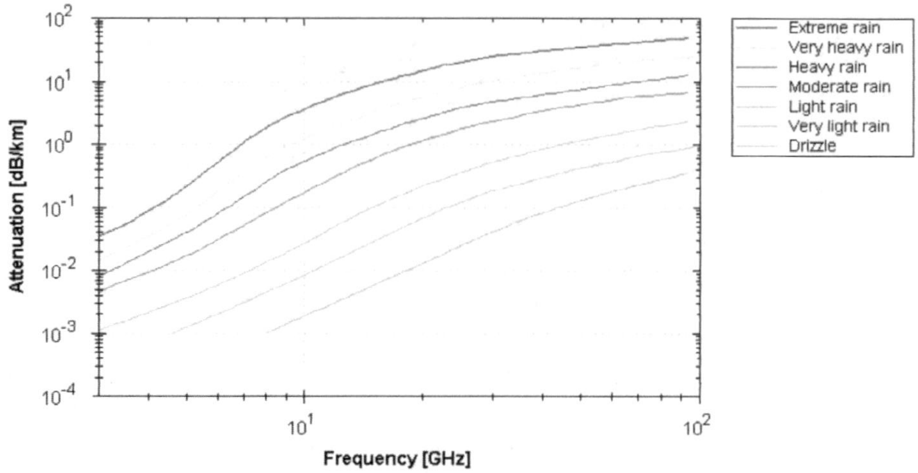

Figure 7. Characteristics of atmospheric attenuation.

Figure 8 illustrates virtual AEBS operation with characteristics of real radars during snowfall. Collision was not avoided, collision has occurred at driving speed of 44.52 km/h. Warning and partial braking TTC values of the second test were lower (late detection and action). In the second case we see malfunction of AEBS control algorithm (the graph of deceleration in figure 8 – it suddenly stopped braking at the stage of partial braking). Perhaps, this situation has occurred because of target tracking fail. In any case, we see that the results of virtual tests are essentially sensitive to the adequacy of ADAS control algorithms and sensors characteristics.

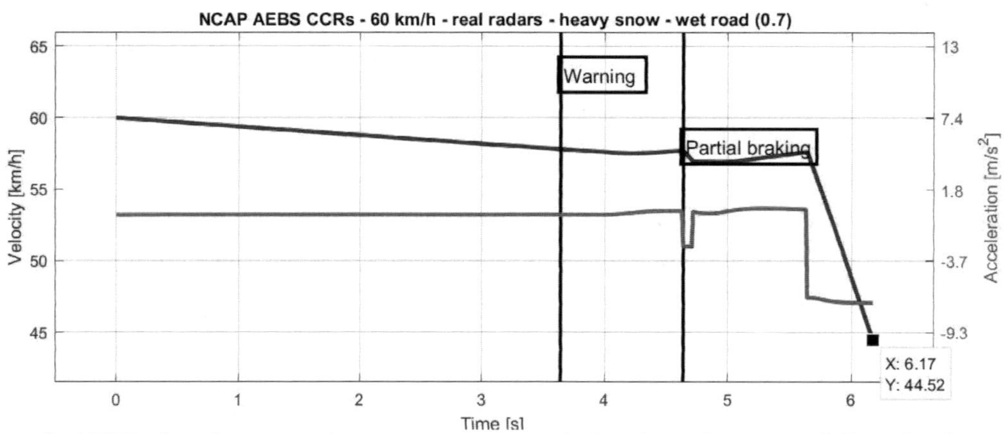

Figure 8. AEBS virtual test – stationary target, wet asphalt, winter time, snowfall, real radars.

The information of exact ADAS control algorithms and vision recognition system characteristics is very important for virtual tests and critical scenarios detection. This type of information can be provided by car manufactures or their OEMs only. Of course this information is essentially confident, but it can be provided as a hardware or software "black boxes" for testing purposes.

4. Virtual Proving Ground

Virtual proving ground is a very important instrument in the task of complex simulations conduction. Of course it is better to have a digital copy of a real proving ground than some imagined one, because it will be possible to validate mathematical models used in simulation. One of such examples is a virtual Mcity proving ground (University of Michigan) integrated in CarSim software [14].

IOP Conf. Series: Materials Science and Engineering **819** (2020) 012016 doi:10.1088/1757-899X/819/1/012016

We have made a detailed virtual proving ground which is the copy of Dmitrovskiy automotive proving ground for automated and autonomous vehicles (FSUE 'NAMI'). Figures 9-11 illustrate some results of our work. This virtual proving ground was designed in Prescan software and it is geodetically attached to the real proving ground in WGS-84 coordinates. Such an approach simplifies the process of virtual simulation results comparison with road tests.

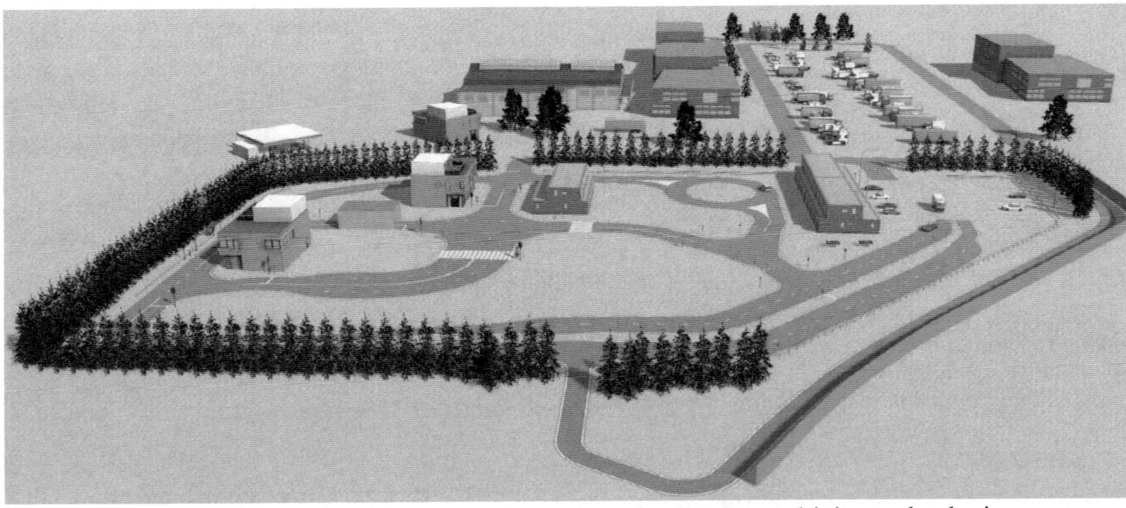

Figure 9. Virtual proving ground for testing of autonomous driving technologies.

Figure 10. Virtual proving ground for testing of autonomous driving technologies.

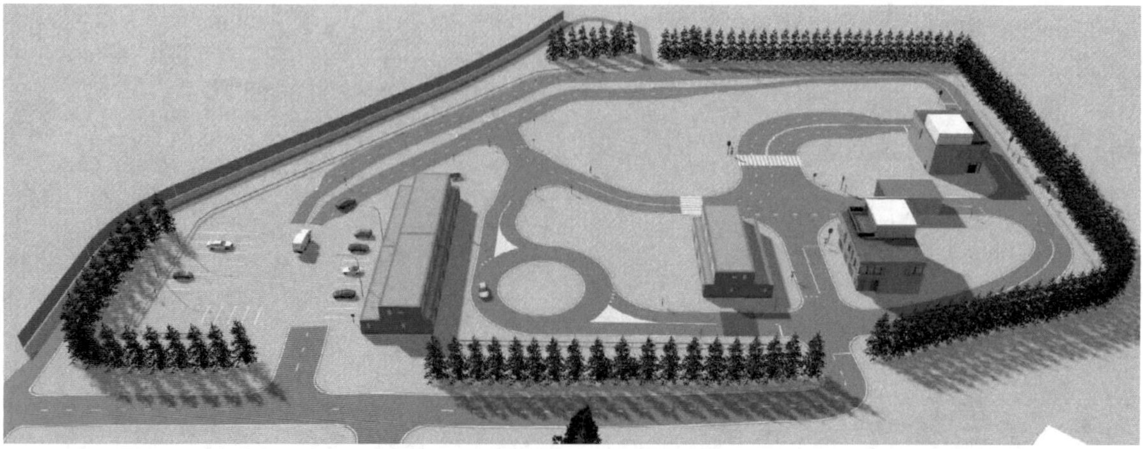

Figure 11. Virtual proving ground for testing of autonomous driving technologies.

Our virtual proving ground consists of roads, crossings, road marking, buildings, traffic lights, road signs, guardrails, street lamps, bus stops, cars, trucks, pedestrians, etc. Complex virtual tests conduction is the subject of our future work and we'll present it later.

5. Conclusion

The overview of UNECE document 'Proposal for the Future Certification of Automated/Autonomous Driving Systems' (ECE/TRANS/WP.29/GRVA/2019/13) was presented. Our vision of autonomous driving technologies' testing system development for the purposes of technology authorization on local market with respect to the local operational peculiarities and road safety aspects was introduced.

We have proposed a method for addressing the issue of critical testing scenarios localization and initial states for each scenario detection.

Results of AEBS virtual testing with respect to the road and weather-climatic conditions in Russia were presented. The information of exact ADAS control algorithms and vision recognition system characteristics is essentially important for virtual tests.

We have introduced a detailed virtual proving ground which is the copy of Dmitrovskiy automotive proving ground for automated and autonomous vehicles (FSUE 'NAMI'). This virtual proving ground was designed in Prescan software and it is geodetically attached to the real proving ground in WGS-84 coordinates. Such an approach simplifies the process of virtual simulation results comparison with road tests.

A comprehensive approach with the use of virtual and field tests is needed to assess the safety of autonomous road vehicles with respect to the peculiarities of road and weather-climatic conditions in Russia. AVs legal permission to access public roads should be done after comprehensive tests only.

References

[1] 2016 *SAE J3016: Taxonomy and Definitions for Terms Related to Driving Automation Systems for On-Road Motor Vehicles* (U.S.A.: SAE) p 30

[2] Ivanov A M and Shadrin S S 2018 Development of autonomous vehicles' testing system *IOP Conf. Ser.: Mater. Sci. Eng.* **315** 012011 (doi:10.1088/1757-899X/315/1/012011)

[3] Technical regulations of the Customs Union "On the safety of wheeled vehicles" TR CU 018/2011

[4] Auerswald R 2018 *Controlled field-testing of automated and connected driving in urban environments* (Stuttgart: Autonomous Vehicle Test & Development Symposium 2018)

[5] Abdellatif H 2018 *Complete Approach for Testing Automated Vehicles on a Testing Ground* (Stuttgart: Autonomous Vehicle Test & Development Symposium 2018)

[6] Fogarty K 2018 *How To Test Autonomous Vehicles* (URL: http://semiengineering.com/how-to-test-autonomous-vehicles/#.W0t1V9iFXyc.linkedin)
[7] UNECE 2019 *Proposal for the Future Certification of Automated/Autonomous Driving Systems* (ECE/TRANS/WP.29/GRVA/2019/13)
[8] Shadrin S S and Ivanov A M 2016 Technical Aspects of External Devices into Vehicles' Networks Integration *International Journal of Applied Engineering Research* **11**(10) pp 7003-06
[9] AEB C2C Test Protocol v2.0.1 (URL: https://cdn.euroncap.com/media/32278/euro-ncap-aeb-c2c-test-protocol-v201.pdf)
[10] Ivanov A M, Kristalniy S R, Popov N V, Toporkov M A and Isakova M I 2018 New testing methods of automatic emergency braking systems and the experience of their application *IOP Conf. Ser.: Mater. Sci. Eng.* **386** 012019
[11] Gaevskiy V V and Ivanov A M 2018 Problems of the application of intelligent driver assistance systems on a single-track vehicles *IOP Conf. Ser.: Mater. Sci. Eng.* **386** 012021
[12] 2018 *PreScan User's Manual*
[13] Continental. Solutions for Passenger Cars. Radars (URL: https://www.continental-automotive.com/en-gl/Passenger-Cars/Chassis-Safety/Advanced-Driver-Assistance-Systems/Radars)
[14] Mechanical Simulation Releases CarSim, TruckSim, and BikeSim 2017.0 (URL: https://www.carsim.com/publications/pressreleases/2016_12_22.php)

IOP Conf. Series: Materials Science and Engineering **819** (2020) 012017 doi:10.1088/1757-899X/819/1/012017

Novel Adaptive Control Method for BLCD Drive of Electric Bike for Vietnam Environment

Chuong Nguyen Khanh[1], **Sachintha Balasooriya**[1], **I Kavalchuk**[1], **A Kolbasov**[2], **K Karpukhin**[2] **and A Terenchenko**[2]

[1] School of Science and Technology, RMIT University Vietnam, 702 Nguyen Van Linh, District 7, Tân Phong, Ho Chi Minh City, Hồ Chí Minh, Vietnam
[2] FSUE "NAMI", 2, Avtomotornaya Str., Moscow 125438, Russian Federation

Email: aleksey.kolbasov@nami.ru, ilya.kavalchuk@rmit.edu.vn

Abstract. Electric bikes are a rising mode of transportation in developing countries as their use reduce the use of hydrocarbon-based energy sources as well as increases convenience for the commuters within the city by reducing traffic. Furthermore, the electric bikes have low noise emission and no exhaust gases produced, which improves the environmental footprint. Due to the Vietnam regulations, most electric bikes use brushless wheel-hub motors as the main drivetrain solution paired with various types of batteries. The key challenges in the design of the drivetrains are related to the efficiency and controllability areas, so the response of the drivetrain can be improved. The key area of improvement lies in the control system design to optimize the performance and energy consumption. This paper presents the research done on developing of optimization PID control algorithms by using GA (Generic Algorithm) and PSO (Particle Swarm Optimization) for driving a synchronized brushless DC motor for the electric bikes.

1. Introduction

The use of electric vehicles has rapidly increased with the further development of cheap and reliable battery technologies for the needs of the transportation industry. Personal transportation has transitioned from the fossil-fuel driven solutions to fully or partially electric-powered vehicles for their lesser carbon footprint with the steady raise of the fuel prices [1]. Electric vehicles (EVs) also come with low maintenance costs since the transmission system is simpler than internal combustion engines [2]. These benefits have given rise to a demand for electric bikes which also has the added benefit of reducing traffic on the roads specially in highly populated areas as shown in [3], a study done on use of electric bikes in China over the last decade.

Due to the low power output and low driving range on a charge, electrification was raising in the applications of the motorbikes and scooters in the developing countries. The key technical benefit is the 95-98% efficiency of the wheel-hub electric drivetrain, in comparison with 25-30% efficiency of the small ICE in the scooters [4, 5]. As the higher efficiency is followed by reduction of the emissions, the government in China and Vietnam have been in favour of supporting electrification of the transport.

IOP Conf. Series: Materials Science and Engineering **819** (2020) 012017 doi:10.1088/1757-899X/819/1/012017

There are two common types of the electric motors used in EVs: synchronous AC motors and asynchronous AC motors, also known as inductive motors [6]. Brushed DC motors are not typically considered for this use case due to the sparking issues and lower efficiency. The synchronous brushless motors have either induced or pre-existing magnetic fields in the form of permanent magnets, especially for the low power applications. This requires the motor's stator to provide the magnetic field that drives the motor in an ordered frequency/sequence, so the rotor can pick up the pace intended by the inverter, operated through the controller [7]. This requires the motor controller to have information on the speed of the rotor's magnetic field, for the proper conditioning. Therefore, large scale synchronous AC motors have embedded hall sensors to track the magnetic field of the rotor [7]. Small brushless motors don't require additional sensors, as their operation is based on the prediction of the rotor's position by tracking it from the starting time by microcontroller, as shown in [8].

Brushless Direct Current Motors (BLDC) have several advantages in comparison with AC Induction motors for low-power electric bikes applications. Firstly, for a same power output, BLDC motor is 40-50% smaller in dimensions. For the two-wheeler application the drivetrain dimensions are the important consideration, especially for the wheel-hub motors, since it might springless masses reduction. Secondly, BLDC motor provides higher torque at low to middle speed range (0 – 2500 rpm) while AC Induction motor has larger torque at higher speed range (2500 – 3600 rpm). Considering city-based usage and that most residential areas in Ho Chi Minh city have the speed limit at 50 km/h, BLDC has low number of the drawbacks for the conditions. In addition, BLDC has 1-3% higher efficiency than AC Induction motor since its rotor is constructed with permanent magnets, and therefore, it possesses no copper losses (I^2 R) in the rotor [9, 10].

This paper analyses the performance requirement of a typical BLDC motor for electric bike application and presents an innovative control system design. Section two describes the system requirements and its mathematical model. Section three presents the developed control algorithm. Section four introduces methodology for optimization of the control algorithm for the city use of the vehicle. Section five gives the results of the developed system. Finally, the conclusion about the new system is given in section six.

2. System Definition and Modelling

Figure 1 shows the reference image of the proposed drivetrain architecture for the electric bike.

Figure 1. Developed Architecture of the Electric Bike.

The BLDC motor is installed at the rear wheel of the bike in order to improve the springless masses and the durability of the solution in the wet conditions, in comparison with the in-wheels architecture.

In order to develop adaptive control system, the mathematical model of the motor was derived in form of a block diagram. Figure 2 shows the mathematical model of the BLDC in form of a block diagram.

IOP Conf. Series: Materials Science and Engineering **819** (2020) 012017 doi:10.1088/1757-899X/819/1/012017

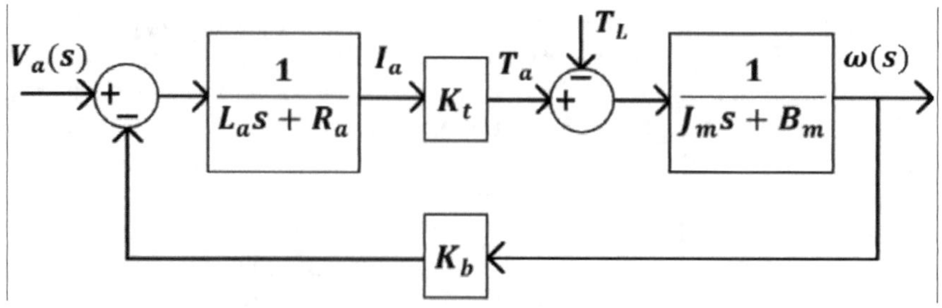

Figure 2. Block Diagram of a BLDC Motor.

Based on Figure 2, the transfer function was obtained for the speed control versus supplied voltage:

$$\frac{\omega(s)}{V_a(s)} = \frac{K_t}{L_a J_m s^2 + (L_a B_a + R_a J_m)s + R_a B_a + K_t K_b} \quad (1)$$

Table 1 shows the parameters for the transfer function from equation (1). Their values are deduced based on the choice of using Nc5000w BLDC motor and the weight of our bike of 100 kg, including batteries.

Table 1. Parameters of BLDC Motor

Nomenclature	Parameter	Value
K_t	Torque constant	$1.75\ Nm/A$
K_b	Back EMF constant	$1.75\ V/rad\, s^{-1}$
R_a	Electric resistance	1.38 Ω
L_a	Electric inductance	0.19 H
B_m	Friction coefficient	0.7 $N.m$
J_m	Moment of inertia	0.17 $kg.m^2$
ω	Output angle	N/A
v_a	Input voltage	N/A

For the final mode, the values were substituted to the equation, so the final s-domain function was derived, as presented on (2):

ω (s)/(V_a (s))=54.18/(s^2+11.38s+124.72) (2) (2)

The key requirements belong to the transient performance, as the specifications to react on the changing environment and input signals and keep the bike in the challenging traffic of the Ho Chi Minh city. Three identified parameters were chosen as overshoot (OS), raise time (T_r) and settling time (T_s) of 2% to ensure high accuracy to be achieved by the controller. The summary of the requirements for the BLDC motor used in the proposed motorbike is presented in Table 2. These parameters will be used to evaluate performance of the developed control system.

IOP Conf. Series: Materials Science and Engineering **819** (2020) 012017 doi:10.1088/1757-899X/819/1/012017

Table 2. - Transient Requirements of the system.

Parameter	*%OS*	T_r *(s)*	$2\%T_s$ *(s)*
Values	≤ 2.5	≤ 0.005	≤ 0.01

3. Development of Adaptive PID Controller for BLDC Motor

Fig. 3 shows the block diagram of the developed controller design. Here, the aim was to optimize the traditional PID controller by using adaptive control algorithms – either GA or PSO, as the optimizing algorithms with the low computational cost and its tunability. The output error is passed to the optimizer and computed in the cost function. The PID gains- K_P, K_I, K_D- that result in the least cost are substituted constantly to the real PID controller so that the system is accustomed to external disturbances.

This allows you to achieve the change in the performance of the drivetrain, based on the request from the accelerator and provide energy consumption optimization at the same time. Furthermore, the dynamics of the low-power solution allows it to stay in the traffic flow with the vehicles with higher power.

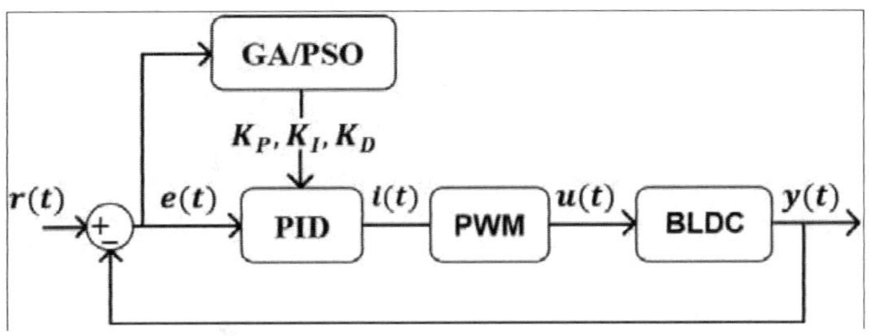

Figure 3. Block Diagram of the Developed Adaptive PID Controller for BLDC Motor.

A performance index or fitness function is an indication to performance of a system. In the context of GA and PSO optimizer, cost functions are usually used as performance indices. As mentioned previously, the error from motor's output is accumulated in the cost function to be fed to the optimization block. Therefore, the smallest accumulated error – fitness value is equivalent to the best values of PID parameters which are used.

There are four commonly used performance indices for PSO and GA: Integral Square Error (ISE), Integral Absolute Error (IAE) and Integral Time Absolute Error (ITAE), Mean Squared Error (MSE), which can be calculated using equations (3)-(6) [11]:

$$ISE = \int_0^T e(t)^2\, dt(3) \tag{3}$$

$$IAE = \int_0^T |e(t)|\, dt(4) \tag{4}$$

$$ITAE = \int_0^T t|e(t)|\, dt(5) \tag{5}$$

$$MSE = \frac{1}{n}\sum_{i=1}^{n} e(t)^2\ (6) \tag{6}$$

Where $e(t)$ is the output error at time t.

4. Design of the Optimisation Functions- Genetic Algorithms and Particle Swarm Optimisation

A. Genetic Algorithm.

As mentioned previously, GA algorithm aims to achieve the best solution in a population by revolutionizing individual within a number of iterations. Here is how GA is used to find the optimal PID values:

Step 1 – Initialisation:

Firstly, GA initializes a random population with a defined amount of individuals (solutions). Each solution carries a set of genes. Within the context of PID tuning, a set carries K_P, K_I, K_D as genes.

Step 2 – Evaluation:

After that, GA assess the fitness/quality of each solution using the cost functions. Within the context of PID tuning, the gene set of an individual – K_P, K_I, K_D – is updated in the real system. After that, the output speed of motor generates a certain amount of error. Refer to fig. 3, the error is then accumulated into the fitness value. The solutions with best fitness value have higher chance to be selected for mating process.

Step 3 – Selection:

Next, the breeding pairs from the population are chosen to generate new offspring – new solution and revolutionize the whole population. There are multiple methods to do this. Here, we choose to implement the most common method – *Roulette Wheel* method. To visualize, this method places all the solution on a wheel and spin it to randomly pick one, similar to how we play a Roulette wheel. The area that each solution occupies is proportional to their fitness value. Therefore, better individual possesses larger area on the wheel to get more chance to be picked [2].

Step 4 – Crossover:

After choosing breeding pairs, we go to crossover. This process is exchanging the genes of two parents to produces two offsprings – mating them. Within the context of PID tuning, GA simply swap/mix the K_P, K_I, K_D values of two chosen solutions so that we have two new ones with different set of genes. We also have a parameter called *"crossover probability"* – how often the crossover is performed [3].

Step 5 – Mutation:

In order to avoid repetition, we also randomly tweak the gene set to ensure the diversity of the population and expand the search space. In mutation, we use "*mutation probability*", to adjust the frequency of mutation process [4].

After that, the process goes back to step 2 and loop until one of *termination conditions* is met [5]:

- A defined fitness value is achieved.
- No major improvement within a number of iteration.
- The defined number of iteration is achieved.
- All of the above conditions are met at the same time.

The full process is shown on Figure 4.

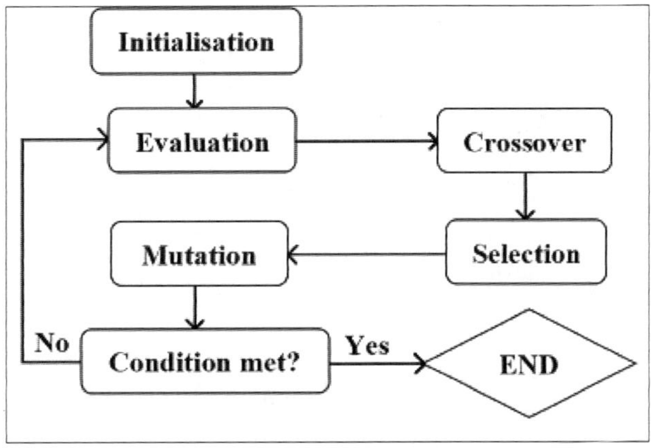

Figure 4. Block Diagram of Genetic Algorithms Optimisation Flow.

B. Particle Swarm Optimisation

As mentioned previously, PSO algorithm searches for the optimum solution (particle) in a population by letting each one goes around their search space and report their best finding after each iteration. The movement of each solution in an iteration is defined by the following steps:

- The solution's previous speed.
- The solution's personal best known location – *pbest*.
- The best known location of the swarm – *gbest*.

Therefore, each solution moves toward pbest and gbest while maintaining its inertia. Because of that, it achieves a new speed in the next iteration, calculated with (7):

$$v_i(t+1) = wv_i(t) + c_1r_1[\hat{x}_i(t) - x_i(t)] + c_2r_2[g(t) - x_i(t)] \quad (7)$$

Where w is the inertia coefficient; acceleration constants c_1 and c_2 are cognitive and social parameter, respectively $(0 \leq c_1, c_2 \leq 2)$; r_1 and r_2 are random numbers from 0 to 1, generated after every velocity update; $v(t)$ is the particle's current speed; $x_i(t)$ is the particle's current position; $\hat{x}_i(t)$ is pbest and $g(t)$ is gbest.

Also, the updated position of each particle is:

$$x_i(t+1) = x_i(t) + v_i(t+1)(8) \quad (8)$$

The PSO optimisation can be presented with the following steps:

Step 1 – Initialisation:

Firstly, we creates a random population with a defined number of particles (potential solutions). Each solution includes a random position and speed.

Step 2 – Evaluation:

Then, PSO assesses quality of each solution using a performance index (cost function).

Step 3 – Assigning pbest and gbest:

If the particle's location is better than pbest, then it will replace pbest. After that, the best values of pbest is updated to gbest.

Step 4 – Update speed **and position of the swarm:**

These updates can be calculated using functions from (7) and (8).

After step 4, the process goes back to step 2 and loop until the termination condition is met, similar to GA. The block diagram and the information flow of the PSO process are depicted in Figure 5.

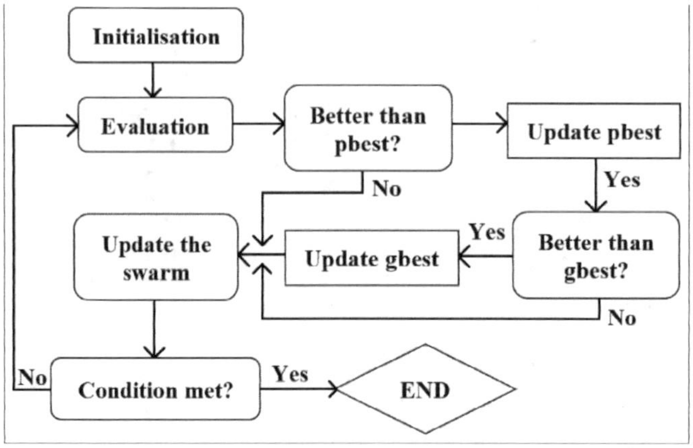

Figure 5. Block Diagram of PSO Controller.

As it can be seen from Figures 4 and 5, the GA and PSO have similar outcomes, but the optimization functions might differ, but the complexity of the processes is comparable.

5. Results of the Implementation

Table 3 shows the parameters, designed for GA optimization system.

Table 3. Parameter of GA.

GA PARAMETERS	VALUE/METHOD
Population size	60
Number of generations	80
Bound values	[0 0 0] – [500 500 500]
Fitness function	ITAE
Selection method	Roulette Wheel
Crossover method	Arithmetic Crossover
Crossover probability	0.8
Mutation method	Uniform Mutation
Mutation probability	0.08

Using GA optimization, the following gains for PID controller were obtained: $K_P = 493.4472, K_I = 489.2545$ and $K_D = 499.9636$.

Table 4 shows the optimization parameters, utilised for PSO-based system.

Table 4. Parameters of PSO.

PSO PARAMETERS	VALUE/METHOD
Population size	60
Number of generations	80
Bound values	[0 0 0] – [500 500 500]
Fitness function	ITAE
Acceleration constant c_1	1.2
Acceleration constant c_2	1.2

Using PSO optimization, the following gains were calculated for the given conditions: $K_P = 45.2444, K_I = 495.8561$ and $K_D = 3.9758$.

Applying the PID parameters to the adaptive PID controller for BLDC Motor (Fig.3), we deduced step responses for both systems, with GA and PSO. Fig 6 presents the step response for both control systems using the same plant model of BLDC motor.

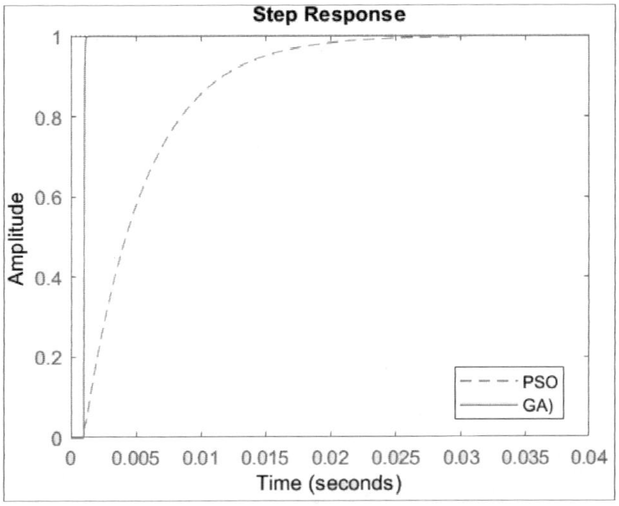

Figure 6. Step Response of the Developed Solution.

Table 5 presents the numerical values of the transient performance indicators, achieved using both systems in comparison with the derived requirements.

Table 5. Simulation Result Table.

Optimizer/ Requirements	%OS	T_r (s)	T_s (s)
GA	0	$8 * 10^{-5}$	$1.5 * 10^{-4}$
PSO	0	0.0102	0.0182
Requirements	≤ 2.5	≤ 0.005	≤ 0.01

The results can be analysed with the following conclusions. Firstly, both optimizing algorithms achieve perfect overshoot - 0% with no steady state error, as it is important for the steady-state performance evaluation. This outcome proves the superior performance in both speed and quality of using adaptive control compared to traditional methods. However, PSO failed to adapt both rise time and settling time requirements while GA not only succeeds on overcoming them, but also achieves exceptionally good results.

In particular, for rise time, PSO gets 0.0102 (s), which is equal to 210% of the requirement. The rise time of GA, on the other hand, is only to 1.6% of the requirement.

For settling time, T_s of PSO equals 185% of the requirement while it is only 0.82% for GA.

In conclusion, due to much superior performance, GA is selected to optimize PID parameters in our BLDC motor drivetrain for the future use due to its performance.

6. Conclusion and Future Work.

The paper presented the optimisation control algorithms design for the BLDC drivetrain of the electric bike. The model of the typical BLDC motor was developed to do the initial tuning of the system. Then, parameters of the function were deduced from the chosen motor type and the framework of the bike. Two optimizing algorithms for the controller were discussed: GA and PSO. Both systems showed no overshoot with zero steady state error. However, it is shown that GA has better overall performance indicators compared to the requirements. PSO, on the other hand, failed to adapt such requirements with larger values of the transient response parameters.

For the future work, we will apply more cost functions and adjust the parameters of PSO to achieve better step responses. In addition, we will do experiments to apply cloud computing for PID optimizer in order to save hardware cost with a reasonable performance. Finally, we will implement this system on the real prototype to justify the results.

References

[1] Kolbasov A, Karpukhin K, Terenchenko A, and Girutskiy O "Efficiency of photoelectric converters intellectual system application on ground electric transport," in *IOP Conference Series: Materials Science and Engineering*, 2019, p. 012011
[2] Kolbasov A, Karpukhin K, Terenchenko A, and Kavalchuk I "Concept of intellectual charging system for electrical and plug-in hybrid vehicles in Russian Federation," in *IOP Conference Series: Materials Science and Engineering*, 2018, p. 012013
[3] Cherry C and Cervero R "Use characteristics and mode choice behavior of electric bike users in China," *Transport Policy,* vol. 14, pp. 247-257, 2007
[4] Waide P and Brunner C U, "Energy-efficiency policy opportunities for electric motor-driven systems," 2011
[5] Sierens R, E. J. T.-A. S. o. M. E. J. o. E. f. G. T. Rosseel, and Power, "Variable composition hydrogen/natural gas mixtures for increased engine efficiency and decreased emissions," vol. 122, pp. 135-140, 2000
[6] Kavalchuk I, Seyedmahmoudian M, Horan B, Oo A T, and Stojcevski A, "Design and Control Algorithms for Power Electronic Converters for EV Applications," *World Academy of Science, Engineering and Technology, International Journal of Electrical, Computer, Energetic, Electronic and Communication Engineering,* vol. 9, pp. 1221-1226, 2015
[7] Pillay P and Krishnan R J. I. T. o. i. a., "Application characteristics of permanent magnet synchronous and brushless DC motors for servo drives," vol. 27, pp. 986-996, 1991.
[8] Iizuka K, Uzuhashi H, Kano M, Endo T, and Mohri K J. I. T. o. I. A., "Microcomputer control for sensorless brushless motor," pp. 595-601, 1985
[9] Uygun D and Solmaz S, "Design and dynamic study of a 6 kW external rotor permanent magnet Brushless DC motor for electric drivetrains," in *2015 IEEE 5th International Conference on Power Engineering, Energy and Electrical Drives (POWERENG)*, 2015, pp. 87-92.
[10] Kavalchuk I, Arisoy H, Oo A T, and Stojcevski A, "Challenges of electric power management in hybrid and electric vehicles," in *Power Engineering Conference (AUPEC), 2014 Australasian Universities*, 2014, pp. 1-7
[11] Hussain K M, Zepherin R A R, Kumar M S, and Kumar S M G, "Comparison of PID Controller Tuning Methods with Genetic Algorithm for FOPTD System," *International Journal of Engineering Research and Applications* vol. 4, pp. 308-314, 2014
[12] Mirjalili S, "Evolutionary Algorithms and Neural Networks: Theory and Applications," in *Evolutionary Algorithms and Neural Networks* vol. 780, J. Kacprzyk, Ed., 1 ed. Gewerbestrasse 11, 6330 Cham, Switzerland: Springer, Cham, 2018, pp. 43-55
[13] Zhao W, Lin Q, Shi Y, and Fang X "Mining the Role-Oriented Process Models Based on Genetic Algorithm," presented at the Advances in Swarm Intelligence, Shenzhen, China, 2012
[14] Namboothiripad M K and B. R., "International Journal of Emerging Technology and Advanced Engineering," *International Journal of Emerging Technology and Advanced Engineering,* vol. 2, pp. 310-314, 2012
[15] Wali M "PROPORTIONAL-INTEGRAL(PID) CONTROLLER DESIGN USINGGENETIC ALGORITHM (GA)," *Al-Qadisiya Journal For Engineering Sciences,* vol. 4, pp. 75-92, 2011

The performance assessment of low-cost air pollution sensor in city and the prospect of the autonomous vehicle for air pollution reduction

Le Thi Phuong Linh[1], **Ilya Kavalchuk**[1, 3], **A Kolbasov**[2], **K Karpukhin**[2], **A Terenchenko**[2]

[1] School of Science and Technology, RMIT University Vietnam, 702 Nguyen Van Linh, District 7, Tân Phong, Ho Chi Minh City, Hồ Chí Minh, Vietnam
[2] FSUE "NAMI", 2, Avtomotornaya Str., Moscow 125438, Russian Federation

E-mail: ilya.kavalchuk@rmit.edu.vn

Abstract. Urban air pollution is clearly a constantly growing problem and the high level of urban air pollution has been shown to pose a significant risk to city dwellers. It is necessary to have low-cost sensors for data collection, ready data source allowing normal citizen to access to and gain information, and have prospect solutions (e.g. autonomous vehicle) for air pollution reduction in the city. The aim of the study is to illustrate the drivers behind the use of low-cost sensors and to review the performance of sensors. In addition, autonomous vehicle is expected to reduce pollution; therefore, the paper analyzes the benefits and adoption of autonomous vehicles in the future. The challenges and outlook for both low-cost sensor deployment and the adoption of self-driving vehicle will also be discussed. A literature review is used to obtain these aims. The study indicates that the main driver of low-cost sensor is to provide high-density spatiotemporal pollution data, assisting in creating emission inventories of pollutants and detecting pollution hotspots without capital investments. A number of performance aspects considered include the coefficients of determination (R^2), variance (CV), repeatability, reproducibly and stability of the sensor. Self-driving vehicle is promising in the change of travel patterns and having impacts on the health of society. Technology and economic challenges, the willingness to use the sensor and the autonomous policy approach are the major challenges. In terms of the outlook, standard guidelines and the calibration methods for low-cost sensors, energy consumption savings and policy supporting for autonomous vehicle should be further investigated.

1. Introduction

The ever-growing population of cities leads to increasing motorization that is one of the major sources of outdoor air pollution. Outdoor air pollution is a major problem attributing to million premature deaths globally. Air pollution is associated with a range of diseases, symptoms and conditions that impair health and quality of life [1]. In addition, air pollution is also responsible for global climate change and environmental problems. Developing countries experiencing rapid urbanization are facing acute challenges in abating air pollution. Many efforts have been carried out but overall levels of air pollution remain hazardous in many emerging cities in the Global South.

In most cities, automobile transportation has been found to be a significant factor in the deterioration of the urban environment and it is closely associated with urban air pollution. In general, more than 60 percent is attributed to motor vehicles. Recently, the data on air pollution have an important role in the great effort of government to reduce air pollution. It can be used to agree policy decisions and can be presented to the public to help them make informed travel choices. Conventionally, air pollution concentrations are monitored by government agencies using accurate static monitoring stations equipped with certified reference instruments for measuring regulatory pollutants, such as carbon monoxide (CO), ozone (O3), nitrogen dioxide (NO2), sulphur dioxide (SO2), and particulate matter (PM) [2]. However, instruments at fixed-site air quality monitoring stations require frequent maintenance and calibration. In addition, static stations are less apt in capturing spatial distribution of pollutants.

Recent advancements in technology have enabled of collecting air quality data beyond fixed monitoring stations. By using a network of low-cost sensors for monitoring real-time air pollution concentrations, the data could be used to inform policymakers for developing better policies and measures that would improve public health. A review of recent article of emerging low cost sensors suggests the majority of studies focus on the benefits and the application of sensors. However, limited attention is paid to performance assessment. Furthermore, many argue that the adoption of autonomous vehicle can reduce the air pollution in the city. Driverless vehicle, also referred to as self-driving vehicle or autonomous vehicle, is vehicle that does not require a person to manually control them [3]. However, many are concerned that autonomous vehicle could cause more air pollution in the city because it increases the travel demand as the time spent in the car will never again seem wasted, and driving will become less tiring and stressful [4].

Table 1. List of commercially available low-cost PM sensors and monitors.

Sensor/ Monitor	Detectable Range	Measurement Unit	Weight	Price
Sharp - GP2Y1010AU0F	0 to 0.8 mg/m^3 approx.	mg/m^3	16g	\$40
Shinyei - PPD42NS	≥ 1.0 µm	pcs/m^3	24g	\$38
Amphenol SM-PWM-01C	1 to 2 µm and 3 to 10 µm	pcs/m^3	n/a	\$40
Dylos DC1100 PRO	n/a	pcs/m^3	1.2lb	\$260
OPC-N1	% Concentration at 10^6 Pcs/l 0.24	pcs/l	< 70g	\$650
SYhitech DSM501	0 to 1.4 mg/m^3	mg/m^3	25g	\$7
M-DUST II	0 to 80 mg/m^3	mg/m^3	750g	300 Eur
Shinyei AES-1	300 to 300000 Pcs/ft^3	Pcs/ft^3	120g	\$1100

Sharp GP2Y1010AU0F *Dylos DC1100 pro* *Shinyei AES-1*

Figure 1. Examples of the Low-Cost Sensors Used in the project.

The objective of the paper is to comprehensively review the usage drivers and performance of low-cost sensors in gathering information high-resolution tempo-spatial pollutant concentration and the adoption of autonomous vehicle as a future solution for urban mobility and air pollution mitigation. The challenges and outlook for the technology deployment and further researches will also be discussed.

2. The Drivers for the Use of Low-Cost Sensors Usage

Urbanization, industrialization and population growth have led to the severe air pollution, in particularly in urban areas. In addition, motorization with the increase in the number of vehicle releases a substantial atmospheric emission, causing significant adverse human health impacts on the global scale [5]. The World Health Organization estimates 4.2 million premature annual deaths globally are linked to ambient air pollution, mainly from heart disease, stroke, chronic obstructive pulmonary disease, lung cancer, and acute respiratory infections in children.

Although the air pollution is recognized among the top ten risks of human development [6], yet many of the world's cities are still unable to comply with the prescribed concentration limits of air pollutant. It is even more acute for developing countries that are experiencing industrialization - about 4.3 million people deaths are attributed to household air pollution and another 3.7 million are from ambient air pollution.

Air quality varies over a relatively small scale [7]; however, in the urban area, the flow of air masses is more turbulent due to built environment and difficult to predict without sophisticated numerical modeling approaches. In order to develop a real-time high-resolution pollutant concentration maps for urban areas requires a large amount of data that are not available for many cities. This complexity assesses actual human exposure to pollutants challenging [8]. Employing low-cost sensor for air pollution collection is one of the proposed solutions for areas with a shortage of small-scale, high-precision air quality measuring equipment. Despite lower quality data, low-cost sensors can be used in locations simultaneously, which allows for high-resolution exposure assessment mapping of city environments [9].

In order to mitigate the impacts of air pollution on human health, and global environment, cities have put tremendous efforts to air pollution monitoring. For conventional measurements of air pollution, many environmental authorities have set up networks consisting of stationary stations, which are large sized with heavy weight, and require significant investment costs. Another drawback of conventional monitoring instruments is the location of the monitoring stations. The air pollution situation in urban areas is highly related to human activities (e.g., construction activities) and location-dependent (e.g., traffic congestion hotspot). The monitoring stations are generally located away from roadsides and major traffic congestion areas. Therefore, the data collected may not be representative for cities or urban regions [9].

To address this issue, many researchers have developed modeling approaches. However, these methods have problems of inherent aleatoric and epistemic uncertainties due to crude approximations in input

IOP Conf. Series: Materials Science and Engineering **819** (2020) 012018 doi:10.1088/1757-899X/819/1/012018

conditions [10]. In order to increase the spatio-temporal resolution of the air pollution information, researchers have deployed low-cost portable ambient sensors widely [11]. Apart from creating emission inventories of pollutants, detecting pollution hotspots, as well as allowing real-time exposure assessment, the low-cost portable sensors enable the mobility and the feasibility in large-scale deployment of the sensor nodes [12]. To facilitate the elaboration of high-resolution air quality maps, sensor nodes can be deployed as dense networks (ubiquitous monitoring) or mounted on vehicles [13] [14]. The reduced size of the low-cost platforms also allows new research in personal exposure [15]. 'Wearable' sensors are able to consider changes in exposure across different urban environments due to changes in location and activities and provide new capabilities to remotely detect the risk of potential toxic exposure, helping to better understand the nature of the exposure and evaluating the health risk from air pollution [16].

Some programs are already using sensor networks to assess their performance for both fixed-site and mobile monitoring [17]. EPA scientists and engineers advancing air pollution monitoring systems by combining the low-cost portable ambient sensors and the Wireless Sensor Network (WSN) into one system known as the Next Generation Air Pollution Monitoring System (TNGAPMS) [18]. TNGAPMS allows the air pollution information be updated in minutes or even seconds. TNGAPMS also helps researchers to understand the distribution of the air pollutants more efficiently and improve the accuracy of air quality models [18]. Furthermore, there are also community-led sensing network projects, or "citizen science" to involve the public at large in data collection and discussion processes (Air Quality Egg, 2014). Examples include *OpenSense* and *Citi-Sense-MOB* [19] that use mobile platforms to monitor air pollution variation in cities, *Everyaware* that helps citizens collect and share noise and air pollution data. *Citi-Sense* was able to enlist public participation to use low-cost air quality platforms in eight cities across Europe [19].

In comparison with other instruments for measuring air pollutants, the low-cost sensors are easy to deploy, operate and manage. Retrieving data from the sensors is simple and their automatic operation allows for extensive deployment in the urban settings. The advantage of the sensors provides information on the pollution sources and helps support more conclusive studies on the effects of air pollution on a wide range of aspects of human life [20].

Most low-cost sensor network adopted a centralized model - data collected from sensors are managed, processed and analyzed centrally. The data storage networks are Internet connected, as is the dissemination channel. Data is shared with all stakeholders, ranging from policy makers, and researchers to the general public via mobile phone applications. This helps to raise public awareness of air pollution and allows populations, especially those already at risk, to make informed decisions relating to their health by avoiding areas of high pollution level [19].

3. Performance Evaluation

Many researchers use the coefficient of determination (R^2) obtained from linear function as a criterion to calibrate the sensor response with the reference measurements. Higher R^2 value implies better sensor performance. However, linear function suffers some drawbacks in capturing pollutant concentration when size distribution (i.e., mean size and geometrical standard deviation) and composition of particles are varied. Therefore, other response functions need to be selected in these cases for a particular sensor by calibrating it under the full range of concentrations [21].

In addition, the difference between real world and the laboratory findings may affect sensor performance (small R^2 values) [22]. Although individual site calibration results in time-consuming and increases labor costs, each sensor must be calibrated on-site in order to increase the accuracy and the precision of the measurements. Laboratory-calibrated sensors should not be used for real-world measurements.

Repeatability of measurements of a particular sensor unit and reproducibility of units of the same sensor model [19] [22] can be used to evaluate the precision of the sensors. Repeatability and

IOP Conf. Series: Materials Science and Engineering **819** (2020) 012018 doi:10.1088/1757-899X/819/1/012018

reproducibility are the closeness between successive measurements of the same measure and carried out under identical and non-identical conditions of measurement, respectively [23]. Estimating the repeatability of PM sensors is very difficult in non-laboratory conditions due to the difficulty of maintaining constant particle concentrations [22]. A number of studies evaluated the variability of output signals from copies of the same sensor model (intramodel variability). According to [24], repeatability characteristics for different low-cost PM sensors are between 2 and 28% and the repeatability deteriorated at low PM concentrations (CV lies in the range of 23–26% at ~50 μg/m^3 PM concentration).

To ensure reproducibility, sensors need to be calibrated individually so as to improve their reproducibility characteristics [25] [26]. Prior studies reported CV values ranging from 0.9 to 16%. Sensors with lower CV values signify with higher reproducibility. CV values below 10% are considered acceptable for low-cost sensor testing studies [27]. In addition, when sensors are exposed to larger sized particles, the level of reproducibility could deteriorate due to the accumulation of particles at the sensing zone.

Stability is another sensor characteristic that needs to be investigated. For the purposes of research or knowledge-based decision making of the government, sensors are normally used for a relatively long duration (up to few months). Due to the sensor aging problems and accumulation of dust, the stability of the sensors also deteriorates. [28] reported that R^2 increases for a sensor when the cofounding variable "days of use" was used in the predictive regression model. Further researches need to be conducted to deal with the issue and ensure that the sensor works sufficiently over a long duration.

4. The benefit of autonomous vehicle as a solution for air pollution mitigation

There are many ways to reduce traffic-related air pollution in the city. One of the emerging solutions is the usage of autonomous vehicle. Autonomous vehicle is believed to be the next-generation technology for the sustainable future societies [29]. There are several ways that autonomous vehicles can reduce pollution in the city.

4.1 More efficient driving pattern

By enabling more efficient driving pattern that avoids excessive acceleration and braking, an autonomous vehicle could help to reduce emission in the urban city. In fact, gasoline is burned when braking and re-accelerating. Autonomous mode could change the driver driving style resulting in less burned gas and less consumed battery power, which leads to less air pollution. In addition, autonomous cars could travel closer to each other, which improved aerodynamics [30]. Accelerating is known to be a particularly energy intensive process. Smooth driving will minimalize the necessity of accelerating. One of the advantages of the autonomous vehicle is that it would not violate the traffic regulations, particularly speed limits. The fuel consumption of vehicular traffic on a given road section depends strongly on the velocity profiles of the vehicles. Driving with the imposed limits would make vehicle obtain the best fuel efficiency and pollution at minimum level [29].

4.2 Traffic congestion and fuel consumption

Traffic congestion puts pressure on the urban transport system and leads to the increased environment impacts. In fact, traffic congestion typically leads to the increase in fuel consumption and emission. The level of fuel consumption estimated for the rush hour is 30 to 50% higher than those in in the conditions of an ideal traffic flow [31]. By suggesting an uninterrupted path with the aid of autonomous technology, the vehicle can maintain the green speed that ensures the lowest fuel consumption. A number of studies have investigated the potential for autonomous vehicles to reduce congestion. Studies conducted in the USA stated that delays on motorways would decrease by 60% and fuel consumption by 25% if the autonomous vehicle occupies 90% of the traffic share [32]. Depending on the vehicle-to-vehicle communication and how traffic-smoothing algorithms are implemented, [33] indicated that the adoption of self-driving vehicles could increase fuel economy and congested traffic speeds by 23–39% and 8–13%, respectively, for all vehicles in the freeway travel stream.

4.3 The increase of car-sharing demand

Several researchers have examined the comprehensive impacts of the deployment of self-driving vehicle fleets in the context of car-sharing and ride-sharing. Many expect that the services will lead to an essential lower demand for private cars. A study by [34] suggests that shared self-driving fleet could remove two thirds of the vehicles currently operating in Singapore while still delivering all of the trips currently made by private vehicles [35]. An agent-based model that is set up in such a way that the fleet of shared and self-driving vehicles deliver the same daily trips in Lisbon have proved that there may be even 65% fewer passenger cars in the streets in peak traffic hours than at present [35]. This will lead to a radical reduction of the congestion level and will significantly reduce the emission. [36] developed a model of a fleet of Shared Autonomous Vehicles. The modeling results suggest that environmental impacts of the implementation of such a fleet are positive, with 5.6% less greenhouse gas emissions, 34% less carbon monoxide emitted, and a 49% reduction in volatile organic compound emissions, among others, compared to the traditional US light duty fleet [35] [36].

5. The Challenges

There are still many technological and economic challenges for the use of low-cost mobile sensors. Because most gaseous and particulate matter sensors require independent evaluation under a range of ambient environmental conditions, the greatest barrier against the uptake of low-cost sensor is the reliability of measured air pollution data.

Other issues are the relatively short working time of sensors as well as their durability. This warrants further evaluation of low-cost sensors in various environmental conditions. In terms of economic challenges, in many cases the costs of maintenance and data management/ analysis are much more expensive than the manufacturing cost of sensors. The ability of citizen science approach in raising awareness of people about the impact of air pollution on the health should not be understated. Community-based air pollution projects and a focus on educational effects might lead to a paradigm shift in air pollution monitoring practice.

In order to deal with air pollution in the urban area, several ways have been implemented, ranging from gasoline consumption reduction to electric vehicle usage [37] [38]. Autonomous vehicle is concerned as one of the promising solutions. The environmental impacts of autonomous vehicles have been realized to be important issues, but many are concerned that autonomous vehicles could encourage more driving, more fossil fuels consumption, and more emissions. In addition, it is very difficult to accurately estimate the potential of autonomous vehicles in the reduction of urban emissions due to a series of variable factors that condition the functioning of the future transport system [39]. Furthermore, self-driving vehicles are still years away and it is difficult to reliably predict the future following disruptive paradigm shifts [32].

6. The Outlook

Firstly, low-cost sensors contribute significantly to the body of information that is important for exposure assessment and subsequent hazard mitigation. Thus, for the use in a citizen based observatory network, we need to develop appropriate infrastructure and quality assurance and control systems. Secondly, due to recent development in sensor devices and other related technologies, the initial (manufacturing) cost of low-cost sensor is not a major problem, however the costs involved in their installation, maintenance and data analysis need to be considered. Thirdly, increasing the sensitivity of the particle sensors necessary to be addressed. Because of the limitation to measure low concentrations, the existing sensors cannot measure ultrafine particles that pose greater risk to human health. In addition, further investigation on the sensor's stability are necessary in order to ensure that the sensor works sufficiently over a long duration. Finally, yet importantly, autonomous cars have the potential to dramatically change the transportation network and contribute to the air pollution reduction. However, before autonomous vehicle hits the road, it will be necessary to change the existing international legal regulations [39]. Policy makers should begin supporting research into how self-driving vehicle could affect transportation and air pollution mitigation, and how to best alter our transportation system as well as travel behavior to maximize their benefits while minimizing any negative consequences of the transition to a largely autonomous fleet of motor vehicles [36].

Acknowledgments
This work was supported by the School of Science and Technology, RMIT University, Vietnam.

References

[1] Raaschou-Nielsen O, Beelen R, Wang M, et al. 2016 Particulate matter air pollution components and risk for lung cancer *Environ. Int.* **87** pp 66-73

[2] Kumar P, Morawska L, Martani C, Biskos G, Neophytou M, Sabatino S, Bell M, Norford L, Britter R 2015 The rise of low-cost sensing for managing air pollution in cities *Environ. Int.* **75** pp 199-205

[3] Information on: https://greenerideal.com/news/vehicles/driverless-cars-environmental-benefits/

[4] Information on: https://www.entrepreneur.com/article/334995

[5] Lim S, et al 2013 A comparative risk assessment of burden of disease and injury attributable to 67 risk factors and risk factor clusters in 21 regions, 1990–2010: a systematic analysis for the Global Burden of Disease Study 2010 *Lancet* **380** pp 2224–2260

[6] Information on: https://www.who.int/air-pollution/news-and-events/how-air-pollution-is-destroying-our-health

[7] Britter R, Hanna S 2003 Flow and dispersion in urban areas *Annual Rev Fluid Mech.* **35** pp 469-496

[8] Vardoukalis S, Gonzalez-Flesca N, Fisher B, Pericleous K 2005 Spatial variability of air pollution in the vicinity of a permanent monitoring station in central Paris *Atmos. Environ.* **39** pp 2725–2736

[9] Kumar P, Morawska L, Martani C, Biskos G, Neophytou M, Sabatino S, Britter R 2015 The rise of low-cost sensing for managing air pollution in cities *Environment international* **75** pp 199-205

[10] Kumar P, Ketzel M, Vardoulakis S, Pirjola L, Britter R 2011 Dynamics and dispersion modelling of nanoparticles from road traffic in the urban atmospheric environment - a review J *Aerosol Sci.* **42** pp 580-603

[11] Engel-Cox J, Oanh N, Donkelaar A, Martin R, Zell E 2013 Toward the next generation of air quality monitoring: Particulate Matter *Atmospheric Environment* **80** pp 584-590

[12] Yi W, Lo K, Mak T, Leung K, Leung Y, Meng M 2015 A survey of wireless sensor network based air pollution monitoring systems *Sensors* **15(12)** pp 31392-31427

[13] Hasenfratz D, Saukh O, Walser C, Hueglin C, Fierz M, Arn T, Thiele L 2015 Deriving high-resolution urban air pollution maps using mobile sensor nodes *Pervasive and Mobile Computing* **16** pp 268-285

[14] Schneider P, Castell N, Vallejo I, Vogt M, Lahoz W, Bartonova A 2016 Data fusion of crowdsourced observations and model data for high-resolution mapping of urban air quality *10th International Conference on Air Quality—Science and Applications*

[15] Castell N, Dauge F, Schneider P, Vogt M, Lerner U, Fishbain B, Bartonova A 2017 Can commercial low-cost sensor platforms contribute to air quality monitoring and exposure estimates?

Environment international **99** pp293-302
[16] Bossche J, Theunis J, Elen B, Peters J, Botteldooren D, Baets B 2016 Opportunistic mobile air pollution monitoring: a case study with city wardens in Antwerp *Atmospheric environment* **141** pp 408-421
[17] Mead M, Popoola O, Stewart G, Landshoff P, Calleja M, Hayes M, Baldovi J, McLeod M, Hodgson T, Dicks J et al 2013 The use of electrochemical sensors for monitoring urban air quality in low-cost, high-density networks *Atmos. Environ* **70** pp186–203
[18] Yi W, Lo K, Mak T, Leung K, Leung Y, Meng M 2015 A survey of wireless sensor network based air pollution monitoring systems *Sensors* **15(12)** pp 31392-31427
[19] Mitchell G, Dorling D 2003 An environmental justice analysis of British air quality *Environment and Planning A* **35** pp 909-929
[20] Kelly K, Whitaker J, Petty A, Widmer C, Dybwad A, Sleeth D, Martin R and Butterfield A 2017 Ambient and laboratory evaluation of a low-cost particulate matter sensor *Environ Pollut.* **221** pp 491–500
[21] Rai A, Kumar P, Pilla F, Skouloudis A, Sabatino S, Ratti C, Rickerby D 2017 End-user perspective of low-cost sensors for outdoor air pollution monitoring *Science of The Total Environment* **607** pp 691-705
[22] Taylor B, Kuyatt C 1994 *Guidelines for Evaluating and Expressing the Uncertainty of NIST Measurement Results*
[23] Wang Y, Li J, Jing H, Zhang Q, Jiang J, Biswas P 2015 Laboratory evaluation and calibration of three low-cost particle sensors for particulate matter measurement *Aero- sol Sci. Technol.* **49** pp 1063–1077
[24] Sousan S, Koehler K, Hallett L Peters T 2016 Evaluation of the Alphasense optical particle counter (OPC-N2) and the Grimm portable aerosol spectrometer (PAS-1.108) *Aerosol Science and Technology* **50(12)** pp 1352-1365
[25] Sousan S, Koehler K, Thomas G, Park J, Hillman M, Halterman A, Peters T 2016 Inter-comparison of low-cost sensors for measuring the mass concentration of occupational aerosols *Aerosol Science and Technology* **50(5)** pp 462-473
[26] Badura M, Batog P, Drzeniecka-Osiadacz A, Modzel P 2018 *Evaluation of Low-Cost Sensors for Ambient PM2.5* (Monitoring Journal of Sensors)
[27] Jiao W, Hagler G, Williams R, Sharpe R, Brown R, Garver D, Judge R, Caudill M, Rickard J, Davis M 2016 Community Air Sensor Network (CAIRSENSE) project: evaluation of low-cost sensor performance in a suburban environment in the south- eastern United States Atmos *Meas. Tech.* **9** pp 5281-5292
[28] Zheng Y, Liu F, Hsieh H 2013 U-air: When urban air quality inference meets big data *Proceedings of the 19th ACM SIGKDD international conference on Knowledge discovery and data mining* pp 1436-1444
[29] Fagnant D, Kockelman K 2015 Preparing a nation for autonomous vehicles: opportunities, barriers and policy recommendations *Transportation Research Part A: Policy and Practice* **77** pp 167-181
[30] Tientrakool, Patcharinee 2011 Highway Capacity Benefits from Using Vehicle-to-Vehicle Communication and Sensors for Collision Avoidance *Vehicular Technology Conference (VTC Fall) IEEE*
[31] Igliński H 2009 Kongestia transportowa w Poznaniu i wybrane sposoby jej ograniczenia *Transport miejski i regionalny* **3** pp 1-8
[32] Fagnant D, Kockelman K 2015 Preparing a Nation for Autonomous Vehicels: Opportunities, Barriers and Policy Recommendations *Transportation Research Part A Policy and Practice* **77**

pp 8-10 DOI: 10.1016/j.tra.2015.04.003

[33] Atiyeh, Clifford 2012 *Predicting Traffic Patterns, One Honda at a Time* (MSN Auto, June 25)

[34] Spieser K, Ballantyne K, Treleaven K, Zhang R, Frazzoli E, Morton D, Pavone M 2014 *Toward a systemic approach to the design and evaluation of automated mobility-on-demand systems: A case study in Singapore* (MIT Open Access Articles, Forthcoming in Road Vehicle Automation, Springer Lecture Notes in Mobility series)

[35] Martinez L, Crist P, Viegas J 2015 Urban mobility system upgrade How shared self-driving cars could change city traffic *International Transport Forum Policy Papers* **6** DOI: 10.1787/5jlwvzdk29g5-en

[36] Fagnant D, Kockelman K 2014 The travel and environmental implications of shared autonomous vehicles, using agent-based model scenarios *Transportation Research Part C: Emerging Technologies* **March** 40

[37] Kolbasov A, Karpukhin K, Terenchenko A Kavalchuk I 2018 Concept of intellectual charging system for electrical and plug-in hybrid vehicles in Russian Federation *IOP Conference Series: Materials Science and Engineering* **315 No. 1** p 012013

[38] Karpukhin K, Terenchenko A, Kolbasov A 2018 The Creation of Modern Electric Vehicles with Additional Source of Energy *IOP Conference Series: Earth and Environmental Science* **159 No. 1** p 012043

[39] Igliński H, Babiak M 2017 Analysis of the potential of autonomous vehicles in reducing the emissions of greenhouse gases in road transport *Procedia engineering* **192** pp 353-358

IOP Conf. Series: Materials Science and Engineering **819** (2020) 012019 doi:10.1088/1757-899X/819/1/012019

Current issues of ensuring human safety in process of vehicles operation

S V Gaysin and V F Koutenev

FSUE "NAMI", 2 Avtomotornaya St., Moscow, 125438, Russian Federation

E-mail: vakutenev@nami.ru

Abstract. Based on the new cases of lung cancer registered by the World Health Organization, which number is annually increasing worldwide, development of activities and acceptance of additional emergency measures on reduction of emissions of particulate matters from wear of tyres and roadway in cities and megalopolises are necessary.
The pilot studies results are given revealing the increased level of emissions of particulate matters (PM) less than 10 microns from wear of tyres and asphalt roadway, but not from exhaust gases of the vehicles with diesel engines.

The paper deals with a problem of air pollution from vehicles operation in cities and megalopolises leading to smog formation both in Europe and in the Russian Federation. Creation and development of vehicles in order to support human life and activities is followed not only by improvement of living and transport conditions for the population but also by negative effects connected with highly significant contamination of the air environment in cities to the levels dangerous for the health of the population.

The topic and material of the document are focused on justification of the need to regulate the emissions of hazardous substances and particulate matters from the wear of tyres and roadway, which at present significantly pollute urban atmosphere during the vehicle operation.

Current development of requirements for vehicle design is regulated in the international legislation by UNECE Regulation 148, which, in general, improves all design elements and functional characteristics step by step within the last 50 years increasing overall integrated safety of the vehicles.

The global indicators in improving vehicle design within the current period are quite impressive because of modern digital technologies penetrating into the transport industry more and more deeply. Transport automation becomes the main trend of advanced technology developments creating the competitive advantages of modern automotive products. Improvement of vehicle design has already reached such amazing heights that the vehicles become driverless and no ancient vehicles spreading clouds of smoke are seen on the roads anymore!

However, at present, it is impossible to definitely say that these new technologies fully ensure human safety.

The efficiency of implementing the requirements of the UN Regulations in force is not yet sufficient. According to the WHO data, today, up to 1.5 million people all over the world die annually in road accidents, and up to 1.5 million people of urban population also die annually due to air pollution in cities [1].

As an example, the results of efficient implementation of the UN Regulations in case of the Russian Federation in terms of design safety (passive and active safety) with regards to both road and environmental safety are shown in figures 1 and 2.

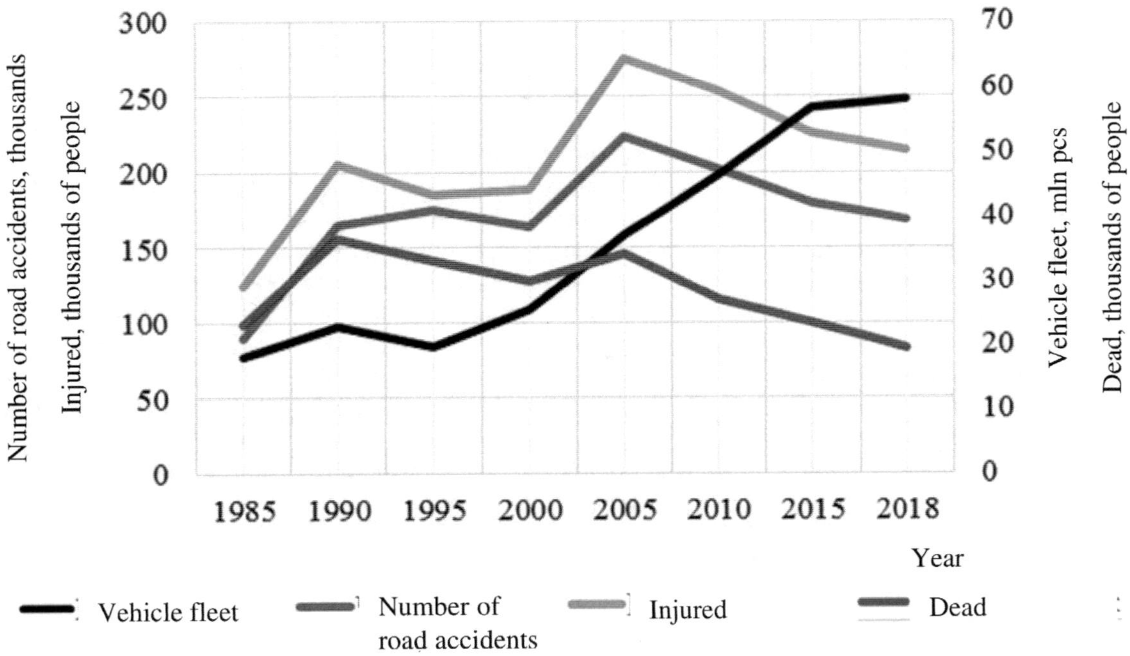

Figure 1. Dynamics of road accidents indicators in Russia for the period of 1985 to 2018.

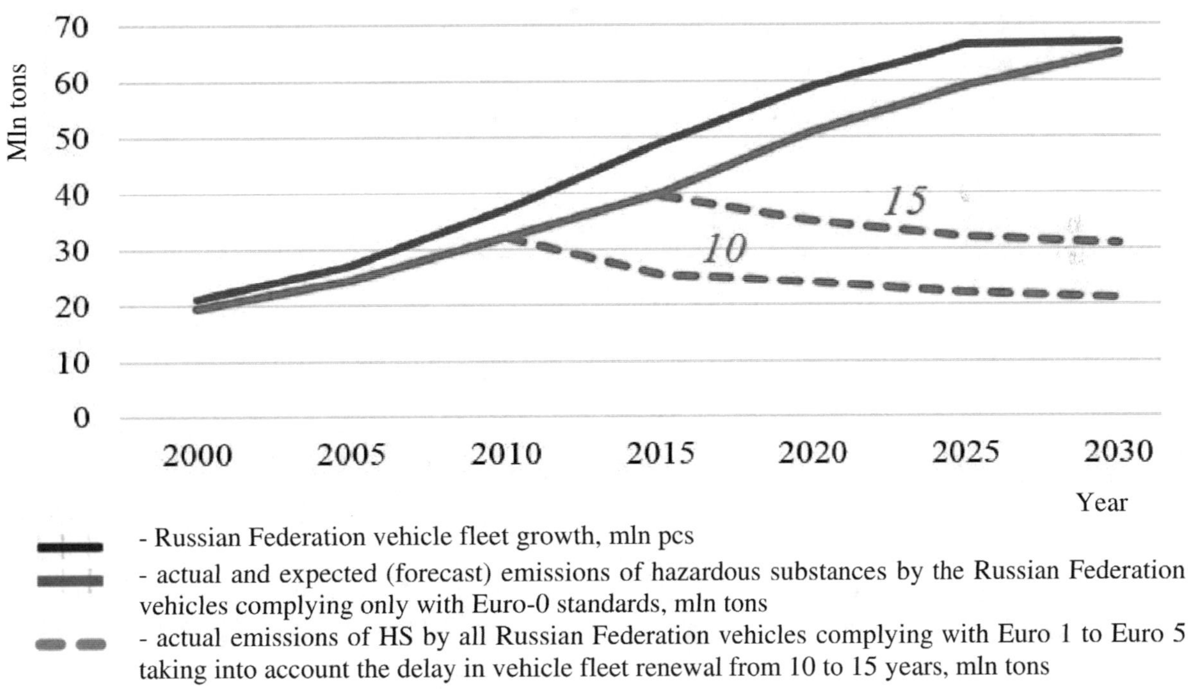

Figure 2. Fleet growth and dynamics of HS emissions by the RF vehicle fleet.

It should be noted that it is the present time period when additional environmental issues were revealed. Back in 2012, the World Health Organization (WHO) proposed to prohibit the usage of diesel-engined vehicles due to the high emission of particulate matters (PM) in their exhaust gases within European cities.

Therefore, today European countries look for effective ways to reduce smog formation in large cities in order to decrease the ambient air pollution there.

According to experts, the program of conversion to vehicle electrification forecasts a share of electric vehicles in the total global output of at least 17% by 2020, of which 75% will be hybrid vehicles, and the total number in the global vehicle fleet will reach 20 mln or only 1,7%[2].

It shall be specifically noted that against the backdrop of intense search (or sometimes its imitation) for the transport ecological problem solution by means of electric vehicles (EV) and alternative fuels, attention to the enormous growth of PM emissions from roadway and tyre wear has been weakened. In our opinion, use of electric vehicles in cities – insignificant at the moment – will not ensure an essential decrease in PM emissions since these vehicles will be driven with the same tyres and on the same asphalt-concrete roadway. This is confirmed by the data of the PM emission surveys carried out in the Russian Federation. Dynamics of the PM emissions from other vehicles systems (braking systems, tyres) and from the roadway wear are shown in Fig. 3 as an example for comparative analysis with regards to the standards of UN Regulations No. 49 and 83 for PM emissions with exhaust gases in case of operation in the megalopolis of Moscow.

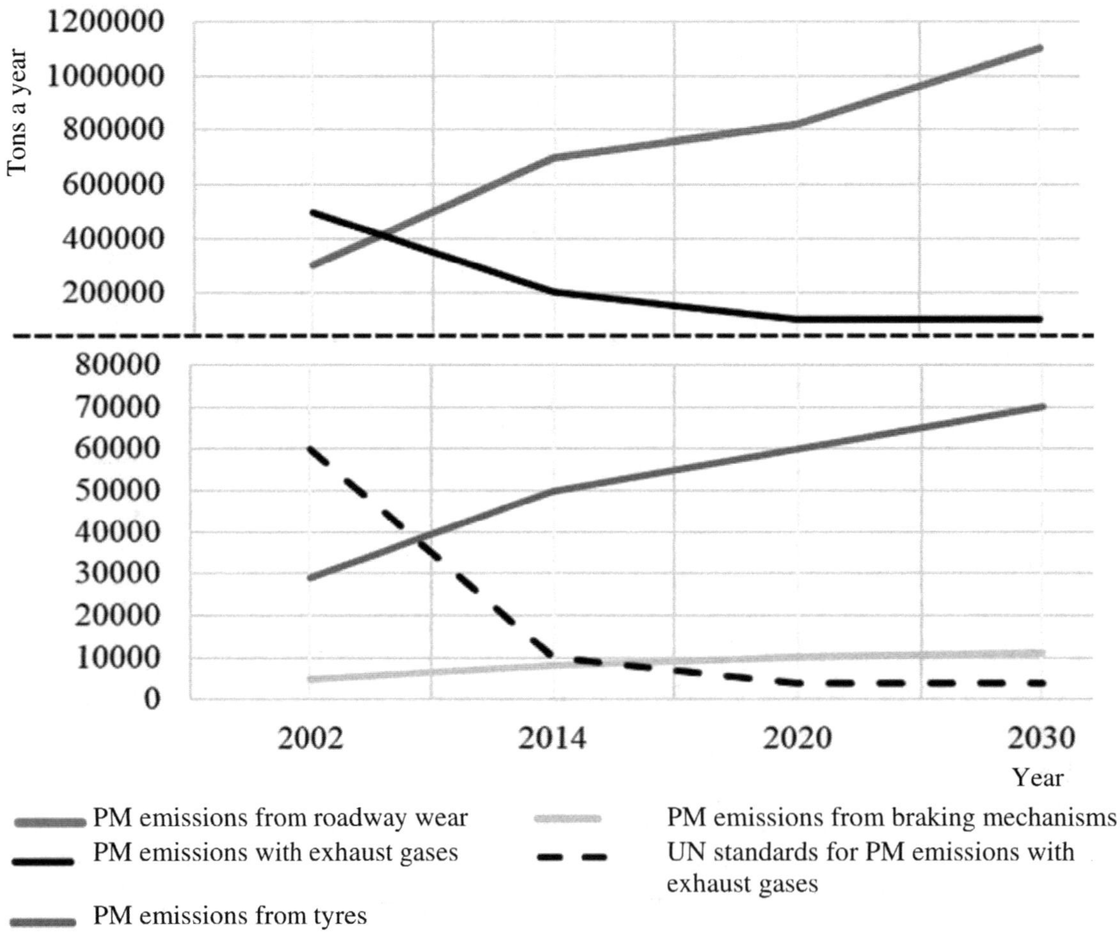

Figure 3. Dynamics and forecast of annual PM emissions from wear of tyres, braking mechanisms and roadway in Moscow, in tons, compared to the emissions of and emissions standards for PM with exhaust gases.

The series of activities with research and development aimed at improvement of design of vehicles' internal combustion engines carried out by manufacturers and legislators all over the world in the past 50 years ensured a significant decrease of PM emissions with exhaust gases, however, in this period, due to the increasing vehicle fleet in all major cities of the developing countries, the air is being polluted because of the dramatical increase in particulate matters emissions, first of all from tyres and roadway wear.

One should pay attention to highly representative materials in the report of the UK submitted in November 2018 at the Working Party on Pollution and Energy (WP.29/GRPE, UNECE Inland Transport Committee) which emphasizes the emissions of particulate matters of 2.5 micrometers and less, which by penetrating into lungs and blood can be transported through the body remaining in human organs and therefore reduce life expectancy [3].

Fig. 4 shows the results of surveys of actual values of emissions of PM less than 2.5 micrometers from various road transport systems in the UK which, by the values of PM emissions from the wear of tyres, roadway and even braking systems, are essentially comparable by level and correspond to the results of the Russian research data shown in figures 3.

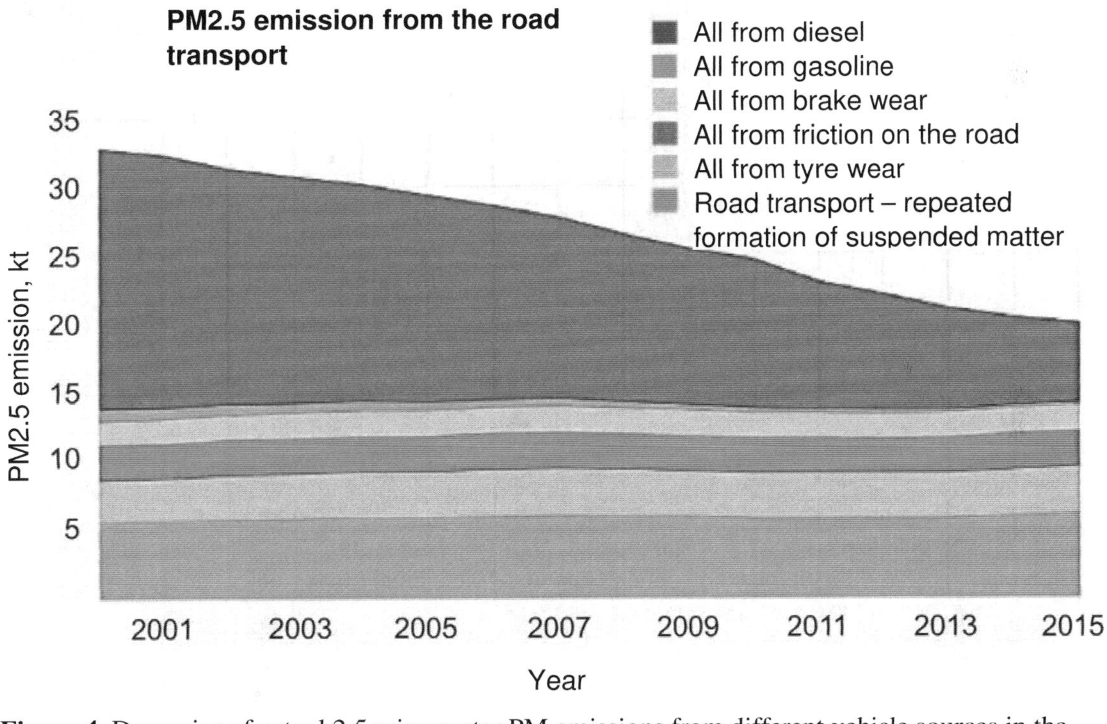

Figure 4. Dynamics of actual 2.5 micrometer PM emissions from different vehicle sources in the UK.

Highly interesting and important data should be especially noted in the document which was provided earlier at GRPE sessions by European Tyre & Rubber Manufacturers Association (ETRMA), in which the results of the PM chemical analysis are shown for the content of polycyclic aromatic hydrocarbons (PAH) in them that contain cancerogenic substances causing human cancer diseases [4].

Table 1 provides information on polycyclic aromatic hydrocarbons (PAH) content in PM. The values are expressed in parts per million (ppm) in the matters mix, sampled behind a moving vehicle.

Table 1

Chemical substance	Road particles (RP)	Tyre wear particles (TWP)	Tyre (tread) particles (TP)
Acenaphthene	4.08	0.04	0.13
Phenantrene	53.4	1.66	1.21
Pyrene	54.84	4.77	0.06
Anthracene	7.36	0.1	0.11
Benzanthracene	38.65	0.18	2.87
Benzo(a)pyrene	12.51	0.28	N.D.
Benzo(k)fluoranthene	7.4	0.02	0.92
Chrysene	17.72	0.36	2.95
Dibenz(a,h)anthracene	2.56	0.1	0.87
Fluoranthene	82.13	0.98	1.62
Indeno-1,2,3(c,d)pyrene	5.36	0.21	N.D.

This list is not an exhaustive one, but it contains the PAHs most widely spread in the environment and considered cancerogenic and able to provoke cancerous diseases of people according to the International Agency for Research on Cancer (IARC) [4].

Due to the differences in evaluation of amounts of PM emissions from the tyre and roadway wear, there has risen a global ecological conflict between the transport and environment in big cities, which by the current time period has become critical both by its level and violent growth.

The comparative results of determination of PM emissions from various vehicle systems in 2015 according to the materials of the Russian Federation (figure 3) and the UK (figure 4) are given in figure 5.

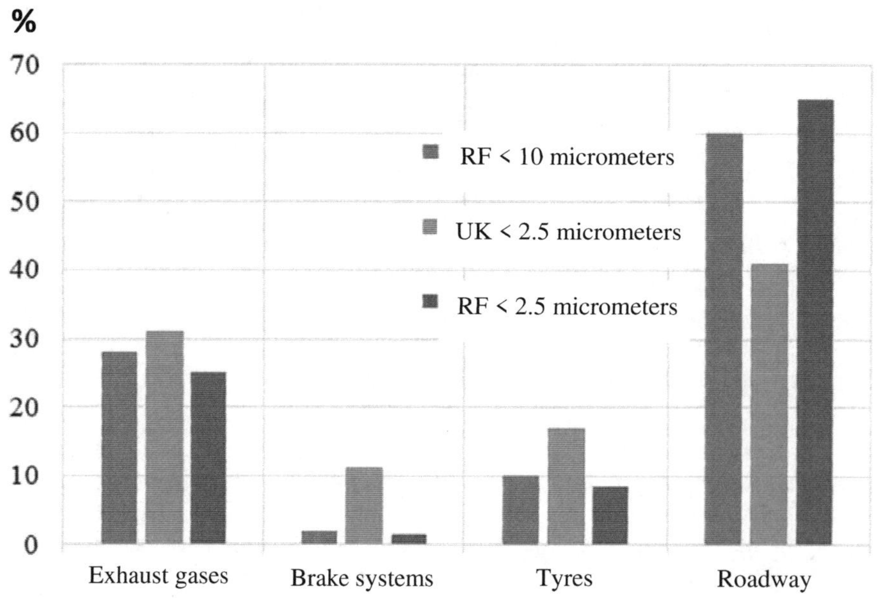

Figure 5. Comparison of PM emissions according to the RF and UK studies.

Thus, the most important issue currently is regulation of PM emissions not only with the exhaust gases, but especially from the asphalt roadway and tyre wear, as the PM emissions from the roadway wear are twice higher than the PM emissions with the exhaust gases, and if considering the tyre and brake system wear, almost thrice, and it is these non-exhaust emissions which influence the air pollution in the urban environment most significantly today.

Based on the abovementioned, there is a need to promote discussion of the following issues of concern by the national and international legislators and vehicle manufacturers:

I. In the current situation, where the urban atmosphere is polluted by hazardous PM, it is necessary for our Governments to initiate and sponsor serious research developments related to reduction of the content of hazardous substances in materials used for production of tyres and roadway, while the international community needs to develop the respective regulatory requirements.

II. Tyre manufacturers and road construction services shall revise the existing technology for production of tyres and roadway considering the potential of ongoing research and development in terms of new materials and technologies.

III. Therefore, we consider it necessary to promote and advance the works on the new problem issues of regulation of integrated environmental safety at vehicles operation within the World Forum (WP-29) starting from 2020.

References

[1] 2016 Sentyabr', *Bulletin of the World Health Organization Ambient (outdoor) air quality and health (Vsemirnoi organizatsii zdravookhraneni: Kachestvo atmosfernogo vozdukha i zdorov'e)*
[2] Azarov V K 2016 *(K voprosu o kompleksnoi bezopasnosti avtomobilei) Izvestiya MGTU «MAMI»* **2(28)** pp 46-54
[3] Emission not related to fuel combustion *UK Department of Transportation point of view*
[4] Kreider M L, Panko J, MacAfee B L and Sweet L I 2010 Physical and chemical characterization of tire-related particles: Comparison of particles generated using different methodologies *Science of the Total Environment* **408** pp 652-9.

IOP Conf. Series: Materials Science and Engineering **819** (2020) 012020 doi:10.1088/1757-899X/819/1/012020

Experience in the development of an effective thermal management system for the high-voltage battery of the vehicle

R Kh Kurmaev, V S Struchkov and V V Novak

FSUE "NAMI", 2 Avtomotornaya St., Moscow, 125438, Russia

E-mail: vasiliy.novak@nami.ru

Abstract. For the effective operation of the vehicle, which includes a high-voltage battery (HVB), it is first necessary to understand what conditions it is designed for. There are a number of criteria for the selection of HVBs for vehicles. One of them is the thermal management system, which ensures the operation of a HVB in optimal temperature mode. In HVBs, it is necessary to maintain an optimum temperature for maximum efficiency and service life. The article discusses the preferred thermal management system of the main manufacturers of vehicles with HVB. The technical solutions, the applied designs and the development trends of the designs of the temperature control of HVBs are described. On the basis of the considered structures, operating conditions and prospects for the development of the market for such vehicles, the concept of creating an effective thermal management system for HVBs for projected vehicles has been proposed. A classification of the thermal management system of HVBs as part of a vehicle, depending on the climatic conditions of operation, is proposed. One of the developed variants of perspective designs of high-voltage batteries of a vehicle operated in a wide range of temperatures is presented.

1. Introduction

Today, based on the main advantages of hybrid electric vehicles (HEVs), associated with efficiency, environmental friendliness and running characteristics, more and more automakers produce new models of these vehicles. Energy efficiency of HEVs is highly dependent on the performance of high-voltage battery, which must meet a number of requirements: capacity, cost, size, weight, operating temperature range, etc. In this case, the temperature mode is one of the main criteria for the functioning of the hybrid.

Temperature mode affects the charging time of a HVB. The results of studies of the dependence of the charge of a HVB on temperature differences show that the full charge of the battery reaches a maximum in the range from plus 20 ºC to plus 40 ºC. [2,3]. Although this range may have deviations depending on the chemical composition of the cells used by the manufacturer, but on average this value corresponds to the main types of battery cells used in HVBs on hybrid vehicles.

Temperature mode also affects the service life of a HVB. When charging and discharging, the HVB must be protected from overheating. Overheating of the battery cells above the maximum value indicated for them leads to cell degradation, which significantly reduces the life of the high-voltage battery, up to 50%. [3, 6].

It should also be noted that when charging a cold battery, cell degradation and loss of capacity also occur. A cold battery must not be used if the vehicle has a recovery mode. Operation of a HEV at low ambient temperatures significantly reduces the power reserve (up to 30 – 40 %). Based on the foregoing, the task of thermal management system of a HVB is very important. [4].

IOP Conf. Series: Materials Science and Engineering **819** (2020) 012020 doi:10.1088/1757-899X/819/1/012020

2. Overview of high-voltage batteries with thermal management systems of serial-produced vehicles on the electric traction

Figure 1 shows the design of the high-voltage battery Chevrolet Volt. The battery is located in the middle part of the HEV and takes place in the central tunnel and under the rear seat. The battery consists of 288 lithium-ion battery cells, which are divided into three modules and outputs a voltage of 360 V with a total capacity of 16 kW*h. Through the inlet tube, the coolant enters the HVB. Inside the battery, there are heat channels that allow coolant to flow through the cooling plates between the flat cells of the batteries. These channels allow cells to be cooled or heated depending on operational requirements. If the temperature of the battery is lower than the operating temperature, the heating element located on the input channel of the cell is activated and powered directly from the HVB [1, 4].

Figure 1. The high-voltage battery of the Chevrolet Volt.

Figure 2 shows the high-voltage battery of the hybrid electric vehicle VW Jetta. The battery is located above the rear axle. Purified air from the cabin is used for cooling. The VW Jetta high-voltage battery has 60 lithium-ion cells with a total capacity of 1.1 kW*h and allows the vehicle to drive exclusively on electric traction over 1.6 km at a speed not exceeding 70 km/h.

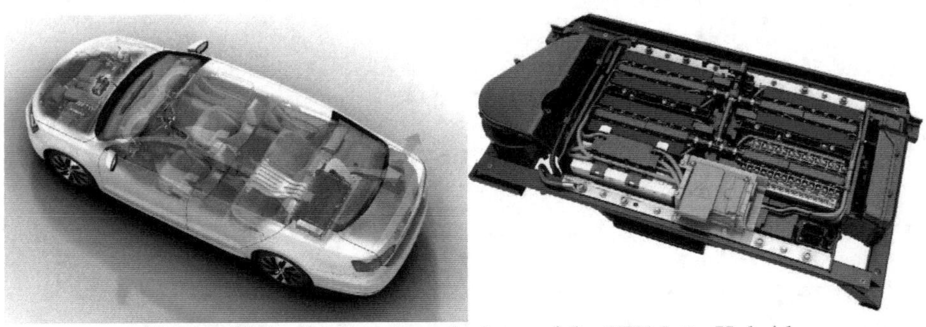

Figure 2. The high-voltage battery of the VW Jetta Hybrid.

Figure 3 shows the high-voltage battery of a BMW i3 eDrive electric vehicle with a Range Extender. The battery has a complex thermal management system, which is a grid of eight flat aluminum tubes 2.0 mm thick, which regulate the temperature of eight modules. To heat the battery, a high-voltage heating strip is used, fixed on each flat tube. Lithium-ion battery is located at the bottom of the electric vehicle. The body of the electric vehicle is made of lightweight materials. Thanks to a well-thought-out design, it was possible to place a HVB with a capacity of 22 kW * h with a voltage of 360 V, which allows the vehicle to travel 130-150 km with a fully charged battery. But the design feature of the "electric" i3 is the version of the Range Extender, which has a fuel tank of 9.2 liters of gasoline for an additional motor-generator. This allows the vehicle to travel an additional 70-80 km.

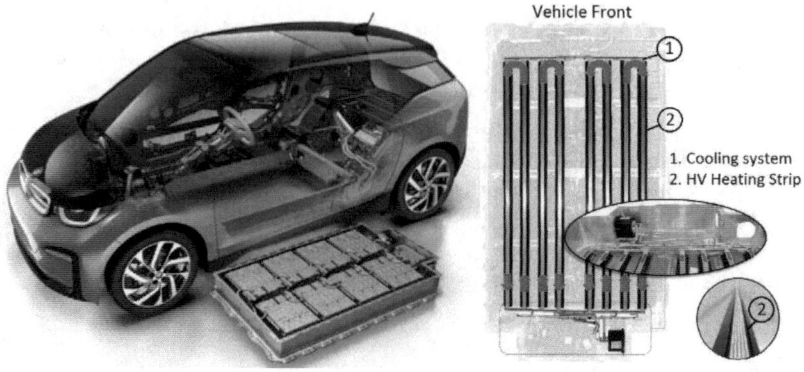

Figure 3. The high-voltage battery of the BMW i3 eDrive Range Extender.

3. Classification of high-voltage battery thermal management systems

There are different methods of cooling and heating high-voltage batteries [5, 2]:

3.1. Cooling and heating with air

Air is used for cooling and heating. Air intake can be direct from the atmosphere or from the cabin, and can also be processed by cooling or heating in the cabin microclimate system. Schemes of air thermal management systems can be passive and active. In both cases, the air is forced through the fan. Figure 4 shows the air schemes of the active and passive battery thermal management system.

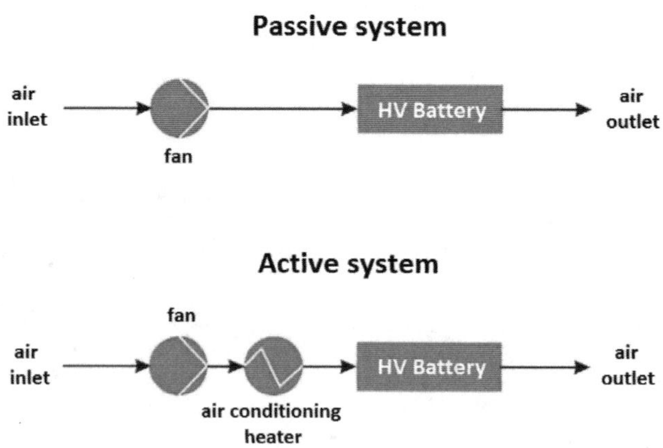

Figure 4. Schemes of the active and passive air cooling system and heating of HVB.

Note that the scheme of the air system provides the functions of heating, cooling and ventilation, and the exhaust air is not returned again. In some cases, a heat recovery system is used, which provides additional potential for energy savings. The forced scheme of the air cooling system with heat recovery is shown in figure 5.

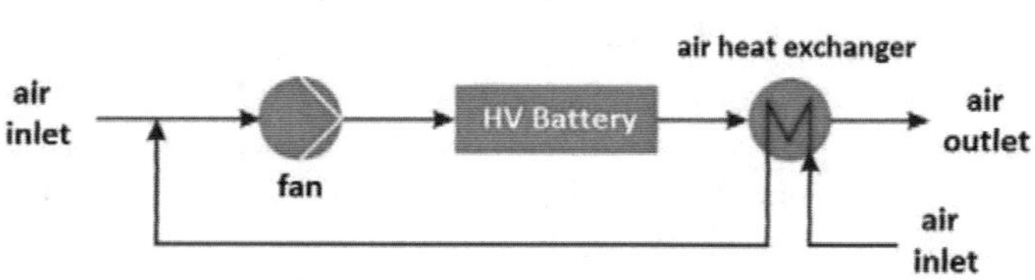

Figure 5. Forced scheme of the air cooling system with heat recovery.

3.2. Liquid cooling and heating

There are two groups of liquids used for thermal control of systems. The first is dielectric liquid that can directly contact battery cells, such as mineral oil. The second is conductive liquid that transfer heat through walls indirectly connected to battery cells, such as a mixture of ethylene glycol and water.

Figure 6 shows the circulating scheme of the passive liquid system. Coolant is circulated by a pump in a closed loop. The circulating fluid absorbs heat from the HVB and releases it through the radiator. Cooling capacity is strongly dependent on the difference in ambient temperature and HVB. Fans behind the radiator can improve cooling performance, but if the ambient temperature is higher than the temperature of the HVB or the difference between them is too small, then the system becomes ineffective.

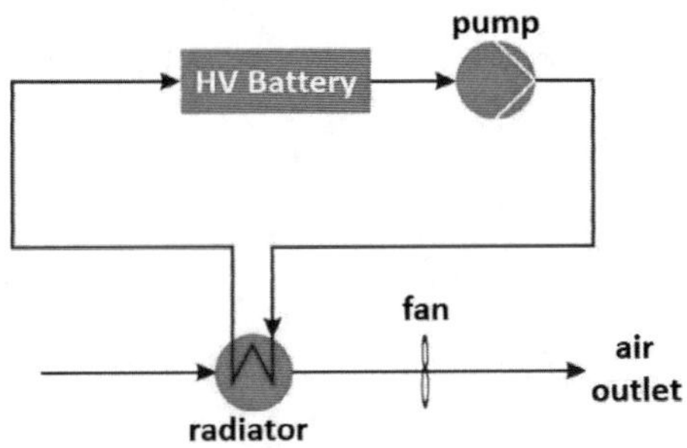

Figure 6. Circulating scheme of a passive liquid system.

In the following scheme, (see figure 7) there are two circuits. Upper circuit - primary, lower circuit - secondary. In the upper circuit of the passive liquid system, the coolant circulates by means of a pump. The lower circuit is from the microclimate system of the cabin. The upper radiator operates as an evaporator for cooling and connects both circuits. When operating in heating mode, the 4-way valve switches and the upper radiator operates as a condenser, and the lower radiator operates as an evaporator.

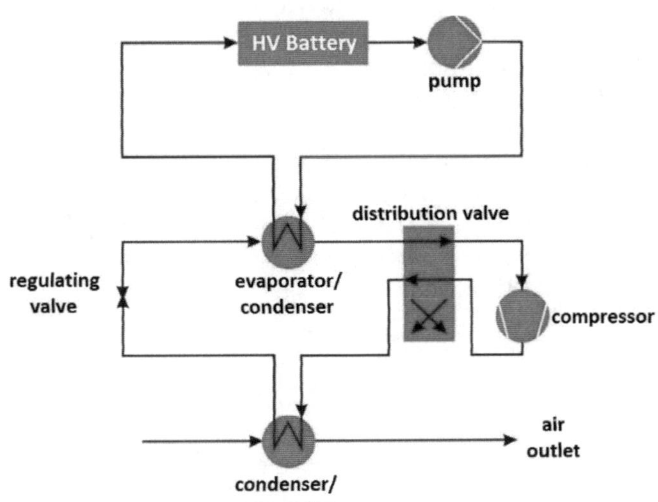

Figure 7. Scheme of the active liquid system.

3.3. Cooling with refrigerant

The principle of cooling with refrigerant is similar to that with cooling fluid. The refrigerant acts as a cooling fluid and circulates through a HVB (see figure 8).

IOP Conf. Series: Materials Science and Engineering **819** (2020) 012020 doi:10.1088/1757-899X/819/1/012020

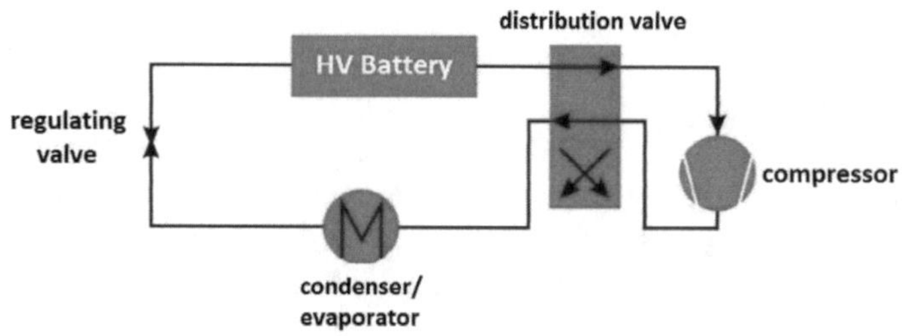

Figure 8. Cooling scheme with refrigerant.

As a result, the following advantages and disadvantages of the considered methods of cooling and heating of electrical components (for example, HVB) can be noted:

- the scheme of the forced air cooling system has a simple structure, high reliability and low cost, as well as ease of maintenance. But not in all cases, the system provides the desired temperature and takes up more space compared to liquid analogues;
- the scheme of the passive cooling system depends on the ambient temperature. Heat from the heat exchanger is dissipated into the environment. In this case, heat dissipation depends on the temperature difference between the liquid and the ambient temperature. At medium temperatures, this scheme operates efficiently, but at high ambient temperatures, it does not. The scheme of the active cooling system has good thermal characteristics, allowing to maintain the battery within the operating temperature, but due to the large number of auxiliary elements, the structure of such a scheme is complex, poorly regulated and has a probability of leaks;
- the scheme of the cooling system using a refrigerant is more efficient, since it uses the refrigerant directly to cool the system, rather than to cool the coolant, which in turn would cool the system. The disadvantages of this cooling system scheme are in the complexity of the structure and maintenance, as well as in the probability of leaks. [5, 2].

4. Experience in developing a thermal management system

The engineers of FSUE "NAMI" carried out research and development of various types of designs of the temperature control system for high-voltage batteries. According to the basic requirements for the system, the design must be efficient, compact, lightweight and meet the safety requirements.

From the point of view of design, layout, manufacturability, pouch cells are the most suitable for heat removal. Their shape allows maximum use of the external surface of the cell case for heat transfer. In addition, we used heat-conducting material, which also provides electrical insulation between the cell case and the cooling system, which can contact other electronic devices of the vehicle through the fluid.

Figure 9 shows the design of the module of 12 pouch cells, between which there are six heat-conducting plates. A cylindrical tube is fixed to each plate through which the coolant flows. The tubes are connected to the collectors of the module.

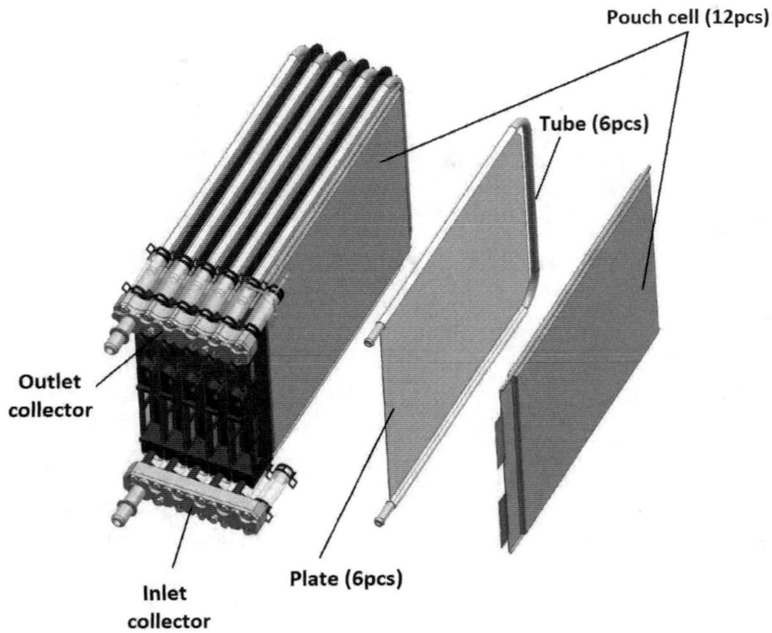

Figure 9. The layout of the module of 12 pouch cells.

The developed high-voltage battery with the capacity of 6.2 kW*h with the voltage of 340 V, has eight modules that are interconnected by common collectors. The model of the developed design of the thermal management system is shown in figure 10.

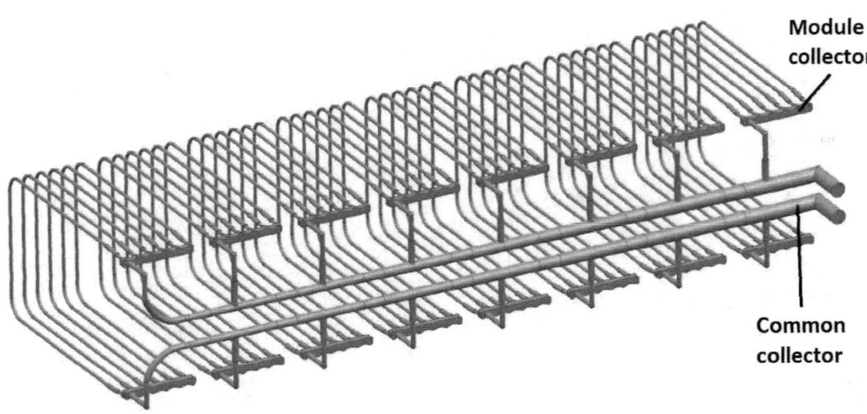

Figure 10. The model of the developed thermal management system for high-voltage battery.

To determine the effectiveness of the thermal management system, hydraulic calculations were carried out.

Design conditions:

- water is selected as the coolant;
- coolant temperature plus 40°C;
- absolute coolant pressure at the outlet of the battery 1.0 atm.

Figure 11 shows the results of the design simulation of the thermal management system for the coolant flow rate of 5 l/min.

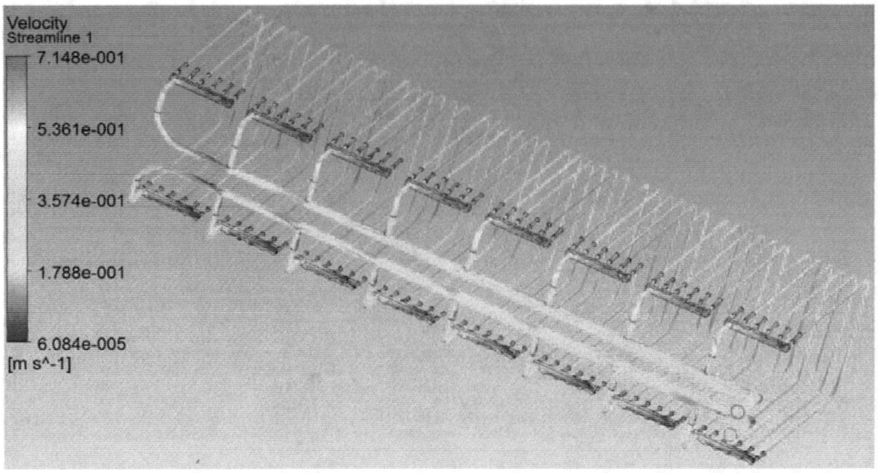

Figure 11. The streamlines and the local velocity of the coolant flow for the design model.

The colors of the streamlines show the velocity of the coolant flow, and also allow you to see the formation of turbulence and the uniform distribution of flow in different parts of the system design. In some sections of the module collectors, which are displayed in blue, it is possible to note a low flow rate of the coolant. To improve the uniformity of flow in these areas, it is enough to change the shape of the collectors.

Figure 12 shows the calculated curve of the loss of total pressure as a function of coolant flow. The graph shows that with an increase in coolant flow rate, the hydraulic resistance increases significantly. This dependence is quadratic.

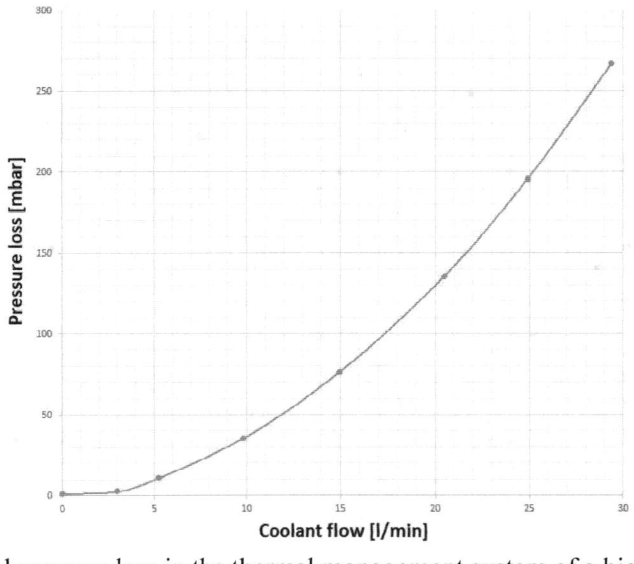

Figure 12. Total pressure loss in the thermal management system of a high-voltage battery.

Figure 13 shows the values of coolant flow rate by modules of HVB. In the presented design of the thermal management system, for its effective operation, the coolant flow must be the same for each module. Based on this, the optimal value of the distribution of coolant flow rates for eight modules will be 12.5% of the total flow. The graph shows that there is unevenness in the distribution of the coolant relative to the optimal value of 12.5% for the version №1. This indicator can be improved by making adjustments to the design of the thermal management system. In version №2, one of the common collectors was changed. Namely, one of the collectors was turned 180 degrees, after which more optimal values of the distribution of the coolant by module were obtained.

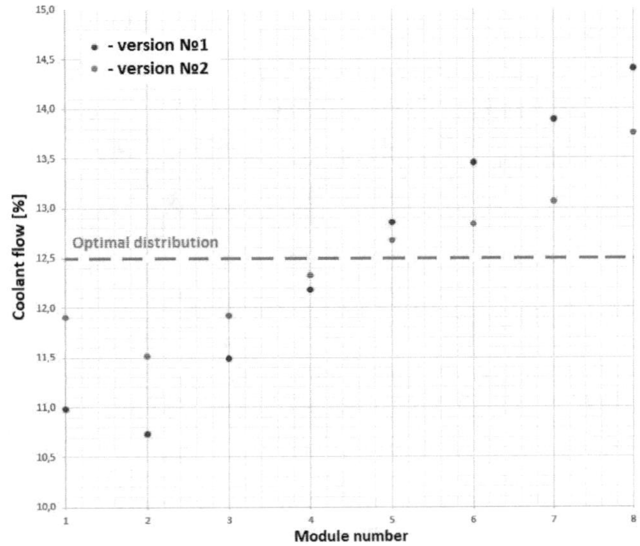

Figure 13. Distribution of coolant flow rates by high-voltage battery modules.

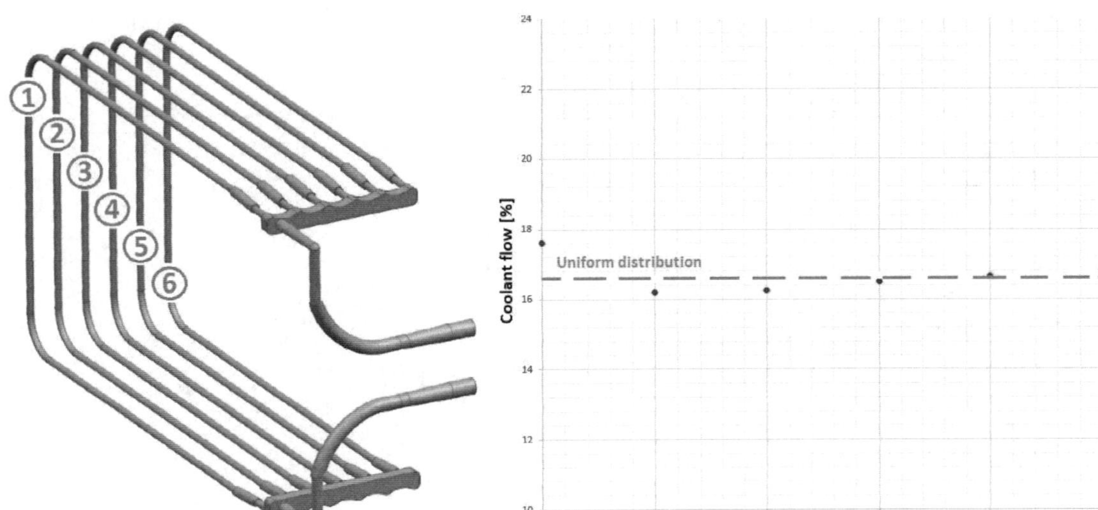

Figure 14. Distribution of coolant flow within the 1-st battery module.

Figure 14 shows the values of coolant flow within the 1-st battery module. For the thermal management system of the module consisting of 6 cylindrical tubes, the uniform distribution of the coolant is 16.6% of the total flow. On the graph, we see that the coolant is uniform distributed along the cylindrical tubes of the module, which indicates the efficiency of the design of the thermal management system of the module.

After the optimization modeling, the design documentation for the high-voltage battery was developed; its appearance and dimensions are shown in figure 15.

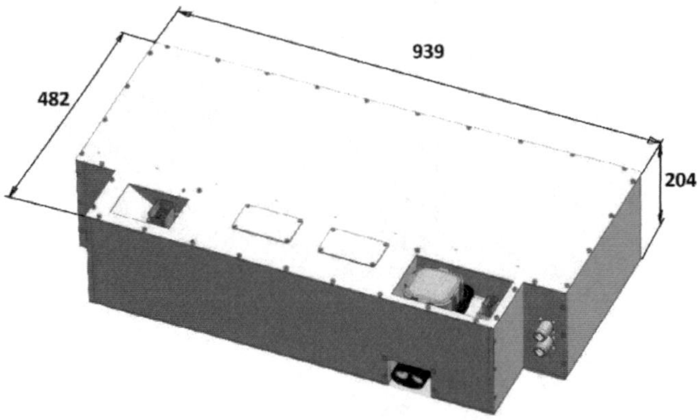

Figure 15. The appearance and dimensions of the high-voltage battery.

The developed design of the high-voltage battery is intended for operation in the HEV at ambient temperatures from minus 40°C to plus 50°C.

5. Conclusion

Thus, when developing a thermal management system, it is necessary to apply the most rational form of collector channels, ensuring minimum turbulence of discharge zones, the smoothest fluid flow without interruptions, which will ensure a more optimal distribution of coolant and the lowest hydraulic resistance and, as a consequence, higher energy efficiency of the heat exchanger.

When developing its own HVB design, the experience of global automakers of batteries for hybrid electric vehicles has been taken into account. The design of the thermal management system of the HVB has good efficiency, since provides quick removal of heat from the battery cells. The efficiency of the design allows you to get an increased power reserve and resource using the required number of cells that are distributed across the modules. The number of modules may be different and depends on the required technical parameters. Improvement of HVBs will allow the use of HEVs in different climatic conditions.

6. References

[1] Guodong Xia, Lei Cao, Guanglong Bi. 2017 A review on battery thermal management in electric vehicle application *Journal of Power Sources* **367** pp 90-105

[2] Jiling Li, Zhen Zhu 2014 Battery Thermal Management Systems of Electric Vehicles *Master's Thesis in Automotive Engineering*, Göteborg p 67

[3] Kurmaev R Kh, Karpukhin K E, Terenchenko A S, Saykin A M, Struchkov V S 2016 Creation of accumulation and storage of electrical energy for driverless electric vehicles of Russian production *International scientific journal "Trans motauto world"* Year I, Issue 1, pp 18-21

[4] Kurmaev R Kh, Umnitsyn A A, Struchkov V S, Karpukhin K E, Liubimov I A 2018 Development and research of temperature control system of a high-voltage battery of a perspective electric vehicle *International scientific journal "Trans motauto world"* Year III, Issue 4, pp 151-4

[5] Kurmaev R Kh, Struchkov V S, Tsimbaluk M A 2015 How to set temperature regimes HEV elements works on the vehicles *Trudy NAMI* **260** pp 69-80

[6] Maan Al-Zareer, Ibrahim Dincer, Marc A Rosen 2017 Novel thermal management system using boiling cooling for high-powered lithium-ion battery packs for hybrid electric vehicles *Journal of Power Sources* **363** pp 291-303

IOP Conf. Series: Materials Science and Engineering **819** (2020) 012021 doi:10.1088/1757-899X/819/1/012021

High-frequency vibration of Z-engine motor

A I Yamanin

Yaroslavl State Technical University, Yaroslavl, Russia

E-mail: a_yam@mail.ru

Abstract. It is shown that the high-frequency vibration of Z-engines with HCCI-process will not deteriorate compared to the four-stroke four-cylinder engine.

1. Main part

The implementation of a promising HCCI process is possible in the so-called Z-engine (Z-engin) [1]. In addition to improving economic and environmental performance, this motor corresponds to another of the current trends in automotive engine construction, known as downsizing (reducing the number and size of cylinders while maintaining the same power) [2]. Distinctive feature the most famous design Z-engin is a two cylinder running on two-stroke cycle with the gas exchange through the usual valves, and one compressor cylinder (possibly larger diameter), blowing air (pressurized to 3 MPa) in cylinders. The engine can be arranged inline or V-type (figure 1).

Figure 1. Downsizing: inline four-stroke four-cylinders motor and V-type Z-engine.

One of the indicators of ecological perfection of the engine is its low vibration activity in the entire frequency range. This question in relation to the Z-engine is practically not studied. In this regard, this study is undertaken. The calculation of high-frequency vibration of the engine is performed by the finite element method according to the well-known proven technique [3]. The basis for the development of the model was adopted by the design of diesel with bore 105 mm and stroke 127 mm with three cylinders, one of which was compressor (maximum air pressure was equal to 1.8 MPa). The perturbing forces were experimentally obtained gas forces [4]; the maximum pressure in the cylinder was 17 MPa.

The calculation was performed in the ANSYS/LS-Dyna software environment (figure 2). The calculation took into account the contact interaction of the engine parts and the variable nature of the loading of the pistons by gas forces.

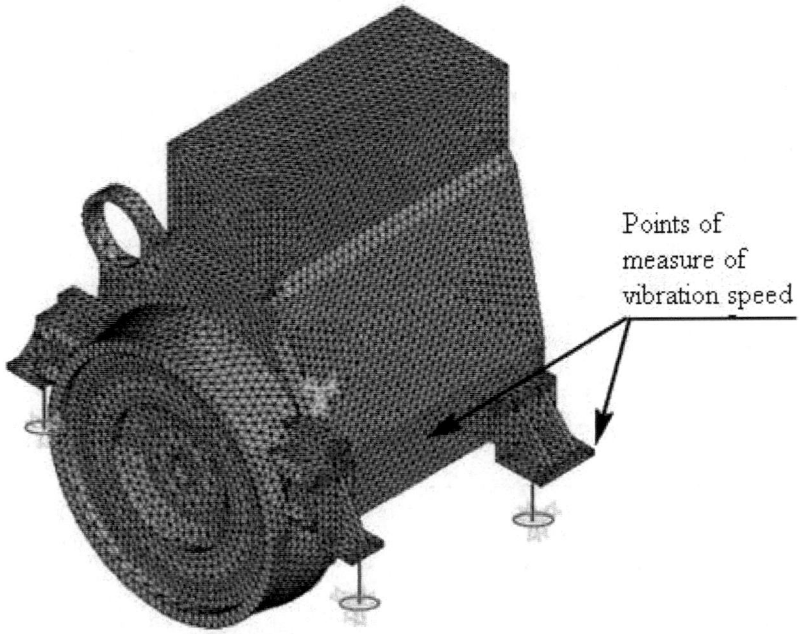

Figure 2. Finite element model of three cylinders Z-engine.

The current values of vibration displacements, vibration velocities and vibration accelerations of the support leg in the vertical direction and points on the lateral surface of the crankcase in the horizontal direction were determined (figure 3). For all the obtained characteristics, the mean square values (MSV) were subsequently calculated. The latter were compared with similar values for an engine running on a traditional four-stroke cycle; the latter were taken as a unit (table).

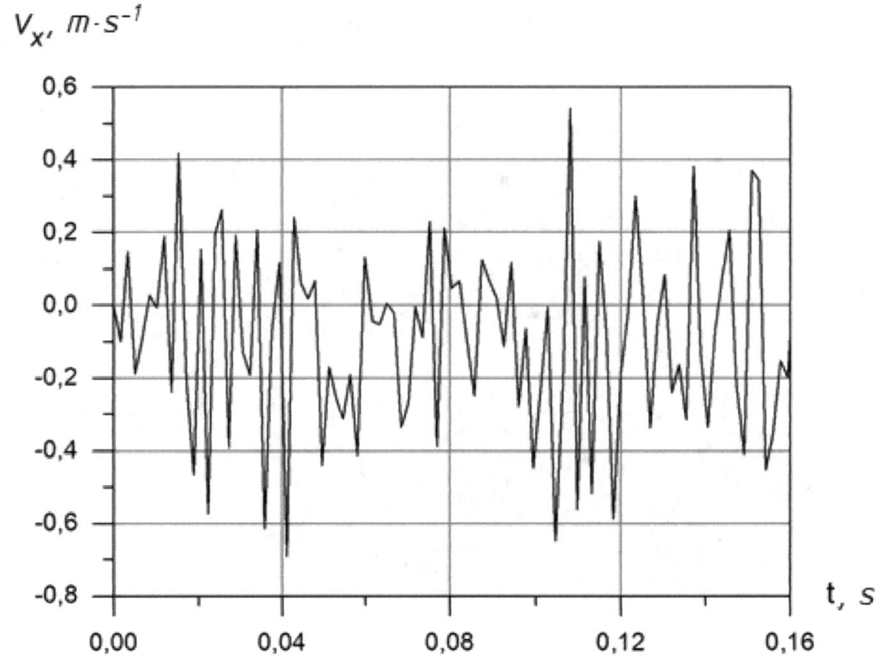

Figure 3. Horizontal vibration speed of the side wall of cylinder block of Z-engine.

Table 1. Ratio of MSV of vibration speed of Z-engine and four-stroke four-cylinder engine.

Support leg	Side wall of cylinder block
1,12	0,841

It can be seen that the implementation of the HCCI-process in the Z-engine does not worsen the parameters of its high frequency vibration. In this regard the level of structural noise compared to a traditional engine, will not increase.

References

[1] Janhunen T T, Akimov V S, Kuleshov A S and Markov V A 2014 Dvigatel` tipa z-engin s HCCI-processom *Gruzovik* **10** pp 33-43

[2] Ter-Mkrtich`yan G G 2015 Daunsajzing v dvigatelyax VAZ s prodolzhenny`m rasshireniem rabochego tela *Avtomobil`naya promy`shlennost`* **11** pp 12-16

[3] Novikov D S, Yamanin A I 2011 Razvitie metoda analiza vibracionnogo sostoyaniya transportny`h dvigatelej *Avtomobil`naya promy`shlennost`* **8** pp 11-13

[4] Akimov V S, Kuleshov A S, Markov V A, Yarovchuk A Yu and Janhunen T T 2015 Raschetny`e issledovaniya vpry`skivaniya i raspy`livaniya topliva v dvigatele s HCCI-processom *Vestnik MGTU im. N. E`. Baumana Ser. «Mashinostroenie»* **6** pp 82-95

IOP Conf. Series: Materials Science and Engineering **819** (2020) 012022 doi:10.1088/1757-899X/819/1/012022

Self-adaptive algorithm for changing braking energy recuperation of an electric vehicle based on neuro-fuzzy inference system

V Medvedev[1] **and I Rudakov**[2]

[1] Research Institute "Special Engineering" of Bauman Moscow State Technical University, Moscow, 105005, Russian Federation
[2] Department of Computer Software and Information Technology of Bauman Moscow State Technical University, Moscow, 105005, Russian Federation

E-mail: irudakov@bmstu.ru

Abstract. This paper describes an expert system prototype that changes the brake energy recuperation coefficient of an electric vehicle. Blackboard is used as the basic architectural pattern, which allowed for working with data obtained from various types of knowledge sources. This approach made it possible to coordinate data from various sources, such as the opinions of drivers, information about the route and the condition of electric vehicles. The expert system is based on the adaptive neuro-fuzzy inference system ANFIS, the rules for which are obtained by mountain clustering data from the training set. Values are averaged using the C-means algorithm.

1. Introduction

There are a number of solutions for integrated monitoring of vehicles. Similar systems are used at various truck and bus depots. However, such systems often handle only the problem of the location determination for the monitored vehicle and in no way provide qualitative adjustment of the operating modes of vehicles.

Modern electric vehicles use braking energy recuperation, which is based on the electric method of kinetic energy recuperation. One of the problems of modern electric vehicles is the inability to “flexibly” adjust the braking energy recuperation. Such a setting will increase the amount of electricity returned to the vehicle power system.

However, there is a point when executing this setting. It depends not only on a specific vehicle, but also on the operating route, as well as on the driving style.

The recovery coefficient is used as one of the parameters describing the intensity of the regenerative braking - the ratio of the maximum generated regenerative braking torque to the maximum possible regenerative braking moment, measured in the range from 0 to 1.

The authors have developed a prototype expert system that, according to available data from various sources (people, vehicles, route information, etc.), adjusts the energy recuperation coefficient of route electric vehicles (hereinafter - electric buses). The work was carried out at BMSTU with the financial support of the Ministry of Education and Science of the Russian Federation under Agreement № 14.574.21.0178 (Unique identifier: RFMEFI57417X0178).

2. Software part design

The difficulty of the problem lies in the need to set the identical energy recuperation coefficients for all electric buses of the same series on the same route, but at the same time individual for each route of operation. It is necessary that the system does not adjust each vehicle for each driver, i.e. there should be an interchangeability of the vehicles. This approach enables to increase the energy efficiency of the vehicle by taking into account the features of the operating route. It is supposed to use 3 types of knowledge:

- information on the state of the electric buses (the number of mechanical brake actuations, the average and maximum speed, total operating time on the route and the time spent at stops);
- information about the route (base route time and energy recuperation coefficient for the route);
- drivers' opinions about the vehicle settings (how hard the vehicle brakes on a scale of 1–10, where 1 is very slow, 5 is normal, 10 is too fast).

Since the system needs to analyze data coming from different types of sources [1], the Blackboard pattern [2, 3] was laid in the basis of the system architecture. The Blackboard pattern is a class interaction shown in figure 1.

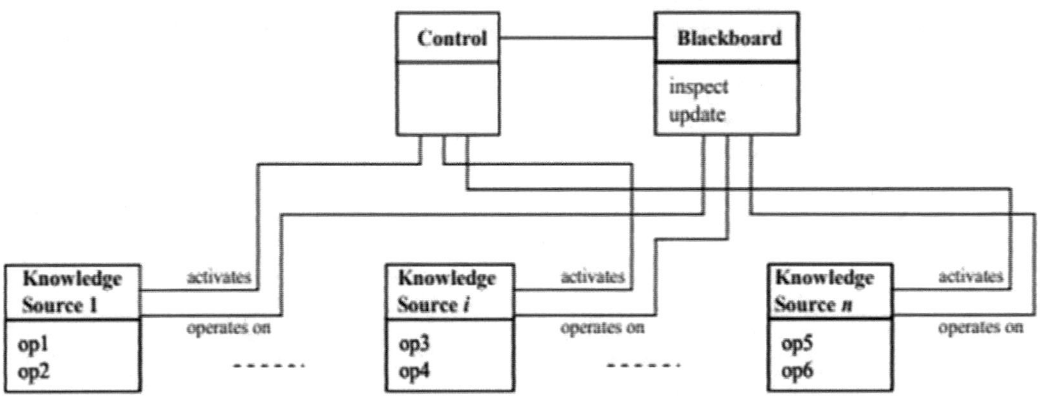

Figure 1. Blackboard pattern.

It will be shown that the pattern consists of the three main components:

- *Blackboard* – a class that contains all the relevant information from sources of knowledge. In fact, blackboard is the base of knowledge of the system. It should be noted that “knowledge” can be different in information fields, but must have a common structure - a common base class.
- *Knowledge Source* – informing entity of the system. Each entity can update its state at any time, therefore the Blackboard pattern actually implements asynchronous interaction of entities.
- *Control* – the control class – is actually an analogue of the *Model* MVP pattern. *Control* processes the data contained on the blackboard, and also generates the command for the knowledge sources to update knowledge on the *blackboard*. Thus, the expert system is implemented in the *control* class.

This approach allows us to simplify the processing of data from external sources due to the compactness of the placement of “knowledge”, and also increases the “flexibility” of the system. Knowledge management is carried out in stages.

At the first stage, the calculation of new energy recuperation coefficient for each electric bus is carried out in accordance with the received data on its status and route information. For calculation,

IOP Conf. Series: Materials Science and Engineering **819** (2020) 012022 doi:10.1088/1757-899X/819/1/012022

the network-based fuzzy inference system ANFIS is used - an artificial neural network based on the fuzzy inference system Takagi-Sugeno [4,5].

3. Algorithm description

ANFIS is one of the first versions of hybrid neuro-fuzzy networks - a feedforward neural network of a special type signal. The architecture of the neuro-fuzzy network is isomorphic to the fuzzy knowledge base. In neuro-fuzzy networks, differentiable implementations of triangular norms (multiplication and serendipitous OR) are used, as well as smooth membership functions. This makes it possible to use fast neural network training algorithms based on the backpropagation for neural-fuzzy networks configuration [7].

ANFIS implements Sugeno's fuzzy inference system as a five-layer feedforward signal propagation neural network.

The layers assignment:

- first layer - terms of input variables;
- second layer - antecedents (premises) of fuzzy rules;
- third layer - normalization of the degree of fulfillment of the rules;
- fourth layer - conclusion of the rules;
- fifth layer - aggregation of the result obtained according to various rules.

At the second stage, the calculation of the “average” selection values is carried out using the C-Means algorithm [6]. Values are calculated for a selection of energy recuperation coefficient and a sample of driver opinions. As a result, this allows us to reduce the selection size to a fixed value, while practically not losing its representativity (a feature of the algorithm is that as a result we obtain points that are the centers of the clusters, i.e. with certain accuracy reflecting the behavior of the entire cluster).

At the third stage, a popular opinion among drivers is selected and the most suitable energy recuperation coefficient is determined taking into account this opinion. After studying the information presented in the articles [8, 9, 10, 11, 12] we designed the update algorithm of the energy recovery coefficient (figure 2).

4. Discussion and results

The prototype of the system was implemented in C#. To conduct the study in accordance with the models describing the motion of LiAZ and PAZ electric buses in Moscow on T36 and T78 routes, a training sample of 1000 records was generated, as well as a test sample of 200 records. As a result of mountain clustering, 493 rules were generated.

When building a neuro-fuzzy network, the choice of membership function (MF) is extremely important. Table 1 shows network learning error obtained for various MF. The study is carried out until the difference between the total network error in the new epoch and the total error of the last epoch is no more than 1E-6.

Table 1. Network learning error obtained with various membership functions.

№	MF name	Network learning error	Epochs quantity
1.	Triangular membership function (trimf)	0,076784	34
2.	Trapezium membership function(trampf)	0,158571	84
3.	General bell membership function (gbellmf)	0,003864	120
4.	Gaussian symmetric membership function (gaussmf)	0,006108	125
5.	Product of two sigmoid membership functions (psigmf)	0,014412	290

IOP Conf. Series: Materials Science and Engineering **819** (2020) 012022 doi:10.1088/1757-899X/819/1/012022

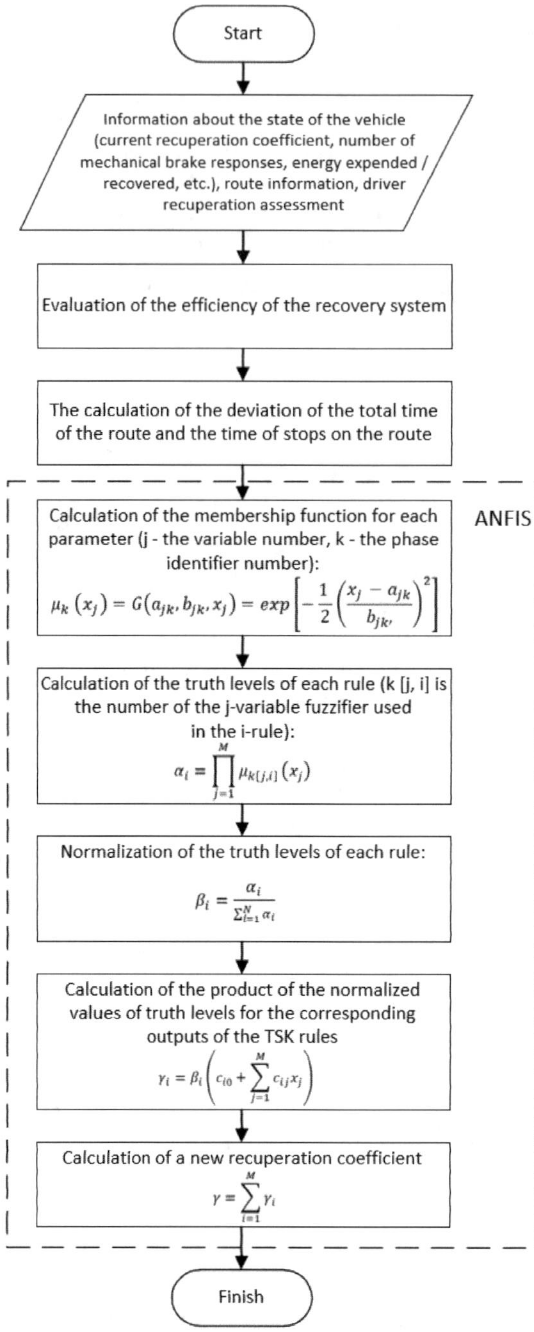

Figure 2. Update algorithm of the energy recovery coefficient.

As we can see, the use of a bell-shaped membership function gives the most accurate result. For further research, this function was chosen.

Table 2 presents the result of training a neuro-fuzzy network with a different amount of source data. In each of the tests, the network was trained in such a number of epochs that the total error of the last but one epoch differed from the total error of the last epoch by no more than 1E-5. The training time was measured when the application was running on the same Intel Core i7-7700 processor with a clock speed of 3.60 GHz.

Table 2. The result of training the neuro-fuzzy network ANFIS.

№	Training set size	Number of rules resulting from mountain clustering	Epochs quantity	Training period (min)	The average value of the error for the test set
1.	100	74	39	67	9,3E-1
2.	200	181	26	69	1,2E-1
3.	400	318	12	61	3,8E-2
4.	700	427	4	78	2,3E-2
5.	1000	609	3	94	8,2E-3

5. Conclusion

As a result of the system's work, the implementation of the dependence of the energy recuperation coefficient on various types of knowledge sources was obtained. Thus, the implementation of this prototype system allowed us to make an assumption about the possible applicability of systems of this class.

References

[1] Giarratano J and Riley G 2007 *Expert Systems: Design Principles and Programming* PWS Publishing Co. Boston MA USA p 1152

[2] Buschmann F, Meunier R, Rohnert H, Sommerlad P and Stal M 1996 *Pattern-Oriented Software Architecture: A System of Patterns* John Wiley & Sons Ltd. England p 467

[3] Ortega-Arjona J L and Fernandez E B 2008 The secure blackboard pattern *Proceedings of the 15th Conference on Pattern Languages of Programs* **22** pp 1-5

[4] Suprata W and Alhasa K M 2016 *Modeling of Tropospheric Delays Using ANFIS* Springer International Publishing. Germany p 109

[5] Walia N, Tiwari K and Malhotra R 2015 Design and Identification of Tuberculosis using Fuzzy Based Decision Support System *Advances in Computer Science and Information Technology* **2** pp 57-62

[6] Sreenivasarao V and Vidyavathi S 2010 Comparative Analysis of Fuzzy C- Mean and Modified Fuzzy Possibilistic C -Mean Algorithms in Data Mining *IJCST* **1** pp 104-106

[7] Matlab Exponenta [Online]. – Available: http://matlab.exponenta.ru/fuzzylogic/book1/1_7_5_7.php (February 8 2019)

[8] Selin P S, Tsurkov V I and Gurchenkov A A 2017 An algorithm for constructing a hereditarily minimax network with predefined vector of node degrees *Herald of the Bauman Moscow State Technical University, Series Natural Sciences* **1** pp 43-58

[9] Vanin A, Aleshin S, Nasirov R, Novikov D and Tulsky V 2016 Investigation of voltage control at consumers connection points based on smart approach *Information* **7** p 42

[10] Korobets B N 2016 Models for technology programmes within an intellectual property management system *Herald of the Bauman Moscow State Technical University, Series Natural Sciences* **6** pp 135-142

[11] Ivashkin Y A, Blagoveschensky I G and Nikitina M A 2018 Neural network and agent technologies in the structural-parametric modeling of technological systems *CEUR Workshop Proceedings* **2098** pp 169-180

[12] Golubkov G V, Manzheliy M I, Berlin A A, Morozov A N and Eppelbaum L V 2018 Problems of satellite navigation and remote sensing of earth's surface *Herald of the Bauman Moscow State Technical University, Series Natural Sciences* **1** pp 61-73

Test bench for experimental research of hybrid powertrain algorithms based on rapid control prototyping unit

D Butarovich, A Smirnov and G Skotnikov

Wheeled Vehicles Department of Bauman Moscow State Technical University, Moscow, 105005, Russian Federation

E-mail: buta73@bmstu.ru

Abstract. The transition from simulation to a real object in case of developing control algorithms for complex systems can be difficult for several reasons. In particular, this process can take a long time due to the lack of hardware of control systems. One of the possible solutions is rapid control prototyping technology. This article discusses the experience of developing a test bench for research of series hybrid powertrain for a bus, in particular, the development of a test bench control system. The use of rapid control prototyping technology is described to the tasks of control system checking and debugging on real objects without hardware developing.

1. Introduction

Today, the traditional initial step of R&D is the stage of mathematical simulation to determine the optimal parameters of the object under research. At this stage, high level control algorithms are being developed, the details of which depend on the assumptions made in the simulation model. Most often, at this stage, issues of integrating the elements of the object under research into the general structure are not considered, the problems of information exchange between system elements are not addressed. Therefore, the next step is to make a prototype of the object or system under research that allows checking the control algorithms developed at the stage of simulation, refine and supplement them taking into account the real behavior of the control object. However, making a prototype requires development of controllers and writing application software. This process takes a lot of time.

It is possible to reduce the time of algorithms debugging on the prototype of the object or system under research by applying the technology of rapid prototyping of embedded control systems. The main idea of the technology is a quick transition from the model of the control system in a simulation environment to controlling a real object or its prototype, while the control system from the development environment is transferred to a rapid control prototyping unit without manual programming. Rapid Control Prototyping Unit (RCPU) is a computer that is designed to perform a wide range of tasks in real-time. The ability to solve various problems is achieved using the input/output modules of analog and digital signals using various data transfer protocols (SPI, I2C, CAN et al.). The real-time problem solution is achieved by a computer running a real-time operating system.

This technology allows solving a complex problem: determine the parameters of the hardware of the control system, check and debug control algorithms. This approach is widely used to solve problems in various fields of science and technology. For example, Lam, Q.L., Bratcu, A.I., Riu, D.,

Boudinet, C., Labonne, A., Thomas, M. used RCPU RT-LAB for development of hybrid power generation system control and analysis of its sustainability [1], Hong, R.L.J., Mahyuddin, M.N., Arshad, M.R., Koo, Y.C. applied RCPU NI MyRIO-1900 for controlling the robot arm for educational purposes [2], Song Hu, Stefano d'Ambrosio, Roberto Finesso, Andrea Manelli made a comparison of models for model-based combustion control in diesel engine using ETAS ES910 [3], Razman Ayop and Chee Wei Tan checked a controller that produces a stable and fast converging operating point for the photovoltaic emulator using RCPU dSPACE ds1104 [4], Davide Tavernini, Mathias Metzler, Patrick Gruber and Aldo Sorniotti tested the developed traction control algorithm for electric vehicle via RCPU dSPACE MicroAutoBox [5]. Rapid Control Prototyping technology has also been used in studies [6 - 10].

2. The goal of the research

Currently, BMSTU (with financial support from the Russian Ministry of Education and Science under Agreement No. 14.574.21.0178, unique work identifier: RFMEFI57417X0178) is implementing research and development project for a hybrid bus powertrain. The project aims to create a suburban bus with the possibility of movement in Zero Emission mode within the city. To ensure competitiveness, this bus have to meet the following requirement:

- possibility of operation without the use of charging stations,
- the ability to move only by battery energy,
- the lowest cost of components.

One of the solutions to achieve the fulfillment of the above requirements is the use of a series type of hybrid powertrain. This type of hybrid powertrain differs in driving wheels only with the help of a traction electric motor (TEM). Thus, there is no mechanical connection between the drive wheels of the bus and the internal combustion engine (ICE), which is used only to drive a generator that charges batteries and provides the necessary power for TEM in heavy traffic conditions. The lack of a mechanical connection between the ICE and the drive wheels also makes it possible to operate the bus in Zero Emission mode.

The key factors in developing of the bus with a serial hybrid transmission that determine its future performance are the right selection of transmission unit parameters and the quality of the control system. In this regard, verifying and adjusting the technical solutions adopted at the early stages of the project is an important task before a full-size prototype of the bus will be made. To solve this problem, it was necessary to make a test bench for researching a prototype of the hybrid powertrain.

3. Methodology

The parameters of the powertrain units, as well as the basic control algorithm, were determined using a simulation mathematical model. A description of the model and an approach to determining the parameters of its subsystems are given in [11-13].

Rapid prototyping technology was used to solve the problem of creating a control system for hybrid powertrain of the bus.

Figure 1 illustrates the transition from a simulation model of the bus with a hybrid powertrain to a test bench. The test bench consists of real powertrain units (TEM, generator, ICE, batteries), the mathematical models of which are highlighted in red in the block diagram. The control system model of hybrid powertrain was transferred to the RCPU.

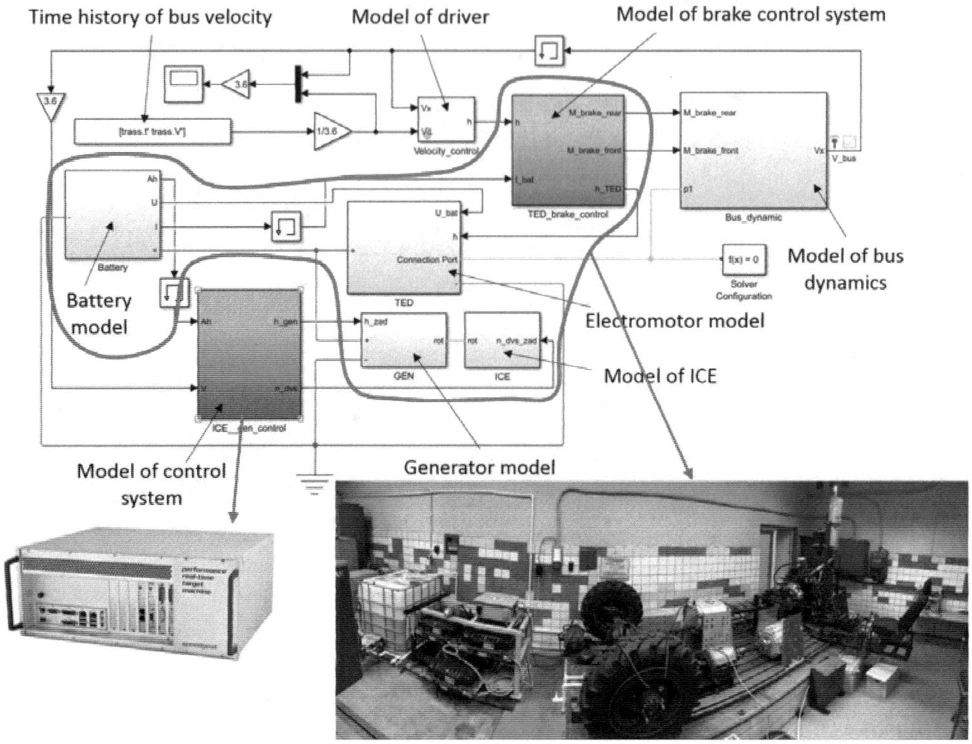

Figure 1. From Simulation Model to System Prototype.

To control the experimental model of the powertrain at the test bench, the control system developed at the stage of the simulation was refined, due to the fact that the bus model did not take into account the methods of data transfer between the unit controllers [12], as well as low-level algorithms for their start-up and initialization. Control system software consists of:

- control algorithms of hybrid powertrain developed using a simulation mathematical model (figure 2),
- algorithms responsible for the initialization and start-up of control systems for the units of the test bench (inverters of the TEM and generator, battery management system, ICE ECU), ICE start and stop algorithm (figure 3),
- algorithm for switching stand operation modes,
- blocks of reception and transmission of analog and digital signals, including CAN signals.

As shown above, in the transition from control algorithms developed based on the simulation model to the control system of the test bench, the amount of software increased several times. Rapid prototyping technology has reduced the time required to develop software for test bench controlling due to the lack of the need to switch to another development environment — test bench control system, as well as hybrid powertrain control algorithms, were developed in MATLAB\Simulink. This allowed avoiding errors that could occur when transferring basic algorithms to another development package.

The drive wheels create resistance on the TED during the movement of the bus. At the test bench, this resistance is simulated using a load device, which is also controlled by the RCPU. To reproduce the movement of the bus along the driving cycle at the test bench the load device creates a moment of resistance, obtained in the form of a time dependence by simulation.

With the help of the developed test bench, studies of the operation of the hybrid powertrain prototype in various modes were carried out, for example, movement with a low battery state of charge (SOC), movement in Zero Emission mode, the battery charge during the operation of ICE in various modes, transient processes, such as ICE start-up via generator, were analyzed. Bellow, there are the graphs of changes in the parameters of the test bench units, obtained in the course of research.

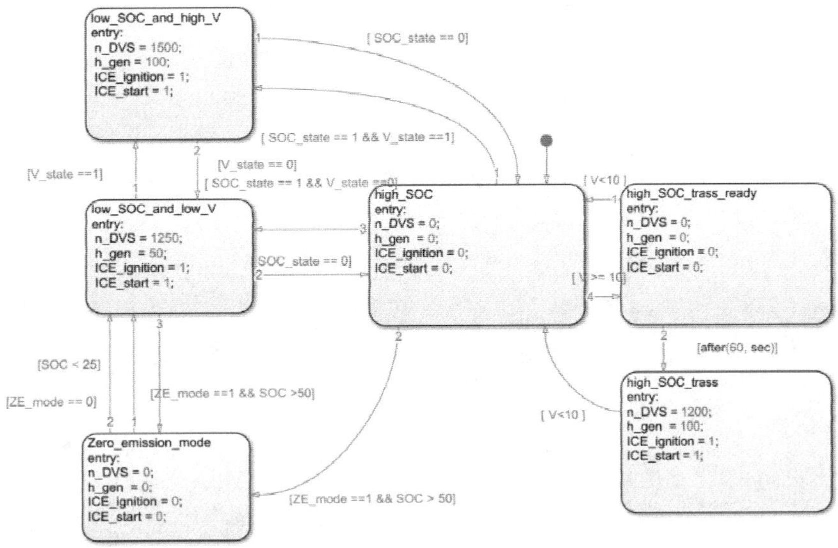

Figure 2. StateFlow Diagram of Hybrid Powertrain Control System.

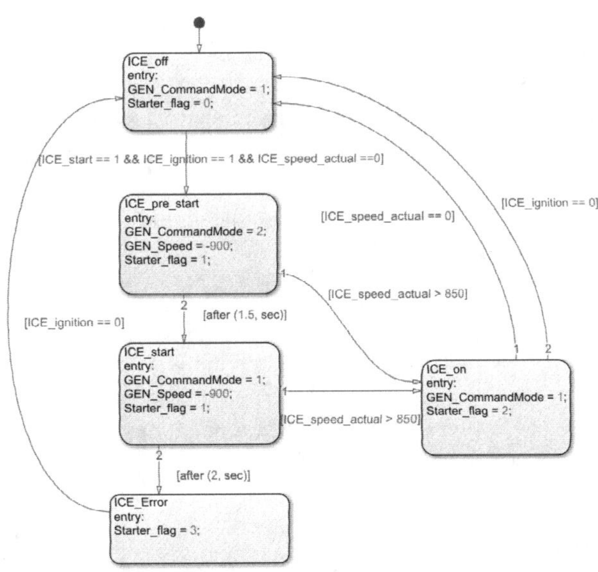

Figure 3. StateFlow Diagram of ICE Start-up via Generator.

Figure 4 shows a graph of the ICE and generator speeds changing in the ICE start mode (155 - 157 s), followed by maintaining the speed when operating under load created by the generator (157 - 211 s) and stopping (212 s).

Figure 5 shows the electric power developed during the start of the ICE, obtained according to the voltage and current sensors installed on generator and battery. Figure 6 shows the change in currents on the TEM, generator and battery:

1) 0 – 1250s Zero Emission Mode,
2) 1250 – 1350s Power Follower Mode,
3) 1350 – 1500s Battery Charging via ICE.

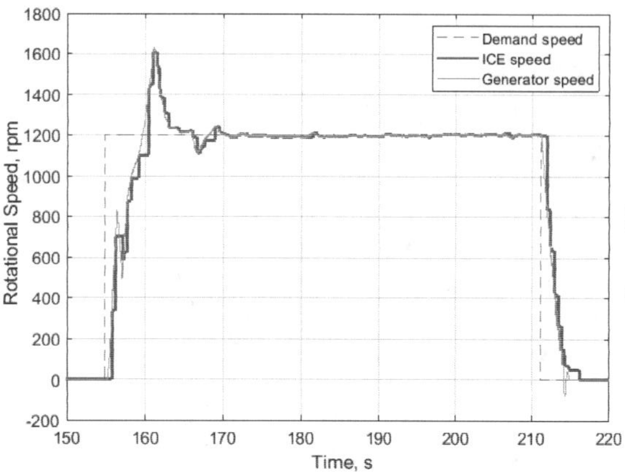

Figure 4. Rotational Speed of ICE and Generator.

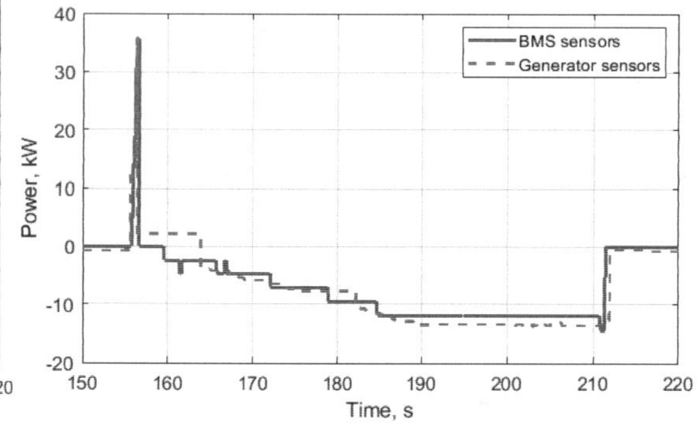

Figure 5. Charge and Discharge Power at ICE Startup and Operation.

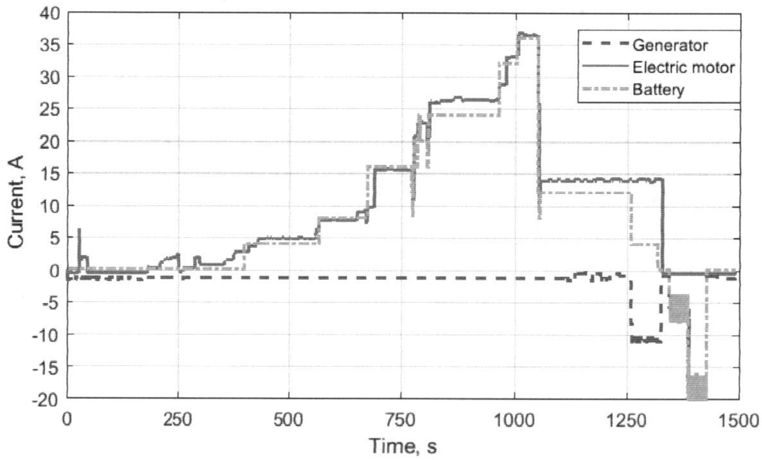

Figure 6. Current of Units in Different Modes of Operation.

4. Conclusion

1. The use of rapid control prototyping technology made it possible in a short time to create an experimental test bench for researching a prototype of hybrid powertrain.
2. Hybrid powertrain control algorithms developed using simulation mathematical modeling were tested on real objects, their performance was proved.
3. The control system of the hybrid powertrain prototype at the test bench developed based on algorithms synthesized at the stage of simulation can be integrated without changes into the general bus control system.

4. Integration of hybrid powertrain units into the general system of the traction drive of the bus using the test bench which was controlled by RCPU made it possible to draw up objective technical requirements for the control system hardware.

Reference

[1] Lam Q L, Bratcu A I, Riu D, Boudinet C, Labonne A and Thomas M 2020 Primary frequency H∞ control in stand-alone microgrids with storage units: A robustness analysis confirmed by real-time experiments *International Journal of Electrical Power and Energy Systems* **115**

[2] Hong R L J, Mahyuddin M N, Arshad M R and Koo Y C 2020 The design and development of an interactive teaching tool for robotics and control system course using rapid control prototyping solution *Lecture Notes in Mechanical Engineering* **2020** pp 515-25

[3] Song Hu, Stefano d'Ambrosio, Roberto Finesso and Andrea Manelli 2019 Comparison of Physics-Based, Semi-Empirical and Neural Network-Based Models for Model-Based Combustion Control in a 3.0 L Diesel Engine *Energies* **12**

[4] Ayop R and Tan C W 2019 Rapid prototyping of photovoltaic emulator using buck converter based on fast convergence resistance feedback method *IEEE Transactions on Power Electronics* **34** pp 8715-23

[5] Tavernini D, Metzler M, Gruber P and Sorniotti A 2019 Explicit Nonlinear Model Predictive Control for Electric Vehicle Traction Control *IEEE Transactions on Control Systems Technology* **27** pp 1438-51

[6] Quintero-Manriquez E, Sanchez E N, Harley R G, Li S and Felix R A 2019 Neural Inverse Optimal Control Implementation for Induction Motors via Rapid Control Prototyping *IEEE Transactions on Power Electronics* **34** pp 5981-92

[7] Trujillo O A, Toro-García N and Hoyos F E 2019 PID controller using rapid control prototyping techniques *International Journal of Electrical and Computer Engineering* **9** pp 1645-55

[8] Broenink T and Broenink J 2019 Rapid development of embedded control software using variable-detail modelling and model-to-code transformation *Proceedings - European Council for Modelling and Simulation ECMS* **33(1)** pp 151-7

[9] Jin L, Sun Z, Liu Z, Li J and Yang M 2019 Simulation Study on State of Charge Estimation of Lithium-ion Battery at Different Temperatures *Qiche Gongcheng/Automotive Engineering* **41** pp 590-598 and 606

[10] Buzunov N V, Kotiev G O and Gorelov V A 2019 Implementation of the interaction of the steering wheel loader control system of the remote-controlled wheeled vehicle operator interface with a real-time simulation model *IOP Conf. Series: Materials Science and Engineering* **534**

[11] Gorelov V A, Butarovich D O, Staduhin A A and Skotnikov G I 2019 Simulation-based identification of the parameters of a minibus hybrid powertrain *IOP Conf. Series: Materials Science and Engineering* **534**

[12] Kotiev G O, Butarovich D O and Kositsyn B B 2018 Energy efficient motion control of the electric bus on route *IOP Conference Series: Materials Science and Engineering* **315**

[13] Kotiev G O, Miroshnichenko A V, Stadukhin A A and Kositsyn B B 2019 Estimating operation modes for the individual wheel electric drive of the all-wheel drive vehicle with the use of the driving simulator *IOP Conf. Series: Materials Science and Engineering* **534**

[14] Astapenko D A, Butarovich D O and Smirnov A A 2019 Development of the upper-layer electronic control unit based on a production-type unit for a commercial vehicle *IOP Conf. Series: Materials Science and Engineering* **534**

Acknowledgments

The research was carried out at BMSTU with the financial support of the Ministry of Education and Science of the Russian Federation under Agreement 14.574.21.0178 (Unique identifier: RFMEFI57417X0178).

IOP Conf. Series: Materials Science and Engineering **819** (2020) 012024 doi:10.1088/1757-899X/819/1/012024

Basic principles of automotive modular battery management system design

A. Zaitsev[1], D. Butarovich[2] and A. Smirnov[3]

Wheeled Vehicles Department, Bauman Moscow State Technical University, Moscow, Russia

E-mail: [1] zaytsevas@bmstu.ru, [2] buta73@bmstu.ru, [3] smr_a@mail.ru

Abstract. Nowadays, manufacturing of electric vehicles remains one of the most dynamically developing industries all over the globe. The issues of battery efficiency improvement by a suitable battery cell structure selection and battery control system enhancement are of the highest priority in the process of the battery design. Battery management systems (BMS) with modular structure have become the most popular as control systems in electric vehicle battery applications. The paper describes design principles of such type of BMS and necessary hardware.

1. Introduction

Nowadays, manufacturing of electric vehicles, mobile energy storage devices and related infrastructure are among the most dynamically developing industries. The transition from internal combustion engine vehicles to hybrid electric vehicles [3, 4, 14, 15] (HEVs) or battery electric vehicles [18] (BEVs) is becoming increasingly apparent. As a result, increasingly stringent demands are made for EV control algorithms [2] and mobile energy storage parameters such as energy efficiency, energy density, charging rate [11], dimensions and mass etc.

As the battery management system is required for the most efficient functioning of the modern mobile high voltage energy source, the selection of optimal parameters, layout and structure [1] of both the source itself and the control system is extremely important. But the factors defining the choice are often not obvious. Further we will use the term "battery" to determine an electrical energy source.

In this article, the principles and basic hardware for the BMS with modular topology design are listed and also solutions of worldwide famous EVs manufacturers are reviewed.

2. Battery layout

Before addressing BMS, we shall give a description of the main structural elements of a battery (see fig. 1). The battery is based on the chemical current source – a lithium-ion cell, which is the simplest structural element of the battery and has a relatively low capacitance and voltage (depending on its chemical composition [5]). Today, battery cells are being mass manufactured in 3 form-factors: prismatic cells, cylindrical cells and pouch cells. The form of the cells largely affects the battery mass and dimensions and also defines the layout of the battery cell assemblies described further.

Several cells connected in parallel form an element. A series of connected elements form a subpack.

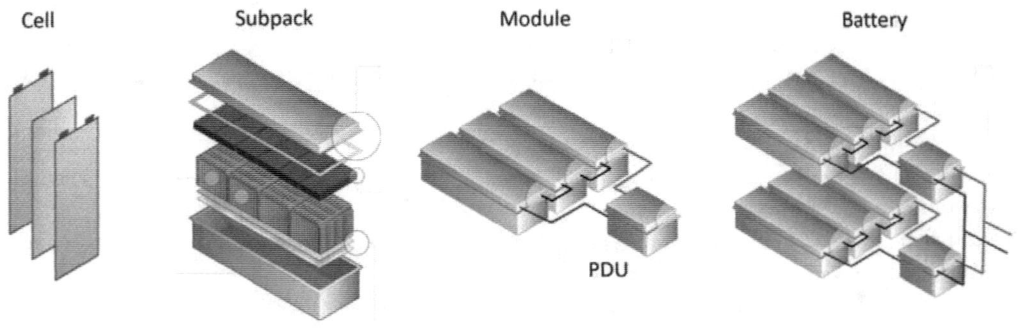

Figure 1. Basic structural parts of the battery.

Subpacks are usually connected in series and placed in a common housing to achieve the voltage required by the traction motor. A possible traction battery configuration of the electric vehicle is shown in figure 2.

As a rule, the module consists of a subpack series, thermal management equipment [13, 17] and a power distribution unit (PDU), which performs the functions of safe load connection and disconnection, precharge of high voltage circuit in the process of connection and communication with the vehicle control unit (VCU).

Any of the structural components described above can be expressed by the formula NpMs, where N is the number of cells connected in parallel in each element and M is the number of elements connected in series (for example 3p4s, 3p88s, 1p10s, etc.). So, NxM equals to the total cell number in the specific structural component of the battery.

Note that the described terms are not universally accepted and vary from manufacturer to manufacturer. Often, additional intermediate structural elements are added. Nevertheless, the described principles of the battery cell layout are general [6].

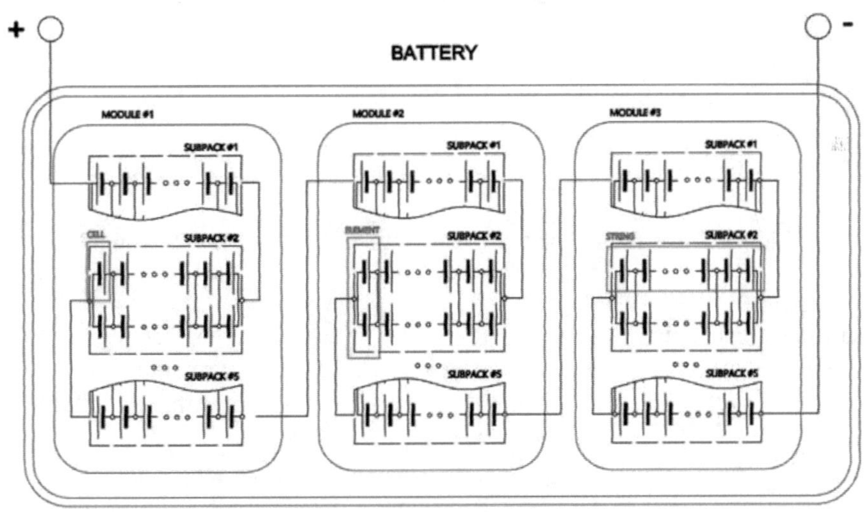

Figure 2. A possible traction battery configuration of the electric vehicle (subpack formula here is 2p10s, module formula is 2p50s, battery formula 2p150s).

3. Modular BMS

Modular BMS [12] have become most popular because of their good expandability (a possibility to connect additional structural elements to the battery system). A modular BMS consists of the following structural elements [7]:

1) master controller,
2) slave controller,
3) additional control units,
4) communication channel "Vehicle-Master",
5) communication channel "Master-Slave".

Possible layout of the listed structural element inside the battery is shown in figure 3.

Usually, each subpack has its own Slave controller and the Master is a part of the PDU [10]. Proximity of Slaves to lithium-ion cells and the Master to the power control devices of the PDU combines well with the functions of these controllers, which are described further.

Note that the shown layout is widely used and easy to understand but it's not the only possible variant. Battery manufacturers often combine different components, for example a Master and a Slave on the same PCB, or use subpacks with different voltages, etc., based on characteristics and form of the battery.

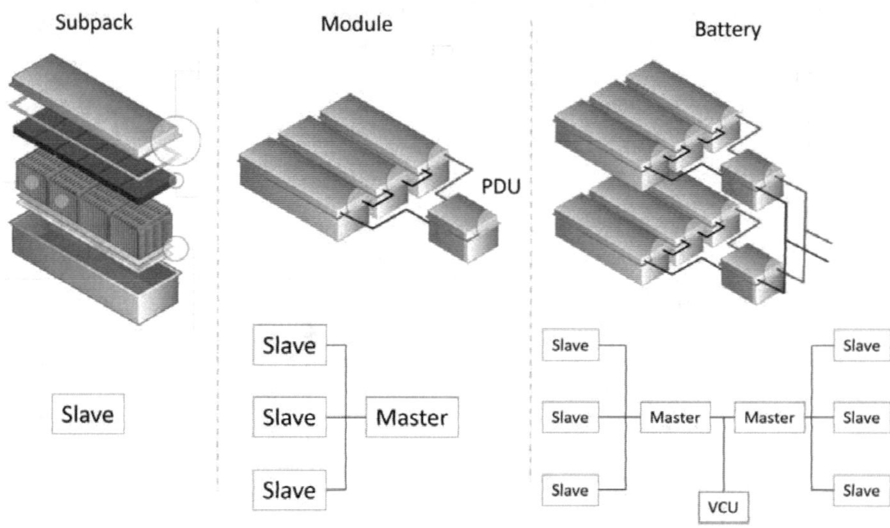

Figure 3. Layout of the main elements of the modular BMS in a battery.

4. Master controller

Master controller implements "high-level" functions. This term means that the controller doesn't work with the battery cells directly but collects and analyses the information received from the Slave controllers. Based on this information, the master invokes algorithms that connect/disconnect the load to the battery, charge the battery, protect it, etc.

To implement the described functions, the master controller is directly connected with the following control devices located in the PDU: main HV switches (contactors) S1 and S2, precharge contactor S3 and current sensor A1 (fig. 4). The figure also shows High voltage measurement units (V1 and V2) and the Insulation Control Unit (ICU). Manual Service Disconnection device (MSD) S4 allows the user to disconnect the HV circuit manually in the process of maintenance.

The process of the battery load connection follows this algorithm:

1) receiving the battery load connection request via the Vehicle-Master channel,
2) battery cells state test (diagnostics of voltage, temperature, etc.),
3) start of precharge session (in case of faults absence),

   – precharge contactor (S3) closing,
   – negative pole main contactor (S2) closing,
4) waiting for 90-95% equalization of battery and load voltage,
5) positive pole main contactor (S1) closing;
6) precharge contactor (S3) opening.

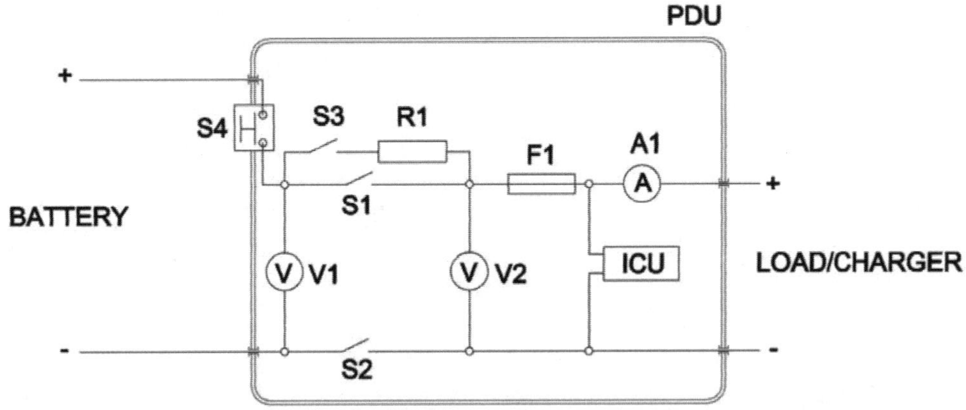

Figure 4. PDU schematics: S1, S2 – main HV contactors, S3 – precharge contactor, S4 – Manual Service Disconnection device (MSD), R1 – precharge resistor, F1 – fuse, A1 – current sensor, V1 – battery output HV measurer, V2 – load HV measurer, ICU – Insulation Control Unit.

Precharge process in this algorithm is necessary. Because of the capacitance of load (of traction invertor, of DC-DC convertors, etc.), main contactor closing without precharge will cause a very high peak current at the first moment. The occurrence of such a current can lead to critical failures of the main contactors (for example, to contact welding).

During the execution of this algorithm, every stage goes with control of Slaves faults, battery current, high voltage circuit insulation and state of switches (S1, S2, S3).

5. Slave controller

A slave controller is a device which implements "low-level" functions. This means that each Slave controls battery cells balancing, measures cell voltage, monitors and calculates the cell parameters such as SOC (State of Charge) and SOH (State of Health), monitors cell faults, etc.

Schematics of the Slave controller is usually more complicated than the schematics of the Master controller because of the connection of large amount of cells (up to several dozens) to one PCB. As a rule, Slave controller power supply is provided directly from the connected cells so that the energy consumption is evenly distributed between the cells, therefore the supply voltage can be up to several tens of volts. Another important problem is that we have to measure voltage differentially for every cell. A conventional controller cannot provide such functionality so special cell monitoring ICs are used in these applications, some of them are listed in table 1 (also see [3]).

That means the microcontroller which implements algorithms of balancing, voltage measurement, etc. is required to work in conjunction with the described ICs. Such ICs have a special interface (SPI, UART, etc.) to exchange the monitoring data with the microcontroller using a special protocol. Also the microcontroller must communicate with other BMS controllers and vehicle devices via automotive interfaces such as CAN, automotive Ethernet or others.

Table 1. Lithium-ion cell monitoring ICs.

Model	Manufacturer	Number of connected cells	Max. number of devices in daisy chain	Possible chemical composition of connected cells
BQ76PL455	TI	6..16	16	All types
BQ76PL536		3..6	32	All types
TC6801	Linear Tech.	up to 12	16	All types
LTC6802		up to 12	16	All types
LTC6804		up to 12	16	All types
LTC6813		up to 18	16	All types
L9763	LG Chem	up to 10	No available data	
ISL78600	Renesas	up to 12	No available data	LCO, LMO, LFP
ISL94202		3..8		
ISL94208		4..6		
MC33771	NXP	7..14	No available data	All types
D15110	(NEC) Renesas	4	32	No available data
T2-A2 PZ38984	Tesla	No available data		

Note that the ICs shown in table 1 don't include any programmed algorithm for the working with the cells (for instance, a balancing algorithm). These ICs can measure the voltage of each connected cell, process readings of the temperature sensors, operate as triggers and signal when the voltage is outside the allowable range on one or several cells, as well as control external balancing circuits.

As a rule, such a connection of the microcontroller with one or several specialty chips is organized as it is shown in figure 5. Each Slave contains one microcontroller and one, two or more daisy chained cell monitors.

As it is shown in table 1, most cell monitoring ICs are able to be connected in daisy chain up to 32 ICs over differential galvanically-isolated interface. In practice topologies with one or two CM are most popular. This choice is based on high reliability requirements for Slave. Obviously, the large number of cell monitors connected in a daisy chain increases fault probability and the time of data reading and writing.

As it was noted before, one of the main functions of the Slave controller is charge balancing [16]. The balancing function provides the most efficient use of the battery energy by keeping the state of charge of the cells approximately equal. The difference between the SOC of the cells in same operation conditions appears because of the inequality of the cells leakage currents [8] due to small cell-to-cell manufacturing process variations. Nowadays, two types of battery cells balancing are used: passive and active balancing.

IOP Conf. Series: Materials Science and Engineering **819** (2020) 012024 doi:10.1088/1757-899X/819/1/012024

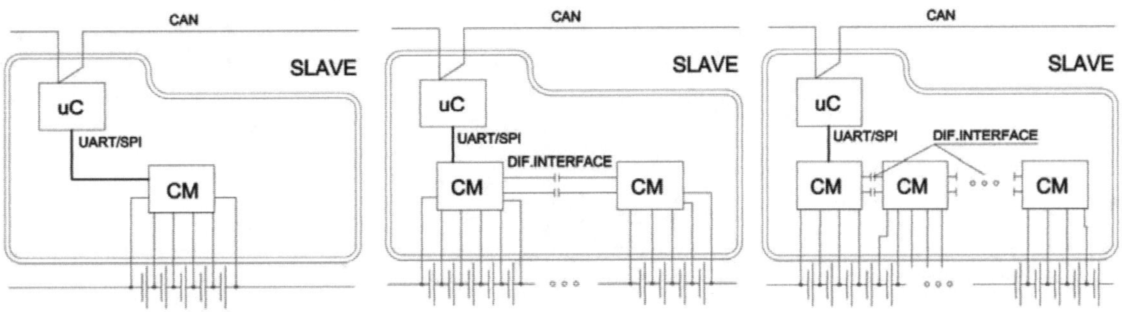

Figure 5. Slave module layout options: a – one cell monitor on each Slave; b – daisy chain of two cell monitors on each Slave; c – daisy chain of several ICs on each Slave.

In case of passive balancing the Slave controller finds the cells having the highest charge and connects them to the balancing load to decrease their SOC by converting the cell energy to heat (fig. 6). Advantages of passive balancing are relative simplicity of design and balancing control algorithms. Main disadvantage is inefficient energy to heat conversion.

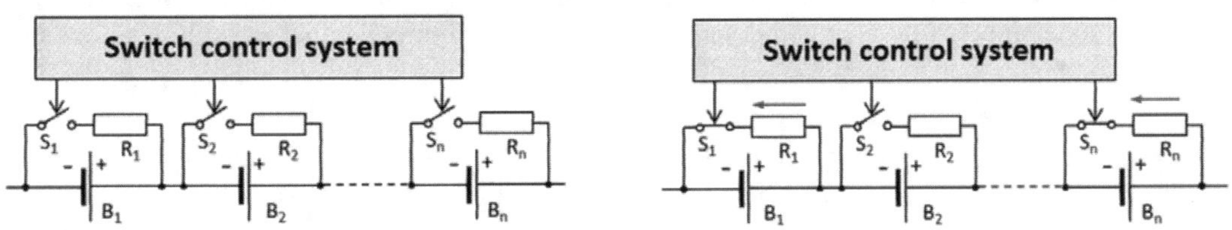

Figure 6. Passive balancing. (a) – electric circuit diagram; (b) – balancing process.

With active balancing the energy can be redistributed inside the battery, while the energy from the cells with a high charge level is transferred to the cells with a low charge level. Active balancing can be used in both charge and discharge processes.

Active circuits are divided into 3 types according to implemented principles of balancing:

1) based on capacitors;
2) based on inductors (transformers);
3) based on DC/DC converters.

The topology of the active balancing system based on capacitors is shown in figure 7. To equalize the charge between the cells the RC circuit is connected to the cell with the maximum charge (for example, B1 in figure 7b), at that the cell energy partially goes into the capacitor. Then the charged capacitor is being connected to the cell with the minimal charge (for example B3, in figure 7c) and the energy stored in the capacitor partially flows to the cell. The resistor R is used to limit the balancing current. The disadvantage of this method is a long time of charge balancing between all the cells.

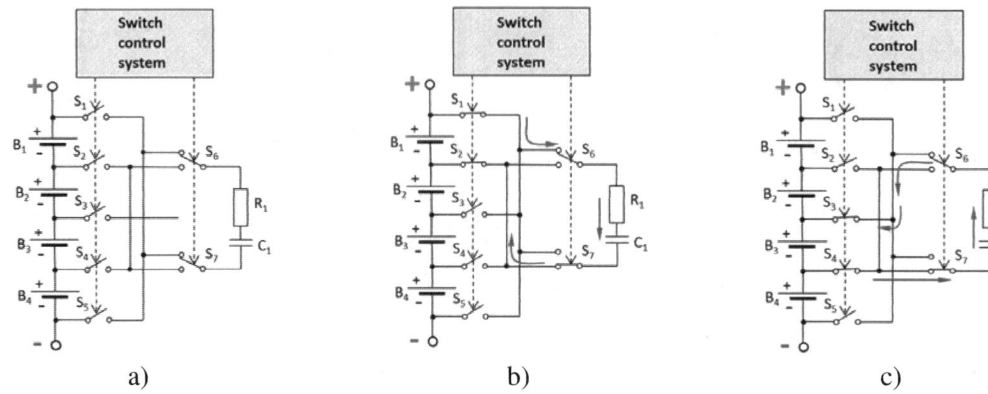

Figure 7. Capacitive active balancing: a – electric circuit diagram, b – capacitor charging process from cell B_1, c – capacitor discharging process to cell B_3.

The general disadvantage of different active balancing systems is the complexity of the circuit design, which increases the time and price of their development.

Obviously, active balancing is more energy-efficient but battery manufacturers often choose the cheaper and easier method of passive balancing. Important criteria of balancing type choice are formed below.

1. Required battery charging time.

If the battery charging time is not critical in a specific application (for example, the vehicle is used in the daytime and is being charged at night), passive balancing is the most reasonable choice. In this case charging requires a long time and balancing can be accomplished during this period with a low current not leading to the BMS elements overheating [9]. However, if the charging time is critical (for example, for public transport, charging several times a day) it's reasonable to use active balancing.

2. Balancing current

To assess the possibility of the passive balancing implementation, it's necessary to define the balancing current, which can be calculated by the following equation:

$$I_b = \frac{(I_{leakage.max} - I_{leakage.min})}{(\frac{t_b}{24})} \tag{1}$$

where I_b is the balancing current, [A]; $I_{leakage.max}$ is the maximal leakage current, [A]; $I_{leakage.min}$ is the minimal leakage current, [A]; t_b is the balancing time during a day, [hours].

Because of one of the main tasks of the BMS while balancing is equalization of the leakage currents for all cells, the data about maximal and minimal leakage currents are necessary. This information can be requested directly from the manufacturers of the cells for accurate calculation. Indirectly, the maximal leakage current can be estimated from the information about the nominal battery storage time. So, for example, if a manufacturer has limited the storage time for a 100 Ah battery by 18 months, then from the following equation:

$$I_{leakage.max} = \frac{C}{t_{storage} \cdot 30 \cdot 24} \tag{2}$$

where C is the nominal battery capacitance [A·h]; $t_{storage}$ is the nominal storage time, [months], we can get the leakage current $I_{leakage.max}$ = 7,7 mA.

The active cooling of the battery cells and BMS elements also must be taken into consideration.

3. Cost

IOP Conf. Series: Materials Science and Engineering **819** (2020) 012024 doi:10.1088/1757-899X/819/1/012024

Cost of each BMS Slave controller is directly connected with the number of cells. So with passive balancing, the addition of one cell requires one switch and one resistor for energy dissipation. With active balancing the possibility of connection of one cell to others must be provided. In this case the cost is determined by specific type of active balancing and includes cost of additional capacitors, inducers, voltage converters and special ICs, intended for switching one cell with any other.

It's important to note that the number of cells in a battery is not a determining factor for balancing type choice. Balancing types used in batteries of big manufacturers are listed in table 2.

Table 2. Solutions of big manufacturers of EVs.

Manufacturer	Product model	Number of cells	Chip model	Balancing current	Balancing type
Tesla	Model S	7104	BQ76PL536 (TI)	9,4 mA	Passive
Tesla	Model 3	2170	T2-A2 PZ38984 (Tesla)	-	Active
BMW	i3	96	LTC6801 (LT)	6,6 mA	Passive
Mitsubishi	i-MiEV	88	LTC6802 (LT)	4,5 mA	Passive
Nissan	Leaf	192	D15110 or D15120 (NEC)	8,6 mA	Passive
Opel	Ampera	-	L9763 (LG Chem)	-	Passive

Conclusion

Optimal choice of BMS topology and parameters is extremely important in terms of the vehicle maximum range and the lifecycle of the battery cells and other electronic devices.

Having analyzed the experience of the world-known manufacturers discussed in the article, we can make a conclusion that the layout of a BMS and the battery itself should be based on the vehicle working cycle, the required mass and, sometimes, the shape of the battery.

The following abbreviations are used in this article:

BMS – Battery Management System
CAN – Controller Area Network
CM – Cell Monitor
DC – Direct Current
ECU – Electronic Control Unit
EV – Electric Vehicle
HV – High Voltage
IC – Integrated Circuit
ICU – Insulation Control Unit
MSD – Manual Service Disconnect
OCV – Open Circuit Voltage
PCB – Printed Circuit Board
PDU – Power Distribution Unit
SOC – State of Charge
SOH – State of Health
TI – Texas Instruments
VCU – Vehicle Control Unit

References

[1] Zaitsev A S, Butarovich D O and Smirnov A A 2018 Obzor topologiy sovremennih sistem upravleniya zaryadom nakopiteley energii dlyz elektricheskich transportnych sredstv [Modern electric vehicle battery management systems topology review] *Zhurnal avtomobilnych ingenerov* **109** pp 29-33

[2] Kotiev G O, Butarovich D O and Kositsyn B B 2018 Energy efficient motion control of the electric bus on route *IOP Conference Series: Materials Science and Engineering* **315**

[3] Gorelov V A, Butarovich D O, Staduhin A A and Skotnikov G I 2019 Simulation-based identification of the parameters of a minibus hybrid powertrain *IOP Conference Series: Materials Science and Engineering* **534**

[4] Yaroslavtsev M V 2014 Determination of parameters of a hybrid car powertrain by modeling of the energy consumption process *Russian Electrical Engineering* **85 (12)** pp 724-728

[5] Semenenko M G 2009 Visual modelling electrochemical processes *Journal of Power Sources* **191 (1)** pp 91-96

[6] Lelie M, Braun T, Knips M, Nordmann H, Ringbeck F, Zappen H nad Sauer D U 2018 Battery management system hardware concepts: An overview *Applied Sciences (Switzerland)* **8 (4)**

[7] Abronzini U, Attaianese C, Monaco M D, Porpora F, Tomasso G, Granato M and Frattini G 2019 Optimal Modular BMS for High Performances NMC Battery Pack *2018 IEEE International Conference on Electrical Systems for Aircraft, Railway, Ship Propulsion and Road Vehicles and International Transportation Electrification Conference ESARS-ITEC 2018*

[8] Redondo-Iglesias E, Venet P, Pelissier S 2018 Global model for self-discharge and capacity fade in lithium-ion batteries based on the generalized eyring relationship *IEEE Transactions on Vehicular Technology* **67 (1)**

[9] Xia B, Zhou J, Yang Y, Wang H, Wang W and Lai Y A 2019 Novel Battery Equalization Method Base on Fuzzy Logic Control Considering Thermal Effect *IOP Conference Series: Earth and Environmental Science* **300 (4)**

[10] Bauer M, Muehlbauer M, Bohlen O, Danzer M A and Lygeros, J 2019 Power flow in heterogeneous battery systems *Journal of Energy Storage* **25**

[11] Wu S.-T, Chang Y.-N, Chang C.-Y and Cheng Y.-T 2019 A fast charging balancing circuit for LiFePO4 battery *Electronics (Switzerland)* **8 (10)**

[12] Liu B, Liu M, Jiang X, Tuo X, Zhou H and Ren J 2018 Design of battery management system based on DSP for BEV *Proceedings of 2017 9th International Conference On Modelling, Identification and Control, ICMIC 2017* pp 857-862

[13] Suthar B, Sonawane D, Braatz R D and Subramanian V R 2015 Optimal low temperature charging of lithium-ion batteries *IFAC-PapersOnLine* **28 (8)** pp 1216-1221

[14] Alfieri L, Bracale A, Caramia P, Iannuzzi D and Pagano M 2019 Optimal battery sizing procedure for hybrid trolley-bus: A real case study *Electric Power Systems Research* **175**

[15] Herrera V I, Milo A, Gaztanaga H, Ramos and Camblong H 2019 Adaptive and Non-Adaptive Strategies for Optimal Energy Management and Sizing of a Dual Storage System in a Hybrid Electric Bus *IEEE Transactions on Intelligent Transportation Systems* **20 (9)**

[16] Feng F, Hu X, Liu J, Lin X and Liu B 2019 A review of equalization strategies for series battery packs: variables, objectives, and algorithms *Renewable and Sustainable Energy Reviews* **116**

[17] Deng T, Ran Y, Yin Y, Chen X, Liu P 2019 Multi-objective optimization design of double-layered reverting cooling plate for lithium-ion batteries *International Journal of Heat and Mass Transfer* **143**

[18] Jung H, Silva R and Han M 2018 Scaling trends of electric vehicle performance: Driving range, fuel economy, peak power output, and temperature effect *World Electric Vehicle Journal* **9 (4)**

Acknowledgments
The research was carried out at BMSTU with the financial support of the Ministry of Education and Science of the Russian Federation under Agreement 14.574.21.0178 (Unique identifier: RFMEFI57417X0178).

IOP Conf. Series: Materials Science and Engineering **819** (2020) 012025 doi:10.1088/1757-899X/819/1/012025

Development of a prototype dashboard for wheeled passenger transport

D Astapenko, D Butarovich and A Smirnov

Wheeled Vehicles Department of Bauman Moscow State Technical University, Moscow, 105005, Russian Federation

E-mail: astapenkoda@bmstu.ru

Abstract. The development of wheeled passenger transport is associated with the need to use many components from various manufacturers. All these components have parameters that should be displayed on the dashboard. In the modern world, the dashboard is the "face" of the vehicle. It should give the driver a complete picture of the vehicle condition. The information shown on the screen should be understandable and not misleading for the user, regardless of the level of his training. World trends show that the analog dashboard is being replaced by fully digital (graphical) ones. When the graphics is added, the software becomes considerably more complicated. In order to reduce the complexity of its development and to design it in parallel, a modular approach is used. For the dashboard, four software modules can be distinguished: drivers for working with the hardware, operating system, top layer software, and a graphical user interface. Modern dashboards use processors with operating systems based on Embedded Linux due to the fact that there are many ready-made drivers for the hardware for this operating system. To accelerate software development, Europian manufacturers use ready-made frameworks for working with graphics, for example, the cross-platform Qt framework. The article presents the results of the development of a prototype dashboard for wheeled passenger vehicles. The device components and necessary functionality of the product for mass production is determined.

1. Introduction

Currently, in Russia and in the world the development of the wheeled passenger electric transport: trolley buses, buses, and electric buses, is under way [1,2,3,4,5]. A lot of attention in the development of the vehicles is given to the dashboard [6,7,8]. The dashboard is the part of the driver's workplace which displays the basic parameters of the vehicle, as a result of which the driver constantly interacts with it. Thus, the dashboard is the "face" of the entire vehicle. At the prototype stage, when the vehicle's functionality is not fully defined, the display of parameters for their control plays a large role. To do this, one must constantly make changes to the dashboard. The article presents the result of using a dashboard development method that solves this problem.

2. Type and composition of the dashboard

The dashboard can be of 3 types: analog (figure 1a), digital (graphic) (figure 1b) and combined (figure 1c).

The analog dashboard is a set of arrow indicators and indicator lights. The digital dashboard is a fully graphical display. The combined dashboard, like the analog one, consists of arrow indicators and indicator lights, but it also uses a small digital display. Comparison of dashboard types is given in table 1.

Table 1. Dashboard type comparison.

Type	Advantages	Disadvantages
Analog	low cost, low complexity software, low complexity hardware	low versatility, impossible to change
Digital	high versatility, easy to change	high cost, high complexity software, high complexity hardware
Combined	middle cost, low complexity hardware	middle versatility, middle software, only the digital part can be changed

At the moment, leading manufacturers of vehicles use fully digital displays. This is due to the fact that:

1) the power of computing devices (microprocessors) has increased significantly, and their cost has decreased,
2) a large number of software tools appeared - frameworks that reduce the complexity of software development,
3) consumer desire to own a car with a graphic display. Due to the mass introduction of smartphones with graphic displays and customization options for users, users are accustomed to extensive personalization options,
4) the graphical interface is a more flexible solution - by changing the software, you can quickly adapt the dashboard to a new vehicle.

a) b)

c)

Figure 1. Dashboards: a – analog, b – digital, c – combined.

Next, the results of the dashboard development for wheeled passenger transport will be presented. The development of a dashboard with a graphical interface includes the following subtasks:

1) collecting data from different communication interfaces,

IOP Conf. Series: Materials Science and Engineering **819** (2020) 012025 doi:10.1088/1757-899X/819/1/012025

2) providing different displays for different vehicles,
3) development of different graphical interfaces for different vehicle models,
4) development of universal top layer software that does not depend on the hardware of the device,
5) quick adaptation of the interface to new vehicle components,
6) hardware development.

Software development for a dashboard is a complex task. To reduce the complexity of the development and to provide development in parallel, a modular approach is used. The software structure used in developing the prototype dashboard for a wheeled passenger vehicle is shown in figure 2.

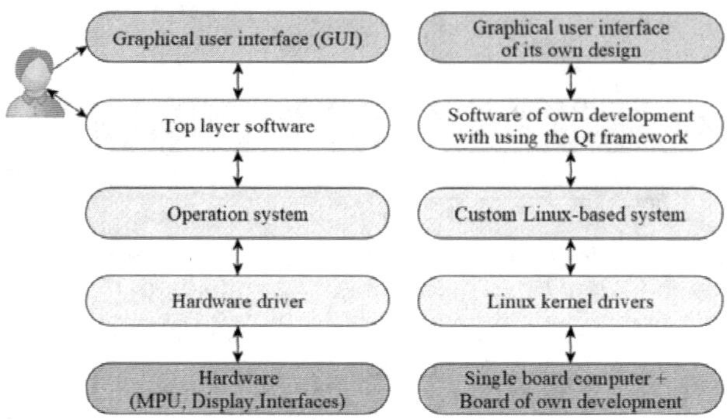

Figure 2. Dashboard software structure.

For the dashboard, four software modules can be distinguished: drivers for working with the hardware, operating system, top layer software, and a graphical interface. Modern dashboards use processors with Linux-based operating systems due to the fact that there are many ready-made drivers for the hardware for this operating system and it does not require large computational resources. To speed up software development, many manufacturers use ready-made frameworks for working with graphics, one of them is the cross-platform Qt framework. It provides ease of interconnection between the operating system, top layer software and a graphical interface.

3. Hardware

Developing hardware from scratch for a prototype product is economically unjustified. It is more cost-efficiently to use ready-made modules and link them together, and when the product is ready for mass production, make new hardware based on the modules used. The hardware of the dashboard is an assembly of three modules: single-board computer Raspberry Pi 3 B+, a display controller board, and a proprietary board (figure 3). The Raspberry Pi single-board computer is used as the computing center of the device, and the software being developed is running on it. The display controller board is used to convert the HDMI interface to LVDS, and to adjust the display settings (contrast, screen calibration, etc.).

The proprietary development board expands the functionality of the Raspberry Pi 3 B+, as it does not have enough peripherals for the developed dashboard. Proprietary elements are located on the board: a voltage converter, an interface for connecting to the CAN bus, a real-time clock, elements for connecting control buttons, a display backlight control module. The 3D model of the board is shown in figure 4.

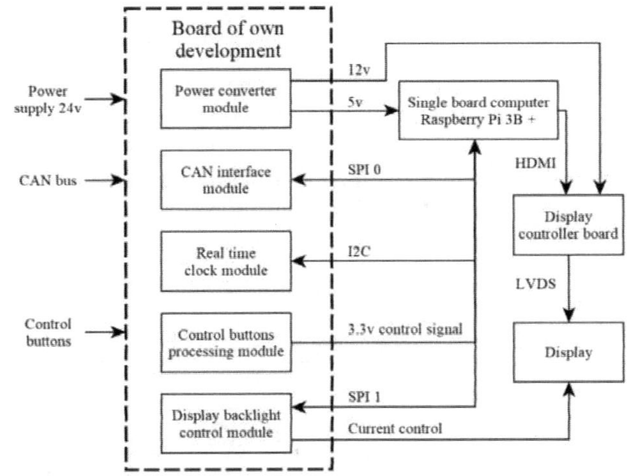

Figure 3. Block diagram of the hardware, proprietary board highlighted by a dashed line.

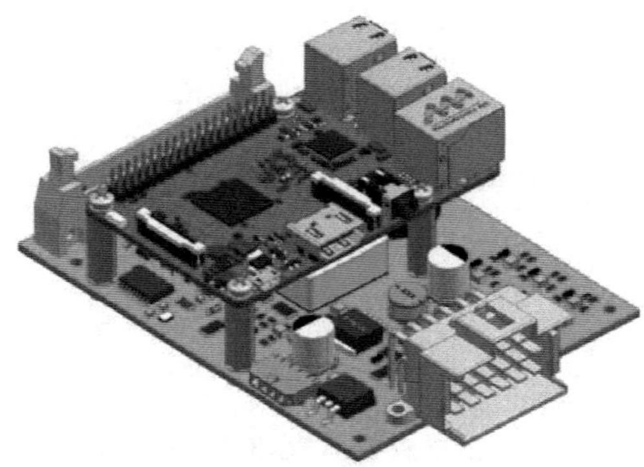

Figure 4. 3D model board by own design and development.

4. Operation system and hardware drivers

A special Linux build using the Yocto project is used as the operating system. The Yocto Project is an open source collaboration project that helps engineers to create custom Linux-based systems for embedded products, regardless of the hardware architecture. The project provides a standard delivering hardware support and software stacks, allowing the interchange of software configurations and builds. The tools allow users to build and support customizations for multiple hardware platforms and software stacks in a maintainable and scalable way. The Yocto Project combines, maintains and validates three key development elements (figure 5) [9]:

1) the set of integrated tools to make working with embedded Linux successful, including tools for automated building and testing, processes for board support and license compliance, and component information for custom Linux-based embedded operating systems,
2) the reference embedded distribution (called Poky),
3) the OpenEmbedded build system, co-maintained with the OpenEmbedded Project.

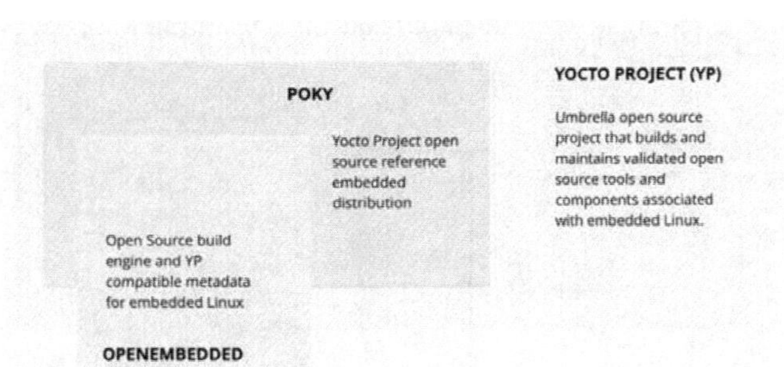

Figure 5. Yocto project structure.

Hardware drivers contain in the Board Support Package. A Board Support Package (BSP) is the collection of information that defines how to support a particular hardware device, set of devices, or hardware platform. The BSP includes information about the hardware features present on the device and kernel configuration information along with any additional hardware drivers required. The BSP also lists any additional software components required in addition to a generic Linux software stack for both essential and optional platform features [10].

IOP Conf. Series: Materials Science and Engineering **819** (2020) 012025 doi:10.1088/1757-899X/819/1/012025

5. Top layer software

Dashboard software was developed in several programming languages. The control logic uses the C++ language with the Qt framework. Algorithms that are responsible for receiving and sending CAN messages, calculating mileage, estimated range, etc. are written in C++. To work with graphics, the qml language using the QtQuick module is used. Animation of arrow movements, font size changes, display color changes is written in qml. Python is used to automatically generate code from a .dbc file for working with CAN messages (figure 6).

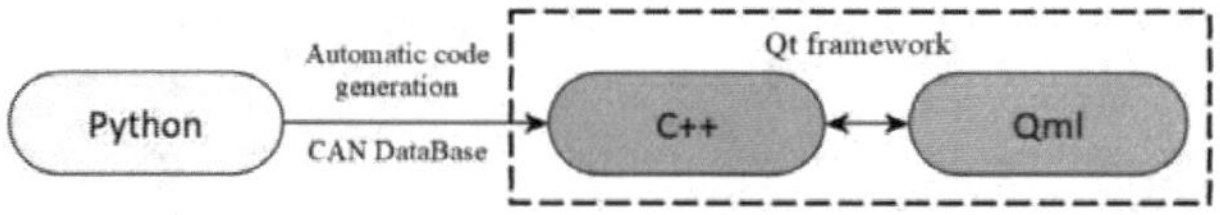

Figure 6. Programming languages used in the project.

Most often, the CAN bus is used to communicate units in a vehicle. A file with the .dbc extension is used to describe messages transmitted over the bus. It contains a textual description of CAN signals and frames. To reduce the adaptation time of the dashboard to new vehicles, and minimize errors when writing code, a script was developed that allows you to create .cpp and .h files from a .dbc file for use in a C ++ project (figure 7).

Figure 7. Generating C++ code from .dbs file.

The basic logic of the dashboard is developed in C++ using the objects of the Qt framework classes, the use of these objects makes the software more universal.

To work with the graphical interface, the Qml language and the QtQuick module are used. QtQuick can work with most modern graphic formats. Source images can be created in any specialized application. The image is exported to QtQuick as special objects. Work with graphics changes the properties of these objects.

To configure the transfer of objects from QML to C++, and vice versa, the Signals and slots mechanism is used (figure 8). Signals and slots are used for communication between objects.

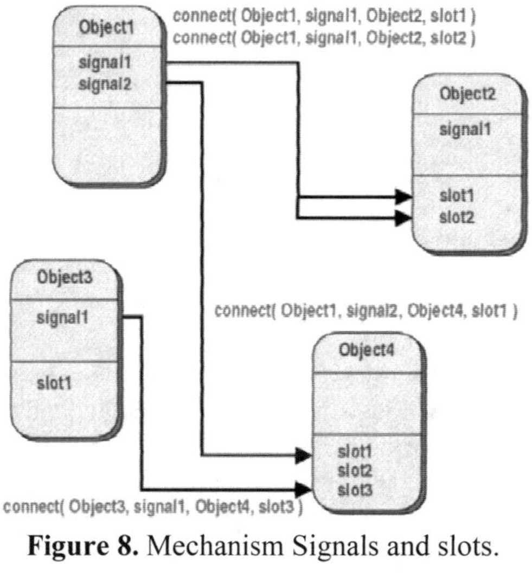

Figure 8. Mechanism Signals and slots.

IOP Conf. Series: Materials Science and Engineering **819** (2020) 012025 doi:10.1088/1757-899X/819/1/012025

The signals and slots mechanism is a central feature of Qt. A signal is emitted when a particular event occurs. Qt's widgets have many predefined signals, custom signals can also be added. A slot is a function that is called in response to a particular signal. Qt's widgets have many pre-defined slots, custom slots can also be added [11].

6. Graphical user interface

The dashboard screen can be divided into several functional parts. The dials are shown on the left and right sides. On the left side the following elements are shown: 3 dial indicators (ammeter, pressure in the first brake circuit, pressure in the second brake circuit) and temperature scale, temperature outside and inside the passenger compartment. On the right side there is one dial indicator (speed), fuel gauge, mileage and current date On the central part the following elements are located: door indication, critical errors, a field for displaying text information and the current time.

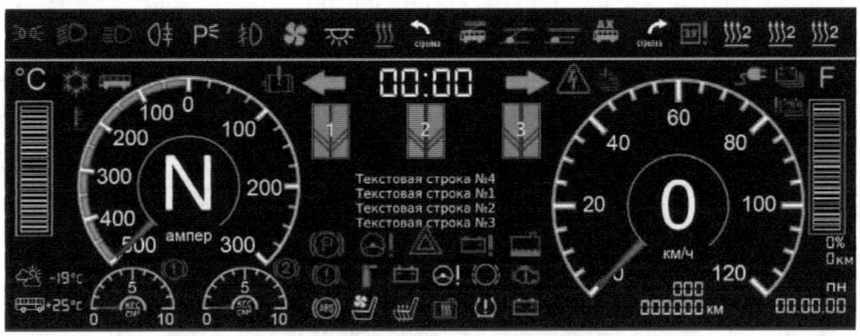

Figure 9. Dashboard interface.

If there are no errors in the system while the vehicle is moving, the logo is displayed on the central part (figure 10). The upper part of the display is reserved for status lights.

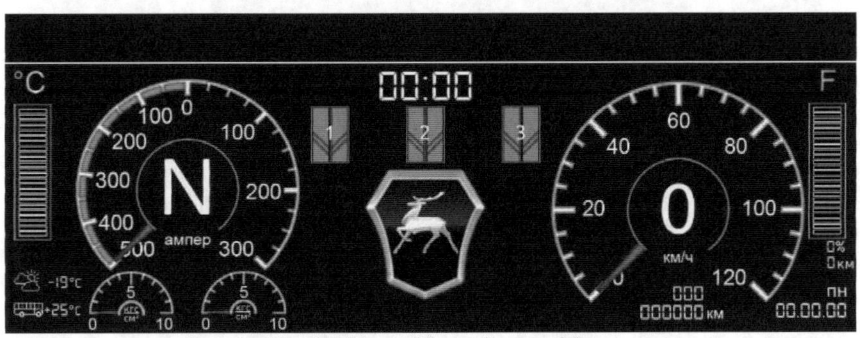

Figure 10. Dashboard interface, without error.

In the future, it is planned to add support for several interface options so that the user can choose it to his liking.

7. Additional configuration for run Dashboard application

To use the device as a dashboard, some additional settings need to be made.

Before starting the main application, you need to start the CAN interface. There are two ways to do this. The first way is to use the QProcess class. To execute a command, it should be passed to the start method and wait for completion:

```
QProcess process;
process.start("Custom command");
process.waitForFinished();
```

IOP Conf. Series: Materials Science and Engineering **819** (2020) 012025 doi:10.1088/1757-899X/819/1/012025

The second way is to run a custom script before running the application. To do that edit the:

/lib/systemd/system/b2qt.service

Under [Service] section you can add:

ExecStartPre=<your_custom_command>.

8. Dashboard testing

A special test bench is used to test the dashboard. The bench is assembled on the debug board from the real peripheral modules (figure 11). By testing one can check the following: the correctness of the device interaction with the CAN bus, conformance of the displayed data with the DBC file, reading and writing date/time from the RTC module, backlight control.

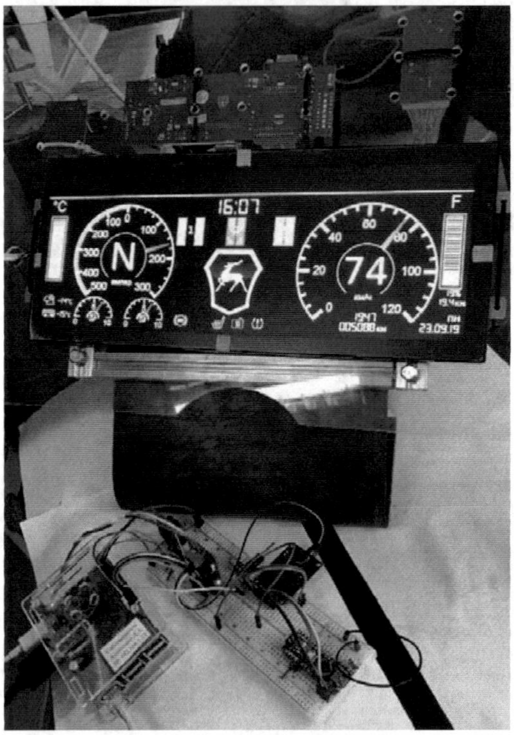

Figure 11. Dashboard test bench.

Using cross-compilation, the project can be assembled on the host computer and only the executable file can be send to the device. Due to this development approach, you can quickly make changes to the algorithms of the device.

9. Conclusion

The results of bench tests of the dashboard have confirmed the workability of using a prototype device on a vehicle. The modular approach to software development has allowed:

1) reduction of the development time, development from scratch to a working prototype took 3 months,
2) achieving software independence from the hardware,
3) using various graphical interfaces.

Further, it is planned to increase the application loading speed from a completely switched off state, now the launch time is from 5 to 7 seconds. In normal mode, it is planned to display more useful information on the central part of the screen: a map with GPS navigation when Drive is in the motion mode, or a rear view camera when reversing.

References

[1] Astapenko D A, Butarovich D O and Smirnov A A 2019 Development of the upper-layer electronic control unit based on a production-type unit for a commercial vehicle *IOP Conf. Series: Materials Science and Engineering* **534**

[2] Buzunov N V, Kotiev G O and Gorelov V A 2019 Implementation of the interaction of the steering wheel loader control system of the remote-controlled wheeled vehicle operator interface with a real-time simulation model *IOP Conf. Series: Materials Science and Engineering* **534**

[3] Buzunov N V, Kotiev G O and Padalkin B V 2019 Wheel vehicle dynamics real-time simulation for on-board stand-alone moving control system realization *IOP Conf. Series: Materials Science and Engineering* **534**

[4] Gorelov V A, Butarovich D O, Staduhin A A and Skotnikov G I 2019 Simulation-based identification of the parameters of a minibus hybrid powertrain *IOP Conf. Series: Materials Science and Engineering* **534**

[5] Kotiev G O, Miroshnichenko A V, Stadukhin A A and Kositsyn B B 2019 Estimating operation modes for the individual wheel electric drive of the all-wheel drive vehicle with the use of the driving simulator *IOP Conf. Series: Materials Science and Engineering* **534**

[6] Muslim E, Moch B N, Lestari R A, Shabrina G and Ramardhiani R 2019 Ergonomic design of electric vehicle instrument panel: a study case on Universitas Indonesia's national electric car *IOP Conf. Series: Materials Science and Engineering* **508**

[7] Pawlenka T, Kulhanek J, Tomcik P and Zapletal R 2019 Design of Digital CAN Based Car Dashboard Unit *Proceedings of the 2019 20th International Carpathian Control Conference*

[8] Ren H, Tan Y and Zhang N 2019 Research on Form Design of Automotive Dashboard Based on Kansei Engineering *IOP Conf. Series: Materials Science and Engineering* **573**

[9] Software – Yocto Project. Yocto Project. Retrieved September 14, 2019, from https://www.yoctoproject.org/software-overview

[10] Yocto Project Board Support Package Developer's Guide. Yocto Project. Retrieved September 14, 2019, from https://www.yoctoproject.org/docs/2.7.1/bsp-guide/bsp-guide.html#bsp

[11] Signals & Slots | Qt Core 5.13.1. Qt. Retrieved September 14, 2019, from https://doc.qt.io/qt-5/signalsandslots.html

Acknowledgments

The research is done in BMSTU with the financial support of the Ministry of Education and Science of Russian Federation under the Agreement № 14.574.21.0178 (Unique identifier: RFMEFI57417X0178).

IOP Conf. Series: Materials Science and Engineering **819** (2020) 012026 doi:10.1088/1757-899X/819/1/012026

The Need for Creation of High-tech Tractors of 0.6-2 Traction Classes with High Degree of Automation

R Solovyev[1], S Cheranev[1], S Karyakin[1], A Kolomeichenko[1], I Gribov[1], V Evgrafov[2], N Mezentsev[2]

[1]Center of Agricultural Engineering, Central research and development automobile and engine institute NAMI, Avtomotornaya St., 2, Moscow 125438, Russia
[2]Center of Intelligent Systems, Central research and development automobile and engine institute NAMI, Avtomotornaya St., 2, Moscow 125438, Russia

Abstract. The article shows the need for the creation of tractors of 0.6-2 traction classes (nominal traction force 5.4-27 kN) with a high degree of automation, which are in demand in the agricultural industry. The authors present the principle of creation of the 4WD modular platform with hybrid powertrain and unification of axles and some technical characteristics for each traction class. It is indicated that for the conditions of digital agriculture it is necessary to provide the possibility to control high-tech tractors in remote and/or unmanned operation modes.

1. Introduction

The main tendencies of development of modern agricultural vehicles are: improvement of operational characteristics (increase of productivity, power, speed, pressure in the hydraulic system; transition to alternative sources of energy; decrease of pressure on the soil; improvement of working conditions of the operator); implementation of automated and remote and/or driverless control, telemetry systems; transition to electric and hybrid drives; development of new vehicles on the basis of modular platforms; reduction of the labour intensity of technological operations; transition to digital agriculture with development of precision farming technologies. Furthermore, agricultural producers are faced with the need to improve the efficiency of the use of machines and farmland, reducing the production costs, and optimizing technological operations during the work.

In connection with the above-stated, the aim of the research is to create high-tech tractors with automated control systems equipped with modern information systems. Development of such tractors will allow engaging modern technologies, digitalize the industry, involve new personnel to direct or remote work in the agricultural industry, including younger people actively exploring information technologies.

2. Problem setting

On the basis of the analysis of tendencies and the forecasted situation in agriculture of Russia to show the necessity of development of high-tech tractors of 0.6-2 traction classes with a high degree of automation.

3. Results

The analysis of the current and forecasted situation in agriculture in Russia shows the following trends (Figures 1-3): there is a steady reduction in the number of employable population in rural areas; it is planned to put into operation unused lands for agricultural production; an increase in the fleet of tractors for agricultural works is forecasted.

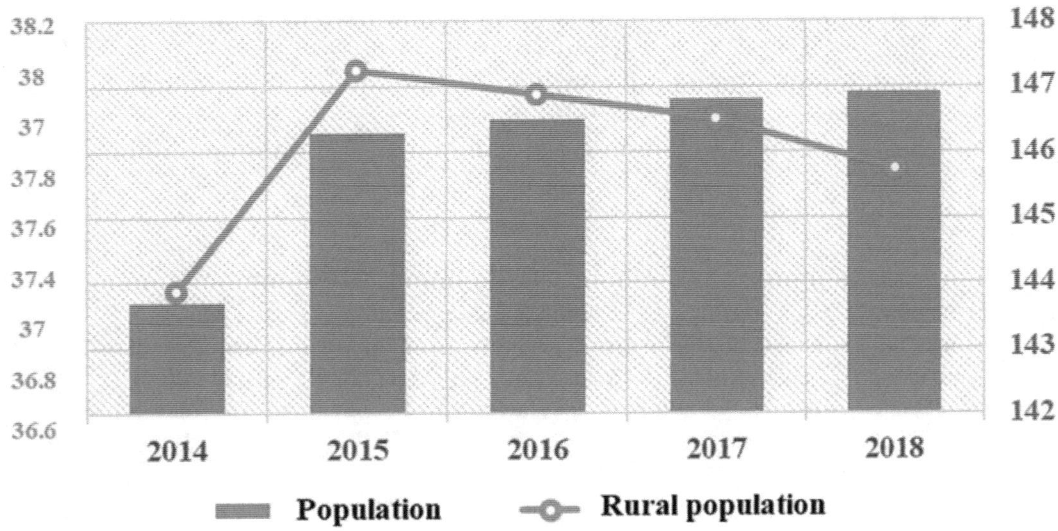

Figure 1. Population change in Russia (million people).

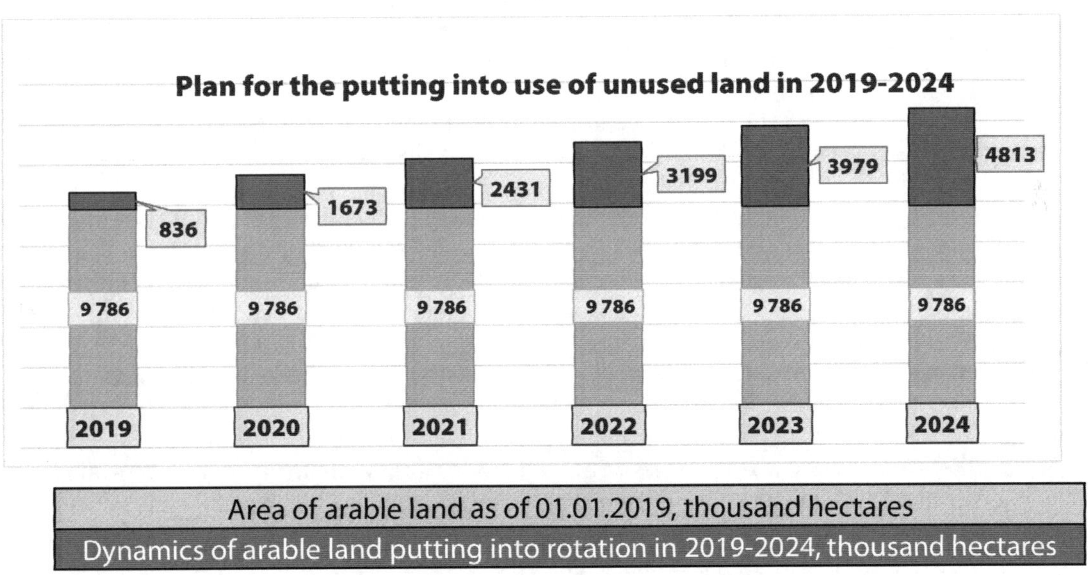

Figure 2. Plan for putting into use of unused land.

IOP Conf. Series: Materials Science and Engineering **819** (2020) 012026 doi:10.1088/1757-899X/819/1/012026

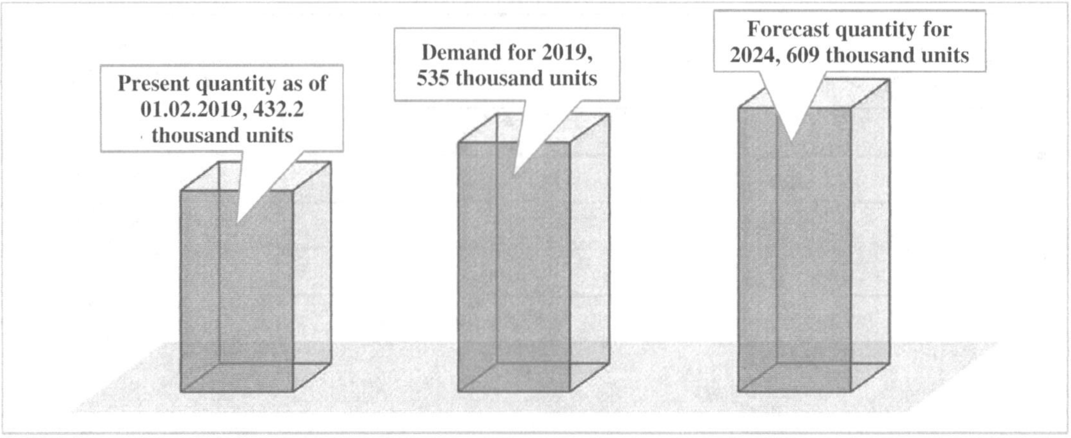

Figure 3. Current and forecasted quantity of tractors.

Tractors of traction classes 0.6-2 can be used in various agricultural operations.

Traction class 0.6 (nominal traction force 5.4-8.1 kN). Tractors of this class are used directly in agriculture and involved in various technological processes: preparation of soil for sowing, sowing; planting of vegetables and roots; planting of decorative and fruit-bearing trees and bushes; care of sowing, care of bushes and trees; harvesting; preparation of feed for animal husbandry; transport, municipal works, drive for various stationary units.

Traction class 0.9 (nominal traction force 8.1-12.6 kN). The range of possible works in this class is slightly wider. Both 2WD and 4WD tractors are presented, and models with low ground clearance for work in mountain areas are developed. It becomes reasonable to install gearboxes with a wide range of gears. The main application of this class is: ploughing light soils; pre-sowing preparation of the soil; interrow hilling; pest control (crop spraying); harvesting; haymaking; transport work.

Traction class 1.4 (nominal traction force 12.6-18 kN). Tractors of this class are designed for operation in difficult road conditions. They are the most effective at performance of works performed by tractors of traction classes 0.6 and 0.9 and, taking into account their exceptional properties, it became possible to use such vehicles for ploughing of light and medium soils, cultivation, harrowing, application of fertilizers in the soil.

Traction class 2.0. This class includes tractors with wheeled as well as caterpillar chassis. This is due to the power of the vehicles and the conditions in which they have to perform their tasks. This class includes garden tractors as well as general purpose and special purpose agricultural tractors and straddle tractors. In addition to the above-mentioned types of works, there are maintenance of vineyards, beet-growing and gardening works, work in mountainous areas.

Modern high-tech tractors of 0.6-2 traction classes have to be developed on the basis of the self-propelled modular platform with a hybrid powertrain, unified axles and variable wheelbase, depending on the traction class (Figure 4). Hybrid powertrain variants: ICE + generator + electric motor or ICE + generator + battery + electric motor. The axles of a modular platform with this type of powertrain have to be equipped with wheel motors. In order to increase manoeuvrability, the tractor has to have crab steering. All units, assemblies and systems of high-tech tractors have to have high reliability [1, 2], because the failure of any element of the vehicle can lead to the interruption of agricultural work.

IOP Conf. Series: Materials Science and Engineering **819** (2020) 012026 doi:10.1088/1757-899X/819/1/012026

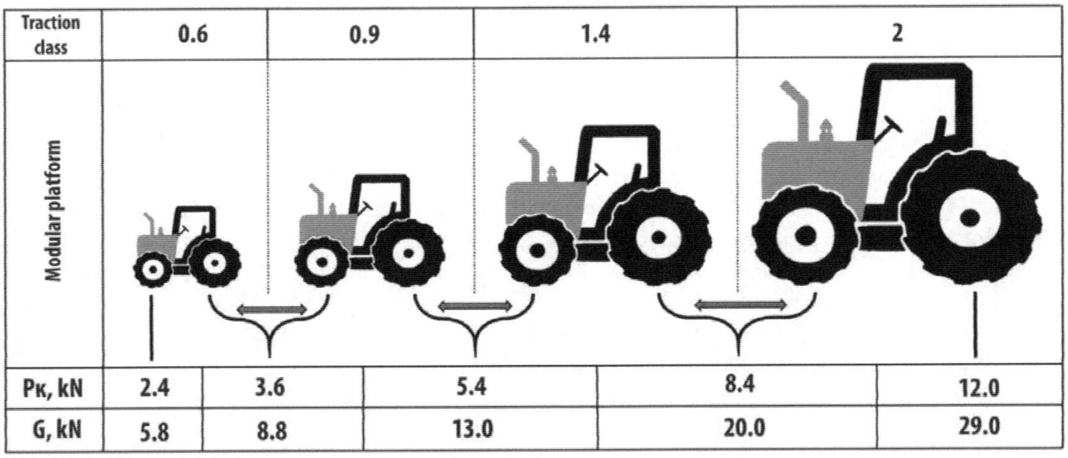

Figure 4. Principle of creation of a 4WD modular platform with a hybrid powertrain and axle unification.

Table 1 shows some required technical characteristics of high-tech tractors of each traction class with the 4WD modular platform.

Table 1. Target indicators for the development of the 4WD modular platform.

Indicator	**Traction class**			
	0.6	0.9	1.4	2
Engine power, kW	24	36	56	80
Completeness of power usage (reference power intensity), kW/kN	1.5	1.5	1.5	1.5
Number of technological spaces for equipment, spaces, not less than	3	3	3	3
Number of technological spaces for containers, spaces, not less than	2	2	2	2
Productivity of the technological process (ploughing), ha/hour	1	1.7	2.5	5
Cost of technological process execution (ploughing), RUB/ha	1106	989	850	843
Maximum wheel ground pressure, kPa	74	80	80	80
Row spacing (adjustment), mm	500	500	500	500
Maneuverability (turning radius), m	2,3	2,9	3,1	3,85
Agrotechnical clearance, mm	700	700	700	700
Transport speed, km/h	30	40	60	60
Load capacity of front linkage, kg	500	1000	1700	2800
Carrying capacity of the rear linkage, kg	1000	1700	2800	4000
Front PTO	+	+	+	+

Development of high-tech tractors of 0.6-2 traction classes for conditions of digital agriculture has to provide possibility of their control in remote and/or driverless operation modes and, accordingly, presence of reliable and safe system of autonomous motion, including at agrotechnical works, with

following functions: recognition of obstacles; automatic stop before an obstacle and/or its bypass. This includes the following:
- gradually introducing into mass production the functions of assisting the operator to stop before an obstacle, pull out of the row, etc., and make them informational at first. At the same time, the development of active assistants to the operator and driving machines for the mass sector has to be carried out (at present they are characterized by high cost and low level of implementation);
- making active operator assistants widely available. At the same time, to work out the possibility of mass implementation of driverless systems;
- determining the necessary and sufficient set of sensors for the computer vision system, considering, the task of minimizing the cost of the object;
- determining the necessary technical equipment of the basic chassis, their level of electrification, control, the possibility of system operating in the tractor's information network.

In addition, it is necessary to develop: a vector map of a field with indication of static obstacles (poles, etc.); the movement of the tractor along the route formed in the field; the detection of obstacles on the route of the tractor using the duplication of sensors of computer vision to improve the reliability of the identification of obstacles; stop or bypass of an obstacle; informing the dispatcher about the presence of obstacles in the lane of the tractor; online transmitting telematic data to the dispatcher's room; interaction of several tractors to work in the same field at the same time.

There is a large number of researches in the field of driverless control of both agricultural vehicles and automobiles [3, 4]. However, all known systems to date have not found mass application primarily because of the high cost of the element base in terms of one unit of equipment. In addition, when moving a high-tech tractor with a high degree of automation or tractor-implement combination, the accuracy of maintaining a given route will be affected by: the probability of different state of the supporting surface under each wheel (humidity, surface hardness), the performance of agricultural work at any time of day, the presence of various natural and atmospheric factors, increased dust content during agricultural operations, leading to contamination of sensors of computer vision sensors. All this has to be taken into account when creating a control system for high-tech tractors in remote and/or driverless operation.

Digital agriculture will allow efficient and flexible redirection of unused power by high-tech tractors of different traction classes (Figures 5, 6).

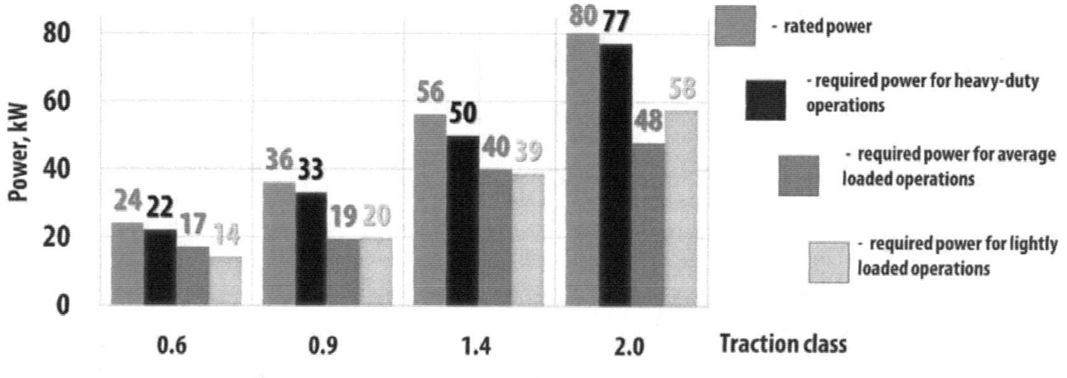

Figure 5. Power requirement for different types of technological operations.

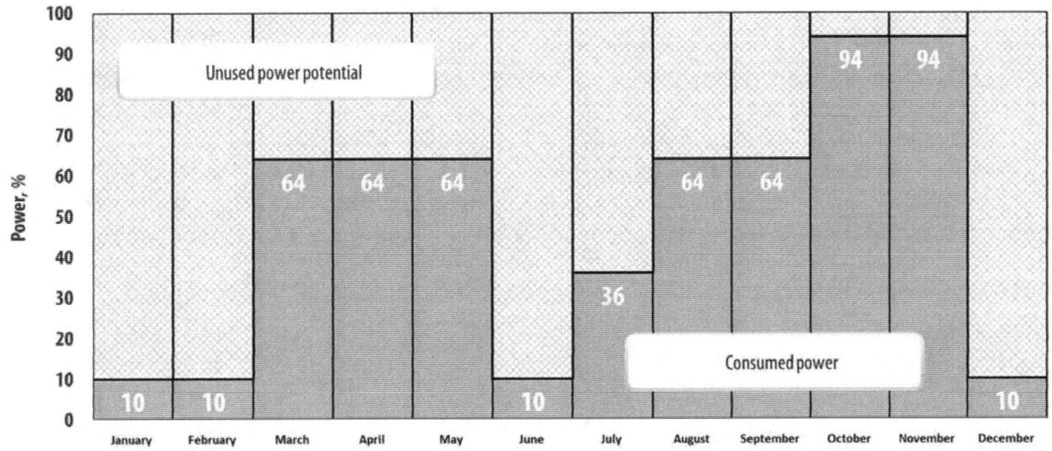

Figure 6. Annual balance of machine and tractor fleet power for the Southern Federal District in grain crop cultivation.

4. Conclusion

Development of high-tech tractors of 0.6-2 traction classes with a high degree of automation would enable reaching the target indicators of the project “Digital Agriculture” - ensuring the digital transformation of agriculture through the implementation of digital technologies and platform solutions for technological breakthroughs in the agro-industrial complex and achieving doubling productivity growth at “digital” agricultural enterprises, as well as another project “Technical modernization of the agro-industrial complex” - providing in 2025, with the state support, renewal of tractors in agricultural organizations to the level of 3.4% [5-10]. Development and use of high-tech tractors of 0.6-2 traction classes with the autonomous control system on the advanced component base with electromechanical and hybrid drive will allow raising essentially efficiency of agriculture and make the transition to its digitalization.

References

[1] Kolomeichenko A V, Kuznetsov Y A, Logachev V N and others Reliability of technical systems. Practical training: tutorial Orel State Agrarian University Russia 2013 p112

[2] Puchin E A,Kolomeichenko A V, Logachev V N Reliability of technical systems. Course design: tutorial Orel State Agrarian University Russia 2012 p 96

[3] Nagaitsev M V “Unmanned” vehicles - stages of development and testing *Journal of automotive engineers* 2012 №5(76) pp 32-39

[4] Yuzayeva A G Unmanned vehicles: dangers and development prospects Actual problems of aviation and cosmonautics 2016 №12 Volume 2 pp 120-122

[5] Decree of the President of the Russian Federation No. 204 of 7 May 2018 “On the national goals and strategic objectives of the development of the Russian Federation for the period up to 2024” URL: http://www.kremlin.ru/acts/bank/43027

[6] Doctrine of food security of the Russian Federation approved by the Decree of the President of the Russian Federation №120 of January 30 2010 URL: https://base.garant.ru/12172719/

[7] Digital Economy of the Russian Federation Program (Order of the Government of the Russian Federation No. 1632-R of July 28 2017) URL: http://static.government.ru/media/files/9gFM4FHj4PsB79I5v7yLVuPg u4bvR7M0.pdf

[8] Resolution of the Government of the Russian Federation of July 14 2012 No. 717 “On the State Program for the Development of Agriculture and Regulation of Agricultural Products, Raw Materials and Food Markets” (as amended and supplemented on February 11 2019 Resolution of the Government of the Russian Federation of February 8 2019 No. 98) URL:

IOP Conf. Series: Materials Science and Engineering **819** (2020) 012026 doi:10.1088/1757-899X/819/1/012026

http://www.gks.ru/free_doc/new_site/import-zam/post717-2012.pdf
[9] Resolution of the Government of the Russian Federation of March 29 2019 No. 377 “On Approval of the State Program of the Russian Federation "Scientific and Technological Development of the Russian Federation”
URL: https://www.garant.ru/products/ipo/prime/doc/72116664/
[10] Decree of the President of the Russian Federation No. 642 of December 1 2016 “Strategy of Scientific and Technological Development of the Russian Federation” URL: http://www.kremlin.ru/acts/bank/41449

IOP Conf. Series: Materials Science and Engineering **819** (2020) 012027 doi:10.1088/1757-899X/819/1/012027

Availability of the agricultural, municipal and road construction vehicle market with Russian diesel engines

R Soloviev, S Cheranev, A Kolomeichenko, M Gerasimov, V Gromov

Center of Agricultural Engineering, Central research and development automobile and engine institute NAMI, 125438, Moscow, Avtomotornaya St., 2

Abstract. The paper describes the enterprises that produce or produced diesel internal combustion engines for agricultural, municipal and road construction vehicles in the Russian Federation. The main emphasis is made on engines with a capacity of up to 4400 cc and a power of up to 150 kW as the most demanded for the country. The main consumers of such combustion engines are presented. It is shown that one of the main drawbacks is the impossibility of manufacturing the component base of the required quality for the organization of production of diesel engines in Russia.

1. Introduction

One of the most important tasks of the domestic industry is the establishment and further development of own production of agricultural, municipal and special vehicles, as well as components for it, including diesel engines, or it is necessary to create conditions for the localization of technologies of foreign manufacturers on the territory of the Russian Federation. Thus it is necessary to establish interaction between large manufacturers of vehicles, on the one hand, and manufacturers of component base (parts, units, spare parts), on the other hand, the absence of which eventually makes it impossible for them to cooperate in the production of vehicles of various purposes in the country. The Russian market of agricultural, municipal and road construction vehicles currently urgently needs a reliable domestic diesel engine with a capacity of up to 4.4 litres. It has to have a high amount of torque, low weight and compact overall dimensions, have high fuel efficiency and meet modern environmental requirements. The current demand for such ICE is up to several thousand units per year with further forecast for a stable annual increasing number of such products due to the renewal of the fleet of vehicles or basically their absence on some necessary for the country's economy positions.

In connection with the above-mentioned, the aim of the work will be to justify the need to establish the production of a range of reliable domestic diesel engines with a capacity of up to 4.4 litres.

2. Problem setting

On the basis of the analysis of tendencies and a developed situation in the engine building and engine component base to show the necessity of manufacture of a range of reliable domestic diesel engines with a capacity up to 4400 cc and power up to 150 kW which is pretty much in demand for a wide variety of agricultural, municipal and road construction vehicles in Russia.

3. Results

We will consider the proposals of Russian manufacturers on diesel engines for agricultural, special and road construction vehicles with a capacity of up to 4400 cc.

Vyatka machine-building Enterprise AVITEK produced small-size single-cylinder four-stroke diesel engines of VSN series (VSN-6D, VSN-7D, VSN-9D). The manufacturer positioned these engines as a power source for driving autonomous small (mini) power plants, welding units, mini-tractors, pumping and compressor units, as well as for road transport and construction machinery and other small mechanization means. At the moment the production of these engines is discontinued [1].

Engines of similar purpose have been developed by Production association TULAMASHZAVOD, these are single-cylinder four-stroke diesel engines TMZ-450D and TMZ-520D [2], having the following characteristics:

- TMZ-450D with a capacity of 454 cc and a power of 8.0 kW;
- TMZ-520D with a capacity of 520 cc and a power of 9.5 kW.

Head Specialized Design Bureau "Transdiesel" of the Chelyabinsk Tractor Plant has developed and produced a 2-cylinder V-shaped diesel engine V2Ch 8.2/7.8, which has air cooling, power of 8.8 kW and a capacity of 0.8 litres [3].

The main consumers of this ICE were such Russian enterprises as:

- Chelyabinsk Tractor Plant (Chelyabinsk);
- Projector Elektrotekhnika (Moscow);
- Electroagregat (Kursk);
- Elkon (Barnaul);
- Sarapul Electric Generator Plant (Sarapul);
- Kurganmashzavod (Kurgan);
- Proletarsky Zavod (St. Petersburg).

Having studied the demand and market requirements, the design bureau of the plant has developed a project of a family of small-size water-cooled diesel engines covering the power range from 18 to 55 hp. The new family of compact V-type liquid-cooled diesel engines consists of:

- 2-cylinder version with a capacity of 1.17 litres;
- 4-cylinder version with a capacity of 2.34 litres;
- 6-cylinder version with a capacity of 3.57 litres.

At the moment such engines are created only in the form of single prototypes [4].

Among domestic engines with a capacity of 1 to 4.4 litres, the products of Vladimir Motor Tractor Plant (VMTZ), which is currently in the bankruptcy stage, were in demand and popular [5]. The company produced the following diesel engines with air cooling:

D120 is a two-cylinder internal combustion engine with a capacity of 2.08 litres, which produces a power of 15.4 - 22.1 kW. It was installed on mini loaders of Uralvagonzavod, forklift trucks of Kalinin Machine Building Plant, electric units produced in Kursk and Vladimir region.

D130 and D130T are three-cylinder engines with a capacity of 3.12 litres and a power of 29.4 - 47.8 kW. It was installed on the following types of machinery:

- tractors Agromash 50 TK, Agromash 50 SSH, T-45, VTZ-2048;
- welding unit of ADD-4004 type;
- cement carrier TC-21.

D144 - four-cylinder internal combustion engine with a capacity of 4.15 litres and a power of 27.2 - 44.1 kW. It was installed on the following machinery:

- T-40, LTZ-55, T28X4M tractors;
- Forklift trucks 4014D, 40811,40261, 40271,40816;
- road rollers DU-63-1, DU-EZ, DU-47B, DU-94, DU-E6, DU-E7;
- Asphalt pavers DS-63-1, DS-155;
- Concrete mixers 581460, 581462;
- Compressor stations of ZIF and PKSD type;
- welding units of ADD type;
- power plants AD-16-T400-1 VP, ED-16-T400-1 VP;
- track machines PRM and MSSHU.

D145T is a four-cylinder engine with a capacity of 4.15 litres and a power of 41.9 - 55.1 kW. The engine was installed on the following types of machinery:

- tractor Agromash 85 TK;
- diesel generators AD-30, AD30-T400-1 VP, ED30-T400-1 VP, HS250-20/4;
- Concrete mixers 58146C, 58147C.

In July 2018, the company underwent mass layoffs of its working personnel. On 30 June 2019, the arbitration court declared the company bankrupt.

Zavolzhsky Motor Plant. The first technical task for the design of the diesel engine was received by the company in the late 1970s. It was planned to produce a turbocharged diesel engine with a cast-iron cylinder block with a capacity of 2.3 litres and a power of 80-90 hp. Over the past 15 years, the engineers created several different prototypes, conducted a number of tests as part of a car at NAMI proving ground, but the engine has not gone into production. Nevertheless, the idea of its creation was not forgotten and in 1993 the project on the development of a diesel engine on the basis of the produced perspective gasoline engine ZMZ-406 was launched. Later in the autumn of 1995, a two-litre 105 hp engine 406D.10 was tested, which served as the basis for the development of the engine ZMZ-514, which after the refinement of the cylinder head (GBTs) by Ricardo specialists, was put into mass production [6].

The first pilot batch of ZMZ-514 engines intended for GAZelle automobiles was assembled in 2002. But as early as 2004 serial production was stopped. The reason for this was the unstable quality of components and the difficulty of maintaining the accuracy of parts processing at the plant. Having carried out works on the refinement of the new engine, having changed the design of the block, rods, valvetrain timing chains and other parts in November 2005 the shop of small series of the enterprise has begun manufacture of diesel engines with an index ZMZ-5143 [7]. Due to the wrong priorities in the engine design, there were claims to the life of this combustion engine throughout its production.

Zavolzhsky Motor Plant upgraded the ZMZ-5143.10 engine by installing Bosch Common Rail fuel injection equipment to meet environmental standards and improve fuel efficiency and traction characteristics. The new ZMZ-51432.10 CRS engine became more demanding to fuel quality, and the cost of repairing its fuel equipment has increased significantly. Many drawbacks in this engine have passed from the old engine in the form of failures of valvetrain drives and high-pressure fuel pump, short life of the oil pump, impulses in the fuel line of the injection pump, cracks in the cylinder head.

The main reason for the decline in demand for ZMZ diesel engines was the Euro-5 emission standard, which was introduced on 1 January 2018. According to manufacturers' estimates, bringing these diesel engines up to Euro-5 (higher fuel injection pressure, the introduction of diesel particulate filters and more efficient catalytic converters, etc.) increases the final cost of the product only at the plant level by 1,000 - 1,500 euros, which will inevitably affect their demand.

A new plant as part of YMZ-Avtodiesel for serial production of engines of the new family was built and launched. This is one of the most advanced production facilities for diesel engines not only in Russia but also in Eastern Europe. Investments in the project amounted to 11 billion rubles, the estimated production volume is up to 50 thousand ICE per year.

The YMZ-530 engine is a new family of medium in-line four- and six-cylinder diesel engines with power from 130 to 330 hp with the potential to meet Euro-6 emission standard. Up to 400 modifications and configurations of engines with a high degree of unification have been developed. The model range was developed by the engineers of YMZ-Avtodiesel and GAZ Group with the support of the world leader in the field of transport engineering AVL List (Austria). The YMZ-530 engines are equipped with advanced design solutions for layout, control and operation of the main systems. Their production began in 2013.

Diesel engine YMZ-534 is in-line four-cylinder, which capacity is 4.43 litres. The power range from 100 to 176 kW, and the torque from 420 to 784 N.m. The design uses: cast iron cylinder blocks, cast iron wet sleeves, oil nozzles cooling the bottom of the piston, connecting rods made of steel. The diameter of cylinders is 105 mm, and the steel crankshaft provides a stroke of 128 mm, which provides the capacity of the internal combustion engine [8]. The engine has a "lower" camshaft located in the cylinder block (Cam-in-Block). The block head has 4 valves per cylinder.

The degree of localization of 530th series ICE components increases every year: in 2015 - 68%, in 2016 - 75%, in 2018 - 80%. GAZ Group has mastered the casting of cylinder blocks, turbo-compressors are manufactured by Turbotekhnika (formerly produced by German Schwitzer), production of pistons has been mastered in Kostroma, KAMAZ supplies crankshafts, and the camshaft is produced by the plant YAZDA, which is located next to YaMZ. They also prepare for the production of high-precision fuel-air pumps. It will produce 2000 bar instead of the usual 1000-1500 bar.

At present, the consumer is inclined to choose an engine which is reliable, efficient, easy to maintain, has high fuel efficiency performance. Due to the fact that the proposals of domestic engine builders are very limited, the scales are tilted towards numerous foreign manufacturers, which are present in the domestic market.

On the territory of the CIS, the factories of Belarus are especially notable in terms of their prevalence. The Minsk Motor Plant (MMZ) with its mass-produced engines due to their cheapness and ease of maintenance, which are in high demand in the post-Soviet space. This is the largest enterprise in Belarus, as well as the leading manufacturer of diesel engines among the CIS countries, it is ranked first in the number of engines for tractors and combines. The market share of this plant is more than 50% of the total fleet of tractors in the CIS. The main products of MMZ are diesel engines D-243, D-245 and their modifications with an in-line arrangement of cylinders and capacity of 4.75 litres. Their number and distribution have led to the fact that the consumer considers these internal combustion engines produced in Russia, but this statement is not true.

Cummins Inc. has localized its production in Russia in the Republic of Tatarstan by establishing a joint venture Cummins Kama to produce diesel engines with KAMAZ in Naberezhnye Chelny [9]. The Cummins "B" series engine assembly line with power from 140 to 300 hp and capacity from 4.5 to 6.7 litres was opened. Since March 2017, Cummins Kama has expanded its product range by launching assembly of L-series engines with a capacity of 8.9 litres and a power of up to 400 hp. Localization of this engine production reaches 60% of its cost.

Unfortunately, the production of Cummins "A" series engines with a power of 31 - 60 hp with a capacity of 1.4 to 2.3 litres and in-line 3- and 4-cylinder series "F" with a power of up to 160 hp with a capacity of 2.8 and 3.8 litres, which are in high demand in the market of light-duty cars, small tractors and municipal vehicles, the company does not plan to move to Russia in the near future. Besides, in the near future, such large foreign manufacturers as John Deere, Caterpillar, Deutz, Perkins do not plan to place their production of diesel engines in Russia yet, but negotiations continue with them [10-13].

At present, the main drawback for the organization of ICE production in Russia is the impossibility to manufacture components of the required quality. Manufacture of such parts for the valvetrain as piston rings, fingers, connecting rods, valves, springs, guide bushings, camshafts, gaskets and glands, causes great difficulties for domestic manufacturers due to high requirements to the quality of manufacturing of component base and ICE. In addition, there are major problems with the manufacture of parts made of aluminium alloys. Even the need for fasteners can be met by up to 50%, but only by "nonresponsible" fasteners. A separate problem is the lagging of Russian enterprises in a wide range of technologies for the production of complex billets by casting from high-strength cast iron and cast iron containing graphite, steel, as well as processing to strengthen the working surfaces of a wide range of parts by chemical-thermal, laser and plasma methods.

In recent years, a lot of attention has been paid to the provision of the agricultural, municipal and road construction vehicle market with Russian-made diesel engines at the highest level. Specialists of the Center of Agricultural Engineering of Central research and development automobile and engine institute NAMI carry out the work to identify the market demand and problem areas for the production of a range of diesel engines, necessary for the needs of the agricultural industry. Together with WDMA (German Engineering Association), a round table on "Development of Component Production and International Cooperation in Russia and the EAU Member States in Agricultural and Road Construction Engineering" was organized at the Agrosalon-2018 in Moscow. The event was attended by key producers of agricultural, construction and road construction machinery and components in Russia and abroad. Together with the National Agency of Social Communications, work was organized at the Strategic Session of the Russian Engine Forum in 2018. A working group "Agricultural piston engine" of leading manufacturers and scientists was formed.

Figures 1 and 2 show the dependence of the engine power on its weight and the dependence of the engine weight on its capacity. The diagrams show the area of piston internal combustion engine parameters, which has an optimal combination of characteristics and is the most demanded by manufacturers of agricultural, municipal and road-construction vehicles.

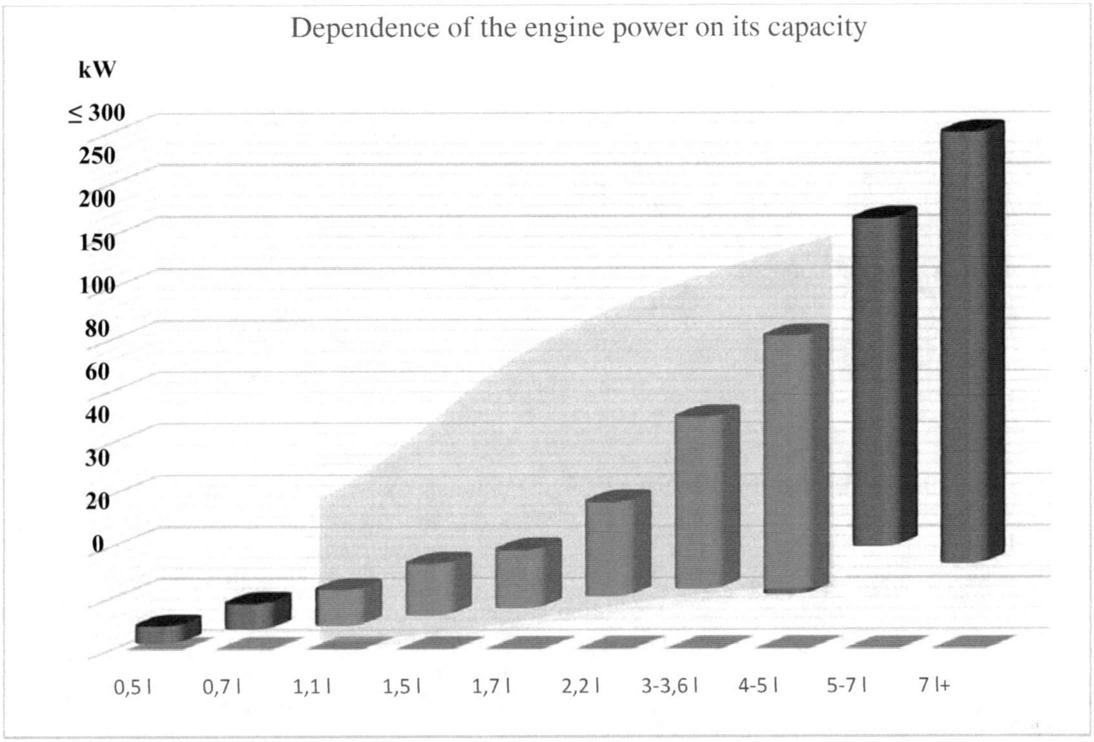

Figure1. Dependence of engine power and capacity.

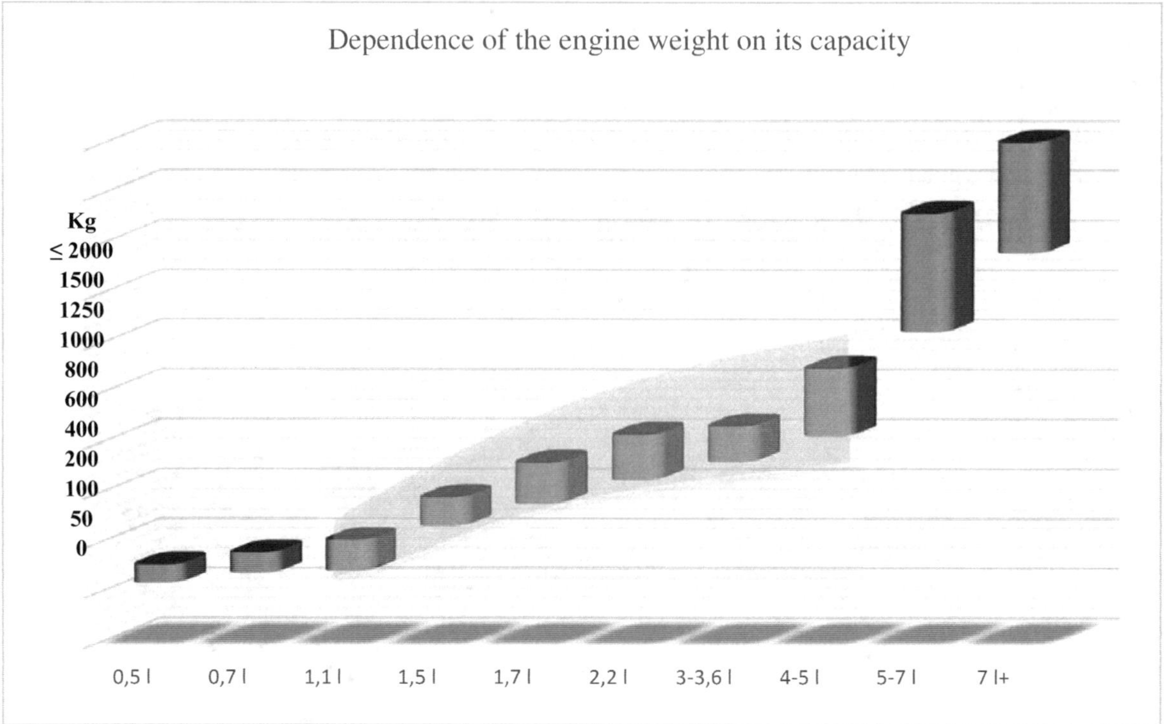

Figure 2. Diagram of dependence of the engine weight and its capacity.

4. Conclusion

When considering the power characteristics of internal combustion engines, it was established that manufacturers in our country need a range of reliable domestic diesel engines with a capacity of up to 4400 cc and a power of up to 150 kW. The analysis of trends and the current situation in the engine manufacturing industry and component base production for them has shown that along with the development of the family of ICE for certain types of vehicles, it is also necessary to develop the production of modern components and systems for them. In that case, machine-building in the Russian Federation will gain significant growth through the development of a considerable number of new advanced technologies. When organizing or localizing production, it will be necessary to pay attention not only to the reliability, efficiency and low noise of engines but also to their environmental friendliness in order to reduce the harmful impact on the environment. Therefore, one of the urgent technologies of the future on the ICE market will be the development and introduction of models operating on both mixed and biofuels and other alternative types of fuel.

Enterprises involved in serial production of diesel engines and components for them will have to master not only new technological processes but also modern methods, equipment and facilities for technical control to improve the quality of production and bring it to conformity with international regulations, requirements and standards. For this purpose, the enterprises to be established will need to employ highly qualified specialists, from workers to engineers and technicians. At present, an active search for partner organizations for these processes is underway. In addition, one of the conditions for consumers is the creation in the country of a widely developed and accessible network of technical support, service and timely delivery of spare parts.

With a proper approach to the problem considered in the work, joint discussion, search and implementation of solutions not only by manufacturers of component base and equipment for various purposes but also by government agencies, scientific institutions, educational institutions, it will be possible to get a quality and competitive product that will allow a breakthrough in the development of domestic engine building, reduce the dependence on imports and provide conditions for the supply of Russian engineering products for export.

References

[1] Website of the Institute of Information Technologies and Computer Aided Design, Federal Information Fund of domestic and foreign catalogues of industrial products "Diesel engines of VSN series". [Electronic resource]. URL: http://xn--80aajzhcnfck0a.xn--p1ai/PublicDocuments/0703989.pdf (accessed 13.09.2019)
[2] Official website of the company Tulamashzavod [Electronic resource]. URL: http://www.tulamash.ru/catalog/4 (accessed: 13.09.2019).
[3] Official website of LLC "Chelyabinsk Tractor Plant - URALTRAC", review of products [Electronic resource]. URL: http://chtz-uraltrac.ru/catalog/categories/1.php (accessed: 13.09.2019)
[4] Official website of LLC "Chelyabinsk Tractor Equipment Plant", overview of products [Electronic resource]. URL: http://www.chztt.ru/index.html (accessed date: 13.09.2019)
[5] Official website of the company "Vladimir Motor-Tractor Plant", review of products [Electronic resource]. URL: http://www.vmtz.tplants.com/ru/company/review/ (accessed: 13.09.2019)
[6] Internet edition of "Za rulem" publishing house [Electronic resource]. URL: https://www.zr.ru/content/articles/7772-molodoj_dizel_zhelajet_poznakomitsa/ (accessed 13.09.2019)
[7] Official website of ZMZ, review of products [Electronic resource]. URL: http://www.zmz.ru/Produktciya/Dvigateli_ZMZ/Semeystvo_ZMZ51410_Dvigatel_ZMZ_5143210_CRS_s_sist (RU) (accessed: 13.09.2019)
[8] Official website of ZMZ, review of products [Electronic resource]. URL: https://www.ymzmotor.ru/catalog/ymz-530/ (accessed: 13.09.2019)
[9] Official website of "Cummins" company, product review, diesel engines [Electronic resource]. URL: https://www.cumminsengines.com/engines (accessed: 13.09.2019)
[10] Official website of John Deere, product overview, industrial diesel engines [Electronic resource]. URL: https://www.deere.com/en/industrial-engines/ (accessed: 13.09.2019)
[11] Official website of the company "Caterpillar", product overview, industrial diesel engines [Electronic resource]. URL: https://www.cat.com/en_US/products/new/power-systems/industrial.html (accessed: 13.09.2019)
[12] Official website of the company "DEUTZ Russia" review of products [Electronic resource]. URL: http://www.deutz.ru/website.ru.html (accessed: 13.09.2019)
[13] Official website of the company "Perkins", review of products, industrial diesel engines Electronic resource]. URL: https://www.perkins.com/en_GB/products/new/perkins/industrial.html (accessed: 13.09.2019)

IOP Conf. Series: Materials Science and Engineering **819** (2020) 012028 doi:10.1088/1757-899X/819/1/012028

Problem of Hazardous Substances Content Reduction in Habitable Compartments of Driverless Vehicles

A M Saikin, S E Buznikov, A V Zhuravlev, E P Zaytseva and A V Lebedev

FSUE "NAMI", Moscow, Russian Federation

E-mail: a.saykin@nami.ru

Abstract. Nowadays, road and environmental safety of people during vehicle operation is one of the most important problems. One of the main ways to solve this problem is development and implementation of driverless vehicles, start of operation of which on public roads is expected since 2020-2025. In the process of development, no due attention is paid to addressing the social problem of ensuring passengers' environmental safety in habitable compartments of driverless vehicles (DLV) and modern driver-controlled vehicles. This fact may lead to the DLV implementation process restraint and negative consequences in the process of their operation as a result of harmful exposure of people's life and health to hazardous substances in the air of habitable compartments. The objective of this research is analysis of the solutions for this social problem.

1. Introduction

Global modern automobile transport development is significantly connected with the toughening of requirements for safety and economic efficiency of cargo and passenger transportation as well as for reduction of vehicle negative impact on human and environment. The priority automobile transport development field is considered to be the driverless vehicles (hereinafter referred to as "DLV") development, which is carried out by all leading automobile manufacturers planning to start the DLV implementation in 2020-2025 [1-6].

It is assumed that the DLV implementation would allow raising the efficiency and reducing the cargo and passenger transportation cost, ensuring traffic safety, using the road capacity more effectively, extending the possibility to use vehicles for people with reduced capabilities, ensuring cargo transportation possibility in danger areas, during natural and technogenic disasters or military operations, raising passengers' comfort and safety, reducing operating fuel consumption and travel time, reducing emissions of hazardous substances (hereinafter referred to as "HS") and greenhouse gases into the atmosphere.

Compared to other vehicle development fields, the DLV, active safety systems (ADAS) and DLV components are evolving at a higher rate. As an illustration of that, the foreign development patenting dynamics for the ground DLV, ADAS systems and DLV components is given in Figure 1.

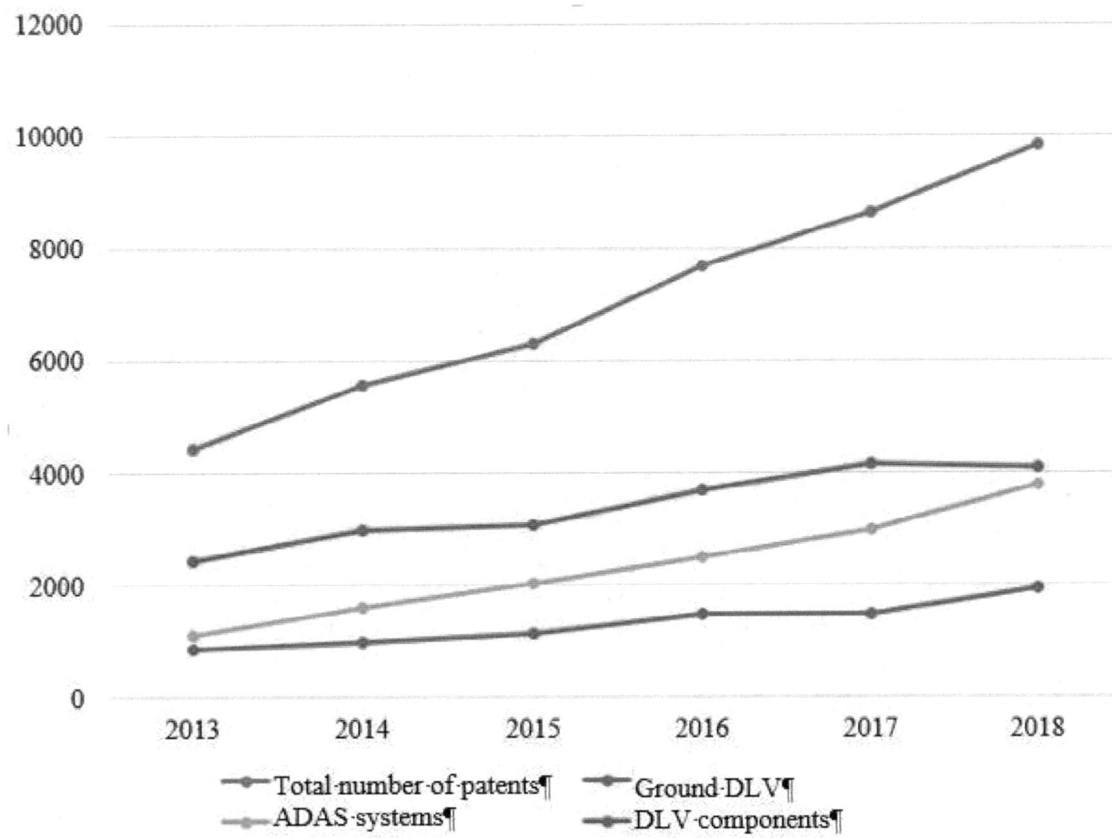

Figure 1. Foreign development patenting dynamics for the ground DLV, ADAS systems and DLV components.

As it appears from Figure 1, the issue of creation and implementation of the advanced DLV is one of the most relevant and prioritized issues of foreign and Russian automobile manufacturers.

Nowadays, in the DLV development, one of the most important issues is addressing the social problem of ensuring the environmental safety for the DLV and modern driver-controlled vehicle passengers in terms of harmful impact of the excessively contaminated air in the habitable compartments.

2. Main part

The lack of solution for this social problem will constrain the DLV implementation process and keep the negative consequences during their operation due to harmful impact on people's health and life. In addressing this problem, it is necessary to consider two circumstances.

The first circumstance is that nowadays the problem of excluding the environmental hazard for the passengers is not solved not only for the advanced DLV, but also for the modern and newly developed driver-controlled vehicles, as the HS content may be hazardous to health under the operating conditions in their passenger compartments, cabins and habitable compartments (hereinafter referred to as "HC"). According to publications in the USA, Germany, Japan, Korea, France, the UK, the Russian Federation and other countries [7-20], the consequences of air contamination in the modern vehicle HC may be premature mortality, common diseases of the whole population, feeling unwell, headache, confusion of consciousness when driving and hence a big number of road traffic accidents.

Nowadays the reduction in average life expectancy (premature mortality) due to vehicle negative impact on human is as follows: 1) about 9 months in the EU countries; 2) 1-2 years in the USA; 3) 25-28 thousand people per year in the Russian Federation [13, 16, 17].

The HS in the HC air are especially dangerous for children, aged people, pregnant women, disabled people, asthmatics, people with bronchopulmonary and cardiovascular diseases. Even small HS concentrations in the air have cumulative toxic effect having an impact on passengers' and drivers' health for all their subsequent life.

Environmentally friendly transport technical regulation and development started about 40 years ago. It related only to the HS specific emissions in exhaust gases (hereinafter referred to as "EG") into the atmosphere and did not provide for elimination of the harmful impact of contaminated air in the vehicle HC on drivers and passengers. However, the actual HS content in the atmospheric air, on the sidewalks and in the modern vehicle HC, especially in traffic jams, heavy traffic flows, tunnels, parking and garages may exceed the hygiene standards from 2-4 up to 10-20 times. In megalopolises, drivers and passengers spend from 0.5 to 1.5 hour on average per day in traffic jams and during this time, being in the HC, they inhale more HS, than during the rest of the day and few times more than people who do not use vehicles [7, 9-13]. As a result, until the considered problem is solved, the negative impact of HS in the driver-controlled vehicle and DLV HC on the population will remain and hold true.

Due to importance and relevance of the discussed problem solution, creation of international Vehicle Interior Air Quality (VIAQ) Informal Working Group was approved at the 70th session of the Working Party on Pollution and Energy (GRPE) of the World Forum for Harmonization of Vehicle Regulations (WP29) of the United Nations Economic Commission for Europe in 2015. This group developed draft Mutual Resolution No.3 (M.R.3) concerning the vehicle interior air quality, which was adopted at session 173 of the World Forum WP29 in November, 2017. However, this resolution includes provisions and harmonized test procedure for the measurement of HS emissions into the HC air from the internal trimming materials only and only for a new vehicle, which came off the assembly line, with the run of at most 80 km provided that the HS entering the vehicle HC from the ambient air are excluded.

The tests according to the harmonized procedure shall be carried out on a static vehicle in a special chamber (box) with measurement of the content of the carbonyl compounds, Volatile Organic Compounds (VOCs) including formaldehyde and acrolein in the vehicle HC air and in the test chamber air. However, the listed HS are not the priority ones in terms of the hazard level of the harmful impact on the population's health. Their share in the general air toxicity in the vehicle HC does not exceed 10-20%.

At present, the VIAQ Informal Working Group started the work on the second stage of development of the requirements for the air contamination by HS entering the HC through the ventilation and conditioning systems, as well as due to vehicle bodies nonhermeticity. At the same time, the EG and fuel and lubrication system evaporations from the test vehicle only are considered to be the contamination sources. The informal working group plans to start developing the requirements for HC air cleaning devices and on-board air quality monitoring systems for the vehicles only [7] from 2021.

It follows that the work plan of the VIAQ group is not aimed at the general solution of the considered problem in the coming years. It is also important to note that the international informal working group focuses only on piloted vehicles controlled by drivers and is not involved in solving the problem regarding passengers and operating personnel in the DLV HC.

Considering the fundamental shortcomings of the test procedures established in the Mutual Resolution M.R.3, the obtained test results when using them are not representative, as they do not allow estimating the air quality in the DLV and driver-controlled vehicle HC, estimating the efficiency of technical and technological solutions for air quality improvement in the HC under vehicle operating conditions at various speeds with excessive atmospheric air contamination, especially in cities.

It is important to note that for modern piloted and advanced vehicles, for instance, electric vehicles and electrically driven DLV that do not emit EG into the atmosphere, the solution of the social problem under discussion is also relevant. Without additional technical solutions, the wear products of road pavement, tyres, brakes and clutch (including PM2.5, PM10, CH2O, VOC, etc.) as well as O3 and other substances hazardous for health contained in the atmospheric air, will keep entering the HC of electric vehicles and advanced DLV.

Table 1 provides general information about the efficiency of removing HS (nitrogen dioxide (NO2), nitrogen oxide (NO), particulate matters (PM10), extra fine particulate matters (PM2.5), ozone (O3), carbon oxide (CO), formaldehyde (CH2O), Volatile Organic Compounds (VOC)) by means of modern air cleaners in the vehicle HC [23-32].

Table 1. Efficiency of modern cabin air cleaners.

Air cleaning methods and systems	Hazardous substances to be removed							Disadvantages of cabin air cleaners
	PM_{10}	$PM_{2.5}$	NO_x	CO	VOC	CH_2O	Soot	
Cabin air filter (paper)	+	-	-	-	-	-	⊥	Removes only coarse particulate matters (PM), does not remove NO_x, CO, CH, formaldehyde
Cabin air filter (paper + carbon)	+	-	-	-	⊥	NI	⊥	Does not remove fine PM, as well as NO_x, CO
Air ionizing air cleaner	-	-	-	+	+	⊥	⊥	Does not remove NO_x, dust; emits O_3 - the HS of hazard class 1 - into the air of habitable compartments
Ozonizer	-	-	-	+	+	⊥	⊥	Does not remove NO_x, PM; emits O_3 - the HS of hazard class 1 - into the air of habitable compartments
Photocatalytic purifier	-	-	-	+	+	NI	⊥	Does not remove NO_x, PM, large weight and overall dimensions
Pulsed corona discharge air cleaner	-	-	⊥	+	+	NI	+	Unacceptably high energy costs, weight, overall dimensions, no PM removal

Remark: + - air cleaning; - - no air cleaning; ⊥ - partial air cleaning; NI - no info.

The most commonly used for air cleaning in the vehicle HC are the **cabin air filters** installed in conditioning systems. The majority of population, and even specialists, have an erroneous idea that they effectively clean the contaminated air in the vehicle HC, ensuring the environmental safety of drivers and passengers.

But the modern **cabin air filters** containing paper (item 1, Table 1) and, at best, carbon filtering material (item 2, Table 1) do not remove nitrogen oxide and dioxide, ozone and carbon oxide from the air. This is their fundamental disadvantage. The cabin air filters are also not very effective regarding other priority HS. Due to the absence of standards for the fineness of air cleaning, paper filtering materials in them remove only coarse PM – larger than 2-5 microns [16] that are much less toxic than finer $PM_{2.5}$ [22, 26-29].

Air ionizing air cleaners (item 3, Table 1) **and ozonizers** (item 4, Table 1) partially remove only CO, some groups of hydrocarbons and soot from the air and reduce the air smell in the HC. However, they do not remove NO_2, NO and dust from the air and are not very effective against formaldehyde. Their major disadvantage is also that they generate and deliver ozone into the air of the cabins, the content of which in the air higher than the hygiene standard is more dangerous than any of the HS, such as NO_x, PM, CH_2O, CO.

In the Russian Federation, vehicle owners widely use air ionizing air cleaners (for instance, automobile air ionizers XJ-801, 802, "Fresh Air To Go", etc.) which are prohibited in the USA. The ozone content in urban atmosphere is typically at the level of or exceeds hygiene standards [24, 31, 32]. For this reason, additional emission of ozone into the air of the vehicle HC by ionizers without using means for diagnosing its content in the air (gas analyzers) is extremely hazardous and is not allowed in accordance with the current sanitary regulations of the Russian Federation. At the same time, air ionizing air cleaners are being successfully certified, which proves the shortcomings of the regulatory framework for vehicle environmental safety.

Photocatalytic purifiers (item 5, Table 1) ensure CO removal up to 70%, are less effective in relation to individual hydrocarbons, formaldehyde and do not remove NO_2, NO and dust from the air, and have a large weight and overall dimensions [25]. For these reasons, photocatalytic air purifiers are not widely used in the vehicle HC.

Pulsed corona discharge installations (item 6, Table 1) [27] along with little knowledge of them, design complexity, large overall dimensions and weight, excessively high power consumption, and limited number of HS that they remove, generate and deliver ozone into the HC air. Therefore, the use of these setups in the vehicle is not promising.

Thus, none of the means presented in items 1 to 6 of Table 1 provides effective air purification in terms of all priority HS. Given this, the world's leading automakers are developing their own technical solutions to reduce air contamination in vehicle HC, however, they are not effective enough. Tesla, for example, applies the control of vehicle HC air bioprotection using HEPA filters; BMW, Toyota and Volvo an apply air recirculation mode in HVAC systems in order to prevent ingress of HS into the vehicle HC from the ambient air [21-23].

Taking into account Russian and foreign experience in solving the social problem under consideration, we can conclude that at present the international community is still far from developing efficient technical and technological solutions to eliminate excessive air contamination in the driver-controlled vehicle and DLV HC.

The second fundamentally important circumstance is that there is no driver in the DLV. Therefore, maintenance of comfortable and safe conditions in terms of air contamination in the DLV HC shall be carried out automatically, regardless of the level of environmental pollution, DLV driving modes, operation parameters of air conditioning systems and applied means to reduce air contamination.

3. Conclusion

Note should be made of the lack of unified standardized assessment methods for the vehicle HC environmental safety and for assessment of technical and technological solutions for its improvement as well as of fragmentary nature of results of investigation of ventilation, conditioning and climate control system operation mode impact on the air quality in the vehicle HC depending on the air contamination level in the habitable compartment air and ambient air, and on the vehicle speed for the whole range of the most dangerous HS as well as CO2.

If using a comprehensive approach, solving the considered problem in the first stage shall begin from the development of an intelligent automatic control system for the HC environmental safety systems and tools operation, which will allow reducing the air contamination levels in the HC of not only DLV but also modern driver-controlled vehicles significantly.

Such control system, based on the gas analyzers and particulate matter counter data on the ambient and HC air contamination during the vehicle operation under the conditions of excessively contaminated air, will allow activating the conditioning system in recirculation mode automatically, thus reducing the HS entering the DLV and driver-controlled vehicle HC from the ambient air, and allow additional automatic activation of autonomous air cleaners to provide auxiliary air cleaning up to the safe levels.

Acknowledgments

This scientific paper was prepared based on the results of the applied scientific research performed with governmental financial support from the Ministry of Education and Science of the Russian Federation, Agreement No. 14.625.21.0043, Unique Project ID RFMEFI62517X0043.

References

[1] SAE J3016. Standard: Taxonomy and Definitions for Terms Related to On-Road Motor Vehicle Automated Driving Systems. Issued: 2014-01-16 – SAE International Inc., 2014

[2] Information on https://venturebeat.com/2017/04/07/meet-t-pod-an-autonomous-electric-truck-that-can-also-be- controlled-remotely/

[3] Information on https://techcrunch.com/2017/04/06/einrides-electric-self-driving-t-pod-is-a-new-kind-of-freight- transport-vehicle/

[4] Saikin A M, Buznikov S E and Karpukhin K E 2016 The Analysis of Technical Vision Problems Typical for Driverless Vehicles *Research Journal of Pharmaceutical Biological and Chemical Sciences*, Vol 7 (4) pp 2053-2059

[5] Kiss D and Papp D 2017 Effective navigation in narrow areas: a planning method for autonomous cars IEEE 15th *Int Symp on Applied Machine Intelligence and Informatics* (DOI: 101109/SAMI 20177880346)

[6] Jeong E 2017 *Evaluating the effectiveness of active vehicle safety systems Accident Analysis & Prevention*, Vol 100 pp 85–96

[7] Kozlov A V and Terenchenko A S 2018 Development of new requirements for air quality in cabins of passenger vehicles *Journal of Automobile Engineers*, no 4 pp 58-60 (In Russian)

[8] Shah A S V et al Global association of air pollution and heart failure: a systematic review and meta- analysis *The Lancet* – 2013 – T 382 – № 9897 – p 1039-1048

[9] Maximum permissible concentration of pollutants in the atmospheric air of populated areas http:// docscntdru/document/901865554 – RF

[10] Yakunova E A and Saykin A M 2016 Air quality in the passenger compartment *Engineering Bulletin* No 2 URL: wwwengsiru/doc/834293html

[11] Liu W T et al Effects of commuting mode on air pollution exposure and cardiovascular health among young adults in Taipei, Taiwan *International journal of hygiene and environmental health* –2015 – T 218 – № 3 – p 319-323

[12] Kubesch N J et al Respiratory and inflammatory responses to short-term exposure to traffic-related air pollution with and without moderate physical activity *Occupational and environmental medicine* –2015 – T 72 – № 4 – p 284-293

[13] *Environmental Science: Processes & Impacts* – 2014 – T 16 – № 6 – p 1309-1317

[14] Jung, HS, et al, Simultaneously reducing CO2 and particulate exposures via fractional recirculation of vehicle cabin air 2017 *World Electric Vehicle Journal*, 160 pp 77-88

[15] Hudda N and S Fruin 2018 Carbon dioxide accumulation inside vehicles: The effect of ventilation and driving conditions *Science of The Total Environment*, 610 pp 1448-1456

[16] Suarez-Bertoa R, A Zardini and C Astorga 2014 Ammonia exhaust emissions from spark ignition vehicles over the New European Driving Cycle Atmospheric Environment, 97 pp 43-53

[17] Sacks J D et al Particulate matter-induced health effects: who is susceptible? *Environmental health perspectives* – 2010 – T 119 – № 4 – p 446-454

[18] Tartakovsky L et al In-vehicle particle air pollution and its mitigation Atmospheric environment – 2013 – T 64 – p 320-328

[19] Tartakovsky L, Baibikov V, Czerwinski J, Gutman M, Kasper M, Popescu D, Veinblat M, Zvirin Y, 2013 In-vehicle particle air pollution and its mitigation Atmos Environ 64, pp 320-328
[20] Tesla Putting the Tesla HEPA Filter and Bioweapon Defense Mode to the Test 2016 Available from: https://wwwteslacom/blog/putting-tesla-hepa-filter-and-bioweapon-defense-mode-to-the-test
[21] BMW Automatic Air Recirculation (AAR) Available from: http://wwwbmwcomkh/asia/en/insights/technology/technology_guide/articles/automatic_air_recirculationhtml?source=index&article=automatic_air_recirculation
[22] Toyota, Owners Manual for Prius MY2017 2017
[23] Li, C, et al, Reducing Mobile Air Conditioner (MAC) Power Consumption Using Active Cabin-Air-Recirculation in A Plug-In Hybrid Electric Vehicle (PHEV) World Electric Vehicle Journal, 2018 9(4) p 51
[24] Sager U Automotive air quality: Adsorption performance of a cabin air filter during a life cycle //Filtration and Separation – 2012 – T 49 – № 4 – pp 38-41
[25] Laumbach R J Outdoor air pollutants and patient health *American family physician* – 2010 – T 81 – № 2 – p 175
[26] Dominici F et al Protecting human health from air pollution: shifting from a single-pollutant to a multi-pollutant approach *Epidemiology* (Cambridge, Mass) – 2010 – T 21 – № 2 – C 187
[27] Ponizovskiy A Z, Ponizovskiy L Z, Kryuchkov S P, et al 2012 Experimental-industrial testing of installations for removing nitrogen oxides and organic compounds from exhaust gases Moscow, Nauka - proizvodstvu, no 3, pp 22-25 (In Russian)
[28] Wang S, Hao J Air quality management in China: Issues, challenges, and options *Journal of Environmental Sciences* – 2012 – T 24 – № 1 – pp 2-13
[29] Shiraiwa M, Selzle K, Pöschl U Hazardous components and health effects of atmospheric aerosol particles: reactive oxygen species, soot, polycyclic aromatic compounds and allergenic proteins *Free radical research* – 2012 – T 46 – № 8 – pp 927-939
[30] Thain W Monitoring Toxic Gases in the Atmosphere for Hygiene and Pollution Control: Pergamon International Library of Science, Technology, Engineering and Social Studies – Elsevier, 2013
[31] UNECE Regulations No49, 83, 96
[32] Izrael Y A Ecology and control of the natural environment – Springer Science & Business Media, 2012

IOP Conf. Series: Materials Science and Engineering **819** (2020) 012029 doi:10.1088/1757-899X/819/1/012029

Test Methods and Equipment for Assessment of Passenger Compartment Environmental Parameters of Modern and Advanced Driverless Vehicles

A M Saikin, S E Buznikov, A V Zhuravlev, E P Zaytseva, A V Lebedev and D A Kuznetsov

FSUE "NAMI", Moscow, Russian Federation

E-mail: a.saykin@nami.ru

Abstract. At the moment, no due attention is paid to the problem of ensuring environmental safety of passenger compartments of modern and advanced vehicles, including the driverless ones. The objective of this research is to analyze solutions to this social problem by developing, at this stage, of a new test method and equipment for assessment of parameters of passenger compartments of modern and advanced driverless vehicles.

1. Introduction

The problem of creation and implementation of advanced driverless vehicles (DLV) is one of the most relevant and prioritized ones [1-13]. In 2016, the European Union (EU) adopted Regulations (347/2012 & 351/2012) on traffic safety. The regulations prescribe that all vehicles registered or used within the borders of the EEC shall be equipped with special systems that ensure higher traffic safety level. Those are advanced driver assistance systems (ADAS) ensuring pedestrian recognition, warning the drivers of an emergency, traffic jam and brake assists, systems for vehicle interaction with other road users and road infrastructure.

According to the expert opinion, implementation of DLV and ADAS will allow increasing of traffic safety, efficiency and will allow reducing of cargo and passenger transportation cost, will allow increasing of ride comfort for passengers, will allow reducing of operating fuel consumption and travel time, will allow reducing of emission of hazardous substances (HS) and greenhouse gases into the atmosphere, etc.

One of the unresolved issues in the development and implementation of DLV in the Russian Federation and abroad is the solution of the social problem of ensuring environmental safety of passengers in the passenger compartments, cabins and habitable compartments (hereinafter referred to as "passenger compartments") of modern and advanced vehicles.

IOP Conf. Series: Materials Science and Engineering **819** (2020) 012029 doi:10.1088/1757-899X/819/1/012029

2. Main part

Under the operating conditions of modern driver-controlled vehicles and DLV that undergo trial operation, the HS content often exceeds the hygienic standards [14-32]. As a result, modern vehicles continue to have a dangerous influence on life and health of the population in the USA, Germany, Japan, Korea, France, the UK, the Russian Federation and other countries [14-32].

Statistically, the USA had the following negative consequences due to excessive air contamination in passenger compartments: - in the 2000s, about 500 people died each year while driving in cars, and about 1,500 people in garages (not as a result of suicide); - the number of prematurely dying because of the effect of particulate matters (PM) in the air is tens of thousands of people, and the number of poisonings with carbon oxide (CO) reaches 10 thousand cases per year. The US experts have established a connection between the increased mortality and exposure to higher ozone concentrations in the air, especially for people over 65 years old. The number of visits to hospitals due to air contamination in the vehicle passenger compartments is estimated at hundreds of thousands of people, while the majority of the population does not associate their diseases with poisoning by contaminated air.

The air in the modern vehicle passenger compartments contains the following (in the decreasing order of harmful impact on health): ozone, nitrogen dioxide, nitrogen oxide, fine particulate matters, formaldehyde, volatile organic compounds (VOC), aldehydes, carbon oxide and other HS. The solution to the problem of reducing air contamination in the DLV and driver-controlled vehicle passenger compartments to meet the hygienic standards shall begin with the stage of developing test methods and benches for assessment of the environmental safety of DLV passenger compartments and the efficiency of technical and technological solutions to reduce air contamination in the driver-controlled vehicle and DLV passenger compartments.

Both for modern driver-controlled vehicles and advanced vehicles, for example, electric vehicles and driverless electric vehicles which do not emit exhaust gases to the atmosphere, the problem of air contamination in the passenger compartments will remain relevant, as modern cabin air filters of ventilation and air conditioning (HVAC) systems intended for air purification (cleaning) are not actually efficient, especially in relation to such most hazardous HS as nitrogen dioxide, nitrogen oxide, particulate matters (especially the fine ones), ozone, carbon oxide, and are not very efficient in relation to formaldehyde and acrolein. The wear products of road pavement, tyres, brakes and clutch, ozone and other HS contained in the air will keep entering passenger compartments of electric vehicles and advanced electrically driven DLV. Therefore, the problem of improving the quality of air in passenger compartments, cabins and habitable compartments will be relevant for the automotive industry in the long term as well.

The main causes of air contamination in the passenger compartments (in the decreasing order of significance) are as follows:

- Low efficiency of air cleaning by means of cabin air filters of ventilation and conditioning systems.
- Passenger compartment nonhermeticity.
- Absence of control systems for the HVAC systems operation modes depending on the air contamination level in the passenger compartments.
- Absence of diagnostic systems as to the level of air contamination in the passenger compartments.
- Absence of a systematic approach when developing ecological regulations for passenger compartment environmental safety. In particular, no comprehensive work is being done on developing standards for the range of Restricted Hazardous Substances (RHS), their limit content in passenger compartment air, requirements for air cleaning efficiency of cabin air filters and cleaners, requirements for body habitable compartment nonhermeticity, HVAC, HS content diagnostics systems for passenger compartment air, testing methods, etc.
- Lack of public awareness of the problem and of actions to be made in case of excessive air pollution.

IOP Conf. Series: Materials Science and Engineering **819** (2020) 012029 doi:10.1088/1757-899X/819/1/012029

Taking into account the importance and relevance of the discussed problem solution regarding the vehicle drivers and passengers, creation of international Vehicle Interior Air Quality (VIAQ) Informal Working Group was approved at the 70th session of the Working Party on Pollution and Energy (GRPE) of the World Forum for Harmonization of Vehicle Regulations (WP29) of the United Nations Economic Commission for Europe in 2015. For two years, this group had been developing draft Mutual Resolution No.3 (M.R.3) concerning the vehicle interior air quality, which was approved and adopted at session 173 of the World Forum WP29 in November, 2017. This resolution includes provisions and harmonized test procedure for the measurement of HS emissions into the passenger compartment air only from the internal trimming materials of a new vehicle, which came off the assembly line, with the run of at most 80 km provided that the contamination of the vehicle passenger compartment air from outside is excluded.

The tests according to this procedure shall be carried out on a static vehicle in a special chamber (box) with measurement of the content of carbonyl compounds, volatile organic compounds including formaldehyde and acrolein in the vehicle passenger compartment air and in the chamber air. However, the listed HS are not the priority ones in terms of the hazard of the harmful impact on the population's health. Their share in the general air toxicity in the vehicle passenger compartment and vehicle chamber does not exceed 10-20%.

At present, the VIAQ Informal Working Group has started working on the second stage of development of the requirements for the air quality in the vehicle passenger compartments regarding the passenger compartment contamination by the substances coming through the ventilation and air conditioning systems, as well as due to vehicle body nonhermeticity. Only the exhaust gases and evaporations from the fuel and lubrication system of the test vehicle are considered the contamination sources. The informal working group plans to start developing the requirements for air cleaning devices and on-board air quality monitoring systems in the piloted vehicles [14] beginning in 2021.

Thus, the VIAQ Informal Working Group formed a limited work plan for the considered problem solution and did not even start developing the test procedure (methodology) for the environmental safety assessment of the vehicle passenger compartments. It is important to note that the international VIAQ Informal Working Group considers only the piloted vehicles controlled by the drivers and does not plan to solve the same problem for the driverless vehicles.

The basic and main drawbacks of the VIAQ group work are as follows:

the tests according to the adopted test procedure are carried out only on a static vehicle, i.e. they do not include simulation (modeling) of the vehicle motion, which makes the main contribution to the air contamination in the driver-controlled vehicle passenger compartments and cabins;

the tests are carried out on a test vehicle under the clean air conditions, i.e. they do not include simulation (modeling) of the traffic flow and contaminated urban atmosphere, which make the main contribution to the air contamination in modern driver-controlled vehicle passenger compartments;

the tests are carried out measuring the following contaminating HS: carbonyl compounds, volatile organic compounds, including formaldehyde and acrolein, while the most dangerous passenger compartment contaminants are particulate matters, nitrogen dioxide, nitrogen oxide, ozone, carbon oxide; the total hazard of the last-named ones exceeds 90% of the total air toxicity in the driver-controlled vehicle passenger compartments and cabins.

The current standard of the Customs Union, GOST 33554-2015, developed in the Russian Federation and compulsory in the countries of the Customs Union, being an integral part of the Technical Regulation On Wheeled Vehicles Safety, practically duplicates the test procedure of the UNECE Regulations and has the same drawbacks. The test procedures applied according to Mutual Resolution No.3 (M.R.3) and Customs Union standard GOST 33554-2015 regarding the vehicle passenger compartment air quality shall be designated as the static ones, which may be conducted under the standard conditions with the specified parameters of the environment and the test vehicle, which may be performed and reproduced by various test laboratories, and that is their advantage.

However, taking into account the basic drawbacks of the test procedures adopted in Mutual Resolution M.R.3 and in the Customs Union standard, the obtained test results are not representative, as they provide assessment of the vehicle passenger compartment environmental safety only at idle and do not allow assessing the air quality in the passenger compartments of intelligent driverless and driver-controlled vehicles as well as the efficiency of technical and technological solutions for the air quality improvement in the passenger compartments under all vehicle operating conditions at different speeds with the excessive atmospheric air contamination, especially in cities and traffic jams.

There are dynamic (road) methods of testing the passenger compartments for environmental safety in the process of vehicle motion on public roads, which are used by foreign test subdivisions, in particular, by Emissions Analytics (Great Britain) with subdivisions in the United Stated, Germany and South Korea. However, the known dynamic test methods for assessment of the environmental safety of vehicle passenger compartments and impact of the ventilation and conditioning system operation modes on the environmental safety [28-32] allow obtaining only qualitative results due to incommensurable test conditions (different climatic conditions (air temperature and humidity, wind, pressure, etc.), road parameters, vehicle speed, ambient air contamination, etc.).

The same drawbacks are also characteristic of the stage of dynamic (road) tests set forth in GOST 33554-2015, which are used only for vehicle certification testing and not suitable for scientific research when addressing the discussed social problem.

Summing up the domestic and foreign experience related to the status of solving the social problem of ensuring the population's environmental safety, the following conclusions can be made.

The problem of creating modern driver-controlled vehicles and advanced DLV passenger compartments safe for human health has not received adequate attention and thus the solution thereof for the advanced DLV is of top priority and relevancy.

Nowadays there are no test methods for quantitative assessment of the vehicle passenger compartment environmental safety, for studying the causes of air contamination in the DLV and driver-controlled vehicle passenger compartments as well as for assessment of the technical and technological means and methods to reduce the air contamination level in the passenger compartments to meet the hygiene and health standards.

The solution for the considered problem regarding the driver-controlled vehicles and, in particular, driverless vehicles shall be comprehensive and shall include development of:

Test methods for assessment of harmless content of HS in the driver-controlled vehicle and DLV passenger compartment air.

Unique experimental setups and facilities to test the technical and technological solutions being developed for the guaranteed compliance with the hygienic requirements for the HS content in the driver-controlled vehicle and DLV passenger compartment air.

Standards for the range and maximum allowable concentration of the most hazardous substances that shall ensure environmental safety of the driver and passengers in the driver-controlled vehicle and DLV passenger compartments.

Environmental safety assessment criteria for the driver-controlled vehicle and DLV passenger compartments.

Technical specifications:

for instrument and equipment to measure the HS content in the air of driver-controlled vehicle and DLV passenger compartments;

for the nonhermeticity level of driver-controlled vehicle and DLV passenger compartments;

for emissions of HS in exhaust gases to the atmosphere;

for the design and health-safe operating modes of the driver-controlled vehicle and DLV passenger compartment HVAC;

for the design of cabin air filters and air cleaners including autonomous ones, for efficiency of air cleaning by means of these systems and devices in the driver-controlled vehicle and DLV passenger compartments;

for materials applied for tyres, clutches, brake systems and road pavement, as well as for the amount of their wear products;

for methods and means of diagnostics of air quality in the driver-controlled vehicle and DLV passenger compartments;

for test equipment and measuring instrumentation.

Due to the absence of a driver in a driverless vehicle, maintenance of comfortable and safe conditions in terms of air contamination in the DLV passenger compartments shall be ensured automatically under any driving conditions of the DLV, regardless of the environmental pollution, DLV driving modes, operation parameters of air conditioning systems (climate control) and means to reduce air contamination in the DLV passenger compartments.

Compared to the well-known methods, the test method being developed shall provide testing under stationary conditions with simulation of operating conditions such as excessive contamination of the air over the traffic way, especially in cities, driver-controlled vehicle and DLV speed, and shall ensure measurement of priority HS content (nitrogen oxide and dioxide, particulate matters, ozone, formaldehyde, carbon oxide and other HS) in the DLV passenger compartment air and ambient air during testing.

3. Conclusion

The test method being developed shall provide simulation of operating conditions (ambient air contamination, vehicle motion speed) by contaminating ambient air in a new specially created unique experimental facility performed in a controlled way by DLV exhaust gases (or by supplying corresponding HS to the test facility air from vessels), as well as by simulating vehicle motion speed, which can be performed by means of a fan blowing air at the DLV with a set or specified speed. In the course of testing, the content of priority HS shall be measured in the DLV passenger compartment air and in ambient air.

The results obtained in the course of the testing can form a basis for the following:

- identification assessment of the environmental safety parameters of intelligent DLV and driver-controlled vehicle passenger compartments;
- development of effective technical and technological solutions to ensure environmental safety of the intelligent DLV and driver-controlled vehicle passenger compartments;
- benchmarking or comparative testing of DLV and driver-controlled vehicle passenger compartments regarding their environmental safety parameter;
- benchmarking or comparative testing of technical and technological solutions to remove HS from the air in DLV and driver-controlled vehicle passenger compartments;
- certification testing of DLV and driver-controlled vehicles regarding their environmental safety parameter.
- The creation of integrated intelligent traffic control systems based on new principles opens the possibility to implement the projects, starting with the operation of unmanned cargo vehicles with electric drive in limited areas and their subsequent introduction to public roads, including the northern regions and the Arctic zone of the Russian Federation. The conceptual core of such systems is mathematics and software, which requires concentration of the developers' efforts in this area.

Acknowledgments

This scientific paper was prepared based on the results of the applied scientific research performed with governmental financial support from the Ministry of Education and Science of the Russian Federation, Agreement No. 14.625.21.0043, Unique Project ID RFMEFI62517X0043.

References

[1] SAE J3016. Standard: Taxonomy and Definitions for Terms Related to On-Road Motor Vehicle Automated Driving Systems. Issued: 2014-01-16 – SAE International Inc., 2014.

[2] Information on https://venturebeat.com/2017/04/07/meet-t-pod-an-autonomous-electric-truck-that-can-also-be- controlled-remotely/

[3] Information on https://techcrunch.com/2017/04/06/einrides-electric-self-driving-t-pod-is-a-new-kind-of-freight- transport-vehicle/
[4] Information on http://www.ixbt.com/news/2017/07/05/einride-t-pod.html
[5] Information on https://hightech.fm/2017/10/09/driverless-fuel-cell-platform
[6] Information on https://www.popmech.ru/vehicles/news-390862-bespilotnaya-gruzovaya-platforma-na-toplivnyh- elementah/
[7] Saykin A M, Bakhmutov S V, Terenchenko A S, Endachev D V, Karpukhin K E, Zarubkin V V 2014 Tendency of Creation of "Driverless" Vehicles Abroad Biosciences *Biotechnology Research Asia*, Vol 11 pp 241-246 (DOI: 10.13005/bbra/1469)
[8] Shadrin S S, Ivanov A M, Karpukhin K E 2016 Using Data From Multiplex Networks on Vehicles in Road Tests, in Intelligent Transportation Systems, and in Self-Driving Cars *Russian Engineering Research*, Vol 36 pp 811-814 (DOI:103103/S1068798X16100166)
[9] Saikin A M, Buznikov S E and Karpukhin K E 2016 The Analysis of Technical Vision Problems Typical for Driverless Vehicles Research Journal of Pharmaceutical, *Biological and Chemical Sciences*, Vol 7 (4): pp 2053-2059
[10] Kiss D and Papp D 2017 Effective navigation in narrow areas: a planning method for autonomous cars IEEE 15th *Int Symp on Applied Machine Intelligence and Informatics* (DOI: 101109/SAMI 20177880346)
[11] Jeong E 2017 *Evaluating the effectiveness of active vehicle safety systems Accident Analysis & Prevention*, Vol 100: pp 85–96
[12] Buznikov S E 2018 System analysis of vehicle active safety *problem IOP Conf Series: Materials Science and Engineering*, Vol 315 (DOI: 101088/1757-899X/315/1/012005)
[13] Kozlov A V, Terenchenko A S 2018 Development of new requirements for air quality in cabins of passenger vehicles *Journal of Automobile Engineers*, no 4, pp 58-60 (In Russian)
[14] Shah A S V et al Global association of air pollution and heart failure: a systematic review and meta- analysis *The Lancet* – 2013 – T 382 – № 9897 – pp 1039-1048
[15] Yakunova E A, Saykin A M 2016 The problem of perspective vehicles ecological safety IASF, October 18-19, Moscow, R20161045
[16] Information on https://wwwwhoint/mediacentre/factsheets/fs313/en/ - WHO
[17] Information on https://wwweeaeuropaeu/publications/air-quality-in-europe-2015 - Eu
[18] Information on https://wwwepagov/criteria-air-pollutants/naaqs-table - USA
[19] Maximum permissible concentration of pollutants in the atmospheric air of populated areas http:// docscntdru/document/901865554 - RF
[20] Information on https://wwwairparifassofr – Paris
[21] Yakunova E A, Saykin A M 2016 Air quality in the passenger compartment *Engineering Bulletin* No 2, URL: wwwengsiru/doc/834293html
[22] Liu W T et al Effects of commuting mode on air pollution exposure and cardiovascular health among young adults in Taipei, Taiwan *International journal of hygiene and environmental health* –2015 – T 218 – № 3 – pp 319-323
[23] Kubesch N J et al Respiratory and inflammatory responses to short-term exposure to traffic-related air pollution with and without moderate physical activity *Occupational and environmental medicine* – 2015 – T 72 – № 4 – pp 284-293
[24] De Nazelle A et al A travel mode comparison of commuters' exposures to air pollutants in Barcelona *Atmospheric Environment* – 2012 – T 59 – pp 151-159
[25] Suárez L et al Personal exposure to particulate matter in commuters using different transport modes (bus, bicycle, car and subway) in an assigned route in downtown Santiago, Chile *Environmental Science: Processes & Impacts* – 2014 – T 16 – № 6 – pp 1309-1317
[26] Saykin A M 2013 New concept of environmentally friendly transport *LAP LAMBERT Academic Publishing* -M, p 103- ISBN 978-3-659-39217-7
[27] Saykin A M, Ter-Mkrtich'yan G G, Karpukhin K E, et al Air quality within vehicles *Russian Engineering Research* – 2017 – T 37 – № 5 – pp 424-427

[28] Li C, et al, Reducing Mobile Air Conditioner (MAC) Power Consumption Using Active Cabin-Air- Recirculation in A Plug-In Hybrid Electric Vehicle (PHEV) *World Electric Vehicle Journal,* 2018 9(4): p51

[29] Lee E S and Y Zhu 2014, Application of a high-efficiency cabin air filter for simultaneous mitigation of ultrafine particle and carbon dioxide exposures inside passenger vehicles. *Environmental science technology* 48(4) pp 2328-2335

[30] Jung H S, et al, 2017 Simultaneously reducing CO2 and particulate exposures via fractional recirculation of vehicle cabin air 160 pp 77-88

[31] Hudda N. and S Fruin 2018 Carbon dioxide accumulation inside vehicles: The effect of ventilation and driving conditions. *Science of The Total Environment*, 610 pp 1448-1456

[32] Suarez-Bertoa, R., A. Zardini, and C Astorga, Ammonia exhaust emissions from spark ignition vehicles over the New European Driving Cycle *Atmospheric Environment* 97 pp 43-53

IOP Conf. Series: Materials Science and Engineering **819** (2020) 012030 doi:10.1088/1757-899X/819/1/012030

Selection of the required deceleration for high-mobility wheeled vehicles with wear resistant brake systems

G O Kotiev[1], I F Gumerov[3], A A Stadukhin[2] and B B Kositsyn[1]

[1] Department of the Wheeled Vehicles of Bauman Moscow State Technical University, Moscow, 105005, Russian Federation
[2] Department of the Multi-purpose Tracked Vehicles and Mobile Robots of Bauman Moscow State Technical University, Moscow, 105005, Russian Federation
[3] PAO Kamaz, Naberezhnye Chelny, Republic of Tatarstan, 423827,Russian Federation

E-mail: kositsyn_b@bmstu.ru

Abstract. Analysis of the driving cycles of the high-mobility wheeled vehicles shows that the driver tries to provide a high speed when performing maneuvers. This results in high loads both on the powerplant and the brake system. During quick changes of the driving modes, the service brake system has to dissipate a lot of energy, which may cause overheating of its working elements and changes (deterioration) in their frictional properties or their increased wear, or even a failure of the system. The article reveals the method for selection of the required deceleration of high-mobility wheeled vehicles with wear resistant brake systems providing higher speeds and considerably lower loads on the service brake system and increased life of the frictional elements. The method is based on the generation of the driving cycle, which is close to the real-life operation conditions of the high-mobility wheeled vehicles with the use of stochastic data on the roads. Thus, driving conditions are described as realizations of random functions of such external factors as the trajectory curvature, the friction coefficient between the tire and the road, and the tire rolling resistance coefficient. The driving cycle is generated with allowance for the vehicle stability, the powerplant performance, the physiological capability of the driver to withstand accelerations, and the maximum available deceleration. By varying the vehicle deceleration we can find the dependence of the vehicle average speed on its deceleration. The data obtained during the analysis helps to select the deceleration which can be effectively provided by the wear resistant brake system without the use of the service brake system. A method for selection of the wear resistant brake system characteristics with the use of stochastic data on the driving conditions. The method allows selection of the torque – speed curves of the units of a wear resistant brake system providing required mobility of the vehicle.

1. Introduction

High-mobility vehicles drive on the roads and on the terrain with different surface conditions [1, 2]. As a rule, when a driver performs a transportation task, he tries to provide the maximum speed determined either by the braking/tractive performance of the vehicle or by the condition of the vehicle stability during maneuvering. In this case, the driving cycle in the given road conditions is determined by the said limitations on the vehicle speed rather than by the driver [3, 4].

This assumption is the base of the proposed method for selection of the required deceleration of the high-mobility wheeled vehicles. At the first stage, a driving cycle is generated for the conditions

IOP Conf. Series: Materials Science and Engineering **819** (2020) 012030 doi:10.1088/1757-899X/819/1/012030

described by long realizations of random functions of the external disturbances. Then, the vehicle deceleration is varied and the dependence of its average speed on its deceleration is defined.

The data obtained during the analysis is used for the selection of the required deceleration, which can be effectively provided by the wear resistant brake system only in order to provide high mobility of the vehicle, as well as to reduce the heat loads and increase life of the frictional elements of the service brake system [5, 6, 7].

2. Generation of the vector of external disturbances

The routes for the high-mobility vehicles are generated from the known stochastic parameters of the roads as realizations of random functions of such road conditions as road curvature, maximum friction coefficient between the tire and the road, rolling resistance coefficient, road surface longitudinal slope angle [8 – 14].

The efficient method for generation of the above-mentioned stochastic parameters of the road is the non-canonical decomposition [9 – 12].

Thus, generating random combinations of these parameters for different road types and different terrain we can form a vector of the external disturbances, which is a combination of the long realizations of the random functions of the distance travelled by the vehicle. In order that this combination of the random parameters (road curvature, maximum friction coefficient between the tire and the road, rolling resistance coefficient, road surface longitudinal slope angle) could be used as a driving route, it should meet the following conditions:

- long realizations of the random functions must be generated as a sequence of the road segments with constant road conditions;
- on each of the road segment all the stochastic parameters must be defined: road curvature, maximum friction coefficient between the tire and the road, rolling resistance coefficient, road surface longitudinal slope angle.

The route generated with this method will have discontinuity at the boundaries of the road segments for each of the stochastic parameter function. Bearing this in mind, we have made the assumption that the given speed (maximum available speed for each segment) is obtained in the middle of the segment. At the same time, between the segments the vehicle speed changes according to a linear law, i.e. the vehicle moves with a constant acceleration. To form the equations of the vehicle motion we will assume that during the transition from one road segment to another the values of the parameters describing the vehicle interaction with the road are equal to their mean values derived from the values for the two segments (figure 1).

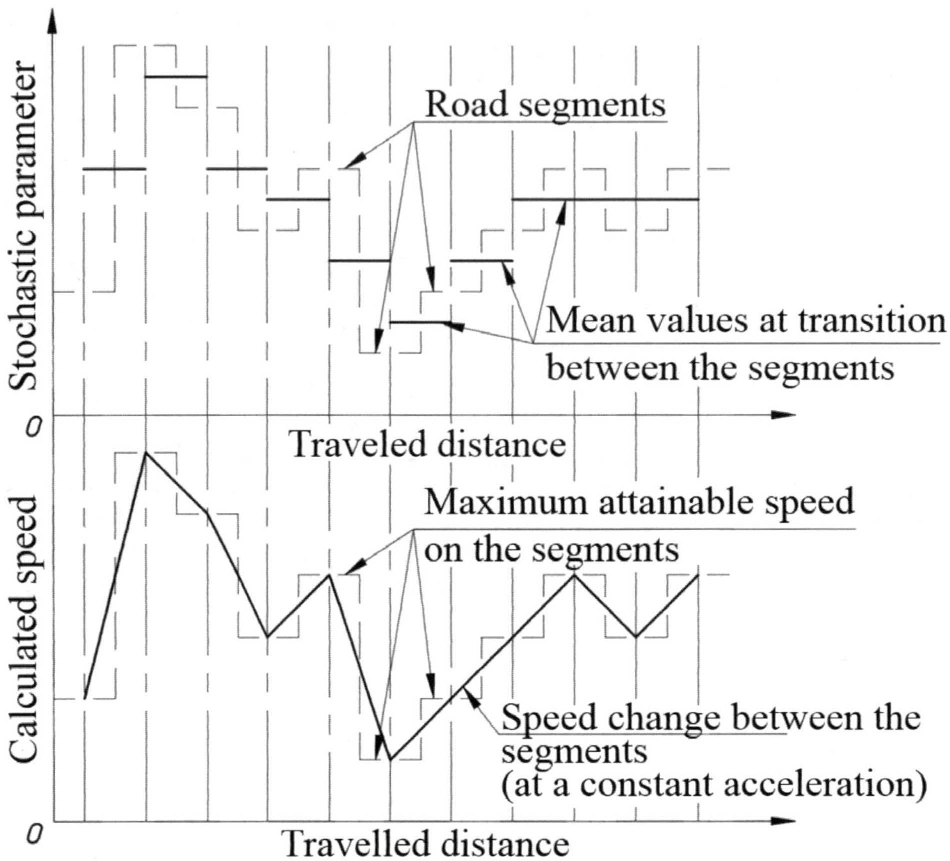

Figure 1. Changes of the road parameters at the transition between the road segments.

3. Generation of the driving cycle for a high-mobility wheeled vehicle with the use of a quasi-stationary model of the vehicle motion

During generation of the driving cycle of the vehicle, first, the upper bound of the vehicle maximum speed on each road segment and the length of the route must be estimated.

This estimate can be efficiently performed with a massless model of the vehicle motion [8]. The following estimates use several assumptions:

- the gearbox output torque – speed curve subject to the energy losses in the drivetrain is a constant power curve (corresponding to the maximum power of the power plant);
- increase in motion resistance at turns is negligible;
- the tire contacts the ground at one point (the turning resistance torque is neglected).

The upper bound of the vehicle maximum speed for each i-th segment of the road v_{lim_i} can be estimated as follows:

$$v_{lim_i} = min\left(v_{skd_i}, v_{roll_i}, v_{Nmax_i}, v_{max}\right),$$

$$v_{Nmax_i} = N_{max}/\left(mg\left(f_i \cos(\alpha_{\text{п}_i}) + sin(\alpha_{\text{п}_i})\right) + k_w F_{front} v_{Nmax_i}{}^2\right), \quad (1)$$

$$v_{skd_i} = \sqrt{\mu_{s\,max_i} g / k_{\text{д}_i}}, \qquad v_{roll_i} = \sqrt{gB/\left(2H_z k_{\text{д}_i}\right)}, \qquad k_w = c_x \rho_w / 2,$$

where f_i is the rolling resistance coefficient on segment i of the route; $v_{skd\,i}$ is the maximum speed limited by the vehicle skidding; $v_{roll\,i}$ is the maximum speed limited by the vehicle roll (the vehicle is assumed to be equipped with an ideal anti-roll bar); $v_{Nmax\,i}$ is the maximum speed limited by the vehicle engine power; v_{max} is the maximum designed speed (the maximum speed according to specification, this speed is usually limited by the engine control system); m is the mass of the high-mobility vehicle; g is the acceleration of gravity; B is the track width of the vehicle; H_z is the height of the vehicle center of mass; N_{max} is the maximum power of the vehicle engine; k_w is the aerodynamic force coefficient; F_{front} is the frontal area of the vehicle; c_x is the aerodynamic drag coefficient; ρ_w is the air density.

The dependence of the maximum speed bound v_{lim_i} on the travelled distance for a stochastically generated route is shown in figure 2 (the vehicle moves over a surfaced road on a flat terrain).

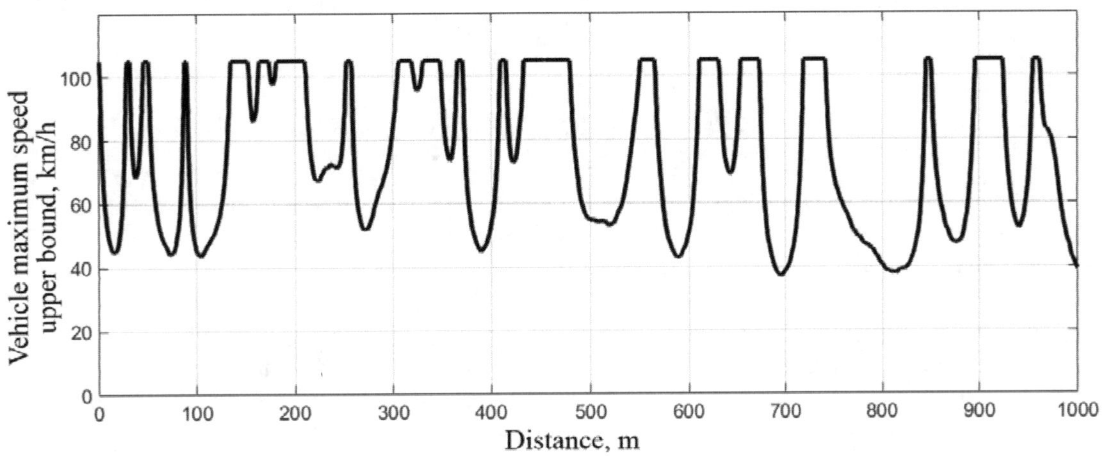

Figure 2. Upper bound of the vehicle maximum speed as a function of the travelled distance.

The length of the route with the given road surface should be estimated under the following condition: the average speed of the vehicle will level off at the end of the route. After that, the simulation can be stopped since the data collected after the speed stabilization will have no effect on the result for the given road. Since the inertia of the vehicle smoothes oscillations of its speed, the length of the route can be estimated with the use of a massless model (using speed upper bound as the average speed [8]). The upper bound of the vehicle average speed v_{mean}^{Max} depends on the travelled distance:

$$v_{lim\ i}^{seg} = \left(v_{lim_i} + v_{lim_{i-1}}\right)/2, \qquad \Delta s_i = s_i - s_{i-1},$$
$$v_{mean_i}^{Max} = \sum_{i=1}^{i} v_{lim\ i}^{seg} \Delta s_i / s_i, \qquad i = 1,2 \ldots n_s \tag{2}$$

where n_s is the number of the road segments constituting the route (s_i is the longitudinal coordinate on the i-th segment, s_{n_s} is the length of the route); $v_{lim\ i}^{seg}$ is the upper bound of the maximum speed of the vehicle during transition from segment i-1 to segment i of the route; Δs_i is the length increment between coordinates i-1 and i of the route (the length of the route segment).

Figure 3 shows the upper bound of the vehicle average speed as a function of the travelled distance. The example presented in figure 3 demonstrates that the required length of the route is 5 km (the discrepancy between the speed at 5 km and the final value at 10 km is less than 2%).

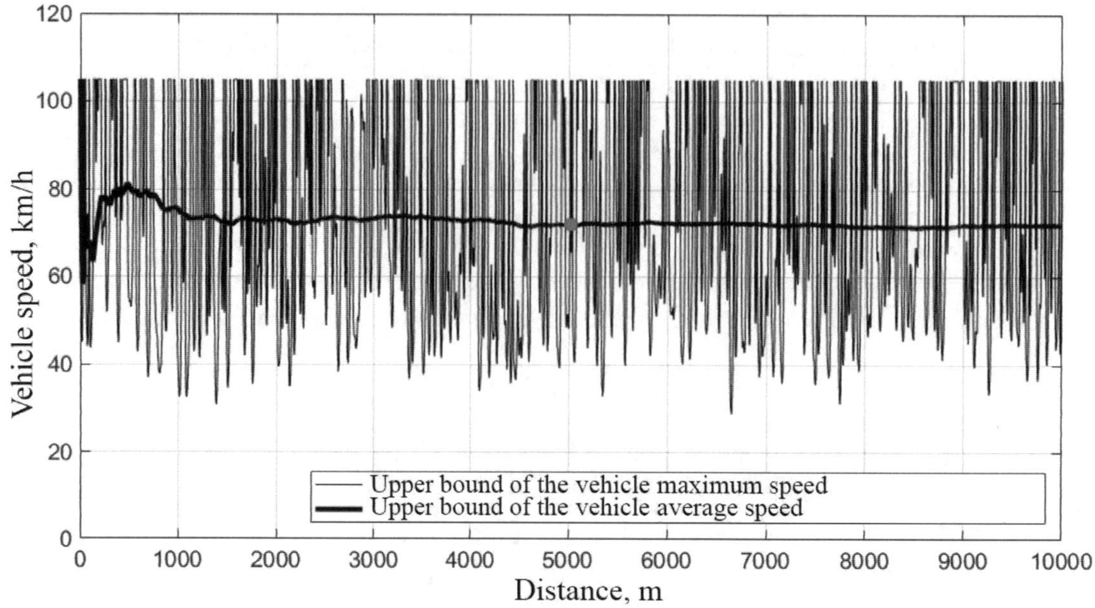

Figure 3. Upper bound of the vehicle average speed as a function of the distance.

On the next stage, limitations on the longitudinal acceleration/deceleration of the vehicle must be imposed on the driving cycle generated with the use of the above mentioned massless model (see figure 2). These limitations are determined by the tractive performance of the engine (including inertia forces) and by a certain maximum attainable deceleration of the vehicle. At that, by varying the latter we can estimate the dependence of the vehicle speed on the said deceleration.

In the driving mode, if the vehicle acceleration a_{lim_i} during transition from elementary segment *i*-1 to segment *i* of the route is higher than the limit value determined by the tractive characteristics of the road a_{μ_i} or by the engine performance a_{eng_i}, the driving cycle needs a correction:

$$v_{tr_i} = \begin{cases} v_{lim_i}, & if\ 0 \le a_{lim_i} \le \min\left(a_{\mu_i}, a_{eng_i}\right) \\ v_{tr_{i-1}} + \dfrac{\min\left(a_{\mu_i}, a_{eng_i}\right)\Delta s_i}{v_{tr\ i}^{seg}}, & if\ a_{lim_i} > \min\left(a_{\mu_i}, a_{eng_i}\right) \end{cases} \qquad (3)$$

where v_{tr_i} is the upper bound of the maximum speed of the vehicle on segment *i* of the route subject to longitudinal acceleration limit; $v_{tr_i}^{seg}$ is the upper bound of the maximum speed of the vehicle during transition from segment *i*-1 to segment *i* of the route subject to longitudinal acceleration limit;
At that a_{lim_i}, a_{μ_i} and $v_{tr\ i}^{seg}$ are defined as:

$$a_{lim_i} = \left(v_{tr_i} - v_{tr_{i-1}}\right) v_{tr_i}^{seg} / \Delta s_i, \qquad (4)$$

$$a_{\mu_i} = \mu_{s_{max_i}}^{seg} g, \qquad \mu_{s_{max_i}}^{seg} = \left(\mu_{s_{max_i}} + \mu_{s_{max_{i-1}}}\right)/2, \qquad (5)$$

$$v_{tr_i}^{seg} = \left(v_{tr_i} + v_{tr_{i-1}}\right)/2 \qquad (6)$$

where $\mu_{s_{max_i}}^{seg}$ is the maximum friction coefficient between the tire and the road during transition from segment *i*-1 to segment *i* of the route.

The following assumptions have been made:

• the gearbox output torque – speed curve subject to the energy losses in the drivetrain is a constant power curve (corresponding to the maximum power of the power plant);

• moment of inertia of the rotating elements of the engine and the drivetrain is transferred to the wheels;

• additional resistance at turns is negligible;

• there is no slip at the tire contact with the road;

• the tire contacts the road at one point;

• kinematic radii of all the wheels are the same.

Then, the following equation can be used: [15, 16]:

$$a_{eng\,i} = \frac{\frac{N_{max}}{v_{tr_i}^{seg}} - mg\left(f_i^{seg}\cos(\alpha_i^{seg}) + sin(\alpha_i^{seg})\right) - k_w F_{front} v_{tr_i}^{seg\,2} - J_z k_{d_i}^{seg}\frac{\Delta k_{d_i}}{\Delta s_i} v_{tr_i}^{seg\,2}}{m\delta + J_z k_{d_i}^{seg\,2}} \quad (7)$$

$$v_{tr_i}^{seg} = (v_{tr_i} + v_{tr_{i-1}})/2, \quad f_i^{seg} = (f_i + f_{i-1})/2, \quad \alpha_i^{seg} = (\alpha_i + \alpha_{i-1})/2,$$
$$k_{d_i}^{seg} = (k_{d_i} + k_{d_{i-1}})/2, \quad \Delta k_{d_i} = k_{d_i} - k_{d_{i-1}}, \quad \delta = 1 + \frac{\sum_{i=1}^{n} J_{k_i}}{m r_{k0}^2}, \quad (8)$$

where f_i^{seg} is the rolling resistance coefficient during transition from segment i-1 to segment i of the route; α_i^{seg} is the slope angle of the road during transition from segment *i*-1 to segment *i* of the route; $k_{d_i}^{seg}$ is the curvature of the trajectory during transition from segment *i*-1 to segment *i* of the route; Δk_{d_i} is the change of the curvature at transition from coordinate *i*-1 to coordinate *i* along the route; J_z is the vehicle moment of inertia about its vertical axis; δ is the mass factor taking into account the inertia of the rotating components of the driveline; n is the number of the wheels of the vehicle; $\sum_{i=1}^{n} J_{k_i}$ is the sum of the moments of inertia of the engine, the drivetrain, and the wheels, the moments being transferred to the axes of the wheels; r_{k0} is the tire rolling radius at zero slip (this radius can be assumed to be equal to the radius of the tire at its nominal load).

The upper bound of the maximum speed as the function of the distance is shown in figure 4.

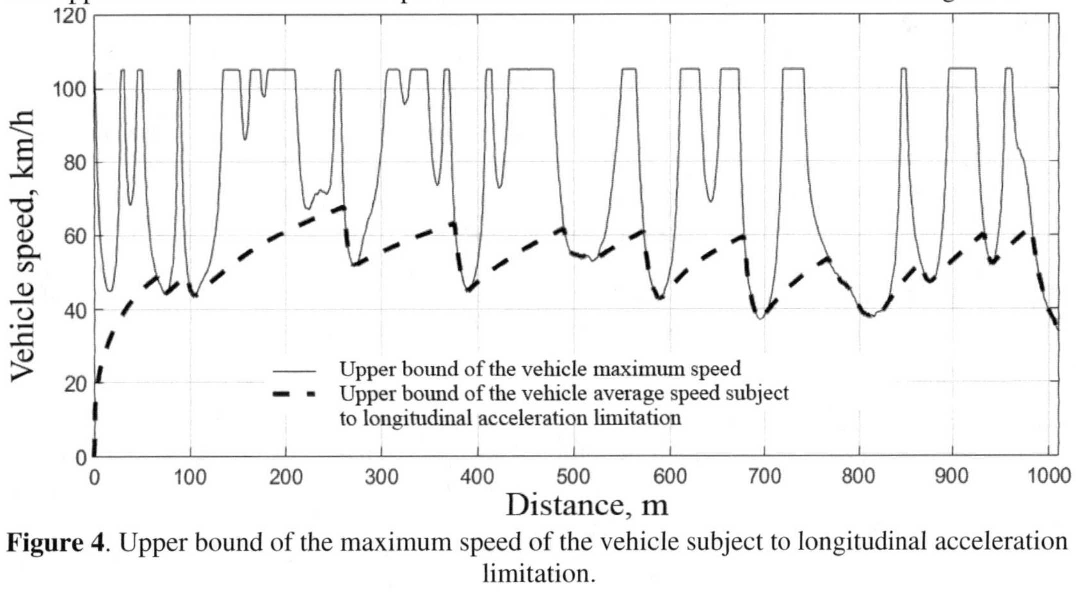

Figure 4. Upper bound of the maximum speed of the vehicle subject to longitudinal acceleration limitation.

In the braking mode, if the vehicle acceleration a_{lim_i} during transition from elementary segment i-1 to segment i of the route is lower than the limit value a_{μ_i} determined by the tractive characteristics of the road or the given deceleration a^*, the driving cycle needs a correction. It is important to note that in this case the segments must be taken in the order from the last one to the first one so that to arrive at the given speed of the vehicle at the end of each segment.

$$v_{brk_{n_s-i}} = \begin{cases} v_{lim_{n_s-i}}, if\ 0 \geq a_{lim_{n_s-i}} \geq \max\left(-a_{\mu_{n_s-i}}, -a^*\right) \\ v_{brk_{n_s-i+1}} - \dfrac{\max\left(-a_{\mu_{n_s-i}}, -a^*\right)\Delta s_{n_s-i}}{v^{seg}_{brk\,n_s-i}}, if\ a_{lim_{n_s-i}} \leq \max\left(-a_{\mu_{n_s-i}}, -a^*\right) \end{cases} \quad (9)$$

where $v_{brk_{n_s-i}}$ is the upper bound of the maximum speed of the vehicle on segment i of the route subject to longitudinal acceleration limit; $v^{seg}_{brk\,n_s-i}$ is the upper bound of the maximum speed of the vehicle during transition from segment $n_s - i + 1$ to segment $n_s - i$ of the route subject to longitudinal acceleration limit; $v^{seg}_{brk\,n_s-i}$ is calculated by the following equation:

$$v^{seg}_{brk\,n_s-i} = (v_{brk_{n_s-i}} + v_{brk_{n_s-i+1}})/2 \quad (10)$$

The speed v_{brk} as the function of the travelled distance is shown in figure 5.

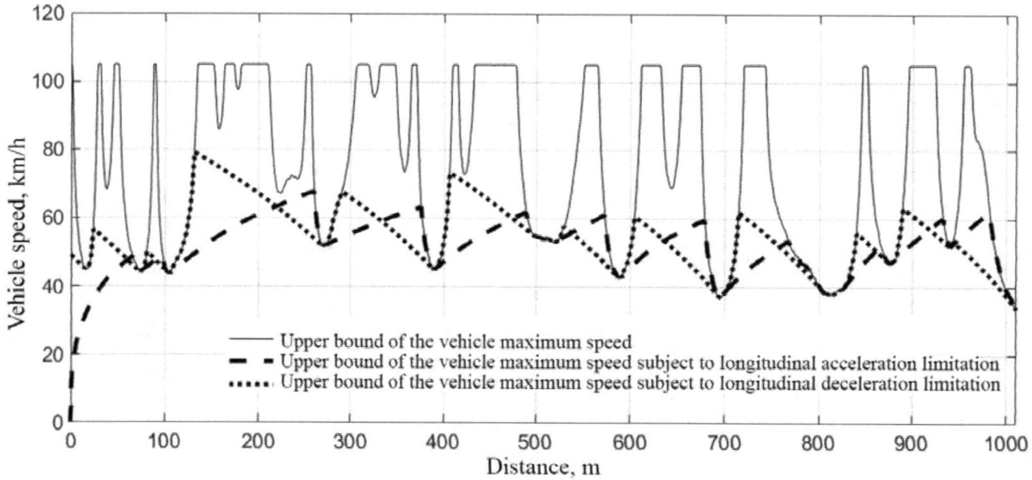

Figure 5. Upper bound of the maximum speed of the vehicle subject to longitudinal deceleration limitation.

The final driving cycle of the vehicle including limitations on the longitudinal acceleration/deceleration is generated as a combination of the curves v_{tr} and v_{brk} (see figure 6).

$$v = \min(v_{tr}, v_{brk}). \quad (11)$$

Thus, the described massless model with additional limitations on the longitudinal acceleration/deceleration provides generation of the driving cycle for a high-mobility wheeled vehicle on a given route. Taking into account the assumption that the driving cycle for a high-mobility wheeled vehicle on a given road is rather determined by the braking/tractive performance of the vehicle or by the condition of the vehicle stability during maneuvering than by the driver, we can argue that the generated driving cycle is close to the real-life operation cycle of the vehicle.

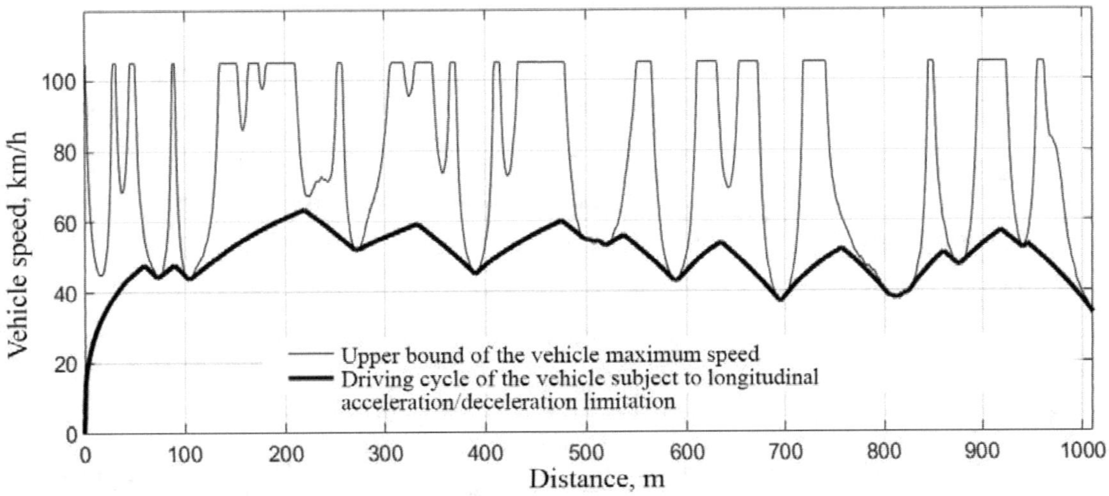

Figure 6. Driving cycle of the high-mobility wheeled vehicle limited by deceleration $a^* = 1\ m/s^2$.

4. Estimation of the deceleration to be provided by a wear resistant brake system of a high-mobility wheeled vehicle

During selection of the deceleration to be provided by a wear resistant brake system we assume that the vehicle speed decreases only due to the operation of the retarders and the deceleration is not higher than a^*.

Then, with the use of the generated driving cycle for a high-mobility wheeled vehicle subject to the longitudinal acceleration/deceleration limitation (see figure 6) the average speed of the vehicle v_{mean} in the given road conditions (for the given acceleration limit a^*) can be calculated as:

$$v_{mean} = \sum_{i=1}^{n_s} v_i^{seg} \Delta s_i / s_{n_s}, \qquad v_i^{seg} = (v_i + v_{i-1})/2 \tag{12}$$

By generating driving cycles for different a^*, we can form the vehicle average speed – deceleration limit curve. By way of example, figure 7 shows the vehicle average speed – deceleration limit curves for a high-mobility wheeled vehicle with gross weight 34 t with different power-to-weight ratios (15 kW/t and 22 kW/t) on a bituminous-concrete surface.

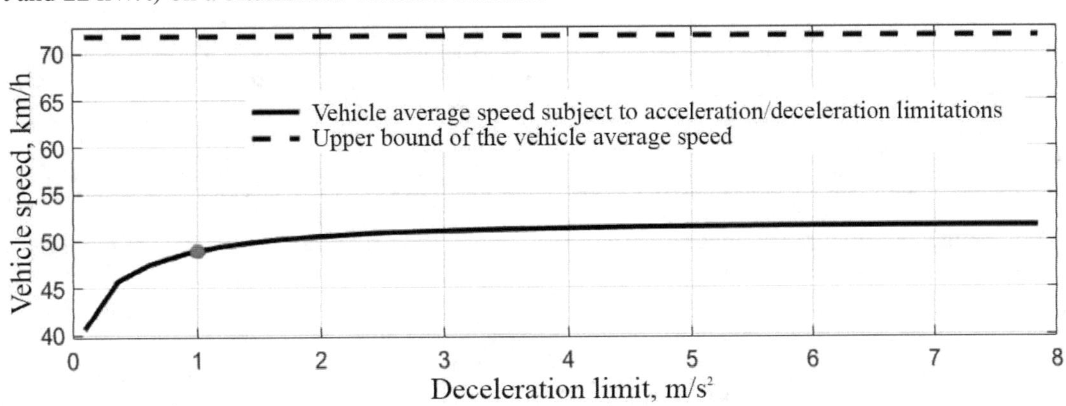

a)

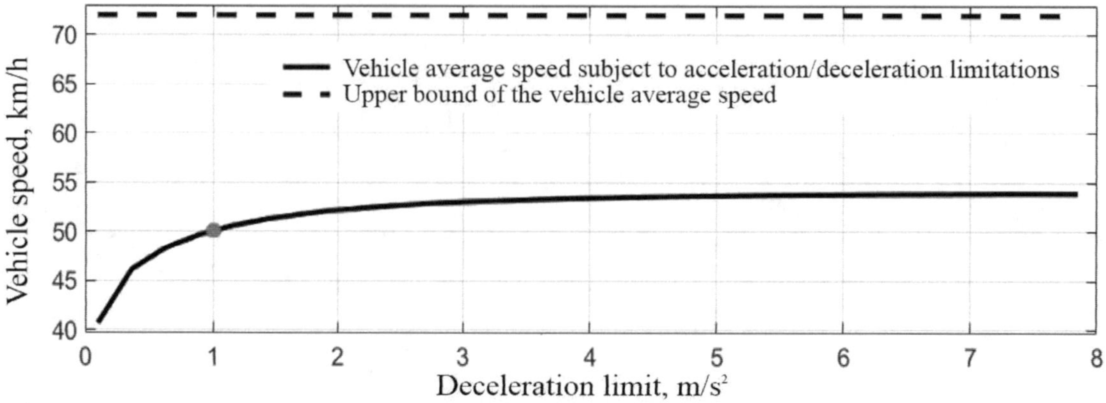

b)

Figure 7. Average speed – deceleration limit curves for a a high-mobility wheeled vehicle: a) power-to-weight ratio 15 kW/t; b)power-to-weight ratio 22 kW/t.

Figure 7 shows that the vehicle deceleration higher than 1 by the units of the secondary brake system (retarders) is not efficient, since it doesn't increase the average speed neither for the vehicles already in operation (with power-to-weight ratio 15 kW/t) nor for the advanced vehicles (with power-to-weight ratio 22 kW/t).

The upper bound of the vehicle average speed did not change with the change of the power-to-weight ratio because in this case the vehicle maximum speed is limited not by the maximum power of the engine, but by the vehicle specification (the control system limits the maximum speed by the value 105 km/h).

The driving cycle of the vehicle with the power-to-weight ratio 15 kW/t for the selected deceleration limit 1 is shown in figure 6.

This deceleration correlates with the data from [17] (deceleration at service braking does not exceed 0.8 – 1.7g) and corresponds to the boundary of the comfort zone of the longitudinal accelerations [18] (the limit longitudinal acceleration for a comfortable ride is 0,1g, table 1).

Table 1. Limit values for the root mean square of the accelerations acting on the people onboard a car [14].

Limit values for the root mean square of the accelerations for	Vertical accelerations	Longitudinal accelerations	Lateral accelerations
comfort boundary	0,1g	0,06g	0,05g
comfortable ride	0,25g	0,1g	0,07g
short term accelerations	0,4g	0,2g	0,1g

The driving cycle shown in figure 6 was generated without limitation on the lateral acceleration acting on the driver (lateral accelerations when the vehicle moves with the roll limit speed are $\mu_{s_{max}} g$) People are not able to withstand driving with these accelerations for a long period of time. Therefore, the driving cycle of a high-mobility vehicle should account for the lateral acceleration limit value $a_y = 0{,}1g$ [18], which corresponds to the limit value for the short term accelerations (see table 1).

To this end, the upper bound of the vehicle maximum speed for each *i*-th segment of the road must be calculated as:

$$v_{lim_i} = min\left(v_{a_y}, v_{skd_i}, v_{roll_i}, v_{Nmax_i}, v_{max}\right), \qquad v_{a_y} = \sqrt{a_y/k_{d_i}}, \tag{13}$$

where v_{a_y} is the vehicle maximum speed subject to lateral acceleration limitation.

Then, the vehicle speed v_{mean} as the function of the deceleration limited by the value of the lateral acceleration can be presented by the curve shown in figure 8.

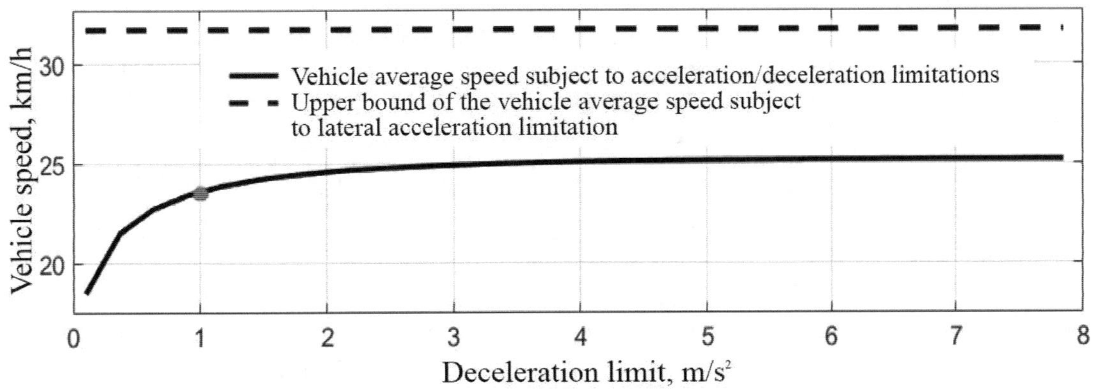

a)

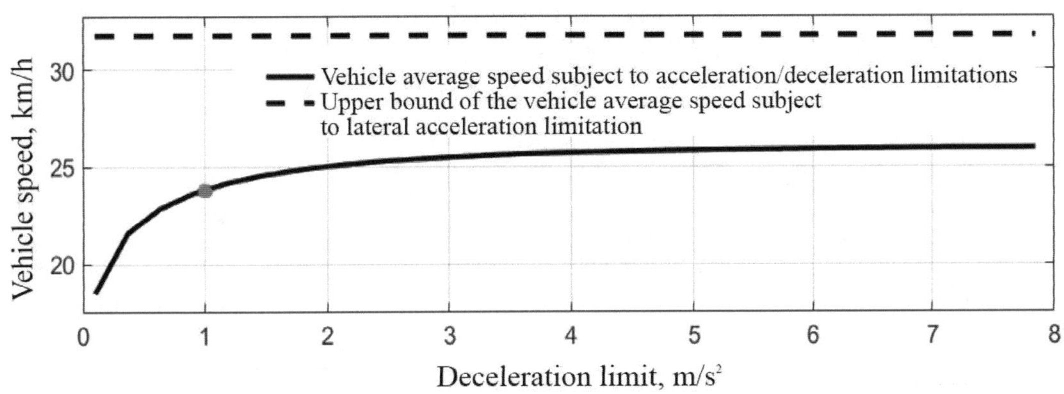

b)

Figure 8. Average speed – deceleration limit curves for a a high-mobility wheeled vehicle: a) power-to-weight ratio 15 kW/t; b)power-to-weight ratio 22 kW/t.

The driving cycle of the vehicle with the power-to-weight ratio 15 kW/t for the selected deceleration limit 1 m/s^2 limited by the value of the lateral acceleration $a_y = 0{,}1g$ is shown in figure 9.

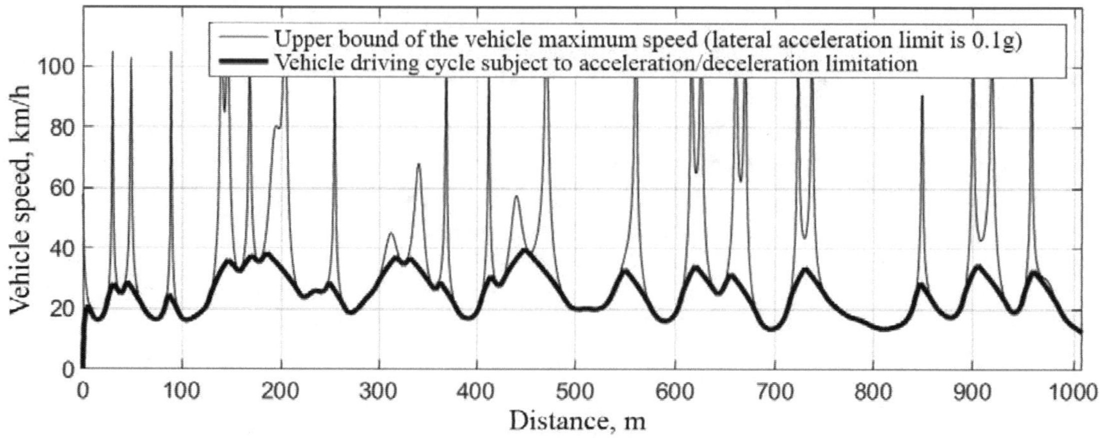

Figure 9. Driving cycle of the high-mobility wheeled vehicle limited by deceleration $1\ m/s^2$

The curves of figures 7 and 8 show that the required deceleration of the high-mobility wheeled vehicle (with limitation of the lateral acceleration) provided by the retarders is $1\ m/s^2$, which is the same value as in the case without limitation of the lateral acceleration.

Thereby, we can conclude that regardless of the engine power and accelerations acting on the driver the required deceleration which must be provided by the units of the wear resistant brake system of the high-mobility wheeled vehicle does not exceed $1\ m/s^2$.

It is easy to see that the lateral acceleration limit $0{,}1g$ corresponds to the roll limit acceleration on an ice-covered road $\mu_{s_{max}} = 0{,}1$ [19]. It means that the results shown in figures 8 and 9 can be regarded as the results obtained for driving on ice, and the deceleration which must be provided by the wear resistant brake system does not depend on the tractive characteristics of the tire and is also not higher that $1\ m/s^2$.

5. Conclusion

The analysis has shown that the deceleration which must be provided by the units of the wear resistant brake system of a high-mobility wheeled vehicle for obtaining the efficient average speed does not depend on the engine power and driving conditions, and is not higher than $1\ m/s^2$.

The described method for the selection of the required deceleration of the high-mobility wheeled vehicles at the design stage provides development of the specifications and torque – speed curves for the units of the wear resistant brake system.

References

[1] Keller A and Aliukov S 2016 Efficient Power Distribution in an All-Wheel Ground Vehicles *SAE Technical Paper* 2016-01-1105

[2] Aliukov S, Keller A, and Alyukov A 2015 On the Question of Mathematical Model of an Overrunning Clutch *Lecture Notes in Engineering and Computer Science: Proceedings of The World Congress on Engineering* pp1319-1324.

[3] Keller A , Aliukov S and Anchukov V 2017 Mathematical model of the truck for investigation of stability and control of movement *Lecture Notes in Engineering and Computer Science* pp. 711-716.

[4] Keller A , Shelepov A and Istomin D 2016 Estimation of Screw-ball Differential on Vehicle Dynamics *Procedia Engineering* 150 pp 1131-1136.

[5] Nosko A L, Safronov E V and Soloviev V A 2018 Study of Friction and Wear Characteristics of the Friction Pair of Centrifugal Brake Rollers *Journal of Friction and Wear* **39** No. 2 pp 145–51

[6] Gekker F R 1998 Generalized dynamic model of 'dry' friction units *Trenie i Iznos* **19** Issue 2 pp 165-70

[7] Yudin E G, Vartanyan V A and Vasilieva E I 1998 Improvement of quality and tribological performance of transmission parts by electrochemical polishing *Journal of Friction and Wear* **19** Issue 2 pp 81-6

[8] Kositsyn B B, Miroshnichenko A V and Stadukhin A A 2019 Modeling of random functions of characteristics of road-ground conditions in the study of the dynamics of wheeled and tracked vehicles at the design stage *"Izvestiya MGTU "MAMI"* - M. **3(41)**. pp 36-46. DOI: 10.31992/2074-0530-2019-41-3-36-46

[9] Shalygin A S and Palagin Yu I 1986 *Prikladnye metody statisticheskogo modelirovaniya*. L.: Mashinostroenie. Leningr. otd-nie. p 320

[10] Kotiev G O and Sarach E B 2010 *Kompleksnoe podressorivanie vysokopodvizhnykh dvukhzvennykh gusenichnykh mashin*. M.: Izd-vo MGTU im. N.E. Baumana. p 184

[11] Rasshcheplyaev Yu S and Fandienko V N 1981 *Sintez modeley sluchaynykh protsessov dlya issledovaniya avtomaticheskikh sistem upravleniya*. M.: Energiya. p 145

[12] Chernetskiy V I 1968 *Analiz tochnosti sistem upravleniya*. M: Mashinostroenie p 247

[13] Sarach E, Kotiev G and Beketov S 2018 Methods for road microprofile statistical data transformation // *MATEC Web of Conferences*. **224**. article 04009.

[14] Kotiev G O, Miroshnichenko A V, Stadukhin A A and Kositsyn B B 2019 Estimating operation modes for the individual wheel electric drive of the all-wheel drive vehicle with the use of the driving simulator *IOP Conf. Series: Materials Science and Engineering* **534** 012004

[15] Savochkin V A and Dmitriev A A 1993 Statisticheskaya dinamika transportnykh i tyagovykh gusenichnykh mashin. M.: *Mashinostroenie*. p 320

[16] Kositsyn B B 2017 *Metod opredeleniya energoeffektivnogo zakona dvizheniya elektrobusa po gorodskomu marshrutu*: diss. …kand. tekhn. nauk: 05.05.03. M.: MGTU im. N.E. Baumana p 165

[17] Kravets V N 2014 *Teoriya dvizheniya avtomobilya*. Nizhniy Novgorod: Nizhniy Novgorod p 696

[18] Rotenberg R V 1972 *Podveska avtomobilya*. Izd 3-e pererabot. i dop. M.: Mashinostroenie p 392

[19] Platonov V F and Leviashvili G R 1986 *Gusenichnye i kolesnye transportno-tyagovye mashiny*. M.: Mashinostroenie p 296

IOP Conf. Series: Materials Science and Engineering **819** (2020) 012031 doi:10.1088/1757-899X/819/1/012031

Selection of the torque – speed curves for the units of a wear-resistant brake system of high-mobility wheeled vehicles

G O Kotiev[1], I F Gumerov[3], A A Stadukhin[2] and B B Kositsyn[1]

[1] Department of the Wheeled Vehicles of Bauman Moscow State Technical University, Moscow, 105005,Russian Federation
[2] Department of the Multi-purpose Tracked Vehicles and Mobile Robots of Bauman Moscow State Technical University, Moscow, 105005, Russian Federation
[3] PAO Kamaz, Naberezhnye Chelny, Republic of Tatarstan, 423827, Russian Federation

E-mail: kositsyn_b@bmstu.ru

Abstract. With the growth of the power-to-weight ratio of the modern high-mobility wheeled vehicles, the capacity of their working brake systems is no longer sufficient for operation at the high speeds owing to high thermal loads of the brakes. In order to increase the average speed of the vehicle during the long-term downslope decelerations, as well as at the intensive use of the wheel brakes it is necessary to use additional devices – exhaust and retarder brakes. This kind of devices include the exhaust brake installed inside the internal combustion engine and the transmission hydro- and electrodynamic brakes (retarder brakes). The use of an electric machine operating in generator mode for storing the braking energy in an electric storage device is also a promising option. The combination of such additional devices is called "a wear-resistant brake system". The research is focused on the selection of the torque – speed curves of a wear-resistant brake system providing higher average speeds of the wheeled vehicles and lower loads of the working brake system. Selection of the torque – speed curves of the units of a wear-resistant brake system is carried out by the analysis of the wheeled vehicle operation in different driving cycles which are close to its real operation cycles. These cycles are usually generated for a stationary case with the use of a quasi-stationary mathematical model of the vehicle motion. In the model, the vehicle – terrain interaction is described by means of the equations of the vehicle motion including stochastically generated roads. The analysis of the vehicle operation in different driving cycles provides stochastic data on the loads of the wear-resistant brake system as a whole, which allow selecting the torque – speed curves of its units depending on their operational features. The proposed method provides selection of the required torque – speed curves of the wear-resistant brake system units: the internal combustion engine (operating in the brake mode), the electric machine, and the retarder. Selection of the parameters of the electric machine is the first step for the solution of the next important problem – selection of the required capacity of the electric energy storage device. The results of the analysis provide selection of such combination of the characteristics for each unit of the wear-resistant brake system that will ensure the required deceleration of the vehicle.

1. Introduction

The proposed method of the selection of the torque – speed curves for the units of a wear-resistant brake system of high-mobility wheeled vehicles is based on the following assumption. When a driver performs a transportation task he tries to provide the maximum speed determined either by the braking/tractive

IOP Conf. Series: Materials Science and Engineering **819** (2020) 012031 doi:10.1088/1757-899X/819/1/012031

performance of the vehicle or by the condition of the vehicle stability during maneuvering [1, 2]. It means that the speed of the vehicle in the given road conditions is rather determined by the driver than by the limitations imposed by the braking/tractive performance of the vehicle or by its stability during maneuvering [3, 4].

Thus, a statistically average driving cycle of the vehicle can be obtained with the use of stochastically generated routes which are realizations of random functions of external disturbances such as: the trajectory curvature, the maximum friction coefficient between the tire and the road, rolling resistance coefficient and the slope angle of the road surface [5, 6, 7, 8, 9, 10].

The obtained driving cycle of the vehicle allows selecting the required torque – speed curves of the retarders providing the given deceleration. At the same time, in order to increase the average speed of the vehicle and also decrease the loads on the working brake system it is reasonable to have the vehicle deceleration due to the operation of the wear-resistant brake system the same as the one realized at the service braking (and not higher $a^* = 1\ m/s^2$ [11, 12]).

2. Results of the quasi-stationary simulation

The required torque – speed curves of the wear-resistant brake system are selected from the driving cycles of the vehicle obtained by the quasi-stationary model of the vehicle motion and stochastically generated roads taking into account the limitations on the longitudinal and lateral accelerations determined by the tractive performance of the tire, the trajectory curvature, the vehicle power plant performance, and physiological capability of the driver to sustain accelerations (see figure 1 and figure 2) [13]

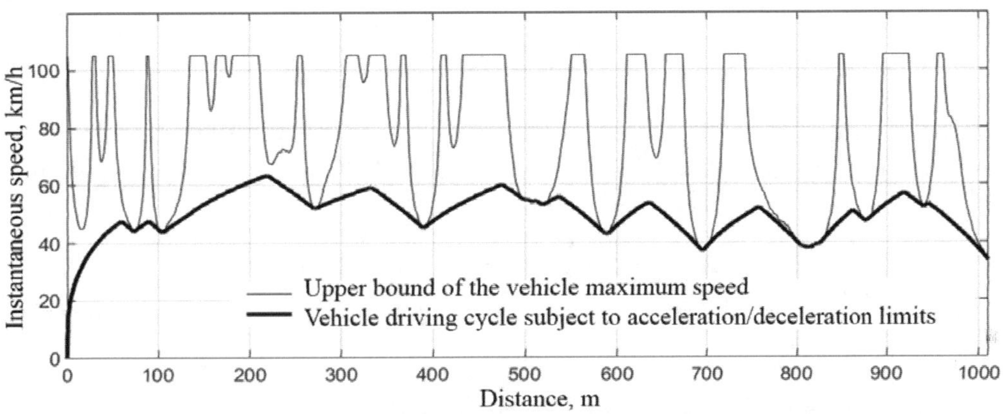

Figure 1. A high-mobility vehicle driving cycle limited by the deceleration $a^* = 1$.

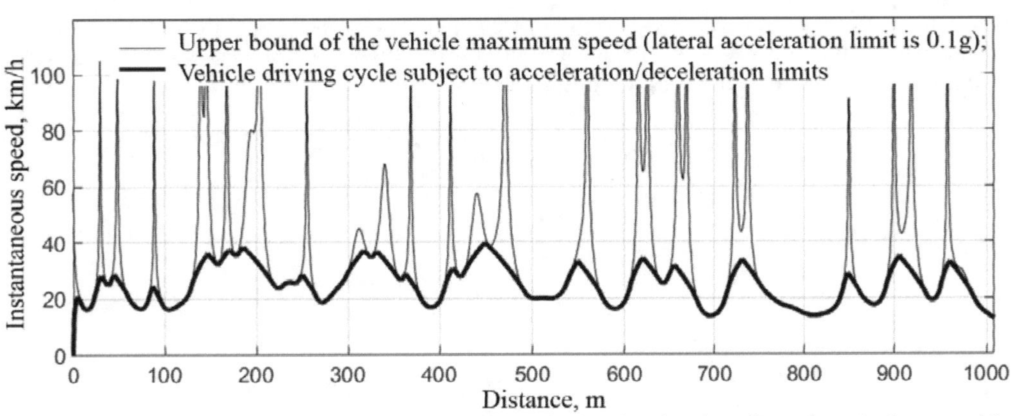

Figure 2. A driving cycle of the high-mobility vehicle for the deceleration $a^* = 1$, the speed being limited by the physiological capability of the driver to sustain acceleration (0,1g).

IOP Conf. Series: Materials Science and Engineering **819** (2020) 012031 doi:10.1088/1757-899X/819/1/012031

Stochastically generated routes used for the generation of the presented driving cycles represent a consecutive set of segments. Each segment features its own slope, curvature, coefficient of rolling resistance and friction coefficient, these parameters remaining constant within the segment.

Thus, the dependence of the magnitude of the traction/braking force F_{reqi} required to maintain a given law of motion (on each i-th segment of the route) can be obtained using a quasi-stationary mathematical model of the vehicle motion and road parameters [14, 15].

$$F_{reqi} = \left(m\delta + J_z {k_{di}}^2\right)a_i + mg\left(f_i\cos(\alpha_i) + sin(\alpha_i)\right) + k_w F_{front} {v_{tri}}^2 + J_z k_{di}\frac{\Delta k_{di}}{\Delta s_i}{v_{tri}}^2, \quad (1)$$

where m is the mass of the high-mobility vehicle; J_z is the vehicle moment of inertia about its vertical axis; δ is the mass factor taking into account the inertia of the rotating components of the drivetrain; a_i is the acceleration of the vehicle during transition from segment *i-1* to segment i of the route; f_i is the rolling resistance coefficient during transition from segment *i-1* to segment *i* of the route; α_i is the slope angle of the road during transition from segment *i-1* to segment *i* of the route; k_{di} is the curvature of the trajectory during transition from segment *i-1* to segment i of the route; Δk_{di} is the change of the curvature at transition from coordinate *i-1* to coordinate *i* along the route; v_{tri} is the upper bound of the maximum speed of the vehicle during transition from segment *i-1* to segment *i* of the route subject to limit longitudinal acceleration; g is the acceleration of gravity; k_w is the aerodynamic force coefficient, $k_w = 0.5\ c_x\rho_w$; c_x is the aerodynamic drag coefficient; ρ_w is the air density; F_{front} is the frontal area of the vehicle.

The value of a_i can be calculated with the following formula:

$$a_i = \frac{({v_i}^2 - {v_{i-1}}^2)}{2\Delta s_i} \quad (2)$$

To simplify the analysis, the tractive/braking force F_{reqi} and the velocity v_i of the vehicle can be transfered to the required torque M_{reqi}^{prop} and angular speed ω_{reqi}^{prop} of the gearbox output shaft (taking into account the assumptions of the dynamic model [15]):

$$M_{reqi}^{prop} = F_{reqi} r_{k0}/u_{mg}, \qquad \omega_{reqi}^{prop} = v_i u_{mg}/r_{k0} \quad (3)$$

where r_{k0} is the wheel radius without slipping; u_{mg} is the gear ratio of the drivetrain from the gearbox output shaft to the drive wheels.

3. Selection of the characteristics of a wear-resistant brake system

Distribution of the operation cycles of the high-mobility vehicle brake system (by way of example the power-to-weight ratio was selected 15 kW/t) on the surfaced road and a flat terrain is shown in figure 3 (only the braking cycles are shown, here and further in the article the characteristics of the wear-resistant brake system are transferred to the gearbox output shaft):

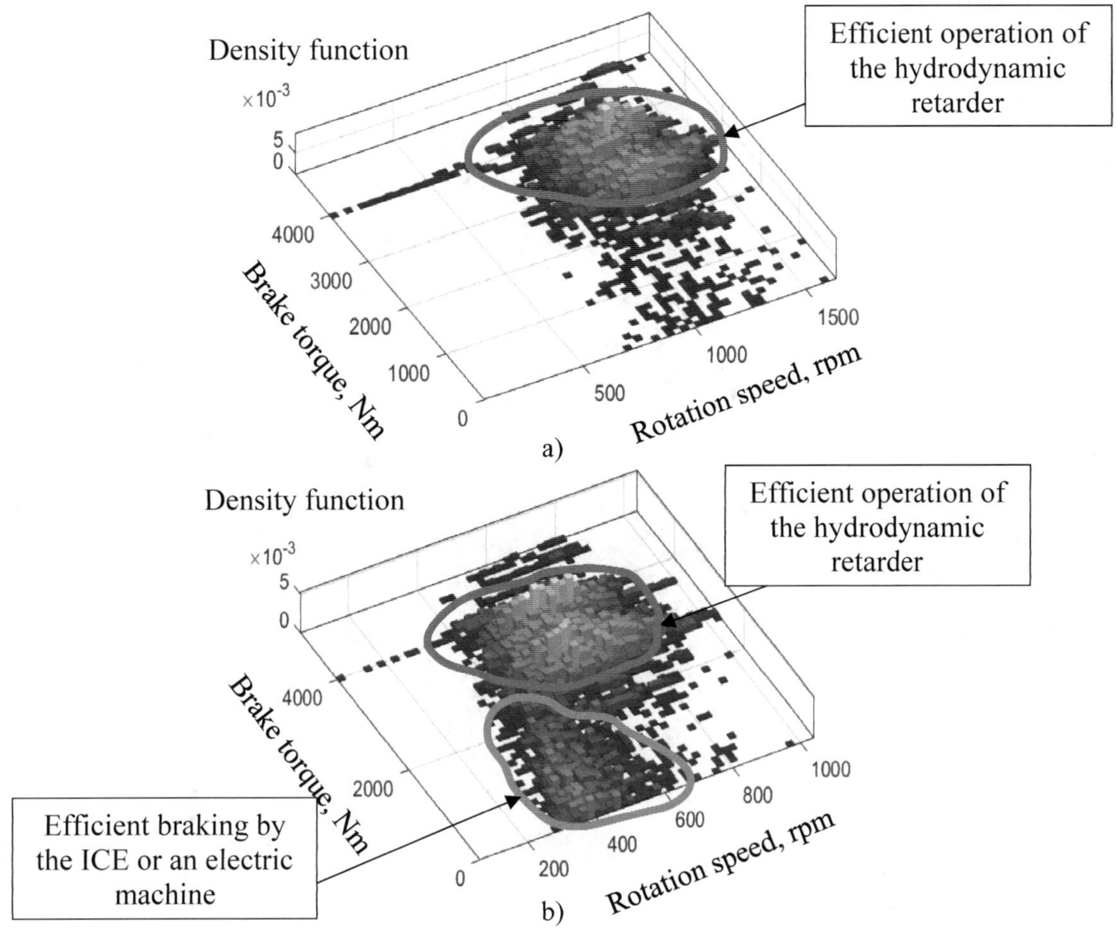

Figure 3. Distribution of the operation cycles of the high-mobility vehicle brake system on a flat terrain: a) without limitations on the lateral acceleration; b) with limited lateral acceleration ($a_y = 0{,}1g$) (the characteristics of the brake system are transferred to the gearbox output shaft).

The histograms show that there are two regions of the brake system operation where braking by two different types of retarders is effective:

• a region of the high torques and rotation speeds of the gearbox output shaft (effective braking by the hydrodynamic retarder);

• a region of the low torques and rotation speeds of the gearbox output shaft (effective braking by the internal combustion engine (ICE) or an electric machine operating in the generator mode [16] to make the high-mobility vehicle more energy efficient.

Similar analysis of the operation on a surfaced road on an undulating terrain has shown the following (see figure 4).

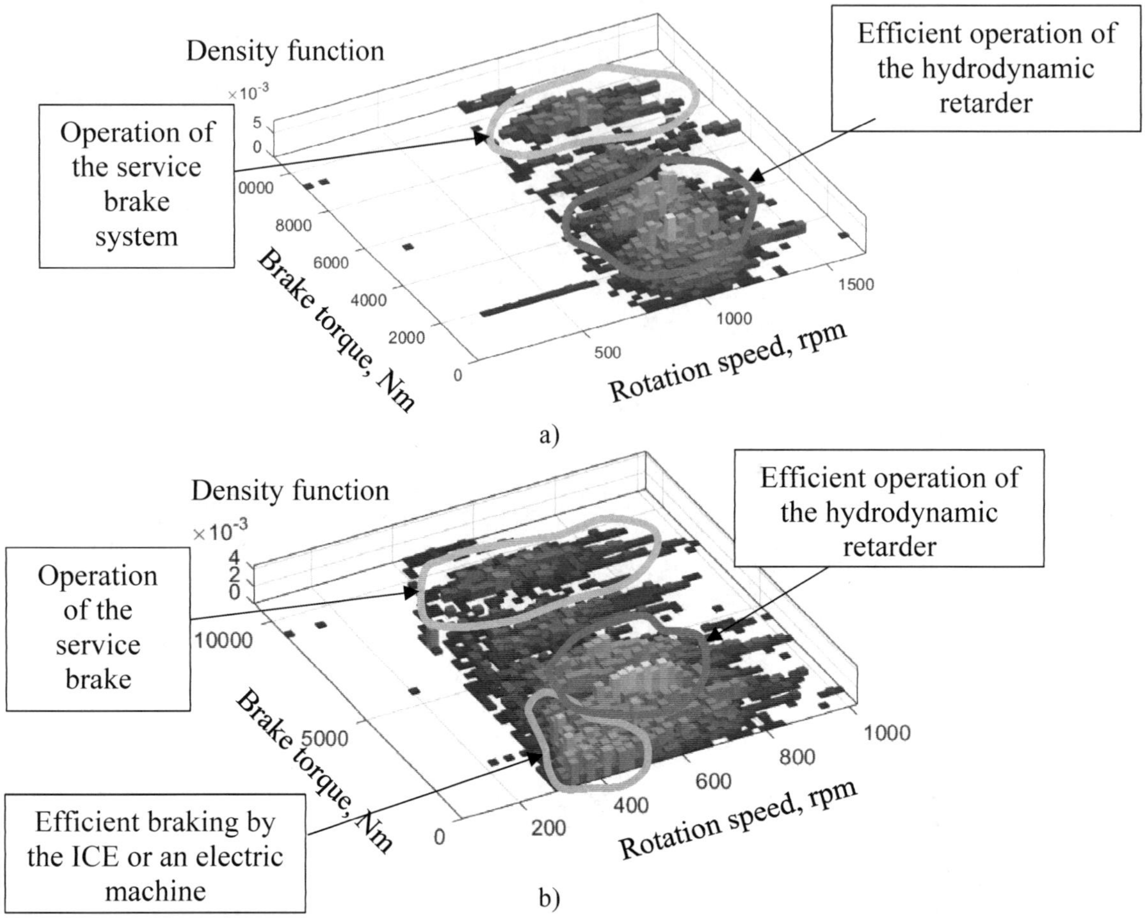

Figure 4. Distribution of the operation cycles of the high-mobility vehicle brake system on an undulating terrain road: a) without limitations on the lateral acceleration; b) with limited lateral acceleration ($a_y = 0{,}1g$) (the characteristics of the brake system are transferred to the gearbox output shaft).

The figures 3 and 4 show that in the case of an undulating terrain there are regions with high braking torque, which should be generated by the service brake system [[17], [18], [19]] since the retarder suitable for such operation cycles would be too bulky to be used on a high-mobility vehicle. At that, the retarder operation regions remain on the same "position" on the histogram.

Hence, further selection of the torque – speed curves of the retarders can be effectively performed from the data obtained on a flat terrain.

For this end, the cumulative distribution functions for the required power of the brake system $N_{req} = M_{req}^{prop}\omega_{req}^{prop}$ and the torque on the gearbox output shaft in the operation cycles in question have been obtained: one (see figure 2) taking into account lateral acceleration limits determined by the physiological capabilities of the driver (see figures 5 and 6) and one (see figure 1) without these limitation.

These results show that the effective power of the retarders with probability 95 % will not exceed $N_{95\%}$ and the gearbox output torque – $M_{95\%}$.

These two parameters are used for description of the "external curve" of the wear-resistant brake system as a whole.

IOP Conf. Series: Materials Science and Engineering **819** (2020) 012031 doi:10.1088/1757-899X/819/1/012031

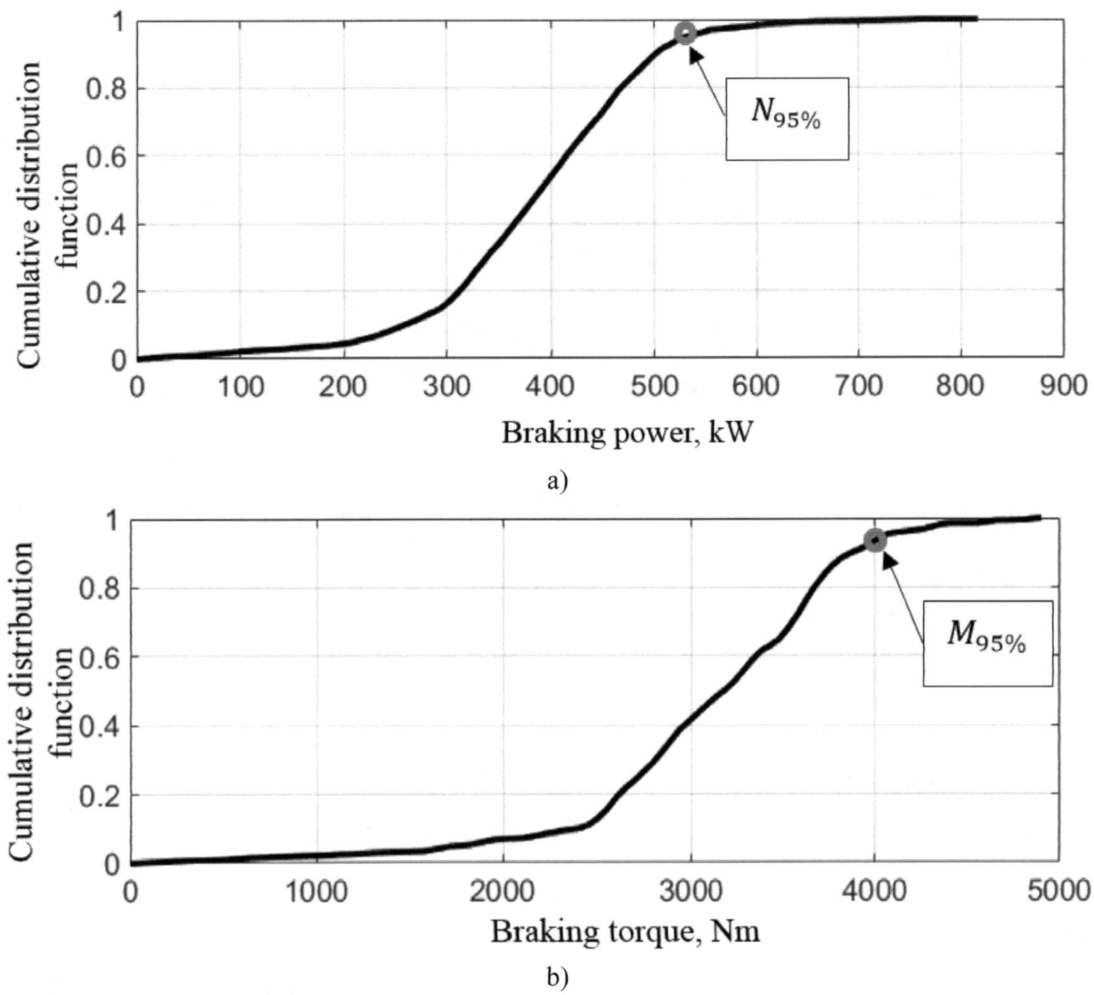

Figure 5. Cumulative distribution functions of the high-mobility vehicle brake system parameters obtained without limitations on the lateral acceleration: a) power; b) braking torque (transferred to the gearbox output shaft).

In order to select the required power of the electric machine N_{ED}^{max}, used as a retarder we need to analyze the density and cumulative distribution functions of the braking power for the driving cycle with limitations on the lateral acceleration (see figure 2). Since, in this particular case according to figure 3 the electric machine is supposed to be used most often (see figure 6).

The density function has two distinctive maximums, the first of which corresponds to the region of the effective use of the electric machine. Its operation in that region will make the high-mobility vehicle more power effective by means of storing the electric power in a storage device and using it later at the acceleration.

Besides, according to figure 6c, the gearbox output shaft torque having probability 95% of not to be overrun corresponds to the same value $M_{95\%}$ as in the case without lateral acceleration limits (see figure 5b). This result means that regardless of the driving cycle the given deceleration $1\ m/s^2$ can be provided by the wear-resistant brake system torque of $M_{95\%}$ for all rotational speeds of the gearbox output shaft.

IOP Conf. Series: Materials Science and Engineering **819** (2020) 012031 doi:10.1088/1757-899X/819/1/012031

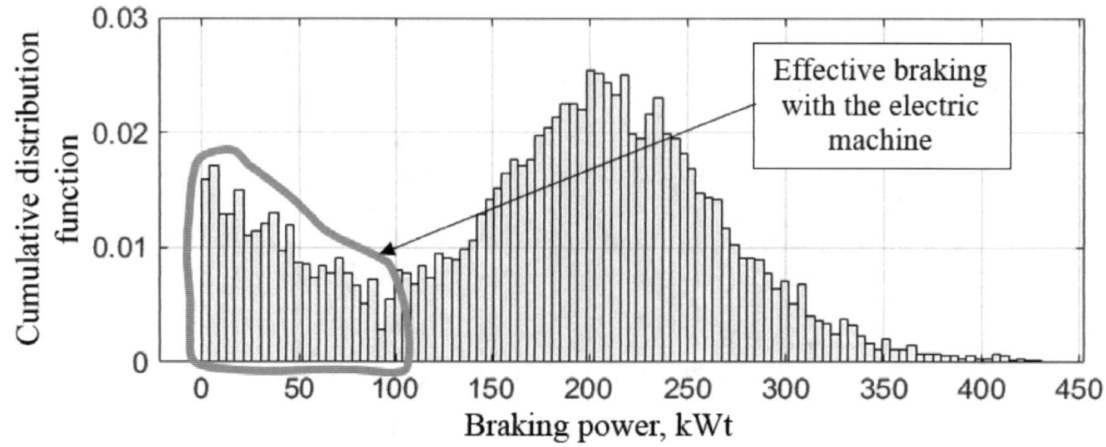

a)

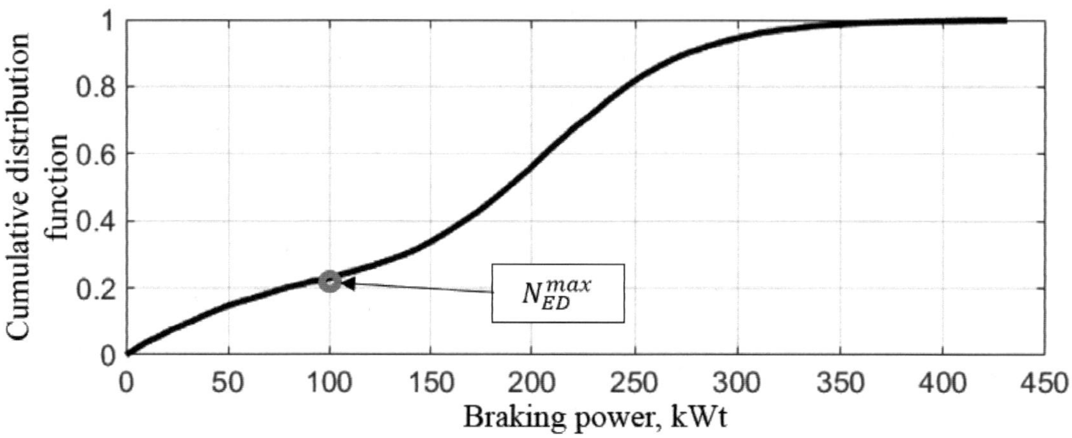

b)

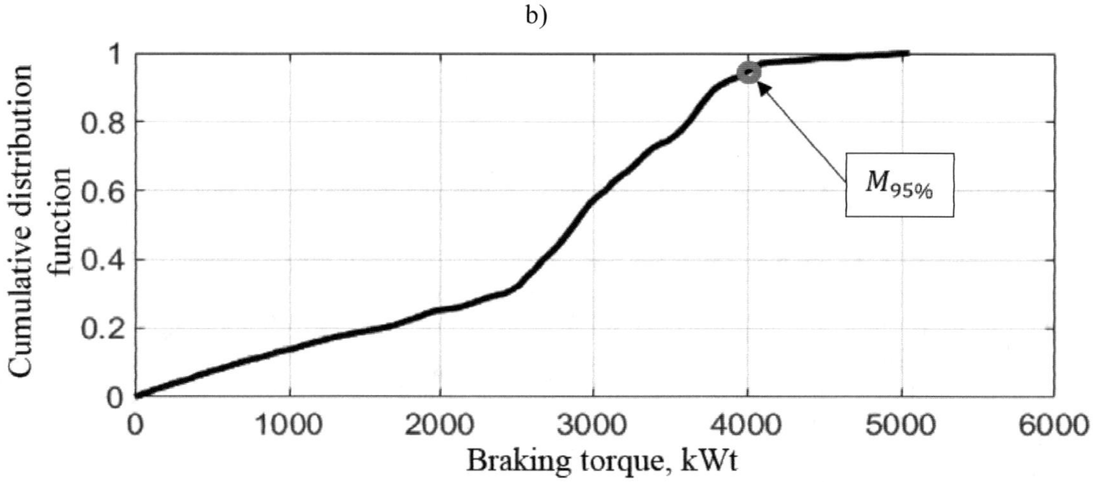

c)

Figure 6. Density function (a) and cumulative distribution function (b) of the effective braking power and cumulative distribution function (c) of the braking torque of the high-mobility vehicle brake system with limited lateral acceleration ($a_y = 0{,}1g$) (the characteristics of the brake system are transferred to the gearbox output shaft).

The selected parameters ($N_{95\%}, M_{95\%}$) of the wear-resistant brake system and the selected power of the electric machine N_{EM}^{max} allow finding the required parameters of each of the units of the brake system.

Since the selection of the engine is based on the required driving performance of the vehicle (obtained by calculation), in the braking mode engine characteristics can be considered to be known and determined by its design parameters: displacement, diameter of the intake/exhaust valves, etc. On the design stage it is reasonable to calculate a combined torque – speed curve of the ICE and its gearbox in the braking mode of the constant power curve. So, by way of example, we will assume that in the braking mode engine power transferred to the gearbox output shaft makes 40% of its tractive power N_{ICE}^{maxT} = 6 kW/t that corresponds to the actuation of a "mountain brake" [16] (the tractive power-to-weight ratio is 15 kW/t).

When an electric machine is used in the drivetrain of a high-mobility vehicle, it is reasonable to install it upstream the gearbox. This solution would provide an extended operating range of the electric machine and operation of the vehicle at a low speed with additional electric drive. Thereby, we will select the vehicle drivetrain layout shown in figure 7. At that, the combined torque – speed curve of the electric machine and the gearbox can be presented as the constant power curve corresponding to the value N_{ED}^{max} .

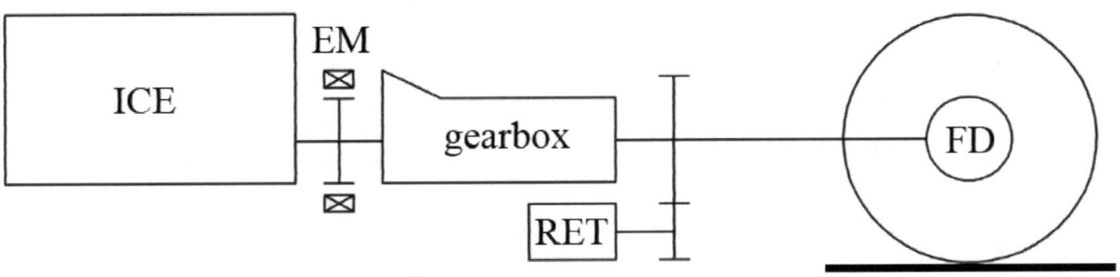

ICE is the internal combustion engine; EM is the electric machine; RET is the hydrodynamic retarder; FD is the final drive.

Figure 7. Layout of the high-mobility vehicle drivetrain.

Thus, the required torque – speed curve $M_{RET}(\omega_{gb})$ of the hydrodynamic retarder transferred to the gearbox output shaft is determined by the difference (the curves of all the units are transferred to the gearbox output shaft as shown in figure 8):

$$M_{RET}(\omega_{gb}) = M_{95\%} - M_{ICE}(\omega_{gb}) - M_{ED}(\omega_{gb}), \tag{4}$$

where $M_{ICE}(\omega_{gb})$ is the dependence of the ICE braking torque (transferred to the gearbox output shaft) from the rotational speed of the gearbox output shaft; $M_{ED}(\omega_{gb})$ is the dependence of the electric machine braking torque (transferred to the gearbox output shaft) from the rotational speed of the gearbox output shaft.

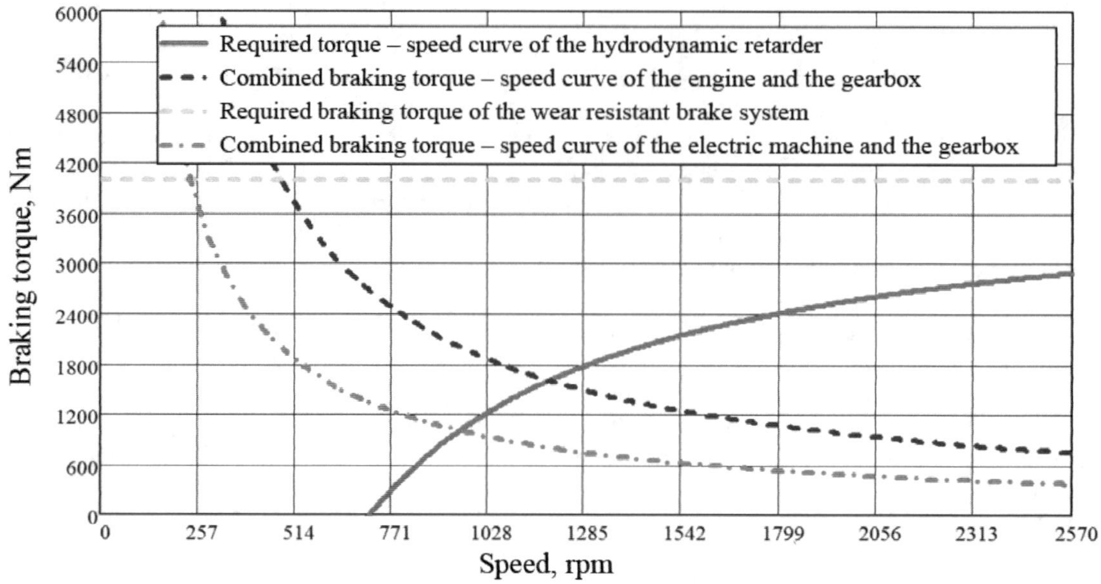

a)

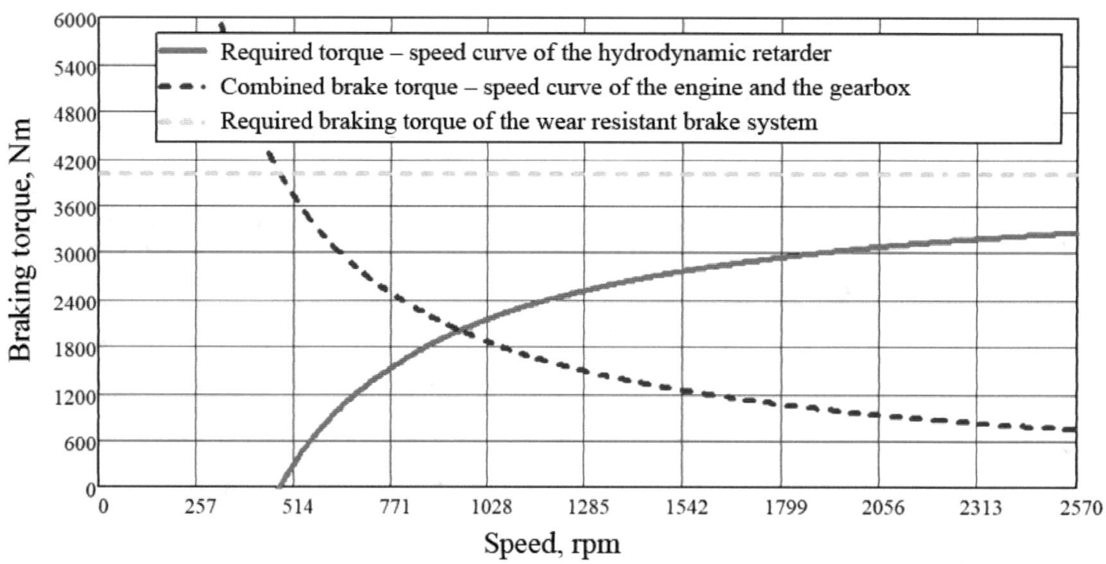

b)

Figure 8. Required torque – speed curve of the hydrodynamic retarder: a) with an electric machine in the drivetrain; b) without an electric machine in the drivetrain.

The next step is the calculation of the hydrodynamic brake external torque – speed curve based on the required one $M_{RET}(\omega_{gb})$.

The torque – speed curve of the filled vane-type hydrodynamic retarder is a parabola passing through the point $M_{RET}(0) = 0$ [21]. Therefore the setting of the hydrodynamic retarder must be designed so that the said parabola would touch the required torque – speed curve of the retarder (see figure 9).

The operation under the torque – speed curve of the filled hydraulic retarder is provided by controlling the volume of the working fluid in the vane section of the retarder, i.e. limit of the torque – speed curve on the right is determined not by the retarder engineering but by its control system operation.

For example, in the case of a long duration braking by the retarder the limit is determined by the maximum heat power that can be dissipated by the cooling system radiator. If this limit is exceeded the coolant and hence the engine will be overheated. At that, in order to get the maximum efficiency of the retarder the heat power dissipated by the radiator must be no less than:

$$N_{RAD} = N_{95\%} - N_{ICE}^{maxT} - N_{ED}^{max}. \tag{5}$$

Since in general case the cooling system of the high-mobility vehicle engine features high thermal inertia, much higher brake power can be utilized at braking (higher than N_{RAD}), but for a short period of time. In this case, the limit for the torque – speed curve of the retarder on the right will be the maximum heat power N_{RET}^{max} at which the coolant is not boiling when passing through the retarder (6). Boiling of the coolant should be avoided since it would lead to a considerable decrease in the working fluid density and a lower effective braking torque.

$$N_{RET}^{max} = c_f \rho_f Q_f \Delta t, \tag{6}$$

where c_f is the heat capacity of the working fluid at the critical temperature; ρ_f is the density of the working fluid at the maximum temperature; Q_f is maximum flow rate of the working fluid in the cooling system; Δt is the maximum temperature difference ($\Delta t \approx 110\ C^o - 90\ C^o = 20\ C^o$).

Thus, the reasonable operational power limit of the hydrodynamic retarder is N_{RET}^{max}. At the same time, the load on the hydro-retarder must be lowered in case the temperature of the coolant reaches the critical value ($\sim 110\ C^o$).

Bearing all this in mind the external torque – speed curve of the retarder can be selected in the form shown in figure 9.

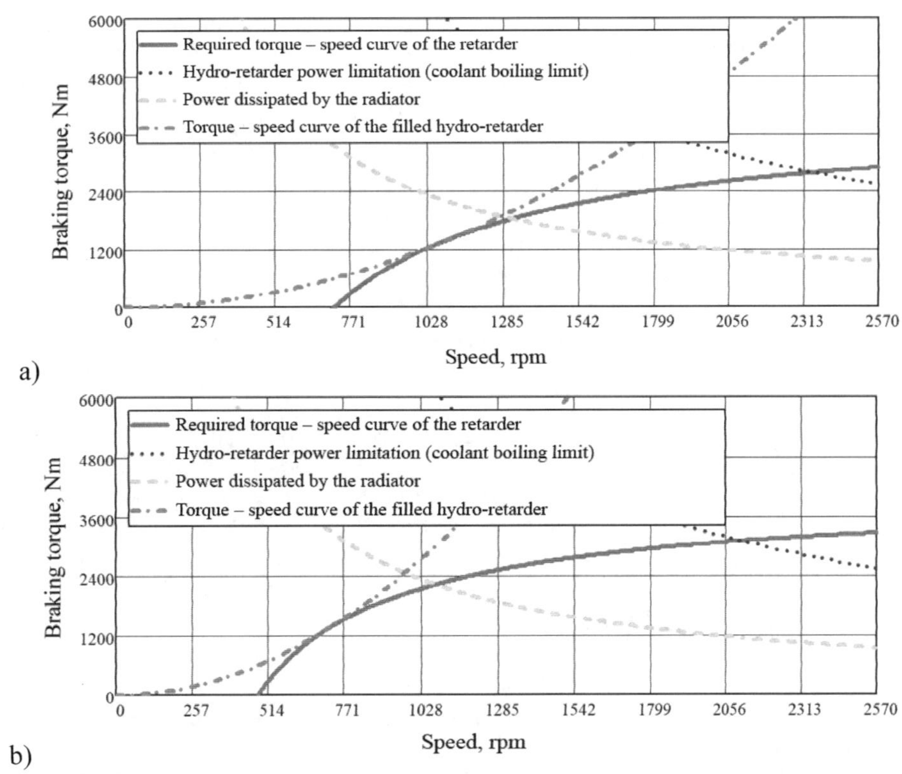

Figure 9. External torque – speed curve of the hydrodynamic retarder: a) with an electric machine in the drivetrain; b) without an electric machine in tthe drivetrain.

IOP Conf. Series: Materials Science and Engineering **819** (2020) 012031 doi:10.1088/1757-899X/819/1/012031

4. Conclusion

The proposed method provides selection of the required torque – speed curves of the units of the wear-resistant brake system of a high-mobility wheeled vehicle: an internal combustion engine (working in brake mode), an electric machine and a retarder the combination of which will ensure the required deceleration of the vehicle.

References

[1] Keller A and Aliukov S 2016 Efficient Power Distribution in an All-Wheel Ground Vehicles *SAE Technical Paper* 2016-01-1105

[2] Aliukov S, Keller A, and Alyukov A 2015 On the Question of Mathematical Model of an Overrunning Clutch *Lecture Notes in Engineering and Computer Science: Proceedings of The World Congress on Engineering* pp 1319-1324.

[3] Keller A , Aliukov S and Anchukov V 2017 Mathematical model of the truck for investigation of stability and control of movement *Lecture Notes in Engineering and Computer Science* pp. 711-716.

[4] Keller A , Shelepov A and Istomin D 2016 Estimation of Screw-ball Differential on Vehicle Dynamics *Procedia Engineering* **150** pp 1131-1136.

[5] Kotiev G O and Sarach E B 2010 *Kompleksnoe podressorivanie vysokopodvizhnykh dvukhzvennykh gusenichnykh mashin.* M.: Izd-vo MGTU im. N.E. Baumana. p 184

[6] Shalygin A S and Palagin Yu I 1986 *Prikladnye metody statisticheskogo modelirovaniya.* L.: Mashinostroenie. Leningr. otd-nie. p 320

[7] Rasshcheplyaev Yu S and Fandienko V N 1981 *Sintez modeley sluchaynykh protsessov dlya issledovaniya avtomaticheskikh sistem upravleniya.* M.: Energiya. p 145

[8] Chernetskiy V I 1968 *Analiz tochnosti sistem upravleniya.* M: Mashinostroenie p 247

[9] Sarach E Kotiev G and Beketov S 2018 Methods for road microprofile statistical data transformation *MATEC Web of Conferences.* **224**. article 04009.

[10] Kotiev G O, Miroshnichenko A V, Stadukhin A A and Kositsyn B B 2019 Estimating operation modes for the individual wheel electric drive of the all-wheel drive vehicle with the use of the driving simulator *IOP Conf. Series: Materials Science and Engineering* **534** 012004

[11] Tarasik V P 2006 *Teoriya dvizheniya avtomobilya.* Spb. BVKh-Peterburg p 478

[12] Kravets V N 2014 *Teoriya dvizheniya avtomobilya.* Nizhniy Novgorod: Nizhniy Novgorod 696

[13] Rotenberg R V 1972 *Podveska avtomobilya.* Izd 3 pererabot. i dop. M.: Mashinostroenie p 392

[14] Savochkin V A and Dmitriev A A 1993 *Statisticheskaya dinamika transportnykh i tyagovykh gusenichnykh mashin.* M.: Mashinostroenie p 320

[15] Kositsyn B B 2017 *Metod opredeleniya energoeffektivnogo zakona dvizheniya elektrobusa po gorodskomu marshrutu*: diss. …kand. tekhn. nauk: 05.05.03. M.: MGTU im. N.E. Baumana p 165

[16] Krasovsky A, Gorbunova E, Bychkov M, Fedorenko A 2018 Torque Control of Switched Reluctance Drive in Generating Mode *25th International Workshop on Electric Drives: Optimization in Control of Electric Drives (IWED)* Moscow Russia.

[17] Nosko A L, Safronov E V and Soloviev V A 2018 Study of Friction and Wear Characteristics of the Friction Pair of Centrifugal Brake Rollers *Journal of Friction and Wear* **39**, No. 2 pp 145–151

[18] Gekker F R 1998 Generalized dynamic model of 'dry' friction units *Trenie i Iznos*, **19** Issue 2 pp 165-170

[19] Yudin E G, Vartanyan, V A and Vasilieva, E I 1998 Improvement of quality and tribological performance of transmission parts by electrochemical polishing *Journal of Friction and Wear* **19** Issue 2 pp 81-86

[20] Heinz Heisler *Advanced Vehicle Technology.* London: Butterworth-Heinemann. 2002 656 p. doi: 10.1016/B978-0-7506-5131-8.X5000-3.

[21] Gapoyan D T and Iliev P 1972 *Avtomobil'nye elektrodinamicheskie tormoza zamedliteli* p 95

Selection of the capacity of the onboard energy storage device for a high-mobility wheeled vehicle wear-resistant brake system with an electric machine

G O Kotiev[1], **I F Gumerov**[3], **A A Stadukhin**[2] **and B B Kositsyn**[1]

[1] Department of the Wheeled Vehicles of Bauman Moscow State Technical University, Moscow, 105005, Russian Federation
[2] Department of the Multi-purpose Tracked Vehicles and Mobile Robots of Bauman Moscow State Technical University, Moscow, 105005, Russian Federation
[3] PAO Kamaz, Naberezhnye Chelny, Republic of Tatarstan, 423827, Russian Federation

E-mail: kositsyn_b@bmstu.ru

Abstract. With the growth of the average speed of wheeled vehicles, the wear-resistant brake system which would increase their mobility by improving their braking performance and lower the heat loads on the service brake system is gaining interest of the manufacturers. At the same time, a good option for increasing power efficiency of the high-mobility wheeled vehicles is to use an electric machine in the wear-resistant brake system. The electric machine can operate in the generator cycle and provide energy recuperation during braking. Recuperation at braking requires accumulation of high values of electric energy in a storage device (usually a molecular storage device), which brings to the forefront the problem of its efficient capacity selection. The article deals with the selection of the required capacity of an onboard energy storage device providing better power efficiency of the vehicle due to the use of an electric machine in the wear-resistant brake system. The method for the selection of the required capacity of the onboard energy storage device is based on the calculation of the energy spent by the vehicle for maintaining the given speed. The following strategy of the electric machine control was used in the research: the energy accumulated in the storage device after braking will be spent at the next acceleration. Therefore, the power efficiency can be increased only due to such compensation of the high frequency oscillations of the spent/accumulated electric energy in the potential storage device that would prevent reaching its maximum charge level in a typical driving cycle. Based on the results of the research the capacity of the storage device is assessed by the peak-to-peak amplitude of the time history of the accumulated/stored energy when driving along a given route. A method that allows formulation of the specifications for the onboard energy storage device used as an element of the wear-resistant brake system, the method uses statistical data on the road conditions of vehicles operation. The method provides selection of the optimal parameters (required capacity) of the electric energy storage device which would guarantee effective operation of high-mobility wheeled vehicles in typical driving cycles.

1. Introduction

A wear-resistant brake system of a high-mobility wheeled vehicle may have besides the exhaust brake installed inside the internal combustion engine and the transmission hydro- and electrodynamic brakes (retarder brakes) also an electric machine operating in the generator cycle [1, 2] with an onboard energy storage device. The main benefit of combining the electric machine with the energy storage device in the brake system is the increased energy efficiency of the high-mobility vehicle due to the transformation of the vehicle kinetic energy into the electric energy (at recuperative braking) and accumulating it in the storage device for later use at acceleration [3, 4, 5, 6].

When the electric machine operates as an element of the wear-resistant brake system, a high-capacity electrical energy storage device (which can withstand high discharge/charge current) is needed. Storage devices which can operate at high values of electric power and at the same time have high capacity have a large mass and a high price.

In view of this, in order to decrease the required capacity of the storage device, the following strategy for the electric machine control is proposed: the energy accumulated in the storage device after braking will be spent at the next acceleration (deceleration by the electric machine with the use of the braking resistors is not addressed). It means that during the operation of the high-mobility wheeled vehicle the electric energy storage device does not discharge fully in a typical driving cycle. In this case, a molecular energy storage device (a super capacitor) can be utilized for energy recuperation at braking. Storage devices of this kind feature low volumetric efficiency but provide high discharge/charge power, which fact allows us to neglect the limitations on the input current at further analysis.

2. Selection of the capacity of the onboard energy storage device for a high-mobility wheeled vehicle wear-resistant brake system with an electric machine

For selection of the capacity of the onboard electric energy storage device the driving cycle of the vehicle close to the real-life operation cycle is used [7 – 14]. The proposed driving cycle was generated analytically by means of the quasi-stationary model of the high-mobility wheeled vehicle motion subject to the limitations on the longitudinal and lateral accelerations determined by the tractive characteristics of the tires, powerplant performance and physiological capability of the driver to withstand the accelerations (see figure 1) [15]. This driving cycle is also based on the assumption that the deceleration which would provide the reasonable average speed on the route is not higher than $a^* = 1\,m/s^2$ [16, 17].

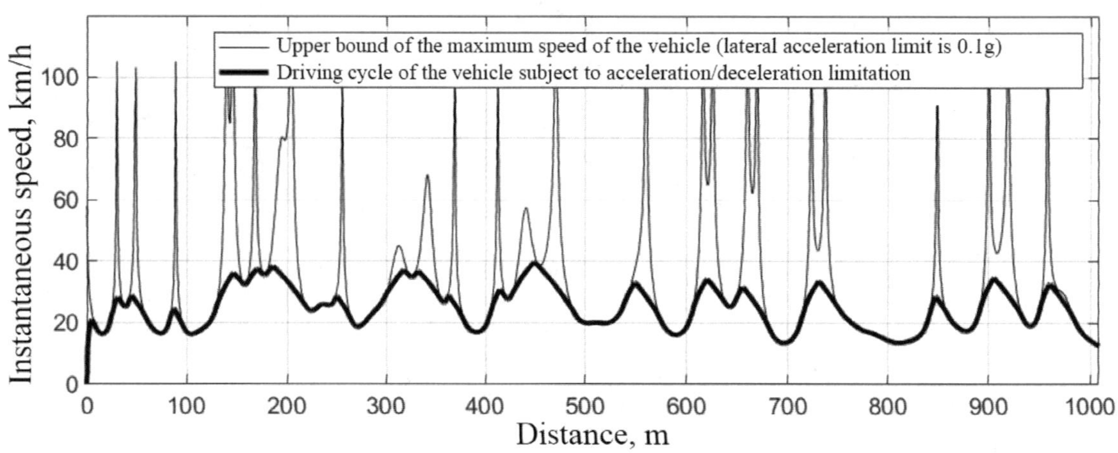

Figure 1. Driving cycle of the high-mobility wheeled vehicle.

Thus, we shall assume that at the acceleration/deceleration with the power less than N_{ED}^{max} (N_{ED}^{max} is the maximum power of the electric machine) the vehicle motion is provided only by the operation of the electric machine in the driving/generator mode, in other cases the extra braking power is provided by the internal combustion engine (ICE), the retarder or the service brake system (see figure 2) [18 – 20].

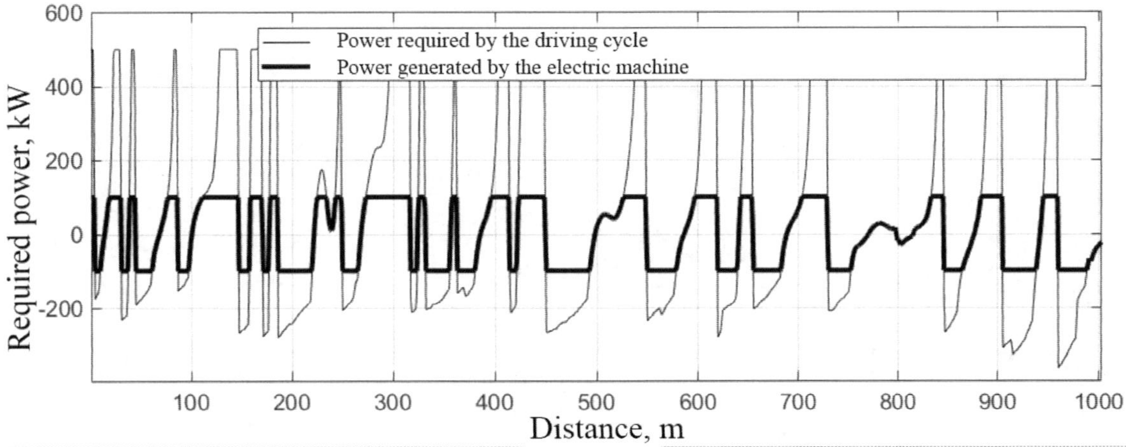

Figure 2. Dependence of the gearbox output shaft power required by the driving cycle of the high-mobility vehicle on the travelled distance.

$$N_{\mathrm{EDreq}_i}^{\mathrm{mech}} = \begin{cases} N_{ED}^{max}, & \text{if } N_{\mathrm{req}_i} \geq N_{ED}^{max} \\ N_{\mathrm{req}_i}, & \text{if } -N_{ED}^{max} < N_{\mathrm{req}_i} < N_{ED}^{max}, \\ -N_{ED}^{max}, & \text{if } N_{\mathrm{treb}_i} \leq -N_{ED}^{max} \end{cases} \tag{1}$$

where $N_{\mathrm{EDreq}_i}^{\mathrm{mech}}$ is the power generated by the electric machine on the gearbox output shaft in the given driving cycle; $N_{\mathrm{req}_i}$is the overall required power.

The dependence of the electric machine power N_{EDreq} required in the given driving cycle can be calculated from the following system of equations:

$$N_{\mathrm{EDreq}_i} = \begin{cases} N_{\mathrm{EDreq}_i}^{\mathrm{mech}}/\eta_{ED}, & if\ N_{\mathrm{EDreq}_i}^{\mathrm{mech}} \geq 0 \\ N_{\mathrm{EDreq}_i}^{\mathrm{mech}} \cdot \eta_{ED}, & if\ N_{\mathrm{EDreq}_i}^{\mathrm{mech}} < 0' \end{cases} \tag{2}$$

where η_{ED} is the electric machine efficiency (for design calculations it is assumed to be constant $\eta_{ED} = 0.9$).

The electric energy E_{ED} spent at driving along the route as a function of time can be calculated as follows:

$$E_{ED\,i+1} = E_{ED\,i} + N_{\mathrm{EDreq}_i}\Delta t_i = E_{ED\,i} + N_{\mathrm{EDreq}_i}\frac{\Delta s_i}{v_{mean_i}}, \qquad t_i = t_{i-1} + \frac{\Delta s_i}{v_{mean_i}} \tag{3}$$

where Δt_i is the time of travelling form route coordinate *i*-1 to route coordinate *i*; Δs_i is the distance increment after travelling form route coordinate *i*-1 to route coordinate *i*; t_i is the time of arriving to the *i*-th coordinate of the route.

The dependence of the spent electrical power E_{ED} on the time is shown in figure 3. The curve shows that the power balance is negative (the spent electric power is growing with time). It means that for the operating cycle of the electric machine shown in figure 2 an additional power source is needed.

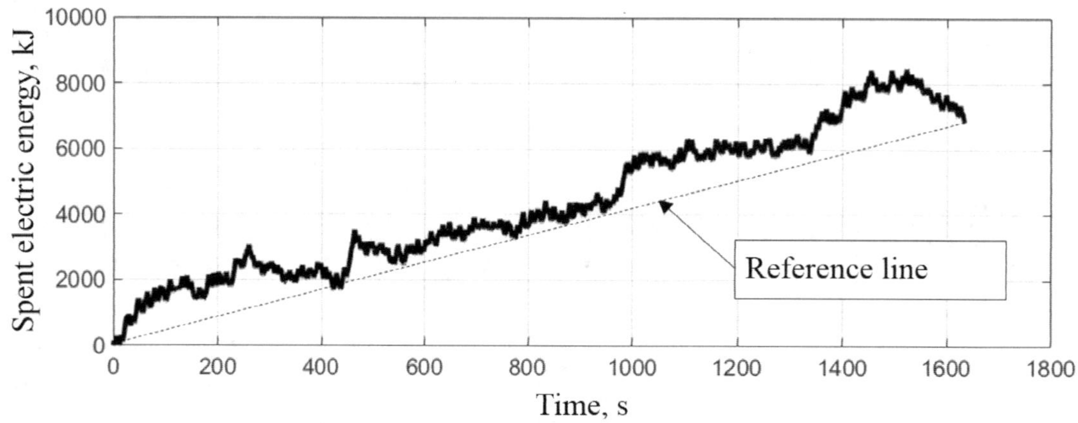

Figure 3. Spent electric energy – travel time curve.

The proposed approach to the electric machine use in the wear-resistant brake system does not require installation of such additional devices, therefore an increase in the vehicle power efficiency can be provided only by compensation of the oscillations of the spent/accumulated energy about the reference line (see figure 3). So, in order to estimate the required capacity of the onboard energy storage device during the calculation of the spent/accumulated energy (3) we must subtract from the value N_{EDreq_i} such power that would provide a neutral power balance in the system (see figure 4):

$$E_{cps\ i+1}^{raw} = E_{cps\ i}^{raw} + \left(N_{\mathrm{EDreq}_i} - E_{ED\,n_s}/t_{n_s}\right)\frac{\Delta s_i}{v_{mean_i}}, \tag{4}$$

where E_{cps}^{raw} is the spent/accumulated electric energy of the onboard energy storage device; n_s is the number of the segments of the route discretization.

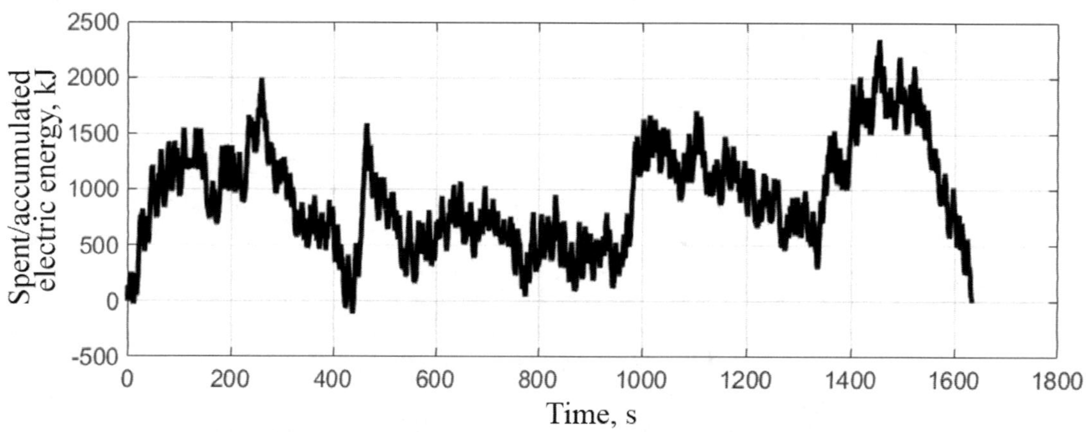

Figure 4. Time history of the spent energy of the storage device.

The generated time history features definite low-frequency oscillations with large amplitude and therefore implies accumulation of much energy onboard. The proposed control strategy and use of the molecular energy storage device do not imply storing the electric energy for a long time in order to reduce the required capacity of the accumulator device (the charge of the storage device is supposed to be used soon after the recuperative braking). For the selection of the energy storage device capacity the low frequency oscillations of the generated curve must be filtered out. The research has shown that the best option for the cut-off frequency is 0.005 Hz (the oscillations with periods lower than 200 s will be discarded). Filtration can be performed, for example, by the 1st order Butterworth filter. The time history of the spent/accumulated energy after filtration is shown in figure 5.

Then, according to the generated time history (see figure 5) the required capacity of the energy storage device can be estimated by the peak-to-peak amplitude:

$$E_{cps}^{max} = max\left(E_{cps}^{f}(t)\right) - min\left(E_{cps}^{f}(t)\right), \quad (5)$$

where E_{cps}^{max} is the required capacity of the energy storage device; $E_{cps}^{f}(t)$ is the filtered time function of the spent/accumulated energy.

Thus, the method allows selecting the required capacity of the onboard energy storage device, which would provide operation of the electric machine according to the selected control strategy preventing the maximum charge in a typical driving cycle.

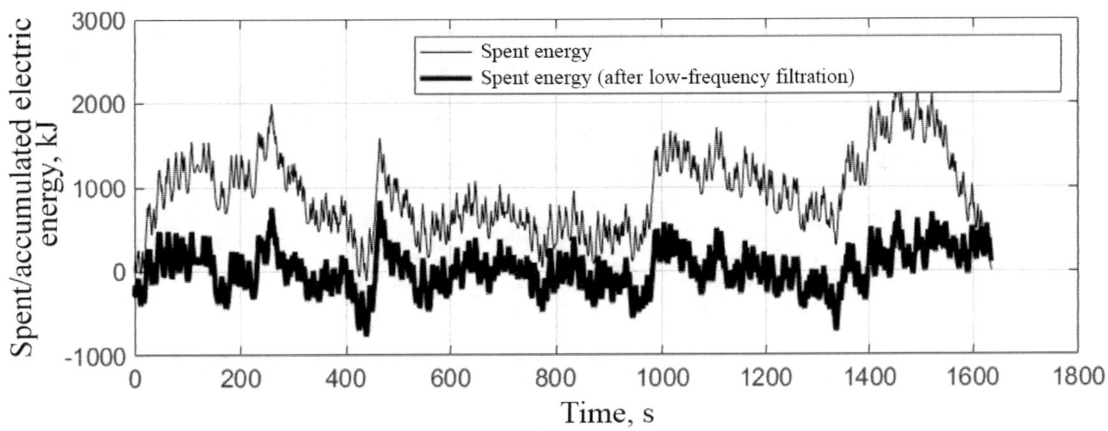

Figure 5. Time history of the spent energy of the storage device.

3. Analysis of the efficiency of the electric machine as an element of a wear-resistant brake system of a high-mobility wheeled vehicle

The most important problem during analysis of the performance of the vehicles equipped with an electric machine as an element of the wear-resistant brake system is the estimation of their efficiency improvement. This estimation can be performed by comparison of the mechanical energy used by the internal combustion engine for keeping the vehicle speed in two cases: with an electric machine and without an electric machine in the brake system.

Braking by the ICE, service brake system, and retarder does not provide energy recuperation. Therefore, when there is no electric machine in the brake system, the mechanical energy of the engine is considered to be spent in all cases when the power required for driving N_{req} is more than 0 (the vehicle operates in the driving mode).

$$E_{ICE\,i+1} = E_{ICE\,i} + \max\left(N_{req\,i}, 0\right) \cdot \Delta s_i / v_{mean_i}, \quad (6)$$

where E_{ICE} is the mechanical energy spent by the engine without an electric machine.

When there is an electric machine in the wear-resistant brake system, the mechanical energy of the engine is considered to be spent when the molecular energy storage device becomes discharged or when there is not enough tractive power for keeping the vehicle speed (when there is not enough power of the electric machine at braking, the mechanical energy of the engine is not spent).

$$E_{cps_{i+1}} = \begin{cases} E_{cps_i} - \min\left(N_{req_i}, N_{ED}^{max}\right) \cdot \Delta s_i / \left(v_{mean_i} \cdot \eta_{ED}\right), & if\ N_{req_i} \geq 0 \\ E_{cps_i} - max\left(N_{req_i}, -N_{ED}^{max}\right) \cdot \Delta s_i \eta_{ED} / v_{mean_i}, & if\ N_{req_i} < 0 \\ 0, & if\ E_{cps_{i+1}} < 0 \\ E_{csp}^{max}, & if\ E_{cps_{i+1}} > E_{cps}^{max} \end{cases}, \tag{7}$$

$$E_{ICE\,i+1}^{ED} = \begin{cases} E_{ICE\,i}^{ED} + \max\left(N_{req_i}, 0\right) \cdot \Delta s_i / v^{\text{ср}}{}_i, & if\ E_{\text{кон}_i} = 0 \\ E_{ICE\,i}^{ED} + \max\left(N_{req_i} - N_{ED}^{max}, 0\right) \cdot \Delta s_i / v_{mean_i}, & if\ E_{\text{кон}_i} \neq 0 \end{cases},$$

where E_{cps} is the effective charge of the energy storage device; E_{ICE}^{ED} is the mechanical energy spent by the engine with the electric machine.

Results of the numerical experiment performed with regard of the described behavior of the system at zero initial charge of the energy storage device are shown in figure 6. By way of example, a high-mobility wheeled vehicle with the following parameters is used: gross weight 34 t, ICE power-to-weight ratio 15 kW/t in driving mode, ICE power-to-weight ratio 6 kW/t in braking mode, electric machine power 100 kW (power-to-weight ratio 3 kW/t), capacity of the onboard energy storage device 0.45.

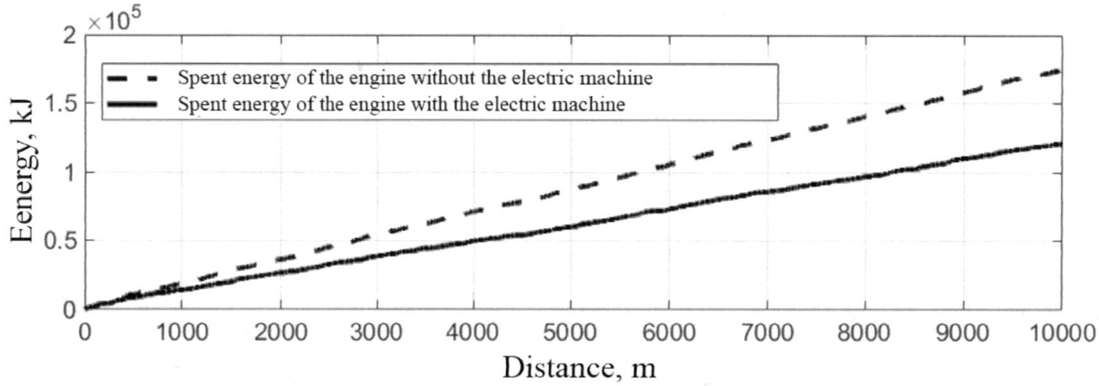

a)

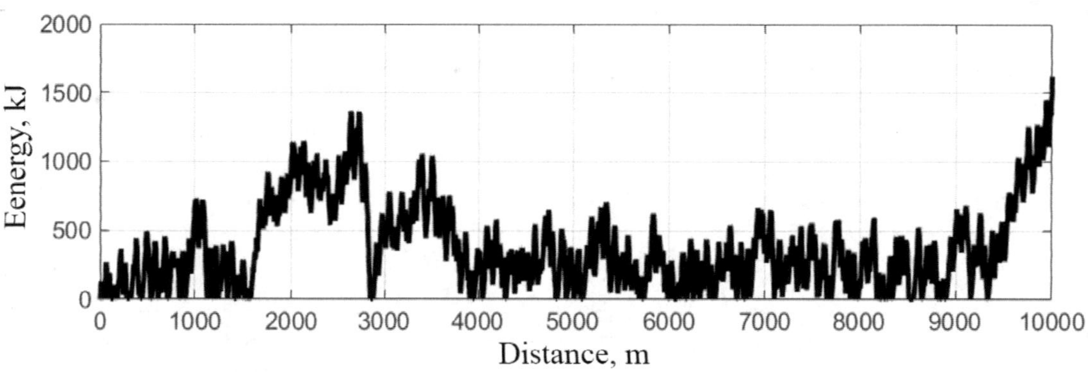

b)

Figure 6. a) mechanical energy of the engine with and without the electric machine; b) energy variation in the energy storage device.

Estimate of the reduction of the energy spent by the ICE in the case of the electric machine installation can be done by the following equation (see figure 7):

$$\eta_{eff_i} = 100\% \cdot \left(E_{\mathrm{ICE}_i} - E^{\mathrm{ED}}_{\mathrm{ICE}_i}\right)/E_{\mathrm{ICE}_i} \tag{8}$$

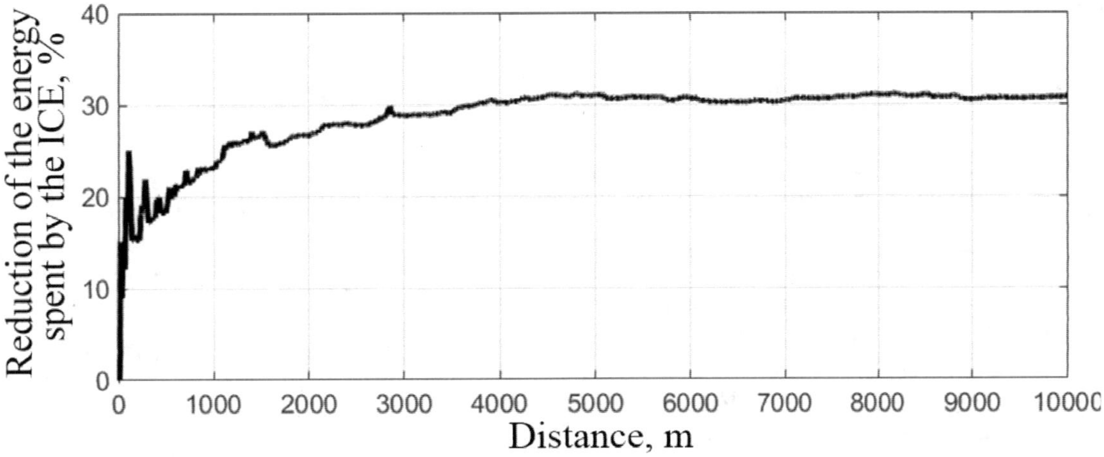

Figure 7. Reduction of the energy spent by the ICE in the case of the electric machine installation.

From the analysis results we can conclude that installation of the electric machine in the wear-resistant brake system for the analyzed driving cycle (see figure 1, the case when the lateral acceleration is limited $a_y = 0{,}1g$ and deceleration $a^* = 1\ m/s^2$) provides reduction of the mechanical energy spent by the ICE by 30.8%.

In order to estimate the effect of the driving cycle intensity (effect of the lateral and longitudinal acceleration limitation) on the reduction of the ICE energy consumption due to the installation of the electric machine in the brake system, the selection of the onboard energy storage device and assessment of the electric machine efficiency in different driving cycles according to the proposed method must be performed under the following conditions:

- without limitation of the lateral acceleration and with limited deceleration – reduction in the energy consumption 15%;
- with limited lateral acceleration, but without limitation on the deceleration – reduction in the energy consumption 13.5%;
- without limitation on the lateral acceleration and with limited deceleration – reduction in the energy consumption 4.3%.

Thus, installation of the electric machine with an onboard energy storage device in the wear-resistant brake system of the high-mobility wheeled vehicle in question (the gross weight 34 t) theoretically provides up to 30% reduction of the mechanical energy consumption.

4. Conclusion

Installation of the electric machine in the wear-resistant brake system provides up to 30% reduction of the mechanical energy spent by the ICE of a high-mobility wheeled vehicle. Reduction in the energy consumption depends on the driving cycle: the more intensive the driving cycle is, the lower the effect of the electric machine installation is. This feature can be explained by the fact that at the intensive deceleration of the high-mobility wheeled vehicle the most part of the energy is not accumulated in the storage device but is dissipated by means of the brake system.

References

[1] Heinz Heisler *Advanced Vehicle Technology* London: Butterworth-Heinemann 2002 p 656 doi: 10.1016/B978-0-7506-5131-8.X5000-3

[2] Krasovsky A, Gorbunova E, Bychkov M and Fedorenko A 2018 Torque Control of Switched Reluctance Drive in Generating Mode *25th International Workshop on Electric Drives: Optimization in Control of Electric Drives (IWED)* Moscow Russia

[3] Kharitonov S A [et al.] 2010 *Analiz i proektirovanie gibridnykh transmissiy transportnykh sredstv na osnove planetarnykh mekhanizmov* M.: MGTU im. N.E. Baumana p 92

[4] Kositsyn B B 2017 *Metod opredeleniya energoeffektivnogo zakona dvizheniya elektrobusa po gorodskomu marshrutu*: diss. …kand. tekhn. nauk: 05.05.03. M.: MGTU im. N.E. Baumana p 165

[5] *Electric Drive Study: Technical report* (final): U.S. Army Tank-Automotive Command Research Development & Engineering Center; General Dinamics Land Systems Division – Warren Michigan 1987 – p 396

[6] Barlow T J [et al.] 2009 *A reference book of driving cycles for use in the measurement of road vehicle emissions* TRL limited p 276

[7] Keller A and Aliukov S 2015 Analysis of possible ways of power distribution in an all-wheel drive vehicle Lecture Notes in Engineering and Computer Science 2218 pp 1154-1158.

[8] Keller A, Aliukov S, Anchukov V and Ushnurcev S 2016 Investigations of Power Distribution in Transmissions of Heavy Trucks SAE Technical Paper 2016-01-1100

[9] Keller A and Aliukov S 2016 Efficient Power Distribution in an All-Wheel Ground Vehicles SAE Technical Paper 2016-01-1105

[10] Aliukov S, Keller A, and Alyukov A 2015 On the Question of Mathematical Model of an Overrunning Clutch Lecture Notes in Engineering and Computer Science: Proceedings of The World Congress on Engineering pp1319-1324.

[11] Shtang A A and Yaroslavtsev M V 2016 Opredelenie osnovnykh kharakteristik kombinirovannoy energeticheskoy ustanovki dlya gorodskogo bezrel'sovogo transporta *Doklady Akademii Nauk Vysshey Shkoly Rossiyskoy Federatsii* **4** doi: 10.172121727-2769-2016-4-111-120

[12] Yaroslavtsev M V 2014 Opredelenie parametrov energoustanovki gibridnogo avtomobilya modelirovaniem protsessa potrebleniya energii *Elektrotekhnika* **12** pp 17-21

[13] Sarach E, Kotiev G and Beketov S 2018 Methods for road microprofile statistical data transformation *MATEC Web of Conferences* **224** article 04009

[14] Kotiev G O, Miroshnichenko A V, Stadukhin A A and Kositsyn B B 2019 Estimating operation modes for the individual wheel electric drive of the all-wheel drive vehicle with the use of the driving simulator *IOP Conf. Series: Materials Science and Engineering* **534** 012004

[15] Rotenberg R V 1972 *Podveska avtomobilya* Izd 3-e pererabot. i dop. M.: Mashinostroenie p 392

[16] Tarasik V P 2006 *Teoriya dvizheniya avtomobilya* Spb BVKh-Peterburg p 478

[17] Kravets V N 2014 *Teoriya dvizheniya avtomobilya* Nizhniy Novgorod: Nizhniy Novgorod p 696

[18] Nosko A L, Safronov E V, Soloviev V A 2018 Study of Friction and Wear Characteristics of the Friction Pair of Centrifugal Brake Rollers *Journal of Friction and Wear* **39** No 2 pp 145–151

[19] Gekker F R 1998 Generalized dynamic model of 'dry' friction units *Trenie i Iznos* **19** Issue 2 pp 165-70

[20] Yudin E G Vartanyan V A and Vasilieva E I 1998 Improvement of quality and tribological performance of transmission parts by electrochemical polishing *Journal of Friction and Wear* Vol **19** Issue 2 pp 81-6

IOP Conf. Series: Materials Science and Engineering **819** (2020) 012033 doi:10.1088/1757-899X/819/1/012033

Legal regulation in the field of automated automobile transport and road tests of autonomous vehicles

V I Kotliarenko

FSUE "NAMI", 125438 Avtomotornaya st., 2, Moscow, Russia

Abstract. Increase in demand for transportation services, on the one hand, stimulates development of the road network and construction of new objects of the transport infrastructure providing necessary transport system capacity, on the other hand, it stimulates development of new technologies allowing increase of human mobility, comfort and safety of vehicle driving. In these conditions, a great significance is acquired by development of intelligent transport systems (hereinafter referred to as the ITS) representing, on the one hand, means of connection, control and management built in the vehicles and objects of transport infrastructure, and on the other hand, means of control and fast decision-making based on the information received in real-time mode and available to transport operators and transport service users. ITS development has resulted in emergence of autonomous (driverless) vehicles. Considering the issues that the transport complex faces, it seems that driverless vehicles will be able to provide substantial help in solving them, for example by eliminating the human factor from transportation process. Creation of driverless vehicles is impossible without development of a unified approach to forming a legal and technical framework for their practical implementation. The abovementioned framework may be divided on a provisional basis into regulatory technical and legal ones. In the process of creating driverless vehicles, development testing on public roads is of great importance. Such tests cannot be replaced by research at closed proving grounds where it is impossible to recreate the whole range of rapidly changing driving conditions of the actual use. The existing legal framework in the Russian Federation and in a number of foreign countries allows testing highly automated and fully automated vehicles on public roads.

1. Introduction

Increase in demand for transportation services, on the one hand, stimulates development of the road network and construction of new objects of the transport infrastructure providing necessary transport system capacity, on the other hand, it stimulates development of new technologies allowing increase of human mobility, comfort and safety of vehicle driving. In these conditions, a great significance is acquired by development of intelligent transport systems (hereinafter referred to as the ITS).

At the same time, the development of ITS has led to the emergence of autonomous (unmanned) vehicles. Considering the tasks set for the transport complex, especially in the field of road safety and the environment, it seems that unmanned vehicles will be able to provide significant assistance in solving them, for example, eliminating the human factor from the transportation process.

2. Normative legal regulation in the field of automated automobile transport

When creating unmanned vehicles in world practice, much attention is paid to the development of a unified approach to the formation of a regulatory legal and technical base for their practical implementation. At the same time, the indicated base can be conditionally divided into regulatory technical and regulatory legal.

The normative technical base establishes technical requirements for vehicles and the rules for their operation, elements of road infrastructure, centers for transport planning and control, as well as for the procedure for transferring data or exchanging data between them.

The regulatory framework governs the rules in the field of road safety, insurance, rules in the field of passenger and cargo transportation and applies to transport entities, officials and personnel associated with the movement of vehicles and maintenance of road infrastructure.

To ensure the safety of the movement of automated cars, it is necessary to solve a large list of regulatory issues and, first of all, to regulate the rules of the Vienna Convention on Road Traffic of 1968 (hereinafter referred to as the Vienna Convention), which require constant driver control of the car, and the rules governing the automatic movement of vehicles funds.

Currently, the World Forum for Harmonization of Vehicle Regulations (hereinafter - the WP.29 Working Party) and the Global Road Safety Forum (hereinafter - the WP.1 Working Party) of the UNECE Inland Transport Committee, to which the Russian Federation is a contracting party, active work is underway to harmonize the requirements for automation systems for driving a car, taking into account the Vienna Convention.

Thus, in 2014, the WP.1 working group approved amendments to Articles 8 and 39 of the Vienna Convention, which entered into force on March 23, 2016. The amendments assume that any vehicle system that affects the way of driving is recognized as complying with the provisions Vienna Convention, if it meets one of two requirements [4]:

1) the system is included in the list of equipment that is allowed to be installed on vehicles and that meets the relevant technical requirements (UN rules annexed to the 1958 Geneva Agreement "Agreement concerning the adoption of uniform technical prescriptions for wheeled vehicles, equipment and parts which can be fitted and/or be used on wheeled vehicles and the conditions for reciprocal recognition of approvals granted on the basis of these prescriptions" [5] and the UN Global Technical Regulations, developed as part of the "Agreement concerning the establishing of global technical regulations for wheeled vehicles, equipment and parts which can be fitted and/or be used on wheeled vehicles" [6]);

2) the driver can at any time disable (deactivate) the above system and / or put the vehicle in manual control mode. For this, a person should always be inside the vehicle and should be ready to take control.

Thus, these changes provide the ability to move along the roads of highly automated and even fully automated vehicles.

At the same time, there are more than 50 provisions in the text of the Convention that directly or indirectly establish the responsibilities of the driver, including those requiring a subjective assessment of the road situation, and for which no specific features of application have been established taking into account the use of automated driving systems. In this regard, the presence of the driver in a highly automated and fully automated vehicle remains mandatory.

In contrast to the process of creating conventional cars, when creating unmanned vehicles it is of great importance to conduct development tests on public roads. Such tests cannot replace studies at closed test sites, on which it is impossible to recreate the whole gamut of rapidly changing conditions, the movement of real operating conditions, although in many countries special test sites are being created for testing unmanned vehicles (figure 1).

Figure 1. Test site of unmanned vehicles near Seoul [11].

As a rule, existing legislation does not allow the use of unmanned vehicles for traffic on roads, but in some countries it is allowed to test them. In particular, with a number of restrictions it is allowed to test autonomous vehicles on public roads in Europe (Germany, Denmark, Spain, Poland, Finland, France), Australia, China, Singapore and the USA.

Figure 2. T-pod unmanned electric truck of the Swedish company Einride on the route warehouse - terminal [12].

In September 2018, a special working group on automated / autonomous and connected vehicles (hereinafter - GRVA) was created within the framework of the WP.29 working group to develop internationally agreed resolutions and guidelines governing the functions of automatic driving. Within the framework of GRVA, four expert groups have been formed in the following areas [9]:

- functional requirements for automated vehicles;
- test driving ability of automated vehicles;
- cybersecurity and software updates;
- data storage systems for automatic driving and event data loggers.

The World Forum for Harmonization of Vehicle Regulations at its 178th session adopted the revised Framework for Automated / Autonomous Vehicles ECE / TRANS / WP.29 / 2019/34 / Rev.1. The main purpose of the document is to identify key safety and security principles for automatic / autonomous vehicles of level 3 and above. This document identifies work priorities for WP.29 and indicates the outcomes and timelines for their achievement.

General principles that will serve as the basis for the further development of unmanned vehicles include, but are not limited to, issues of cybersecurity, safety of the driver and other road users, a fail-safe response (detection of system failures) and the possibility of automatically involving the driver in the driving task [7].

It should be noted that the European Commission has developed a strategy for Connected and Automated Driving (hereinafter - CAD) regarding connected and automated driving, and has also created a special working group that prepared preliminary recommendations on legislative changes in the field of autonomous vehicles. At the same time, the working group concluded that [10]:

- EU directives on liability for defective products (85/374/EEC) and on motor insurance (2005/14/EC) are sufficient for future automated systems;
- there is no need to agree on test requirements (the Vienna and Geneva Conventions are enough to test autonomous vehicles while the driver/operator is in the car);
- it is necessary to clarify the responsibility associated with the storage of CAD data and create a mechanism for regulating access to this data;
- The 2010 Intelligent Transport Systems Directive (2010/40/EU) can be used as the basis for adopting an agreed set of rules at EU level to create a single market for autonomous vehicles.

One of the leading countries in the promotion of unmanned vehicles is the United States. On September 7, 2017, a law was passed in the United States Self Drive Act, which allowed the testing of unmanned vehicles on public roads [8]. If earlier the authorities of individual states had to independently determine and regulate the rules of law relating to unmanned driving, now all acts are subject to uniform standards. At the same time, the manufacturer of an unmanned vehicle must prove the safety of the structure and immediately inform the National Highway Traffic Safety Administration about road incidents that have occurred with their car.

An unmanned vehicle approved for trial operation must meet a number of requirements:

- the autopilot system must be reliably protected from hacker attacks and theft of personal data of passengers;
- supervisory authorities should have full access to information on vehicle parameters, its movement, participation in road accidents;
- unmanned vehicles must safely respond to obstacles, road crashes, software failures.
- autonomous vehicles should switch to manual control without problems, be able to interact with pedestrians.

Several states have adopted separate laws governing the testing and movement of unmanned transport systems. As of 2018, 33 states have introduced autonomous vehicle legislation. Twenty-one states have enacted legislation on autonomous vehicles, and governors in nine states have issued administrative decrees for autonomous vehicles (figure 3 [8]).

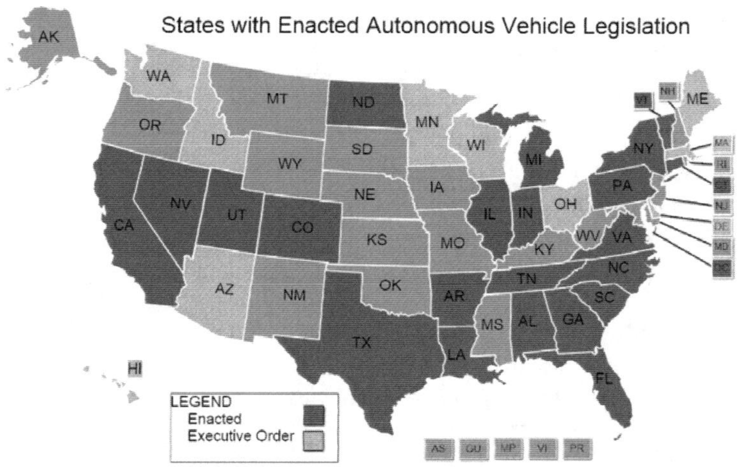

Figure 3. Blue indicates the enacted state legislation, green indicates administrative decrees.

IOP Conf. Series: Materials Science and Engineering **819** (2020) 012033 doi:10.1088/1757-899X/819/1/012033

March 29, 2018 the Government of the Russian Federation approved an action plan ("roadmap") to improve legislation and remove administrative barriers in order to ensure the implementation of the National Technology Initiative in the "Autonet" area.

The revised plan (as amended on May 13, 2019) includes 86 items. This plan is aimed at ensuring the priority positions of Russian companies in emerging global markets, including the development and promotion of products and services in the field of multimodal logistics, including the creation and development of telematics service platforms, interaction platforms for sharing and other platform solutions in the field of connected cars. According to plan should be developed [1]:

- the concept of safe interaction of autonomous vehicles with other road users;
- proposals for the normative consolidation of the features of the organization of the movement of autonomous vehicles in urban conditions, outside the city and on highways;
- proposals for the responsibility of developers, owners and operators of autonomous vehicles;
- proposals for accounting for the use of autonomous and automatic driving systems in the development of programs for the integrated development of transport infrastructure and integrated traffic management schemes;
- proposals on the procedure for allowing autonomous vehicles to be used on public roads;
- training requirements for operators of autonomous vehicles.

To create the possibility of testing autonomous vehicles on public roads, the Government of the Russian Federation issued a resolution dated November 26, 2018 No. 1415 "On the experiment on the operation of highly automated vehicles on highways." In accordance with the said decree, from December 1, 2018 to March 1, 2022, manufacturers of autonomous vehicles can test their cars on public roads in the territories of Moscow and the Republic of Tatarstan [2]. The car is allowed on the road, provided that the driver is the operator, who is ready to intervene in the control process at any time if necessary, and the car must be insured during the trial operation in the amount of 10 million rubles.

Figure 4. Yandex equips serial Prius with its own unmanned modules [13].

In May 2019, State Duma introduced a draft federal law "On the pilot operation of innovative vehicles and amending certain legislative acts of the Russian Federation" (hereinafter - the bill) to the State Duma of the Russian Federation. The bill is aimed at removing legislative barriers and creating favorable legal conditions for the pilot operation of highly automated (innovative) vehicles and involves the introduction of the following changes [3]:

- definition of concepts necessary for regulating the trial operation of innovative vehicles;
- establishing the special legal status of innovative vehicles;
- introduction into the legislation of the Russian Federation the possibility of organizing and conducting pilot operation of innovative vehicles on public roads;
- vesting the Government of the Russian Federation with a number of powers in the field of pilot operation of innovative vehicles;

IOP Conf. Series: Materials Science and Engineering **819** (2020) 012033 doi:10.1088/1757-899X/819/1/012033

- regulation of the procedure for issuing permits for the pilot operation of innovative vehicles on public roads.

After the entry into force of the relevant Federal Law, the Government of the Russian Federation will have to determine the federal executive body vested with the authority to develop state policy and legal regulation in the field of innovative transport and control functions in this area, as well as establish the procedure for the pilot operation of innovative vehicles on public roads and the procedure for issuing the appropriate special permission, requirements for the minimum insurance amount, within which the insurer undertakes to indemnify for damage caused by the operation of innovative vehicles.

The amendments proposed in the bill will increase the degree of investment activity of testing innovative technologies of transport and create conditions for the development of the innovative vehicle market segment.

3. Findings

When creating unmanned vehicles, lapping tests on public roads are of great importance. Such tests cannot replace studies at closed test sites, on which it is impossible to recreate the whole gamut of rapidly changing conditions, the movement of real operating conditions.

The existing regulatory framework in the Russian Federation and a number of foreign countries allows testing of highly automated and fully automated vehicles on public roads.

References

[1] Decree of the Government of the Russian Federation dated March 29, 2018 No. 535-r “On the plan of measures (“roadmap”) for improving legislation and eliminating administrative barriers in order to ensure the implementation of the National Technology Initiative in the conduct of an experiment on operation on roads highly automated vehicles”

[3] Explanatory note to the draft federal law “On the pilot operation of innovative vehicles and amendments to certain legislative acts of the Russian Federation”, Internet: https://sozd.duma.gov.ru/bill/710083-7

[4] Convention on Road Traffic. Done at Vienna, November 8, 1968

[5] Agreement concerning the adoption of uniform technical prescriptions for wheeled vehicles, equipment and parts which can be fitted and/or be used on wheeled vehicles and the conditions for reciprocal recognition of approvals granted on the basis of these prescriptions E/ECE/TRANS/505/Rev.2

[6] Agreement concerning the establishing of global technical regulations for wheeled vehicles, equipment and parts which can be fitted and/or be used on wheeled vehicles ECE/TRANS/132

[7] Framework document on automated/autonomous vehicles ECE/TRANS/WP.29/2019/34/Rev.1

[8] A Policy Proposal to Support Self-Driving Vehicle Accessibility, Internet: http://drivelab.org/pdfs/brinkley_policy_proposal.pdf

[9] Safety at core of new Framework to guide UN regulatory work on autonomous vehicles, Internet: Internet: http://unece.org/trans/main/welcwp29.html

[10] At a Glance: Autonomous Vehicles, Internet: https://www.twobirds.com/en/news/articles/2017/global/at-a-glance-autonomous-vehicles

[11] A testing ground for smart cars is being built in South Korea (AgentVDK.RU) Internet: http://agentmdk.ru/interesnoe/10575-v-yuzhnoy-koree-stroitsya-ispytatelnyy-poligon-dlya-umnyh-mashin.html

[12] Efremova E “The unmanned truck of the future T-pod was shown in Sweden” Internet: https://www.ridus.ru/news/299307

[13] Polishchuk D (Moscow, Vedomosti) “Tesla calls his regime autopilot, but this is marketing guile” Internet: https://www.vedomosti.ru/technology/characters/2019/08/19/809088-dmitrii-polischuk-yandeks-tesla-nazivaet-

IOP Conf. Series: Materials Science and Engineering **819** (2020) 012034 doi:10.1088/1757-899X/819/1/012034

Adaptation of a universal method for assessing the strength and durability of suspension elements applicable to an unmanned vehicle

I Nazarkov, A Bokarev and S V Bakhmutov

FSUE "NAMI", 2 Avtomotornaya St., Moscow, 125438, Russia

Abstract. Ensuring the required vehicle life is one of the most important global engineering challenges. Checking the fatigue life, first of all, is carried out separately for components, electronic and mechanical systems of the car by the bench test method, and the next step is to check the fatigue life of the car assembled by the road test method. The most urgent task is to check the strength and durability of the car chassis in the early stages of design. Currently, the mandatory and classic approach of all leading automakers when designing a car at the stage of a technical project is to apply the methodology for checking the vehicle's steering suspension device and its power structure on multi-channel stands, and the hub assembly verification methodology is separately applied. [1] This paper describes the basic principles of the hub assembly methodology checking to be applied to any motor vehicle. The aim of this work is to demonstrate the versatility of the developed methodology for assessing the strength and durability of suspension elements. Substantiated adaptation (adjustment) of the forced load cycle of a passenger car of class M1, M2 is offered to an unmanned vehicle.

There are several methodologies for checking the chassis (Table 1):

- Methodology of Porsche Engineering (PBA);
- Methodology of Fraunhofer Institute LBF (STAMAS);
- The methodology of FSUE "NAMI".

Table 1. Characteristics of methodologies for testing chassis components.

Chassis Verification Methodologies	Porsche Engineering Methodology (name: PBA)	Fraunhofer Institute LBF Methodology (name: STAMAS)	Proposed Methodology of FSUE NAMI
Applicability:			
cars (up to 3.5 t) --->	OK	OK	OK
trucks (more than 3.5 t) --->	NOK	NOK	OK
Load cycle simulation:			
operation on European roads --->	OK	OK	OK
operation on Russian roads --->	NOK	NOK	OK
the ability to simulate a cycle as required customer (% distribution and mileage) --->	NOK	NOK	OK
Simulation of wheel and transmission forces:			
Fx (traction force) --->	NOK	OK	OK
Fx (brake force) --->	OK	OK	OK
Fy (lateral force) --->	OK	OK	OK
Fz (vertical force) --->	OK	OK	OK
Mx (torque) --->	OK	OK	OK
My (braking torque) --->	OK	OK	OK
My (traction torque) --->	NOK	OK	OK
Mz (torque) --->	NOK	OK	OK
Turn steering wheels --->	NOK	OK	OK
Transmission torque --->	NOK	OK	OK
Duration: --->	**200 hours.**	**200 hours.**	**200 hours.**

It is important to note that the developed methodology is applicable for the categories of cars M1, M2, N1 and N2 (cars and trucks). The load cycle can be considered as a modeling of car operation both on European roads and Russian roads, and it is also possible to optimize the load cycle according to customer requirements and a kilometer resource.

In many countries special research laboratories for road transport are involved in the development of model cycles. In particular, the TRL limited transport research laboratory commissioned by order of the Department of Transportation of England contains information on the road cycles of a wide variety of settlements for different categories and classes of vehicles. [2]

The principle of the proposed ideology (see Figure 1) is to determine the input impact by the scaling method (wheel loads Fx, Fy, Fz, Mx, My, Mz) depending on the suspension parameters, the basic parameters of the car and the operational requirements for the developed car. These parameters are percentage of operation for different types of roads, mileage to reach the full resource, requirements for quasistatic modes (overload coefficients). To force the load cycle (reduce test time) only extreme operating conditions are assumed to be used. The set of these modes for M1, M2 passenger car categories covers the following road events:

- maximum acceleration, braking when moving along a curved path on damaged asphalt with potholes;
- moving along a mountain road with maximum intensity;
- movement on a cobblestone road of various profiles, the Belgian bridge road (resonant modes);
- Maneuver «Eight» with maximum lateral acceleration on a good asphalt surface on road.

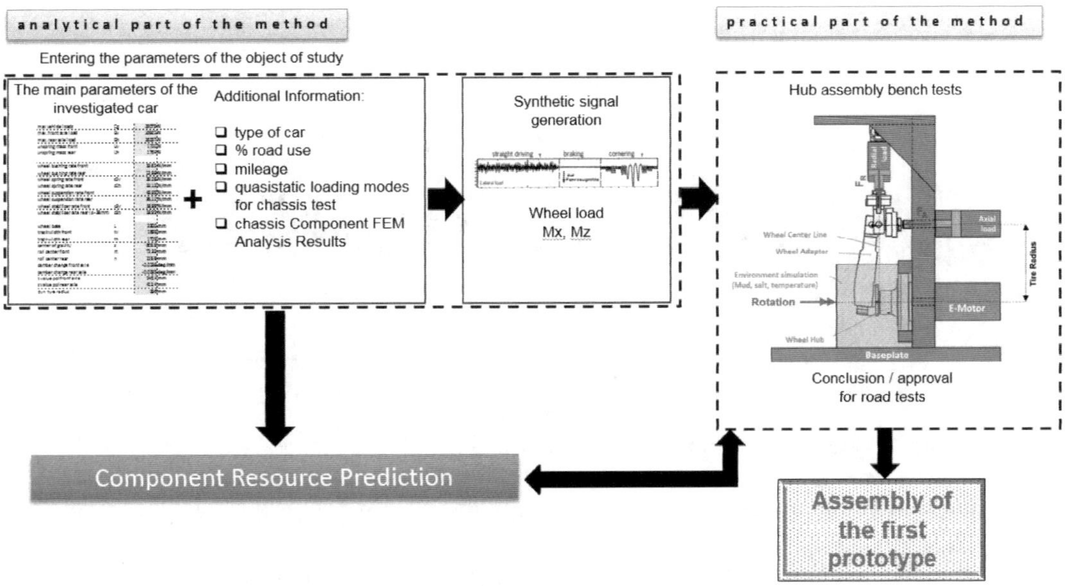

Figure 1. The principle of the proposed methodology.

The above-mentioned road events and modes were implemented at the test site of FSUE NAMI. The combination of these operating modes allows us to load the suspension guide and its power structure most effectively the along the longitudinal, lateral and vertical components of the effort and to provide the required load factors.

Transitions between operational, road and field tests are evaluated using the pseudo-damage values of the wheel loads Fx, Fy, Fz, Mx, My, Mz. The operational coefficient of an everyday test in transition to a field (polygon) test in terms of pseudo damage is found by the general formula:

$$К_{\text{factor i}} = \frac{D_{Fx\_1} + D_{Fy\_1} + D_{Fz\_1} + D_{Mx\_1} + D_{My\_1} + D_{Mz\_1}}{D_{Fx\_2} + D_{Fy\_2} + D_{Fz\_2} + D_{Mx\_2} + D_{My\_2} + D_{Mz\_2}}$$

Where:

$К_{\text{factor i}}$ - conversion factor for the i-th road with reference to the equivalent mileage coefficient;

D_{Fi_1}, D_{Fi_2} - pseudo-damage values for wheel load signals during operation on public roads ("1" index) and polygon roads ("2" index) - Fi forces;

D_{Mi_1}, D_{Mi_2} - pseudo-damage values for wheel load signals during operation on public roads ("1" index) and polygon roads ("2" index) - Mi torgues;

The coefficient of equivalent mileage is determined by the formula:

$$К_{\text{equivalent mileage i}} = \frac{S_{1\,i}}{S_{2\,i}} \cdot К_{\text{factor i}}$$

Where:

$S_{1\,i}$, (km) - mileage on the i-th public road;

$S_{2\,i}$, (km) - mileage on the i-th road and polygon tests.

The transition from field tests to bench tests is carried out by eliminating slightly pseudo-damaging sections (see the example in Figure 2). To implement this method, the calculation of pseudo-damages in the time domain is usually carried out, and by slightly superimposing temporary histories on each other (the investigated temporal history and the pseudo-damages of the investigated temporal history), to identify slightly damaged zones. To work with temporary histories, the FSUE "NAMI" uses RPS pro software (Remote Process Control - feedback signal management software) of MTS company.

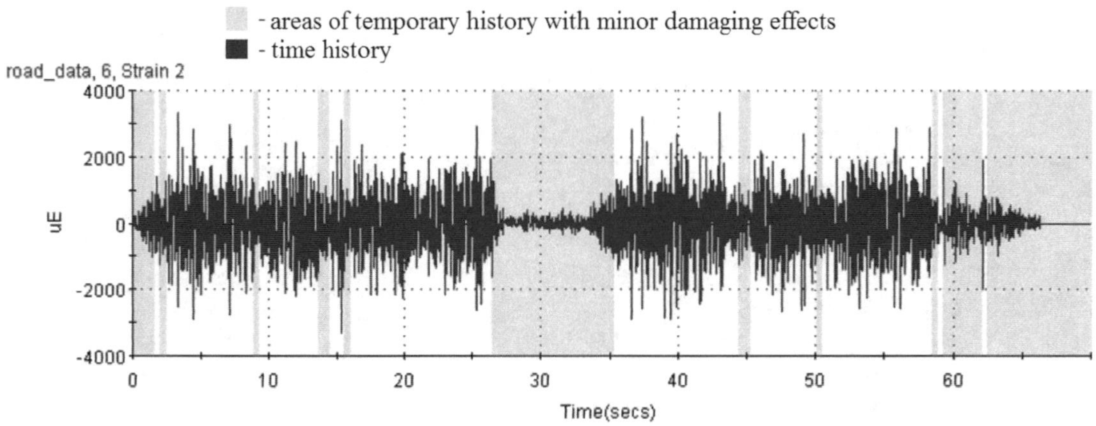

Figure 2. Temporary history with damaging effects.

Processing the weakly pseudo-damaging areas reduces the test time to 1000% while maintaining the reference value of damage / pseudo-damage.

The absolute values of the pseudo-faults Fx, Fy, Fz, Mx, My, Mz are used to evaluate the test groups of signals. Pseudo-damages are calculated using the standard Wöhler line in logarithmic coordinates (Wohler curve, k = 5), shown below in Figure 3. [3]

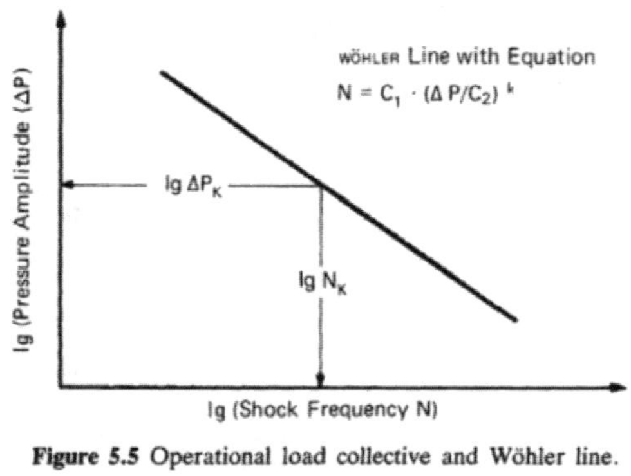

Figure 3. Wöhler line in logarithmic coordinates.

Where:

K - is the slope coefficient of the straight line (in the practice of studying material fatigue experimentally, an average coefficient k = 5 was obtained for the analysis of the pseudo-damage of steel and aluminum structures).

X axis - is the number of cycles in logarithmic coordinates;

Y axis- is the amplitude of the signal under investigation.

To study the hub unit, the force factors Mx and Mz are used. On the basis of road data and requirements for the vehicle's service life, load cycles of the hub are formed taking into account its rotation by the scaling method. The destructive factor for the hub is damage mainly from pulsation of bending moments, as well as the effect of torque / braking moments is negligible. It is proposed to investigate the forecasting of the hub resource by the material fatigue curve under loading with equivalent design load cycles of constant amplitude, which are determined by the method of calculating the pseudo-damage from the total bending moment Mrez [N•m] in the target operating cycle.

A diagram that determines the calculation sequence before the test stageis shown below.

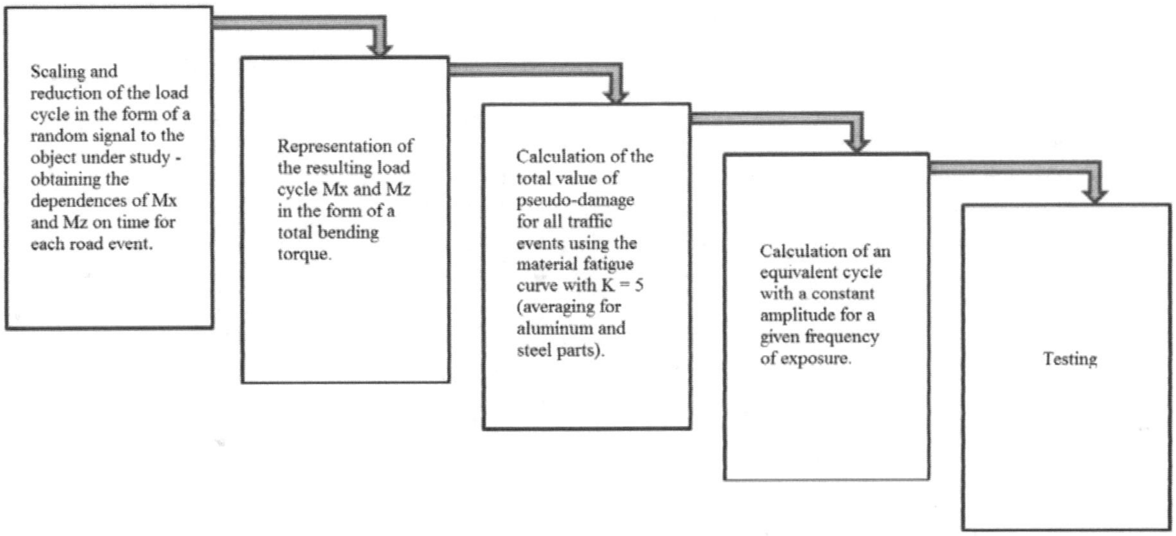

Figure 4. Calculation sequence before the test stage.

The formula for determining the total bending moment Mrez [N•m] taking into account the rotation of the hub unit is:

$$f(x) = \Delta + A\sin(2\pi\omega t + \varphi),$$

where $\Delta = 0,\ A = \sqrt{M_x^2 + M_z^2}$[N•m], $\omega[rps], \varphi = 0.$

rps - revolutions per second.

The final result of the methodology is a software package for the formation of an operational load cycle for prototypes of car categories of M1, M2 with a wheel formula of 4x2 and 4x4 (Figure 5).

This article is considered as an informing about a potential possibility for the implementation of the hub unit resource cycle for M1, M2 vehicles category with a 4 × 2 and 4x4 wheel arrangement. Currently, a software package for the formation of the operational load cycle is being developed. The results and the algorithm for scaling Mx and Mz signals are not be presented in this article.

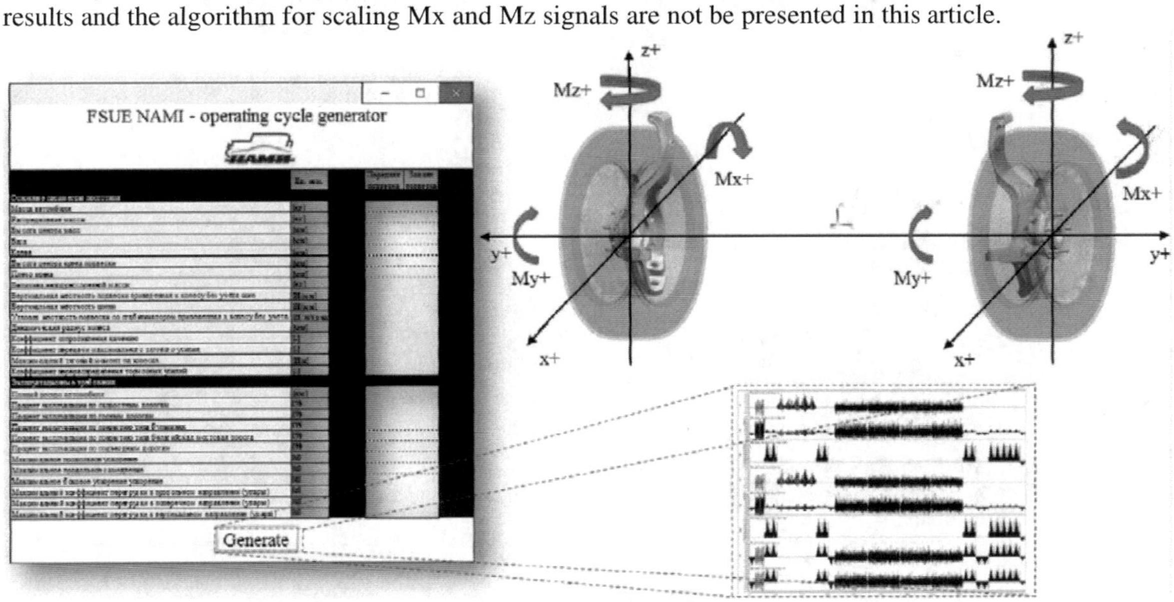

Figure 5. Software package for the formation of the operational load cycle.

A vehicle of a particularly small class of category M2 for closed areas has been taken as an unmanned vehicle. The fundamental difference between the driving modes of an unmanned vehicle and a passenger car of class M1, M2 is the absence of significant lateral loads that arise when driving along a curved path with great acceleration. The maximum speed of an unmanned vehicle designed to carry passengers in closed areas on paved roads is 25-40 km / h (limited by electronics), which, in its turn, minimizes the failure of the hub unit caused by brittle or fatigue destruction of the hub body. Hub assembly failure can most likely occur as a result of wear on the hub bearing components.

Thus, the approach consists in revision of the forced load cycles for an unmanned vehicle, eliminating maximum accelerations, and braking when driving along a curved path as well as adjusting the content of traffic events.

Figure 6 shows the general principle of loading modes formation and the recurrence of load cycle events.

The number of repetitions is related to conversion factors, equivalent mileage coefficients and resource requirements for the hub unit of an unmanned vehicle in accordance with the technical specification (full mileage to failure).

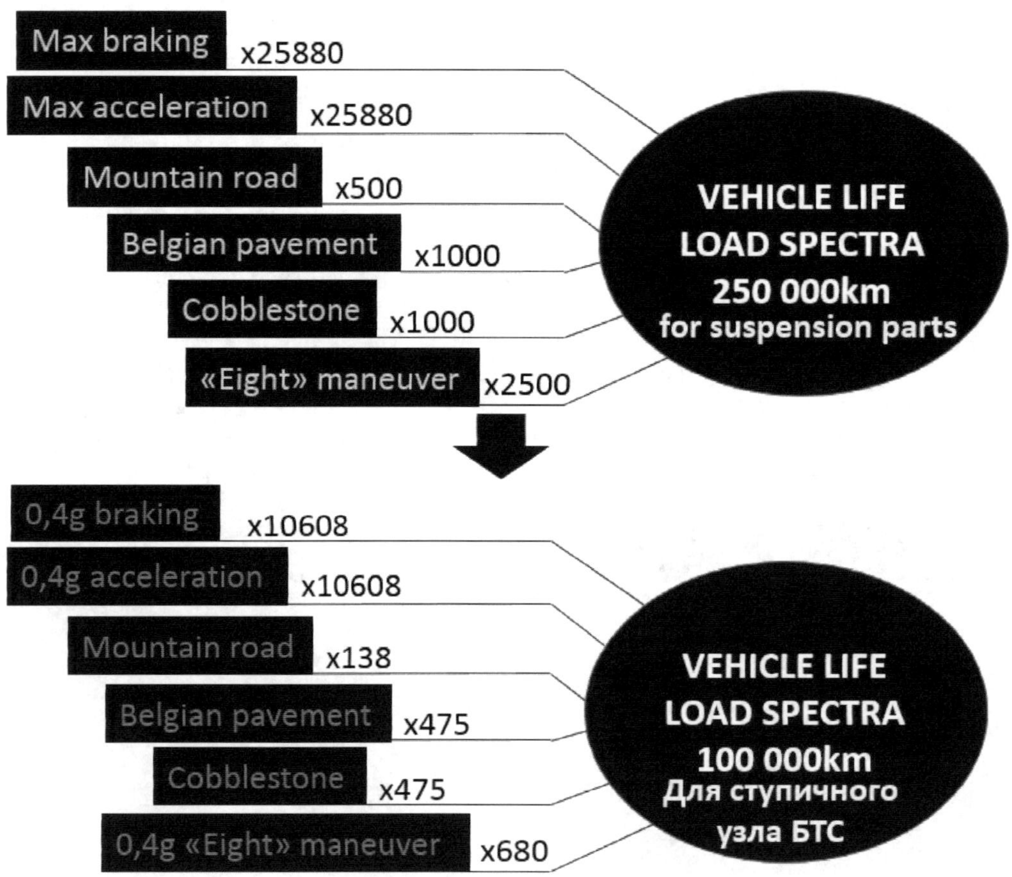

Figure 6. Revised software requirements for the formation of the cycle, taking into account the coefficients for the hub unit of an unmanned vehicle.

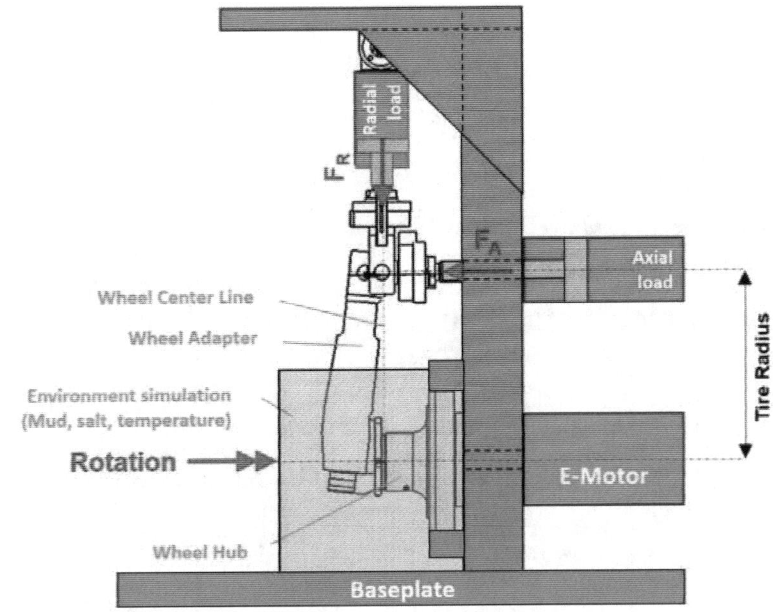

Figure 7. Schematic diagram of a 3-channel test bench for the hub unit.

Conclusion

1. The use of developed by FSUE NAMI methodology for assessing the strength and durability at the early stages of vehicle design allows to eliminate or at least to predict failures as well as weak zones of the power structure in chassis and suspension guide in all subsequent phases.
2. The use of the above methodology allows to reduce the test time to 1000% while maintaining the reference value of damage / pseudo-damage.
3. In this paper, the possibility of adapting the forced load cycle of a M1, M2 passenger car to an unmanned vehicle is demonstrated.

References

[1] A I Bokarev, I A Nazarkov, V A Kulagin 2019 Strategy of developing a bench test methodology for testing the strength and durability of the components of the car chassis at the stage of a technical project *Materials of the International scientific-practical conference* (Minsk: BNTU) ISBN 978-985-583-238-7 (t. 1) 978-985-583-240-0

[2] Bokarev A I 2018 Improving the efficiency of an individual adjustable electric drive of a vehicle *The dissertation for the degree of candidate of technical sciences* p 75

[3] Lee Y.-L, Barkley M E and Kang H.-T Metal Fatigue Analysis Handbook *Code technical Reference Book* v.5.4.

IOP Conf. Series: Materials Science and Engineering **819** (2020) 012035 doi:10.1088/1757-899X/819/1/012035

Comparative assessment of vehicle anti-lock braking system operation using friction brake mechanisms and e-machine in the vehicle with electric drive of traction wheels

S V Bakhmutov[1], A A Umnitsyn[1] and V G Ivanov[2]

[1] FSUE "NAMI", 2 Avtomotornaya St., Moscow, 125438, Russia
[2] Research Professor TU Ilmenau, Ilmenau, Germany

E-mail: artem.umnicin@nami.ru

Abstract. Study of the braking process of the vehicle with an electric machine in the traction wheels drive equipped with an anti-lock braking system (ABS), where the actuators are the electric machine and friction brake mechanisms, is a relevant task. This is related to the worldwide trending increase in the share of production of vehicles with an electric machine within the transmission. In this paper, the "traditional" ABS and two systems with different variants of usage of the electric machine and friction brake mechanisms as actuators are compared: 1. The electric machine fulfills the function of the wheel slip regulator; the friction brake mechanisms maintain constant pressure in the brake line, which depends on the tyre-road friction coefficient. 2. The friction brake mechanisms fulfill the function of the wheel slip regulator; the electric machine maintains the set brake torque. According to the study results, usage of the combined actuator system within the ABS allows reducing the braking distance significantly, raising the vehicle deceleration value and improving the driver's and passengers' comfort during braking.

1. Introduction

The Anti-lock Braking System (ABS) is an important vehicle system having direct effect on traffic safety. One of the latest trends in the development of this system is the use of an e-machine as one of the actuators. In recent years, this has been connected, first of all, with the considerable increase in sales of electric vehicles and vehicles equipped with a hybrid power unit [1, 2]. Studying the properties and methods for control of the modern anti-lock braking system is an important and relevant task of the automotive industry [3, 4]. A comparative analysis of the previously developed anti-lock braking system operation algorithm [5] with continuous control and the traditional system with rule-based control is performed within this paper.

2. Object of investigation

In this paper, the object of investigation is the mathematical model of the rear-wheel-drive electric vehicle of category N2, the key specifications of which are given in Table 1.

Table 1. Specifications of N2 category vehicle.

Fully loaded weight	4200 kg
Weight on front axle	1653 kg
Weight on rear axle	2547 kg
Unladen (kerb) weight	4060 kg
Weight on front axle	1555 kg
Weight on rear axle	2505 kg
Height of the center of mass:	
in the unladen weight condition	998 mm
with fully loaded weight	1001 mm
Tyres	185/75R16C

The characteristics of the electric machine used in this model are shown in figure 1, [6].

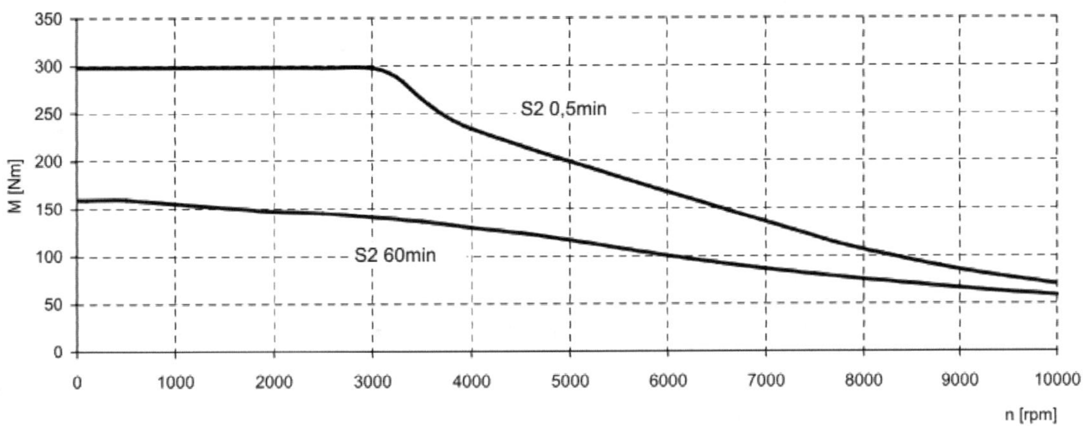

Figure 1. Characteristics of electric machine.

3. Variants of ABS actuator control

At the moment, there are two main types of ABS actuator control:

1. Continuous control that is characterized by constant control of pressure in the wheel brake cylinders, which ensures reduction of wheel slip oscillation amplitude relative to the target value, and also enables application of the electric machine as an actuator.

2. Rule-based control that implies setting the slip boundaries according to which the pressure in the wheel brake cylinders is controlled. Such approach is the most reliable and easy to set up, but at the same time, it does not allow achieving low wheel slip oscillation relative to the target value, as a result, the braking distance may be increased and the driver and passenger comfort worsened, it also makes it difficult to use the advantages of the electric machine as an ABS actuator to the full extent.

3 variants of ABS actuator control are considered within this work.

1. Control type: constant. Actuators: friction brake mechanisms and electric machine. Slip is regulated by the electric machine. The algorithm monitors the electric machine load and, if necessary, smoothly adjusts the pressure in the rear wheel brake lines. The target pressure in the front brake mechanisms and the target electric machine torque are determined by PID controllers.

2. Control type: constant. Actuators: friction brake mechanisms and electric machine. Slip is regulated by the friction brake mechanisms. The algorithm monitors the pressure in the rear brake mechanism brake lines and, in case of exceeding a certain threshold, raises the electric machine brake torque smoothly. In case the pressure falls to the set threshold, the electric machine brake torque decreases. The target pressure in the brake mechanisms is determined by the PID controllers.

3. Control type: rule-based. Actuators: friction brake mechanisms. The minimum and maximum slip thresholds are defined as the optimum one ± 0.01. If the current wheel slip value is above the threshold one (max.), the pressure is decreased. If the current slip is below the threshold one (min.), the pressure is increased. If the current slip is between the minimum and maximum threshold values, the pressure is maintained. The e-machine is not involved in braking.

Definition of the wheel slip target values in all cases is performed by the method specified in paper [5]. The anti-lock braking system layout and general information are presented in paper [7].

4. Virtual test method

The following virtual test conditions are suggested for the evaluation of the operation efficiency of the control variants described above.

1. Initial speed: 60 and 100 km/h.
2. Wheel adhesion coefficient during driving: 0.8 and 0.2.
2. The accelerator pedal is completely released.
3. The brake pedal is pressed with 300 N force.
4. Upon decelerating to 15 km/h, the test stops.

5. Criteria for ABS operation efficiency evaluation

The criteria for the ABS operation efficiency evaluation used in this paper are taken from paper [7]. The ABS operation efficiency is assessed according to 6 criteria:

1. Braking distance.
2. Average deceleration.
3. ABS index of performance (ABSIP) equal to the ratio between the average deceleration during the ABS operation and the average deceleration when braking without the ABS.
4. Average slip.
5. Slip control peak-to-peak value, which is characterized by the percentage ratio of the difference between the maximum and minimum wheel angular speed to the maximum wheel angular speed at the start of ABS operation.
6. ITAE Jerk – integral of the absolute value of the derivative of acceleration times the time.

6. Comparative assessment of anti-lock braking system algorithms operation

Table 2 contains the parameters of the assessment criteria for the ABS operation efficiency during braking at 60 km/h.

Table 2. Summary table of assessment criteria for ABS operation efficiency during braking at 60 km/h.

Control variant	Braking parameters			ABS operation parameters			Jerk ITAE
	Braking distance, m	Average deceleration, m/s^2	ABSIP	Average slip, %	Peak-to peak value, front axle, %	Peak-to peak value, rear axle, %	
			μ=0.2				
Variant 1	75.86	-1.75	1.11	-5.93	60.86	4.401	1.59
Variant 2	76.13	-1.74	1.11	-7.61	61.35	61.48	3.51
Variant 3	81.59	-1.61	1.02	-24.5	14.39	7.8	76.40
without ABS	83.13	-1.58	1.00	-98.74	-	-	-
			μ=0.8				
Variant 1	21.20	-6.47	1.07	-15.59	6.63	31.84	3.42
Variant 2	20.81	-6.55	1.08	-20.87	10.65	17.7	3.80
Variant 3	20.75	-6.48	1.07	-24.66	4.1	2.95	52.50
without ABS	22.21	-6.04	1.00	-92.88	-	-	17.55

IOP Conf. Series: Materials Science and Engineering **819** (2020) 012035 doi:10.1088/1757-899X/819/1/012035

Figure 2 provides circular radar charts with indicators according to the assessment criteria for the ABS operation efficiency during braking at 60 km/h

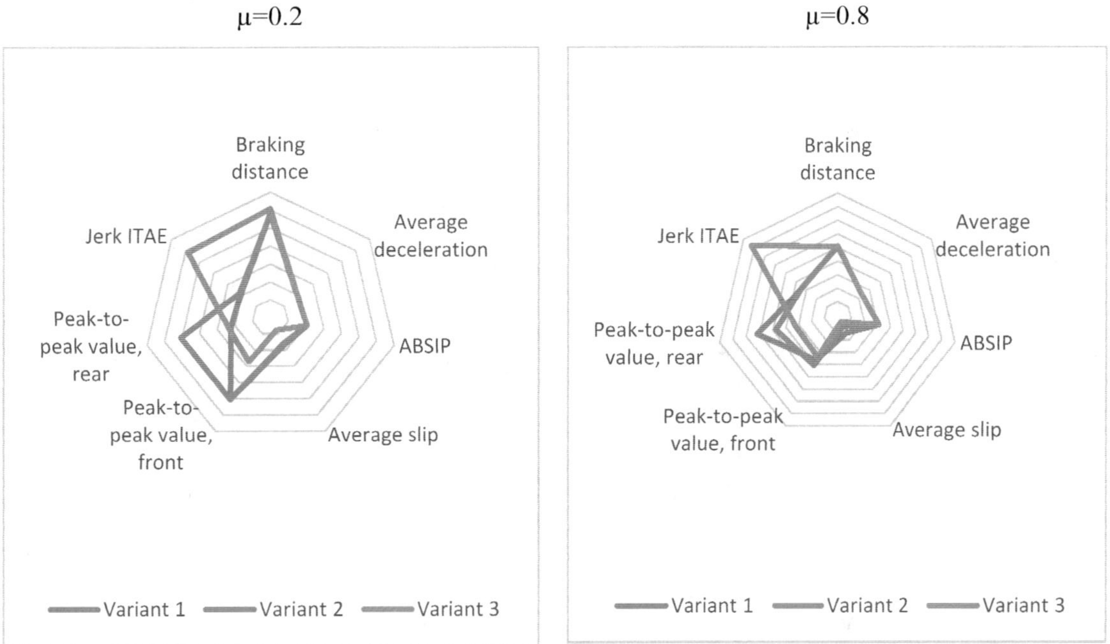

Figure 2. Circular radar charts with indicators according to the assessment criteria for ABS operation efficiency during braking at 60 km/h.

Table 3 provides indicators of the assessment criteria for the ABS operation efficiency during braking at 100 km/h.

Table 3. Summary table of assessment criteria for ABS operation efficiency during braking at 100 km/h

Control variant	Braking parameters		ABS operation parameters				Jerk ITAE
	Braking distance, m	Average deceleration, m/s^2	ABSIP	Average slip, %	Peak-to peak value, front axle, %	Peak-to peak value, rear axle, %	
			μ=0.2				
Variant 1	219.80	-1.73	1.08	-5.93	60.86	4.401	1.59
Variant 2	213.90	-1.78	1.11	-7.61	61.35	61.48	3.51
Variant 3	227.70	-1.65	1.03	-24.5	14.39	7.8	76.40
without ABS	234.90	-1.60	1.00	-98.74	-	-	-
			μ=0.8				
Variant 1	58.85	-6.64	1.07	-15.59	6.63	31.84	3.42
Variant 2	56.88	-6.85	1.11	-20.87	10.65	17.7	3.80
Variant 3	57.60	-6.67	1.08	-24.66	4.1	2.95	52.50
without ABS	62.11	-6.19	1.00	-92.88	-	-	17.55

Figure 3 provides the circular radar chart with indicators according to the assessment criteria for the ABS operation efficiency during braking at 100 km/h.

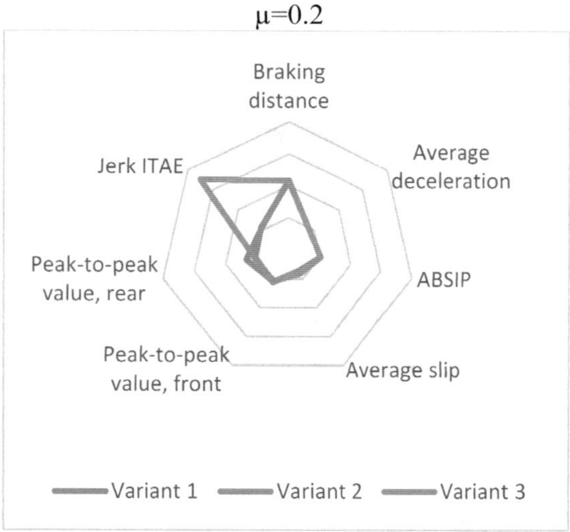

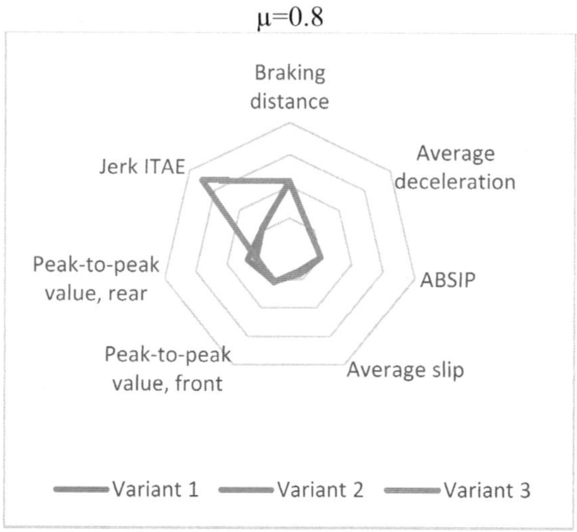

Figure 3. Circular radar chart with indicators according to the assessment criteria for ABS operation efficiency during braking at 100 km/h.

It may be noted that variant 3, currently applied in most vehicles, provides the worst efficiency among the considered variants.

Variant 1 – continuous control where the slip is controlled by the e-machine – compared to Variant 3, in case of braking from 60 km/h on the road with adhesion coefficient μ=0.2, reduces the braking distance by 7%, and in case of braking on the road with adhesion coefficient μ=0.8, the braking distance is increased by 2.2%. In case of braking from 100 km/h speed on the road with adhesion coefficient μ=0.2, this variant reduces the braking distance by 3.5%, and in case of braking on the road with adhesion coefficient μ=0.8, the braking distance is increased by 2.2%.

Variant 2 – continuous control where the slip is controlled by the friction braking mechanisms, and the electric motor only generates constant torque – ensures the best parameters of the anti-lock braking system efficiency. In case of braking from 60 km/h speed on the road with adhesion coefficient μ=0.2, this variant, compared to the third one, reduces the braking distance by 6.7%, but in case of braking on the road with adhesion coefficient μ=0.8, the braking distance is increased by 2.2%. In case of braking from 100 km/h speed on the road with adhesion coefficient μ=0.2, this variant reduces the braking distance by 6.1%, and in case of braking on the road with adhesion coefficient μ=0.8, the braking distance is reduced by 1.3%.

7. Conclusion

Comparing two types of the anti-lock braking system, it can be noted that the constant control has a number of advantages in comparison with the threshold one, especially when used in the vehicles equipped with an electric machine within the vehicle transmission. On the other hand, the presence of the differential located between the electric machine and the vehicle wheels does not allow exploiting all the advantages of the individual wheel slip control by means of electric machines [8]. Thus, it is more efficient to use variant 2 of the algorithm presented in this paper in case of usage of one electric machine within the ABS.

IOP Conf. Series: Materials Science and Engineering **819** (2020) 012035 doi:10.1088/1757-899X/819/1/012035

References

[1] Terenchenko A S, Karpukhin K E, Kurmaev R 2015 Features of Operation of Electromobile Transport in the Conditions of Russia *28th International Electric Vehicle Symposium and Exhibition 2015* **28**

[2] Karpukhin K E, Terenchenko A S 2016 Features of Creation and Operation of Electric and Hybrid Vehicles in Countries with Difficult Climatic Conditions in the Russian Federation *IOP Conference Series: Materials Science and Engineering* 2016 **157** № 1 p 012014

[3] Kamenev V F, Terenchenko A S, Karpukhin K E, Kolbasov A F 2017 The strategy of creating urban electric vehicle: Challenges and solutions *International Journal of Civil Engineering and Technology* **8** Issue 12

[4] Karpukhin K E, Terenchenko A S 2017 The specificity of the popularization of hybrid and electric vehicle in the Russian Federation *EVS 2017 - 30th International Electric Vehicle Symposium and Exhibition*

[5] Bakhmutov S V, Umnitsyn A A, Ivanov V 2019 Creation of electric vehicle abs operation algorithm with possibility of hybrid braking based on slip-slope approach at wheels slip determining, *IOP Conference Series: Materials Science and Engineering*

[6] http://media3.ev-tv.me/Azure300VDC-400Acurve.pdf

[7] Bakhmutov S V, Ivanov V, Karpukhin K E and Umnitsyn A A 2018 Creation of operation algorithms for combined operation of anti-lock braking system (ABS) and electric machine included in the combined power plant *IOP Conference Series: Materials Science and Engineering* **315**

[8] Savitski D, Ivanov V, Augsburg K, Pütz T, Barber P, 2016 The new paradigm of an anti-lock braking system for a full electric vehicle: Experimental investigation and benchmarking *Proceedings of the Institution of Mechanical Engineers, Part D: Journal of Automobile Engineering*

IOP Conf. Series: Materials Science and Engineering **819** (2020) 012036 doi:10.1088/1757-899X/819/1/012036

Providing vehicle running life in case of loss of tyre air pressure

S A Kantorin[1] and S V Bakhmutov[2]

FSUE "NAMI", 2 Avtomotornaya St., Moscow, 125438, Russia

E-mail: [1] semen.kantorin@nami.ru, [2] s.bakhmutov@nami.ru

Abstract. This paper comprises general information on the tyre air pressure loss problem and methods for ensuring vehicle mobility in case of such pressure loss, namely the information on current tyre-wheel system development trend, as well as the use of special internal supports (run-flat) of different design. Furthermore, the paper describes practical aspects of implementation of this system, reveals positive and negative characteristics and contains a conclusion on the system development prospects depending on their application.

1. Providing vehicle running life in case of loss of tyre air pressure

The problem of ensuring traffic (vehicle motion) safety in case of sudden tyre damage leading to complete loss of air pressure is always relevant.

Loss of internal pressure leads to deformation, bead come-off from the wheel bead ledge and tyre destruction. It impairs vehicle stability and leads to an accident in most cases. The research by Dunlop (England) shows that, on the European roads, tyre failures due to mechanical damages amount to 40%. According to the estimation by the Federal Statistical Office (Germany), there are more than 1,500 accidents a year happening because of the tyre defects in Germany. The striving to improve traffic (motion) safety in case of loss of tyre pressure led to the emergence of new tyre and wheel design variants.

Creation of tubeless tyres in 1954-1958 was a major step in this direction. In case of foreign item penetration, tubeless tyres are less prone to a sudden drop in pressure, however, after the damage, the tyre running life reduces, though it is still longer than the one of a tube tyre. This problem is partially solved by application of a rubber-cord spacing ring.

However, it does not solve the safety problem completely. Moreover, a need appeared to create special tyres and systems named "safe" for general-purpose vehicles, and for special vehicles, mainly for the Ministry of Defence and security services – "bullet-proof". The purpose of such structures is ensuring sufficient sealing of the hole in case of a puncture and stable vehicle motion. In order to provide the possibility to drive a vehicle with a flat tyre, the safe tyres shall be highly resistant to side slip allowing vehicle control without drifting on a straight road caused by lateral forces and when turning, as well as when performing road maneuvers, tyres shall keep their position on the rims so that the beads are not displaced and shall not come off from the wheel bead ledge.

The safety improvement problem also includes creating safe tyre-wheel systems and other methods of increasing tyre functionality in case of operational damages. The major tyre and automotive companies of

Europe, USA, Japan, and other countries work on creating designs of tyres and tyre-wheel systems ensuring vehicle motion safety in case of mechanical damage of the tyres.

All methods of increasing the tyre penetration resistance and bulletproof capability and ensuring reliable driving of a vehicle with a damaged tyre can be conditionally divided into two main groups: self-sealing of mechanical punctures on the tyre tread and ensuring long-term driving of a vehicle with a damaged tyre.

Traditional tubeless tyres including various sealing compositions and tyres with special spongy layers in the internal cavity under the tyre tread belong to the first group. The sealing liquids are characterized by the following properties: prevention of air leakage, keeping nominal operating pressure in tyres and reduction of tyre heating. Being inside the tyre during the whole service life, such liquids remedy punctures when these emerge. They do not cause tyre disbalance, steel rusting, disc corrosion or rubber destruction. The sealing liquids maintain their operational properties in a wide range of temperatures (from -20 to +75 °C). Thus, the use of sealing liquids in tyres is of interest in terms of vehicle motion safety.

One of the tyre design options allowing further driving after the tyre damage is a highly self-sealing tyre such as Seal Inside by Pirelli (Italy), Figure 1. The internal surface of the basic innerliner is provided with an additional layer intended for self-sealing in case of small holes. This technology is effective for the majority of punctures, except for the punctures breaking the tyre integrity.

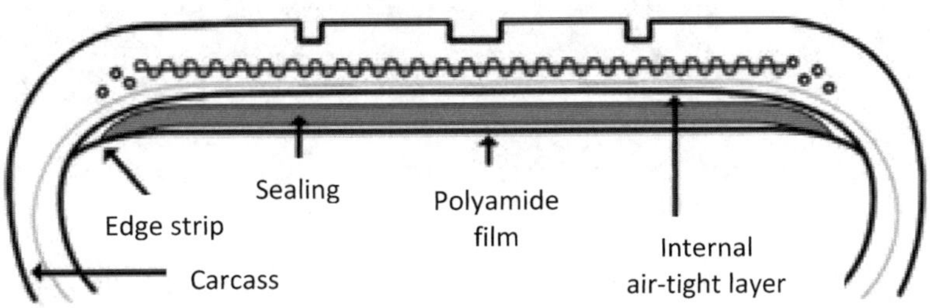

Figure 1. Seal Inside tyre by Pirelli.

For ensuring long-term vehicle motion with a damaged tyre or with all tyres damaged, new requirements shall be added to the abovementioned characteristics. Regarding the safe operation, it is important to ensure safety concerning tyre rupture or wheel failure and to have a reliable contact between the tyre beads and the rim edges; the tyre shall be also of tubeless type. With no air pressure in the tyre, the vehicle shall move without tyre destruction for up to 200 km at no less than 80 km/h, and, in case of 100% tyre damage, special vehicles shall remain mobile and run for up to 50 km at up to 50 km/h with a possibility of subsequent repair of tyres and wheels (and upon that, the through-hole size shall be no more than 8 mm, and the number of such punctures shall be no more than 10). It is also important to have reliable brake and tractive force transmission, to provide sustainable vehicle driving on dry, wet, ice- and snow-covered roads, to have low heat generation, and to sufficiently resist tyre side damages and feature low sensitivity to obstacle crossing.

Additional requirements for such tyres are as follows: low production cost, long service life and running life, recoverability and low rolling resistance. High comfortableness can be ensured by appropriate elasticity and absorption characteristics, optimum cornering resistance, as well as minimum lateral and radial runout.

Rigid wheel supports can be rotatable and non-rotatable. A wheel with rotatable rigid support consists, for example, of a tubeless tyre and sectors separately inserted into the tyre and bolted to each other. The rim is sealed by a ring rubber cord located in a cavity between the rim halves. A spacing ring holds the tyre bead on the bead ledges in case of the tyre air pressure drop. When the pressure drops, the tyre internal surface is set on the support, which reduces tyre radial deflection and bears most of the wheel load. In order to reduce friction between the tyre internal surface and the support, the latter can rotate about the rim on a friction (slide) bearing over the spacing ring external surface.

The two-segment internal additional support (IAS) design (Rodgard, Figure 2) by Hutchinson considerably simplifies installation of the support within the tyre. The Rodgard design also features a pilot (guide) friction bearing mounted on the rim, which the support can rotate about. According to the manufacturer, the structure can be mounted on a standard rim and ensure the run for up to 50 km with a flat tyre at no less than 50 km/h. As a rule, the number of segments is defined by the feasibility of in-tyre assembly.

Figure 2. Rodgard IAS by Hutchinson.

NIIShP (Scientific Research Institute of the Tyre Industry) tested a wheel with IAS of the same design on the asphalted roads and on the soil tank road at 20-40 km/h. After 30 minutes of driving at 40 km/h, the tyre temperature reached 180 °C. With the completely destructed tyre, the wheel with rotatable support did not provide torque transmission and vehicle mobility.

CRF IAS (Figure 3) by Hutchinson is assembled of three segments. According to the producer's declaration, besides the relative simplicity of manufacturing and installation, it is distinguished by the strong bullet and splinter protection as a whole of the wheel center and tubeless tyre.

CRF provides run for up to 80 km maintaining vehicle drivability without air pressure in the tyres. IAS includes composite materials and is assembled by means of bolt joints.

Figure 3. CRF IAS by Hutchinson.

In the assembled internal support connecting (detaching) points, there are connecting design elements, which are the stress raisers. Such stress raisers significantly increase the probability of the structure failure in case of the exposure to bending moments arising during the vehicle turning. Besides, the wheel running gear vibration loads connected with the drop of the vehicle longitudinal stability when driving with flat tyres significantly increase. Failure of the great majority of detachable IAS under test happens because of destructions in the support connecting points.

An illustrative example of the rigid insert is the metal support ring – Continental Support Ring (CSR) IAS (Figure 4) by Continental – which is mounted along with a standard tyre on a regular rim. During normal driving, the CSR insert does not influence the vehicle's operational properties. In case of tyre pressure loss, further vehicle motion is possible with the top speed of 80 km/h for up to 200 km. CSR is generally offered for equipping AWD passenger cars.

The design can be used on the majority of standard rims in combination with many tyre types. After long and comprehensive testing, Maybach and Bently decided to use this system to increase their vehicles' safety. Similar designs became widespread and are patented by many foreign companies, including Yokohama.

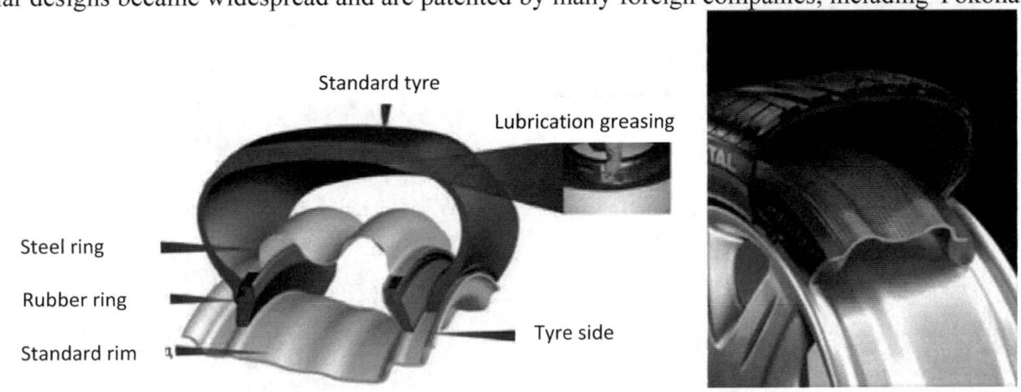

Figure 4. CSR IAS by Hutchinson.

The safe tyre design was considerably simplified by supports made either of rubber only or of combined (rubber and metal) material that increases the stiffness and acting as spacing rings for fixing beads and supports.

The design manufactured since 2002 by Michelin – PAX System (Figure 5) – has become the most widespread one. It was installed on Audi A8, Rolls-Royce Phantom, Audi A4 and Honda Odyssey passenger cars.

In case of tyre pressure loss, an elastic insert helps the tyre sides to keep their shape – it is a ring mounted inside and fixed on the rim.

According to the manufacturer, a vehicle equipped with the PAX system can run no less than 200 kilometers at up to 80 km/h with a flat tyre. Any external force impact on the wheel increases the stress inside the tyre, the beads of which represent a locking gear (beadlock) and can transmit the torque from the rim even if there is no tyre overpressure. In the tyre attachment points, the cord is clamped in the slot, so that the tyre is locked on the rim and cannot come off from the wheel bead ledge even under high dynamic loads. For reducing friction between the support and the tyre, the tyre internal surface is covered with a special greasing layer. At present, tyres with the PAX system are actively used on armoured executive vehicles, as well as on some passenger car models.

IOP Conf. Series: Materials Science and Engineering **819** (2020) 012036 doi:10.1088/1757-899X/819/1/012036

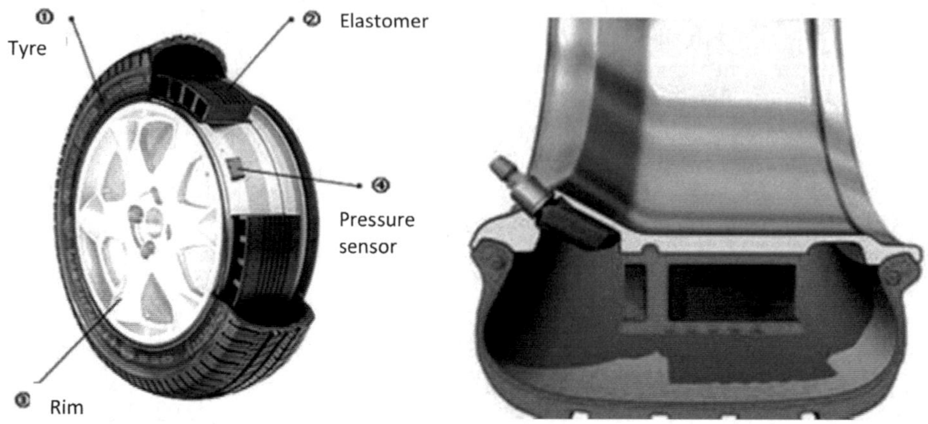

Figure 5. PAX System by Michelin.

In many design versions, the torque is transmitted from the rim to the tyre by means of pressing its beads to the rim edges. In this regard, a need emerged to eliminate friction in the IAS tread contact with the tyre.

Non-detachable VFI IAS for heavy armoured wheeled vehicles are produced by Hutchinson (Figure 6). Despite the need to use a special device pressing the support rubber ring to put it inside the tyre, there is a tendency for such support design application. Due to the absence of connecting elements, the design mass and number of stress raisers in the elastic ring are reduced. According to the manufacturer's declaration, the design provides the run with a flat tyre for up to 50 km at no less than 50 km/h.

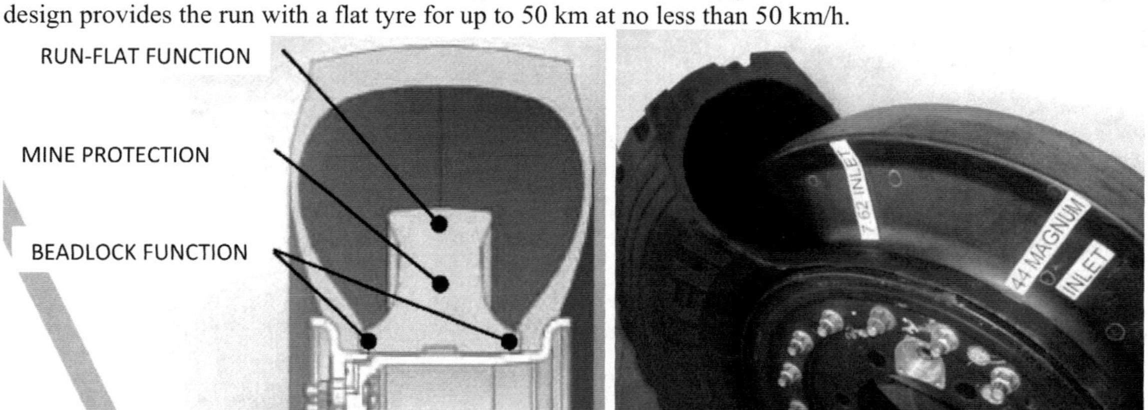

Figure 6. VFI IAS by Hutchinson.

A Russian IAS example is the petal-type support offered by the 3rd Central Research Institute (CRI) of the Ministry of Defence of the Russian Federation (Figure 7) allowing reduction of destruction intensity of the tyre internal cavity in its contact with the support when rolling without overpressure due to reduction of slipping of the bending support petals. Furthermore, the design has also other advantages, such as easy support installation in the tyre due to its lower deformation (folding) resistance and better off-road performance on deformable soils with no air pressure in the tyre due to lower radial stiffness of the support. It should be noted that the internal support stiffness reduction results in bigger internal friction losses which leads to the increase of the wheel rolling resistance and to higher thermal load on the structure.

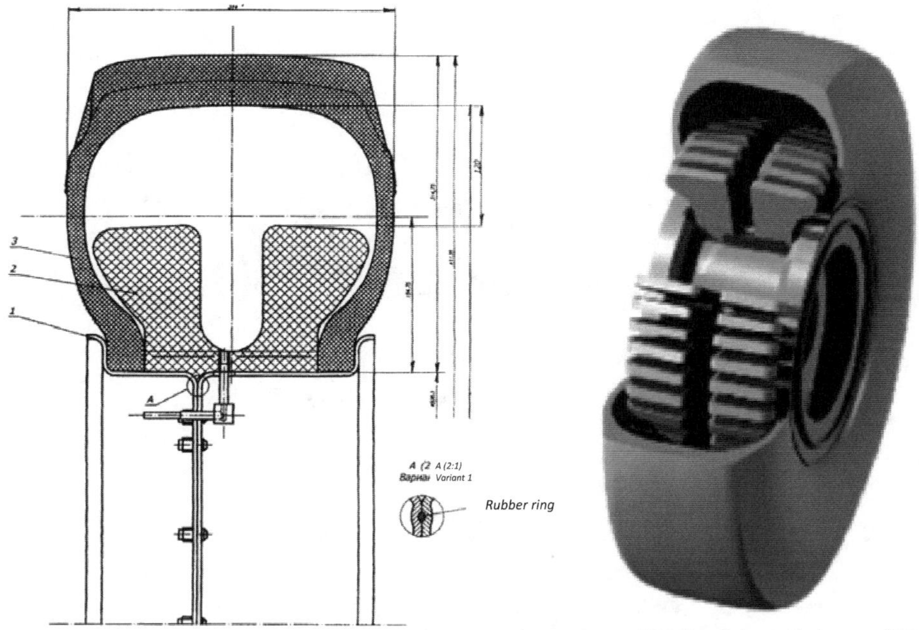

Figure 7. Petal-type IAS by the 3rd Central Research Institute (CRI) of the Ministry of Defence of the Russian Federation.

A special group of tyres designed to ensure running in case of partial and full internal pressure loss comprise the tyres with high own stiffness or with a special tyre side structure. In the Continental concern, such tyre was named SSR, in Michelin – SST, and in Bridgestone – RFT. The cornerstone of the technology is a rather simple postulate: in order to continue moving with a zero-pressure tyre, its sides shall withstand all dynamic and static loads, which are borne by the tyre in the inflated condition. Despite the simplicity of the idea, its implementation took quite a long time and even demanded to join efforts by several tyre concerns to create a special material that could be applied for the run-flat tyre sides (Figure 8). According to the producer's declaration, the technology allows the vehicle to move for no less than 80 km more at up to 80 km/h.

All run-flat tyres require obligatory use of the tyre pressure monitoring system. Using the tyres on vehicles, which are not regularly equipped with such system, is inadmissible. Upon that, the run-flat tyres do not demand application of any special rims. Nevertheless, their installation requires special tyre mounting equipment due to the higher tyre side stiffness.

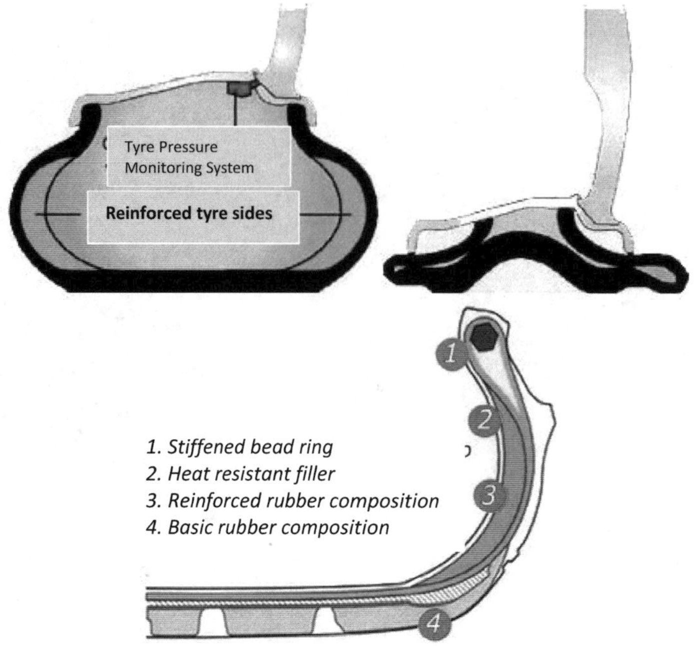

a) b) c)

Figure 8. Tyre with run-flat sides (a, c) and regular tyre (b).

Disadvantages of such tyres are as follows:

1) if a tyre puncture is not noticed in due time, it may cause unexpected effects in an emergency situation;

2) higher vibration and worse steerability on rough roads;

3) these tyres are barely subject to repair;

4) bigger rubber weight that causes increased non-sprung weight, worse steerability and dynamic characteristics.

A safe wheel with a carcass-type tyre without internal overpressure represents a new tyre design direction. The tyre can operate without internal pressure as a result of replacing the carcass threads under tension by elastic sprung wires bearing the tyre radial load. Abroad, great attention is paid to this direction and research activities are conducted very intensively, especially in the Michelin research center. The wheel with the atmospheric pressure tyre was named Tweel. For the first time, the Tweel system (Figure 9, a) was used in traditional wheelchairs where tyre inflating is the number one problem. Then, an improved tyre was tested on Audi A4. In this tyre, the rubber tread is supported by flexible polyurethane spokes. Its testing continues.

At present, Michelin intends to enter the market with the Tweel system (Figure 9, b) for applications with low speed and low weight. Certain progress in creating tyres operating at the atmospheric pressure was reached also in Russia. Originally, the idea of a carcass tyre was implemented on a 3.5-10 tyre. Upon that, the selected safe carcass tyre has shown the values of side and radial stiffness, as well as of uniformity, rolling resistance at the level of similar pneumatic tyre values. When testing in NIIShP, the tyre was damaged by drilling through-holes with a diameter of 12.5 mm (2 on the tyre side and 2 on the tyre tread). According to the bench tests, the drilled holes have no impact on the characteristics of the tyres and their running life on the test bench.

IOP Conf. Series: Materials Science and Engineering **819** (2020) 012036 doi:10.1088/1757-899X/819/1/012036

a)

b)

Figure 9. Tweel tyres by Michelin.
First concept of 2005 (a) and later models (b).

The following development by Michelin is slightly more original – a tyre named Airless (Figure 10) made from elastic rings with pre-determined longitudinal, transverse and vertical stiffness.

IOP Conf. Series: Materials Science and Engineering **819** (2020) 012036 doi:10.1088/1757-899X/819/1/012036

Figure 10. Airless tyre by Michelin.

The abovementioned rings are made of a patented mix including thermosetting resin, fiberglass and polyurethane. The advantage of the French development is that strong impacts and sharp objects damage only separate tyre elements, therefore, the wheel is distinguished by the highest survivability.

A little later, the American company Amerityre offered a design similar to the Michelin Tweel: composite wheel where the rubber tyre tread is connected to the rim through flexible polyurethane spokes (Figure 11).

Figure 11. Airless tyre by Amerityre.

The creators declare lower operating temperature (that means low tyre tread wear) and lower rolling resistance (that means low fuel consumption). The short-term outlook for such wheels is mass production

IOP Conf. Series: Materials Science and Engineering **819** (2020) 012036 doi:10.1088/1757-899X/819/1/012036

and delivery to conveyors as space-saving tyres (donuts). Similar tyre designs were developed by Japanese Bridgestone with AirFree concept (Figure 12, a) and American Hankook with its I-Flex (Figure 12, b).

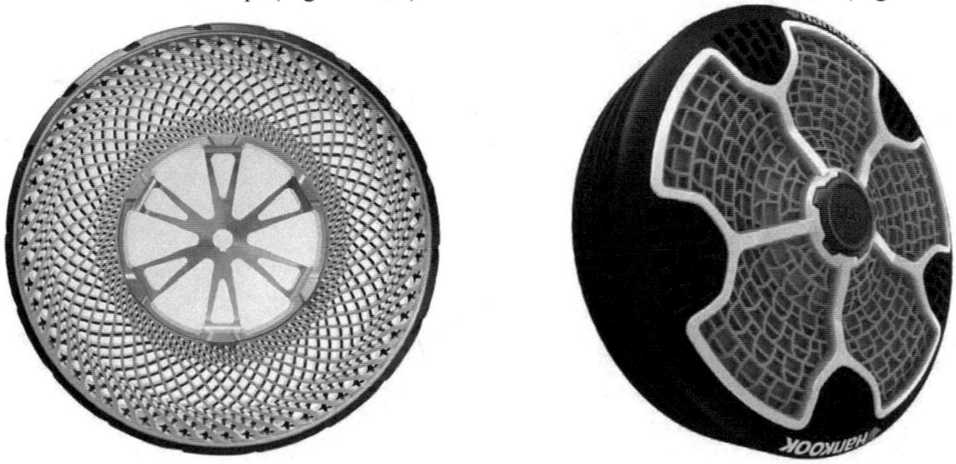

a) b)

Figure 12. AirFree by Bridgestone (a) and I-Flex by Hankook (b).

The Hankook engineers state that the I-Flex technology developed by them allows wheels to operate at up to 130 km/h, and upon that such tyres can be installed on standard rims. A demonstration of the company possibilities was the tyres for Volkswagen Up! city car in the ABT version presented in 2013.

At the moment, different airless tyre variations are not ready to conquer the automobile market, as they perfectly proved themselves on off-road and special vehicles. At the same time, bringing rolling resistance to modern standards, lowering noisiness and improving long-drive heat resistance of the airless tyres can make them peer competitors of traditional tubeless tyres.

Conclusion

Thus, analysis of a number of safe and bullet-proof wheel designs allows determining certain trends in evolution of a structural form of the solutions applied on vehicles including special and military wheeled vehicles.

1. At present, the most promising design of the safe wheels for general-purpose vehicles are carcass-type tyres without internal overpressure.
2. For special and military vehicles, the advantages of safe and bullet-proof wheels with IAS are obvious.
3. In modern designs, thermoplastic polymers or elastomers, mainly polyurethanes, are used as IAS material. These materials have such properties as durability, lightness and elasticity. Damages, high temperatures and dynamic loads have no significant effect on the functionality of the supports made of these materials.
4. The most promising IAS designs are designs compatible with standard wheel rims.
5. The majority of IAS installed on foreign special-purpose vehicles weighing up to 6-7 t are assembled in a tyre of two or three identical segments. This is explained by relative simplicity of design installation in a tyre and a small number of parts.
6. When selecting the IAS segment connecting method, reduction of stress concentration in the connecting points shall be taken into account.
7. The problem of high friction between IAS and the tyre is not finally solved at the moment, in spite of the fact that friction in the IAS-tyre contact is one of the main factors leading to the tyre destruction.

Measures to reduce friction between design elements, for example, greasing with high viscosity coefficient, will allow inhibiting tyre tread deterioration and increasing the vehicle run after the tyre damage.

References

[1] Biderman V L et al. 1963 *Automotive tires — Construction, design, testing and operation* Ed. Biderman V L (Moscow: Goskhimizdat) p 384 (In Russian)

[2] Evzovich V E and Raibman P G 2010*Vehicle tyres, discs and rims* (Moscow: Avtopolis-Plus) p 144 (In Russian)

[3] Bukhin B L 1988 *Introduction to the mechanics of pneumatic tires* (Moscow: Khimiya) p 224 (In Russian)

[4] Bukharin N A 1973 *Cars. Construction, load modes, work processes, strength of car assemblies* (Leningrad: Mashinostroenie) p 504 (In Russian)

[5] 1976 *Theory, design and calculation of wheeled military vehicles. Study guide* Ed. Medvedev V.I. (Moscow: Malinovsky Military Armored Forces Academy, Armoured Personnel Carrier and Vehicle Department, Academy Publishing) p 405 (In Russian)

[6] Babakin A N 2010 Increasing survivability of military vehicles by using modern bullet-proof vehicle tyres *Vehicle and Tractor Construction in Russia: Development Priorities and Training of Specialists: Materials of International Scientific and Technical Conference of Association of Automotive Engineers Dedicated to the 145 Anniversary of MGTU "MAMI"* **1** (Moscow: MGTU "MAMI") pp 22-28 (In Russian)

[7] Seal Inside, Pirelli Available at: http://www.pirelli.ru/tyres/ru-ru/legkovyye-shiny/naydite-svoi-shiny/vse-o-shinakh/tekhnologiya-sealinside (In Russian)

[8] Michelin PAX System, Michelin U.S. website Available at: URL:http://www.michelinman.com/automotive-tires/michelin-pax.page (accessed: 03 July 2015).

[9] RunFlats, Hutchinson worldwide Available at: http://www.hutchinsoninc.com/CMS/index.php?page=Tactical_RunFlats

[10] Tweel, Michelin Available at: http://www.michelintweel.com

IOP Conf. Series: Materials Science and Engineering **819** (2020) 012037 doi:10.1088/1757-899X/819/1/012037

Charging balance management system modeling and implementation in intelligent vehicle with combined power system

V V Debelov, O A Dzhodzhua, K M Sednev and D V Endachev

FSUE "NAMI", 2 Avtomotornaya St., Moscow, 125438, Russian Federation

E-mail: vladimir.debelov@nami.ru

Abstract. The article presents the results of mathematical modeling and implementation of algorithms for ensuring the charge balance of an automotive low-voltage battery as part of an intelligent vehicle control system with a hybrid power system. The aim of the research is to evaluate the effectiveness of the proposed complex of mathematical models in calculating the control parameters of the DCDC converter to achieve a positive charge balance of the low-voltage battery. The work evaluates the boundary conditions of the system and adapts the mathematical models and algorithms implemented in the program code of the controller as part of the current model of an intelligent vehicle with a combined power system. The following parameters are evaluated as criteria for the optimality of the proposed models: SOC - state of charge (%), battery charge (A * h), charge and discharge current of the battery (A), voltage on the battery during its charge, in the absence of a charging current and at rest (V). A full-scale experiment was carried out, experimental data and graphs were obtained. Optimal control coefficients for the processes of charge and discharge of the battery, current characteristics, the dynamics of their changes during operation, experimental time-current dependences during the operation of each component of the low-voltage power supply network were studied. The results of the work make it possible to make adjustments to the charge balance management model and contribute to improving the energy balance of the power grid, increasing the efficiency and service life of batteries, forming requirements for power systems and their subsequent implementation.

1. Introduction

The main objective of modern intelligent vehicles, which is to increase the efficiency of use of fuel and energy resources, is to achieve the greatest efficiency of electrical and electronic systems [1]. This became possible with the use of a combined power system, which contributes to the double conversion of energy: from electrical to mechanical and vice versa. At the same time, important tasks are efficient use of battery power, reducing self-discharge currents, stabilizing the supply voltage of the on-board network, maintaining the battery charge at a high level for restarting systems and a hybrid installation, optimal loading of control systems and increasing the resource of use of components [8].

A hybrid installation is a complex system and consists of many components, such as an electric machine that serves as a starter and generator, a high-voltage battery that accumulates charge for

IOP Conf. Series: Materials Science and Engineering **819** (2020) 012037 doi:10.1088/1757-899X/819/1/012037

subsequent engine start-up and power supply of electronic systems, an inverter energy converter for powering the electric drive at the time of starting and charging the high-voltage battery after starting the engine, DCDC converter, as well as a current sensor and a charge control system [3]. It is believed that the system works efficiently provided that the generated and stored energy is enough for all consumers and there are practically no self-discharge currents.

In hybrid systems, it is customary to distinguish two electrical networks: a high-voltage 300-400V and a low-voltage 12-24V. Due to the fact that the largest number of consumers in the vehicle that use electric energy are part of a low-voltage power supply with a nominal voltage of 12V, it is most important to first achieve the effectiveness of a low-voltage power supply of 12V, obtain a positive charge balance of the battery and provide conditions for the most optimal operating mode DCDC power converters, and timely prevent leakage of current from the battery when DCDC voltage converters open are obtained [5].

The aim of this work is to obtain quantitative characteristics and evaluate the effectiveness of the proposed methods for controlling the charge balance of a car battery in the mains with a nominal voltage of 12V, assess the impact of delays in the operation of the control system on the battery charge, determine the magnitude of the error in calculating the current settings of the converters, compare the control methods of DCDC converters.

To achieve this goal, you need to solve several problems:

1. Assess the reliability of the choice of charge balance control mode;
2. Evaluate the operation of the low-voltage battery diagnostic model;
3. Evaluate the operation of the model for monitoring the residual battery;
4. Evaluate the operation of the model disconnecting powerful consumers;
5. Evaluate the performance of the DCDC converters control model.

The results of mathematical modeling and full-scale experiment will allow us to draw conclusions about the effectiveness of the proposed mathematical models and methods for controlling the charge balance, to evaluate the accuracy and speed of reaction to deviations of the system from the state of a positive charge balance. Subsequently, the results of the work will contribute to the improvement of the proposed methods and form a deeper understanding of the processes of charge and discharge of the battery, affecting its charge balance.

2. A set of models for the positive charge balance of an automotive low-voltage battery

In order to evaluate the proposed method for controlling the charge balance of a car, it is necessary to simulate a set of charge balance models, which includes five mathematical models performed in the MATLAB Simulink environment (figure 1) [7]:

- manager of control modes - “Control Mode Manager” (CMM);
- low-voltage battery diagnostic model - “Diagnostics Model” (DM);
- model for monitoring the residual charge of the battery - “Residual Charge Control” (RCC);
- Consumer Shutdown Model (CSM);
- DCDC control model - “DCDC Control Model” (DCM).

It should be noted that each of the models performs a separate function, which is performed when the conditions determined by the charge balance control mode manager (CMM) are met. It is also worth considering that there are two states when the control model is working: when the control unit is in active mode in one of the operating states and when the unit is in sleep mode and wakes up in order to evaluate the charge balance of the low-voltage battery.

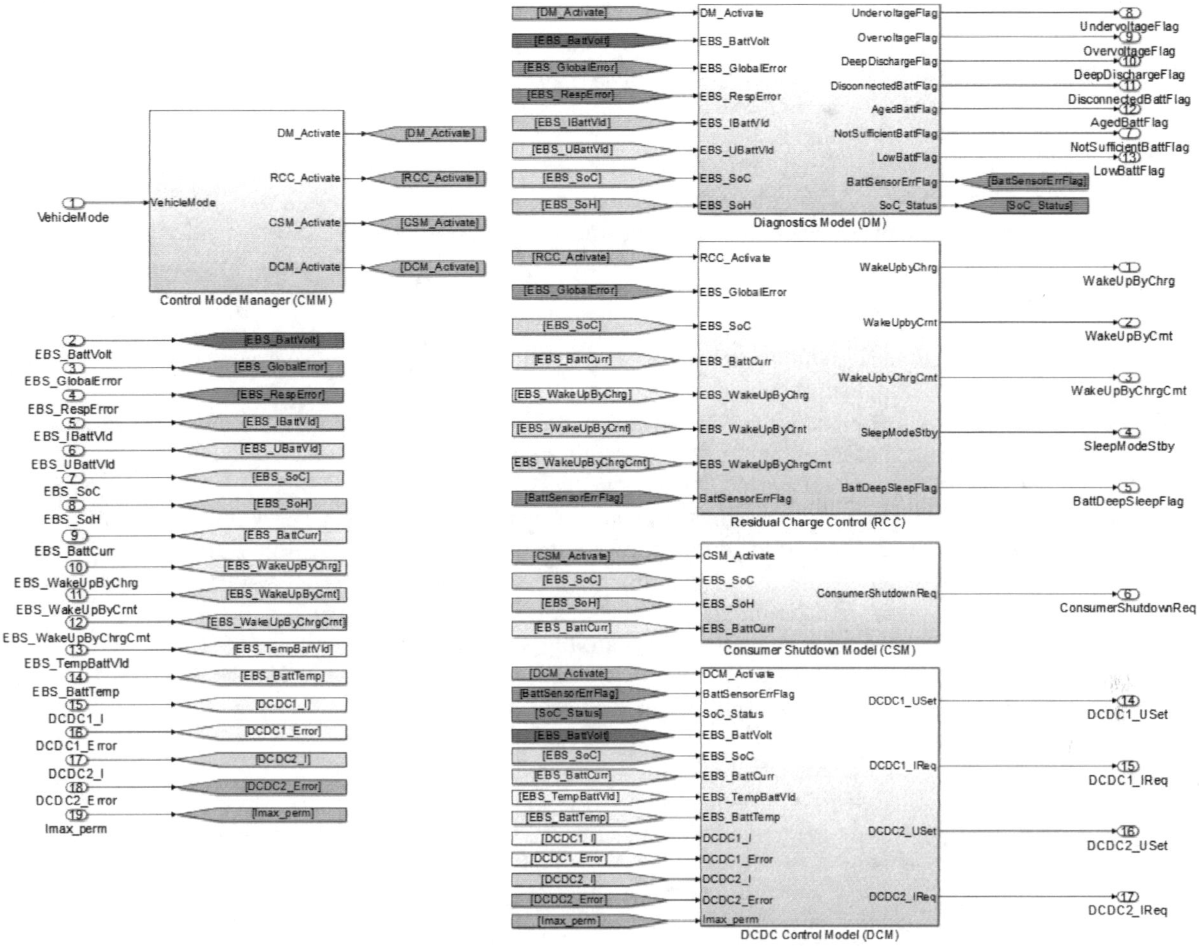

Figure 1. Structure of models for managing the charge balance.

The manager of control modes receives input about the vehicle operating mode - "VehicleMode", which determines the status of the connected terminals and the power supply mode of the vehicle, and based on this information activates the corresponding control models for the following signals:

- "DM_Activate" activates the low-voltage battery diagnostic model;
- "RCC_Activate" activates the model for monitoring the residual charge of the battery in the presence of leakage currents;
- "CSM_Activate" activates a model for controlling the disconnection of powerful consumers to unload the power grid;
- "DCM_Activate" activates the DCDC drive control model.

The mode manager is a simple logical function and does not affect the speed and accuracy of calculation and control of the charge balance of the low-voltage battery.

3. Testing the diagnostic model

Let us consider in more detail the DM model, which is responsible for the diagnosis of a low-voltage battery and contributes to the choice of its charging mode. The input of the DM_Activate model receives signals from an electronic intelligent sensor of a low-voltage battery:

- "EBS_BattVolt" - voltage at the battery terminals;
- "EBS_GlobalError" - status of the global error of the battery sensor;

- "EBS_RespError" - error status of the battery sensor communication;
- "EBS_IBatt" - error status of battery current measurement;
- "EBS_UBatt" - error status of battery voltage change;
- "EBS_SoC" - battery charge;
- "EBS_SoH" - battery life.

Based on the "EBS_BattVolt" signal, the diagnostic model detects and generates the programmed battery failure status, which indicates the following:

- low battery voltage (voltage at the battery terminals is lower than 10.5V for 10s) - "UndervoltageFlag";
- overvoltage (voltage at the battery terminals is higher than 16V for 10s) - "OvervoltageFlag";
- deep discharge of low-voltage battery (voltage at the battery terminals is lower than 9.5V for 60s) - "DeepDischargeFlag";
- loss of sufficient contact between the terminals and the battery poles (caused, for example, by a broken cable, blown fuses, short circuit, oxidation of conductive contact surfaces, change in contact zone resistance) - "DisconnectedBattFlag".

The signal "EBS_SoC" contains the actual charge of the battery, which the battery sensor integrates over the entire period of its operation and, after checking for reliability, it is converted into the signal "SoC_Status". Further, by comparing the current charge level with the set thresholds, the required battery mode is determined.

The following battery level statuses are highlighted in the model:

- Fully discharged battery: "EBS_SoC" $<$Limit1;
- Fast battery charge: Limit2$>$ "EBS_SoC" $\geq$ Limit1;
- Normal battery charge: Limit3$>$ "EBS_SoC" $\geq$ Limit2;
- Fully charged battery: "EBS_SoC" $\geq$ Limit3;
- Error - in case of a prolonged state, "EBS_SoC" = 100% or "EBS_SoC" = 0%.

The described limits in a practical experiment will have the following values that are as close as possible to the real ones:

- Limit1 = 45%;
- Limit2 = 65%;
- Limit3 = 98%.

The signal "EBS_SoH" contains the value of the level of wear of the battery and determines the corresponding status:

- "AgedBattFlag" - battery aging status. It means that there is a deviation of the battery from the declared technical characteristics;
- "NotSufficientBattFlag" - battery unusability status. It means that the battery is unsuitable for further operation as part of the car, which is determined by the model of the battery sensor.

The presence of any of the signals from the intelligent battery sensor: "EBS_GlobalError", "EBS_RespError", "EBS_IBatt" or "EBS_UBatt" - generates a battery sensor error - "BattSensorErrFlag", which is stored in the controller memory and puts the model into emergency operation by default.

Experimental studies showing the reliability of the model are presented in figure 2 and figure 3.

Figure 2 illustrates the definition of signs that indicate the low-voltage battery exiting from the optimal operating mode, in which a deviation of the operating voltage or resource characteristics occurs. Diagrams of determining deviations from the normal mode for reduced (less than 11V) voltage, increased (more than 16V) voltage, critically low voltage (less than 9.5V), aging of the battery and its failure, which is determined by the amount of stored charge, are presented (in percent, %) ceteris paribus compared to a new battery. Figure 3 conventionally shows the boundaries of the battery charge modes, when changing between them, the control strategy of the DCDC converters changes to a new strategy, providing a charge for a low-voltage battery and power on-board network.

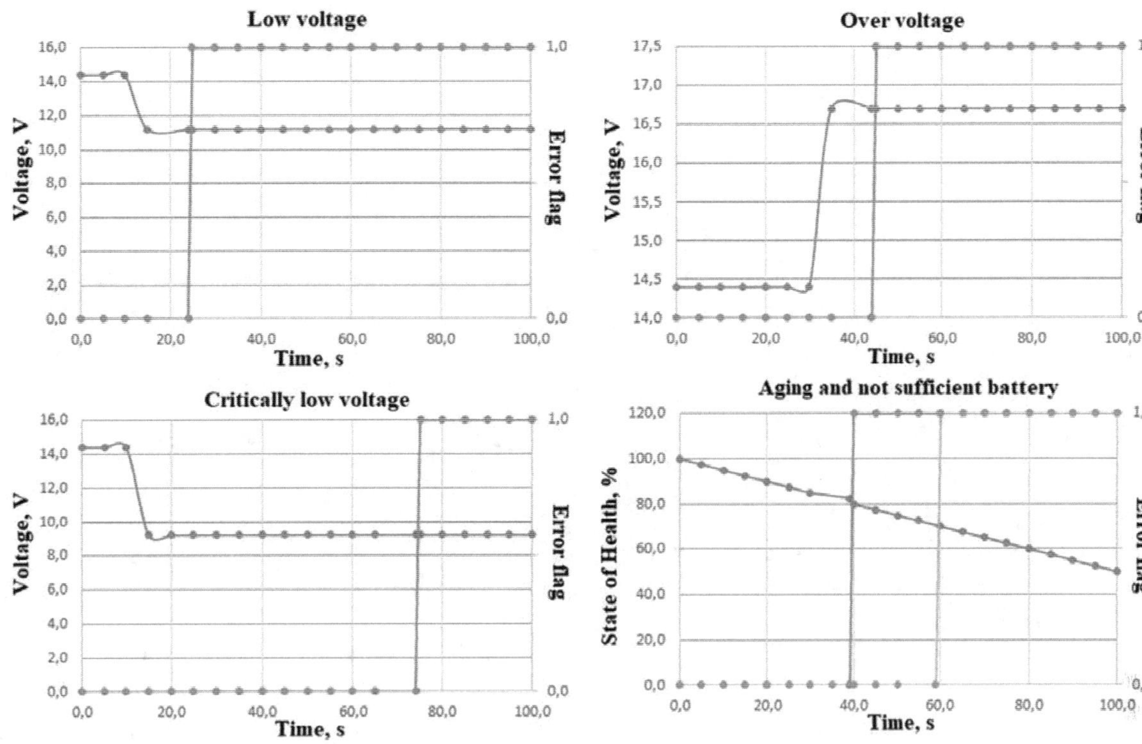

Figure 2. Critical battery conditions.

IOP Conf. Series: Materials Science and Engineering **819** (2020) 012037 doi:10.1088/1757-899X/819/1/012037

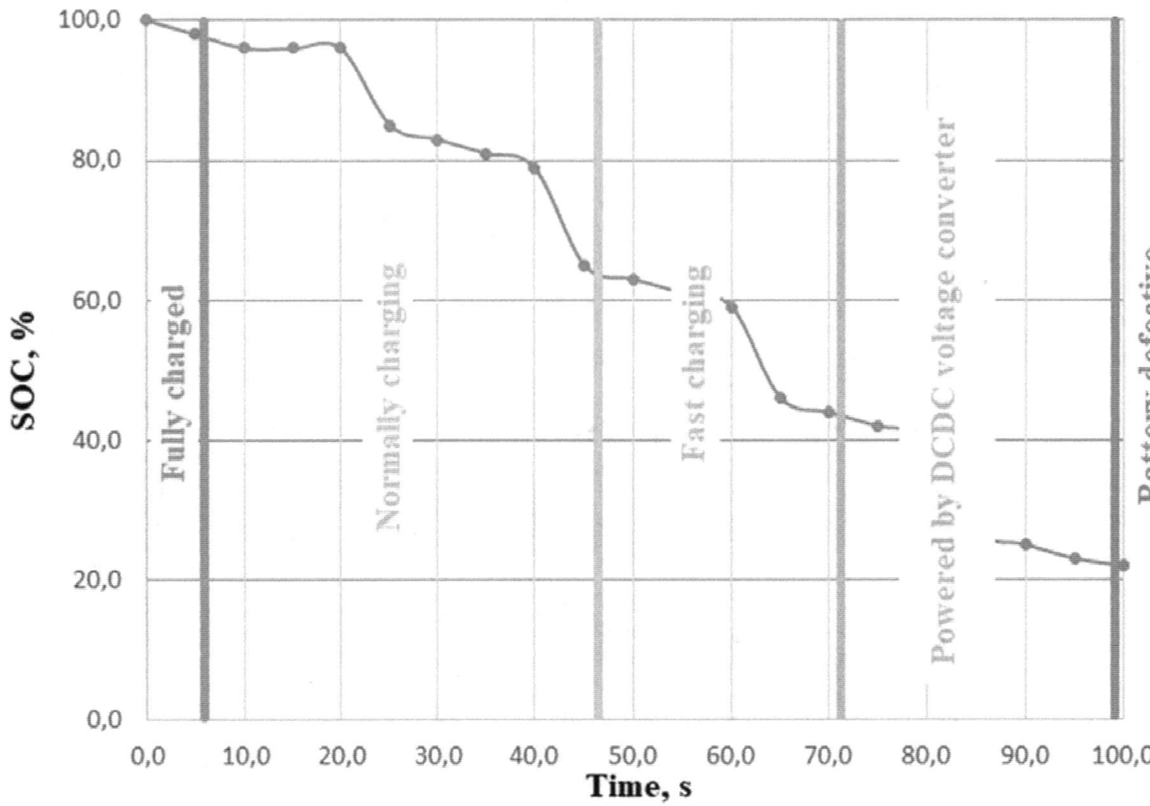

Figure 3. Determining state of charge of the battery.

4. Testing the model for calculating leakage currents

The model, which provides control of leakage currents, has a number of features. This model works both in the active state of the unit, when the DCDC inverters are turned off to take into account the current (5-40A) that flows before all consumers are turned off and switched to low power mode (I <5mA), and in the inactive state, when the control unit wakes up under various conditions related to the settings of the intelligent battery sensor. Conventionally, two modes of operation of the model can be designated as the following:

Prepare to sleep mode - readiness mode for disconnecting consumers;

Awake mode - conditional wakeup mode of a block.

Each of the modes, before the control unit is ready to turn off the power lines and go to sleep, requires sending the configuration to the smart battery sensor, which will indicate that it will subsequently be the source of awakening of the control unit.

The most common and common parameters are:

Wake up time (c) - required for periodic monitoring of the battery charge in order to detect its discharge and turn off consumers that may be active in low power mode;

Wake up by charge (A * h) - the level of charge at which the control unit spills and disconnects active consumers;

Wake up by current (A) - current at which the control unit wakes up in order to detect the source of consumption and limit the discharge current;

Wake up charge current (A) - the current at which the control unit wakes up in order to turn on special systems that must be in an active state while the battery is charging.

The DeepSleep flag is not a current sensor setting, it is a derivative function and serves to ensure that all consumers in other systems are instantly de-energized to prevent self-discharge of the battery.

The following signals are input signals to the residual charge control model:

- EBS_GlobalError (Bool) - a sign of an error during the operation of the battery sensor, which sets the model to default mode: data from the sensor is ignored, and predefined values substitute for them;
- EBS_SOC (%) - the actual battery charge, which is used to compare it with threshold values to which it can fall;
- EBS_BattCurr (A) - battery current, which monitors the presence of current leakage and determines the limits of acceptable consumption for each of the operating modes of the control unit when the DCDC converters are turned off;
- EBS_WakeupByChrg (Bool), EBS_WakeUpByCrnt (Bool), EBS_WakeUpByTime (Bool) - the signs by which the control unit identifies the reason for waking up and, further, activates the necessary script for the vehicle's power supply manager.

Testing the model consists in checking how the numerical data used for the configuration of the current sensor is calculated, as well as checking the correctness of the formation of a sign that allows you to limit the consumption of all active devices and determine the correctness of the formation of error flags when the consumption current does not fall for the calculated time.

The experiment presents a graph (figure 4) for the latter case, in which flags are shown that determine the formation of an error indicator, provided that the consumer current does not drop to a predetermined threshold at a set time, which indicates a deviation in the operation of the systems.

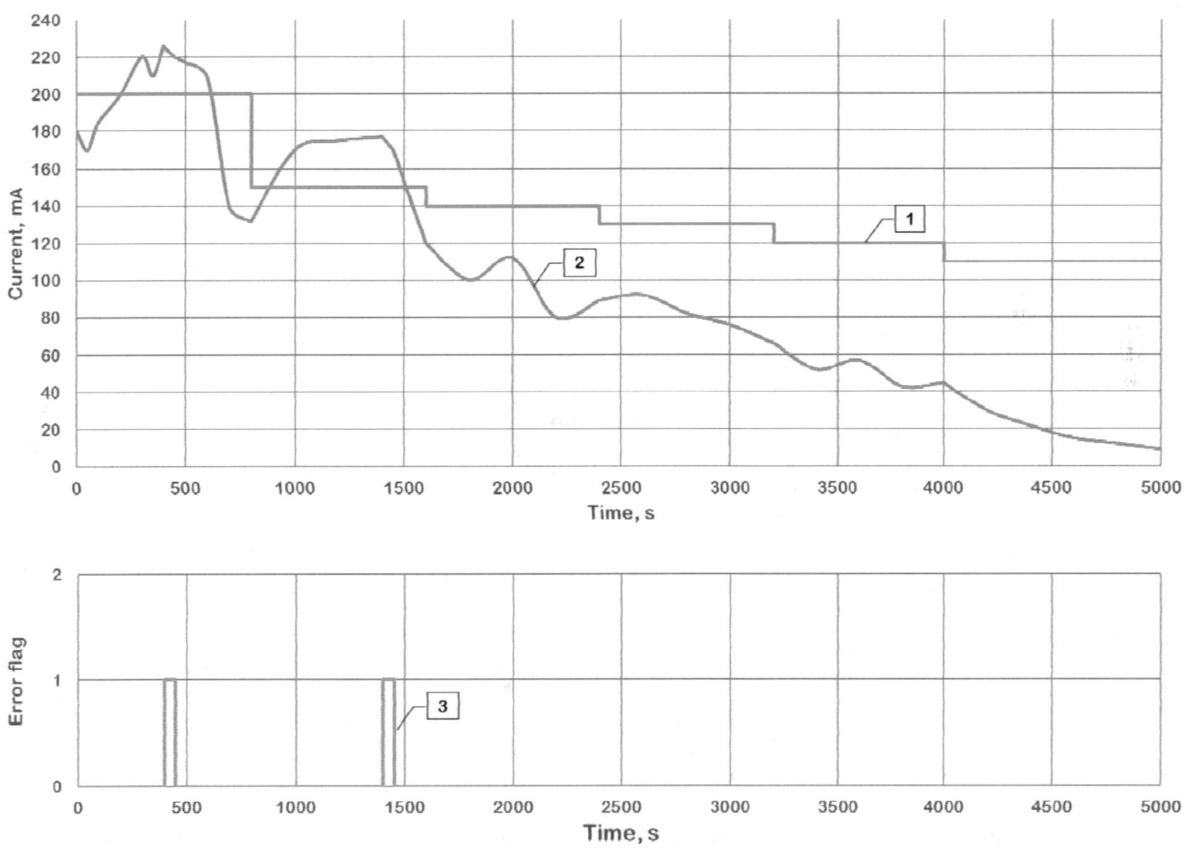

Figure 4. Determination of an error in the operation of systems powered by the on-board voltage network of 12 V, when consumers are disconnected before entering Sleep mode by means of a time-current characteristic.
1 - boundaries of the current consumption, 2 - filtered value of the current consumption, 3 - a sign of an error in the operation of any of the systems when it is necessary to timely reduce the current consumption.

5. Testing the model for disconnecting powerful consumers

The model for disconnecting powerful consumers consists of a mathematical function and a logical one. The mathematical function of the battery charge, the level of wear and the current through the battery allows you to estimate the time during which the battery is charged up to 100% or discharged to 0%. The logical function evaluates the steepness of the graph, which allows us to draw conclusions about which group of consumers should be turned off to improve the situation with current consumption in the on-board network. Consumer groups depend on where the curve is located, leaving 0% (when charging) or 100% (when discharging) relative to the ordinate axis. If there is a charge, then the curve lying between the ordinate axis and the first graph allows you to judge that the battery is charging quickly and does not require disconnecting powerful consumers. If the charge time curve were between 3 and 4 curves, this would indicate that it is necessary to disconnect a certain group of consumers in order to shift the graph to the left to 1 or 2 curves. The discharge is determined in a similar way, but the opposite. Thus, the longer the discharge time, the less consumers need to be disconnected at this moment. But if it turns out that the graph shifts closer to the ordinate axis, closer to 8 or 7 curves, then an immediate reaction is required to prevent the battery from quickly discharging.

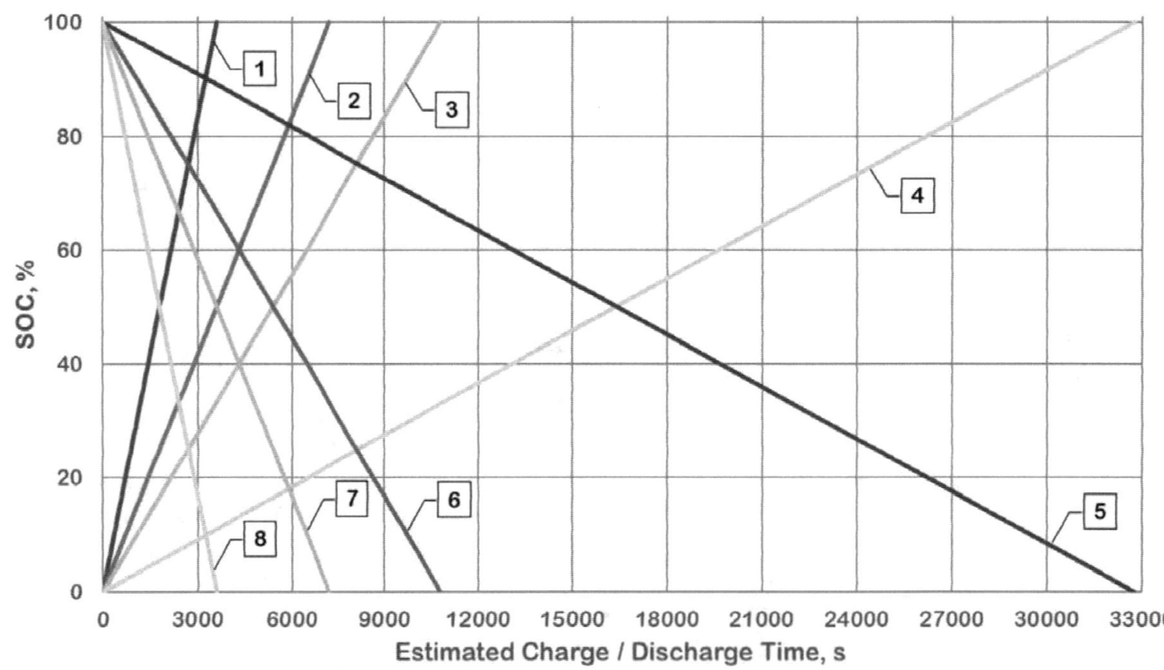

Figure 5. Graph of boundary conditions for which it is necessary to disconnect powerful consumers from the mains.
1 - consumer shutdown level “0”, positive current, estimated charging time up to 100% - 3600 sec;
2 - consumer shutdown level “1”, positive current, estimated charging time up to 100% - 7200 sec;
3 - consumer shutdown level “2”, positive current, estimated charging time up to 100% - 10800 sec;

4 - consumer shutdown level “3”, positive current, estimated charging time up to 100% - 32767 sec;
5 - consumer shutdown level “0”, negative current, estimated discharge time up to 0% - 32767 sec;
6 - consumer shutdown level “1”, negative current, estimated discharge time up to 0% - 10800 sec;
7 - consumer shutdown level “2”, negative current, estimated discharge time up to 0% - 7200 sec;
8 - consumer shutdown level “3”, negative current, estimated discharge time up to 0% - 3600 sec.

The experiment showed that the proposed method allows one to predict in advance the remaining time for which the battery will be charged or discharged when the DCDC converters are turned off, which contributes to a premature decision to disconnect consumers. The time for which you can make a forecast about how the charge will change is about 10-100 ms, which is fast enough for the system under test. And it also allows you to divide all consumers into several groups, due to their power, criticality and safety.

As a result of this, the application of the proposed model allows one to reduce the time of consumers' exposure to the battery and prevent battery discharge, where for each of the modes its own border is calculated and a group of consumers is selected depending on their consumption of electric energy (Ah).

6. DCDC control model for positive charge balance of low-voltage battery

The control model of the DCDC converters calculates the current and voltage limit settings for each of the two converters at the same time that the mode control model permits its operation.

Limitations are calculated based on the following parameters:

- "EBS_BattVolt" - battery voltage (V);
- “EBS_SoC” - battery charge (%);
- "EBS_TempBattVld" - the reliability of the readings from the battery sensor;
- “EBS_BattTemp” - temperature value from the battery sensor (˚C);
- "DCDC1_I" - the actual current of the first DCDC converter (A);
- "DCDC1_Error" - error in the readings of the first DCDC converter;
- "DCDC2_I" - the actual current of the second DCDC converter (A);
- “DCDC2_Error” - error in the readings of the second DCDC converter;
- "Imax_perm" - the maximum total continuous current for the first and second converters, stated in the technical specifications (A).

IOP Conf. Series: Materials Science and Engineering **819** (2020) 012037 doi:10.1088/1757-899X/819/1/012037

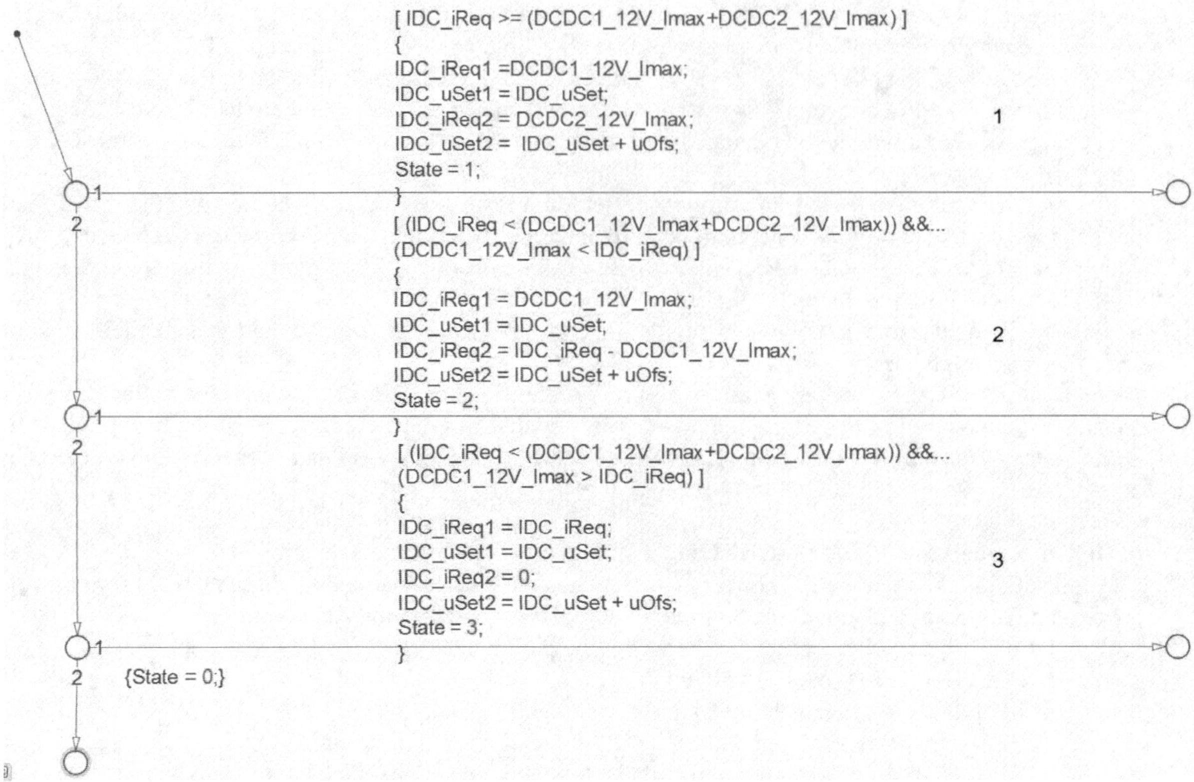

Figure 6. Model for calculating currents and voltages set point for two DCDC converters according to the previously proposed concept.

As mentioned earlier, in this concept, one of the converters is loaded to the maximum value, and, after that, the second converter starts loading. This allows you to minimize the participation of one of the converters [2].

As practice has shown, this method contains several disadvantages due to the fact that only one of the converters is constantly loaded. The second one is at rest at this time, and its power keys are closed. To transfer the system to the active state, a certain time will be required, which will create a lag in the current of the converters from that required by the system, which will briefly load the battery with high currents, instead of charging it [4]. This is clearly shown in figure 7.

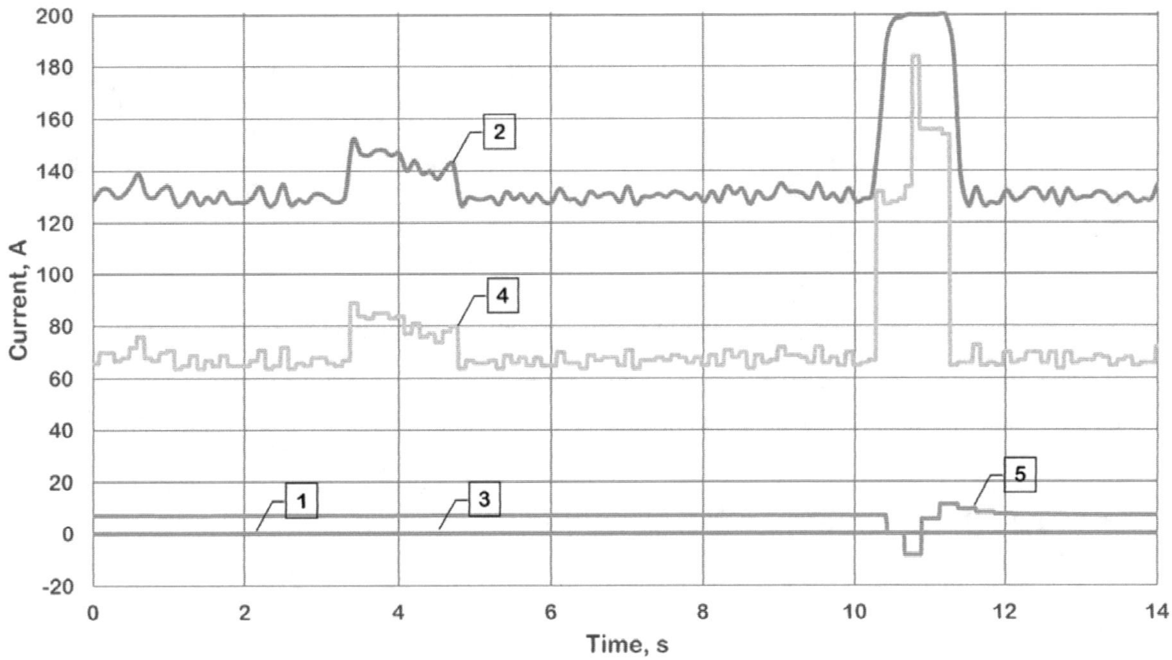

Figure 7. Realization of current request by converters according to the previously proposed concept of calculating the current setpoint by two DCDC converters.

1 - EEPMve_I_maxDcdc1 - requested current setpoint limitation from DCDC1 (A);
2 - EEPMve_I_maxDcdc2 - requested limitation of the current setpoint from DCDC2 (A);
3 - SWINve_I_LoVltgDcdc1 - the actual current of the DCDC1 converter (A);
4 - SWINve_I_LoVltgDcdc2 - the actual current of the DCDC2 converter (A);
5 - SWINve_I_IBS_LvBatCurr - current of the battery when it is charged from two DCDC converters (A).

Looking at the graph obtained, we can conclude that the control of the current setpoint of the converter in a system consisting of two DCDC converters, according to the principle of master and slave, is not effective, and the requests for the current settings are distributed unevenly. While one converter is operating in the range from 60 A to 185 A, the other converter remains unloaded. Connecting a powerful load generates a current of the order of 100 A for a short time interval of 1 s, which brings the converter operation mode to a limit of capabilities of 200 A. Moreover, the current implementation is delayed from the setting by approximately 1 s, which leads to a drop in the battery charge current to 8- 10A. Uneven loading of the converters affects the resource and the thermal mode of their operation, and also does not allow to ensure a stable state of the charging current, which, when the load is connected abruptly, leads to a delay in the operation of the converters, and, as a result, a drop in current on the battery and its short-term discharge [6] . This method can be used in control systems with one DCDC converter, but in systems with two converters it is not very effective.

In this regard, changes were made to the proposed control method and a new control concept for DCDC converters was developed for systems including more than one DCDC converter. It takes into account those factors that each of the converters must be partially loaded, which contributes to a faster response to changes in the load in the system.

The method consists in the fact that it calculates the current balance of the power supply, followed by the determination of the required current limit setting and voltage settings, which vary depending on which temperature and battery charge, how many consumers are on and what the actual load of DCDC converters is.

IOP Conf. Series: Materials Science and Engineering **819** (2020) 012037 doi:10.1088/1757-899X/819/1/012037

Calculation of the current is carried out based on what current is required from the DCDC converters to ensure battery charge, what maximum battery charge current is acceptable at current values of charge and temperature, and what current flows in or out of the battery itself. The method for generating the current setpoint for a system of two DCDC converters was presented by other formulas.

7. Calculation of the current setting for DCDC converters balanced by maximum power and load factor of each converter

As experimental studies have shown, loading one of the DCDC converters is most complete, and compensating for the lack of power due to the second or the remaining converters leads to the fact that the first converter is operating at the limit of its capabilities and prematurely produces its useful resource. This is manifested in the fact that with a large number of consumers, a large loss power is released, which goes to heating the converter itself, which leads to the aging of its components under the influence of high temperatures. At the time of heating of the components, the power supplied by the converter itself is limited. And the lack of power is compensated by the remaining converters. Due to the increased load on the first converter with this concept of power distribution, its resource and reliability are reduced. This requires the use of other methods for distributing the requested power between all converters in order to increase the reliability of the entire system.

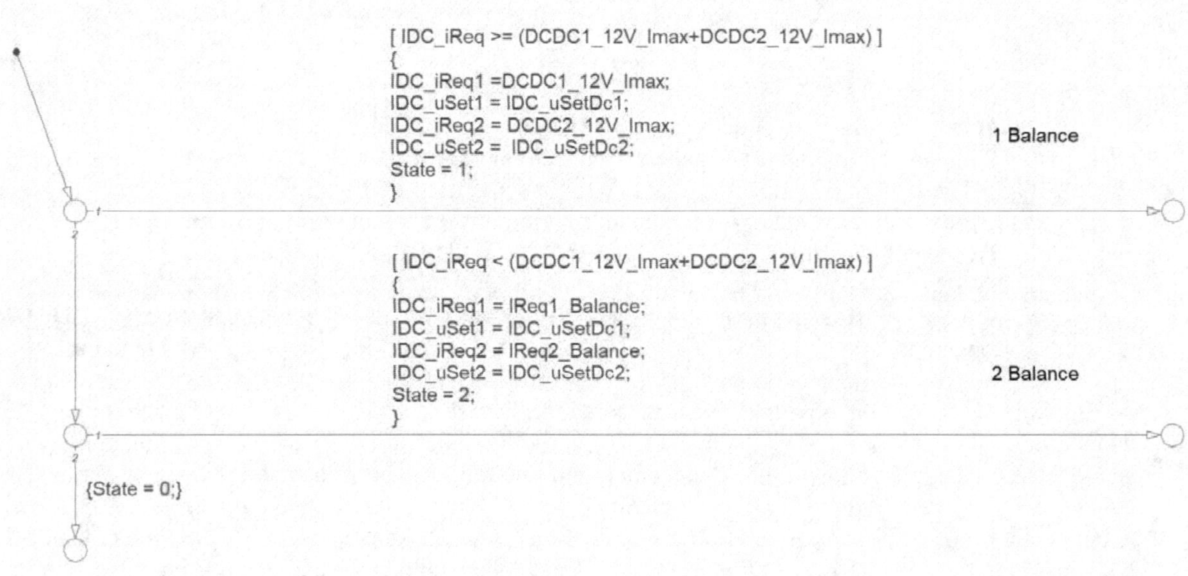

Figure 8. Model for calculating a balanced set point of currents and voltages for two DCDC converters.

In order to achieve the maximum efficiency of all converters, it is necessary to evenly distribute the requested power between all converters in proportion to the maximum power of each of them and the current load, which is determined by the thermal mode of operation of the converters. To do this, it is necessary to perform a series of the following calculations, which will allow you to evaluate the current load, the maximum power output taking into account the current load and calculate the current setting for each of the converters.

Initially, it is necessary to calculate the maximum current that a DCDC converter can provide, taking into account its current load and the temperature mode of operation, according to the formula:

$$I_{max1} = \frac{I_{DCDC1} \cdot 100\%}{K_{load.actual}}$$

$$I_{max2} = \frac{I_{DCDC2} \cdot 100\%}{K_{load.actual}}$$

where I_{DCDC1} is the current of the DCDC converter at the current time (A), $K_{load.actual}$ is the current load factor of the DCDC converter (%).

In this case, the current is limited by the minimum possible and maximum permissible value of each converter.

$$I_{min1} < I_{DCDC1.max} < I_{max1}$$
$$I_{min2} < I_{DCDC2.max} < I_{max2}$$

Next, you need to calculate the load limit of each converter, based on the total requested current for the two converters and the load factor of each of the converters.

$$I_{DCDC1.Req} = \frac{I_{max1} \cdot (\frac{I_{Req} \cdot 100\%}{I_{max1} + I_{max2}})}{100\%}$$

$$I_{DCDC2.Req} = \frac{I_{max2} \cdot (\frac{I_{Req} \cdot 100\%}{I_{max1} + I_{max2}})}{100\%}$$

where I_{Req} is the current, requested from the converters, taking into account the charging of the on-board battery with a nominal voltage of 12V.

The calculated converter load value is a balanced current request for each converter. This allows you to evenly load each of the converters and helps to improve the thermal mode of operation of each of the converters.

In order to evaluate the adequacy of the work of the proposed calculation, an experiment was performed in which, depending on the current load and battery current, the current of each converter was selected, and taking into account that the power of the converters was different, the optimal balance of the current setting for each converter was calculated by the formula.

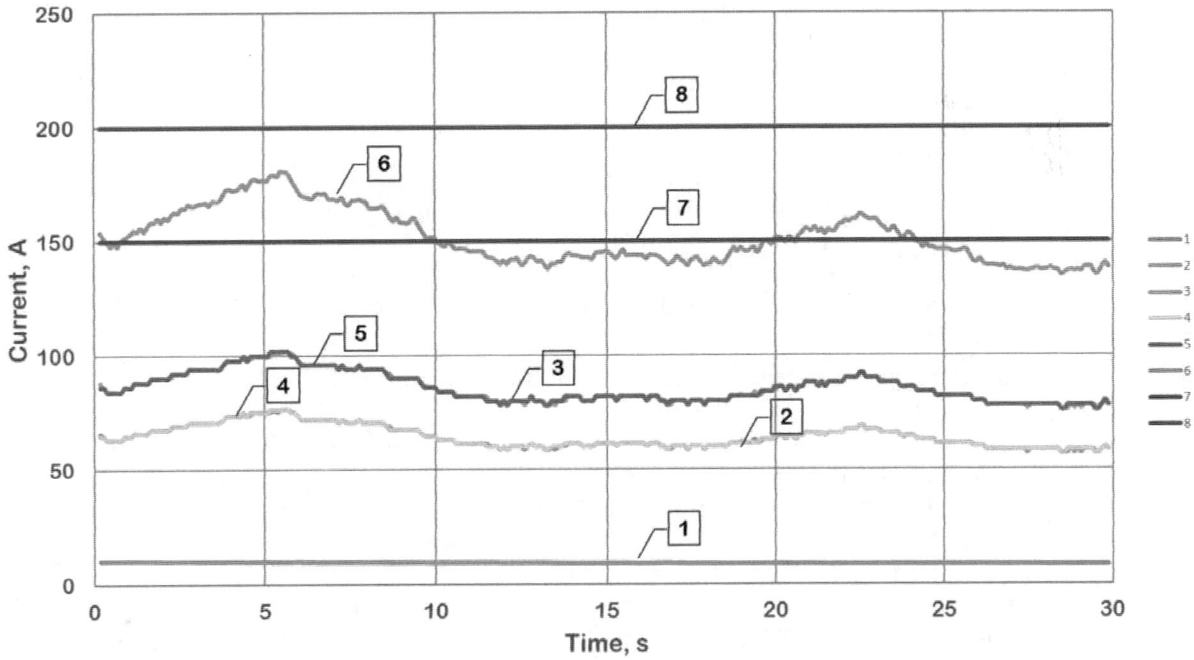

IOP Conf. Series: Materials Science and Engineering **819** (2020) 012037 doi:10.1088/1757-899X/819/1/012037

Figure 9. Graph of the current settings of the DCDC converters for the current load factor, the value of the system current and charge current of the battery.
1– SWINve_I_IBS_LvBatCurr - current of the battery when it is charged from DCDC (A);
2 - EEPMve_I_maxDcdc1 - requested current setpoint limitation from DCDC1 (A);
3 - EEPMve_I_maxDcdc2 - requested current setpoint limitation from DCDC2 (A);
4 - EEPMve_I_IDC_iReq1 - calculated limitation of the current setpoint from DCDC1;
5 - EEPMve_I_IDC_iReq2 - calculated limitation of the current setpoint from DCDC2;
6 - EEPMve_I_IDC_iReq - calculated value of the current setting from two converters;
7 - EEPMve_I_MaxCalcCurDc1 - calculated value of the maximum current setting that the DCDC1 converter is capable of working out based on the actual load;
8 - EEPMve_I_MaxCalcCurDc2 - calculated value of the maximum current setting that the DCDC2 converter is capable of working out based on the actual load.

Consider an example with real values. The maximum current that each DCDC converter can provide, taking into account its current load and temperature conditions:

$$I_{max1} = \frac{I_{DCDC1} \cdot 100\%}{K_{load.actual}} = \frac{64{,}447 \cdot 100\%}{0{,}43} = 150 \text{ A}$$
$$I_{max2} = \frac{I_{DCDC2} \cdot 100\%}{K_{load.actual}} = \frac{25{,}447 \cdot 100\%}{0{,}1272} = 200 \text{ A}$$

where I_{DCDC1} is current of the DCDC converter at the current time, $K_{load.actual}$ is the current load factor of the DCDC converter.

Limitation of the minimum possible and maximum permissible current value of each converter:

$$150 \text{ A} > I_{DCDC.max1} > 0 \text{ A}$$
$$200 \text{ A} > I_{DCDC.max2} > 0 \text{ A}$$

The load limitation of each converter based on the total requested current for two converters and the load factor of the converters:

$$I_{DCDC1.Req} = \frac{I_{max1} \cdot (\frac{I_{Req} \cdot 100\%}{I_{max1} + I_{max2}})}{100\%} = \frac{150 \cdot (\frac{151{,}406 \cdot 100\%}{350})}{100\%} = 64{,}89 \text{ A}$$

Rated current $I_{DCDC1.Req}$ = 64,89 A, experimental current $I_{DCDC1.Req}$ = 64,4375 A. $\Delta I_{DCDC1.Req}$ = 0,4525 A.

$$I_{DCDC2.Req} = \frac{I_{max2} \cdot (\frac{I_{Req} \cdot 100\%}{I_{max1} + I_{max2}})}{100\%} = \frac{200 \cdot (\frac{151{,}406 \cdot 100\%}{350})}{100\%} = 86{,}52 \text{ A}$$

Rated current $I_{DCDC2.Req}$ = 86,52 A, experimental current $I_{DCDC2.Req}$ = 85,9688 A. $\Delta I_{DCDC2.Req}$ = 0,5512 A.

I_{Req} is the current requested from the converters, taking into account the charging of the on-board battery with a nominal voltage of 12V.

The graph shows that the current that is requested on the converters is distributed balanced in relation to the actually requested current and the maximum current that each converter can provide. The current on the battery has an almost linear characteristic, which is due to the fact that all load changes are compensated by the DCDC converters, which perform a damping function, and the battery is charged with current at this moment approximately equal to 8-10% of the nominal battery value of 100 A. The shape of the current curves of the DCDC converter repeats the shape of the curve that describes the current setting for each converter. Thus, the delay in the actual implementation of the current setpoint request tends to zero and is compensated by the battery current, and the uniform distribution of current between the converters minimizes the actual load. The graph shows that the current of the converters is significantly lower than each of the restrictions, but the percentage of the realized current and the maximum are approximately equal for each converter. This suggests that the thermal mode of operation of each of the converters is optimal and allows long-term operation with connected loads. The deviation of the actual value of the converter current from the setting is not more than 1%, the relative constancy of the charge current indicates that the battery is charging optimally and damps extremely sharp surges due to the instantaneous connection of powerful loads.

The proposed change in the previously obtained model allows us to solve the problem of controlling the converters in optimal temperature conditions and to ensure the relative constancy of the low-voltage battery charge current, which increases the efficiency and reliability of the entire system.

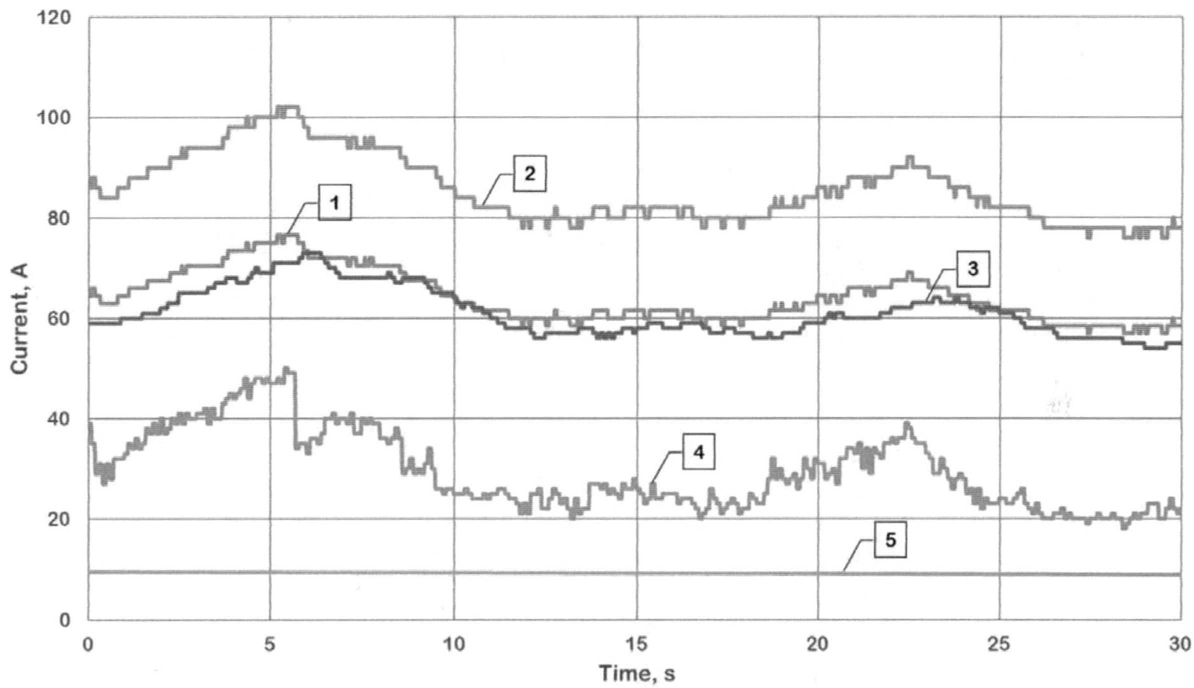

Figure 10. Graph of battery charge in balanced mode.
1 - EEPMve_I_BalanceCurDc1 - calculated value of the current setpoint of the DCDC1 converter, balanced with respect to the current load of each of the converters;
2 - EEPMve_I_BalanceCurDc2 - calculated value of the current setting of the DCDC2 converter, balanced relative to the current load of each of the converters;
3 - SWINve_I_LoVltgDcdc1 - actual current of the DCDC1 converter;
4 - SWINve_I_LoVltgDcdc2 - actual current of the DCDC2 converter;
5 - SWINve_I_IBS_LvBatCurr - battery current when charging from DCDC.

8. Results and conclusions

As a result of the experimental and analytical work, we can come to the following conclusions:

1) The model for diagnosing the condition of the battery is an effective solution that contributes to the timely decision-making on the maintenance of the battery. It allows you to correctly select the charge mode of the battery, which allows you to maintain the battery in a charged state at a time when the dynamics of changes in the current consumption has significant gradients in the characteristics of consumption.

2) The leakage current monitoring model allows detecting and timely servicing those systems that consume electrical energy at a time when their current consumption should be reduced. Monitoring

IOP Conf. Series: Materials Science and Engineering **819** (2020) 012037 doi:10.1088/1757-899X/819/1/012037

leakage currents allows you to more reliably judge the charge and condition of the mains, which is especially true at low temperatures.

3) A method that allows disconnecting powerful consumers is effective and allows you to adjust the current consumption of powerful consumers in the range of 2-80A, which is achieved by reducing the time of their work or by completely disconnecting them from the on-board network of 12V.

4) The model of DCDC converters allows you to maintain battery charge at a given level, which positively affects its operational characteristics. The speed of the original method, where one converter is leading and loaded up to 100%, is small, and delays of up to 2 seconds appear, which, in turn, cause a short-term increase in load up to 100A, which affects the resource characteristics of the battery.

5) An improved balanced control method for DCDC converters is proposed to solve the problem of current delay, which has proved to be much more efficient. The maximum delay in the implementation of the setpoint does not exceed a value of 100 ms, and the temperature regime of each converter becomes optimal, which allows to improve the operational characteristics of the components.

The solution of the tasks was achieved through the implementation of mathematical functions integrated into a single model and implemented in the form of program code as part of the automotive electronic control unit for body electric equipment.

The presented generalized concept for the implementation of the mathematical model for controlling the charge balance of a car battery can be used on various vehicles: in traditional vehicles with a LIN driven generator, in hybrid vehicles, in electric vehicles and electric vehicles, and other vehicles.

It requires an assessment of the applicability of the technology in severe climatic conditions and an assessment of the boundary modes of application.

Testing the model as part of the control unit on the current vehicle model confirms that when it is used, a positive battery charge balance is achieved for standard electronic control systems.

References

[1] Kozlovsky V N, Aidarov D V, Vasiliev M M and Debelov V V 2018 Development of electric vehicles and combined power system projects *Truck* **6** pp 18-21

[2] Debelov V V, Endachev D V, Deev O I and Yakunov D M 2018 The technology for ensuring the charge balance of an automotive low-voltage battery in control systems of a car with a hybrid power system In the collection: Intellectual transport systems technology *Collection of works of the International Automotive Scientific Forum IASF-2018* pp 287-309

[3] Kozlovsky V N, Stroganov V I, Debelov V V and Petrovsky S V 2018 Modeling of energy-intensive drives of a combined automobile power system *Truck* **11**

[4] Debelov V V, Endachev D V, Yakunov D M and Deev O I 2018 Charging balance management technology for low-voltage battery in the car control unit with combined power system *In the collection: IOP Conference Series: Materials Science and Engineering* pp 284-302

[5] Kozlovsky V N, Debelov V V, Deev O I, Kolbasov A F, Petrovsky S V and Novikova A P 2017 Perspective diagnostic systems for managing an autonomous transport object *Truck* **6** pp 21-28

[6] Yakunov D M, Debelov V V and Endachev D V Energy efficiency of electric and hybrid vehicles in conditions of low negative temperatures In the collection: Intellectual transport systems *Collection of works of the International Automotive Scientific Forum IASF-2017* pp 270-78

[7] Dzhodzhua O A, Debelov V V and Endachev D V 2018 Unification of software modules of electronic control units of a vehicle as a way to reduce time and resources for prototyping of varying complexity In the collection: Intellectual transport systems technology *Collection of works of the International Automotive Scientific Forum IASF-2018* pp 279-287

[8] Stroganov V I, Kozlovsky V N, Shakursky M V and Zayatrov A V 2018 Modeling the basic processes of electric vehicles and cars with a combined power system *Electronics and electrical equipment of transport* **2** pp 8-13

Regulatory legal and technical confirmation with the risk analysis while mounting ITS (intelligent transport systems) auxiliary equipment or replacing power units

O I Girutskiy[1] and K A Kirillov[2]

FSUE "NAMI", Moscow, Russian Federation

E-mail: [1] giruzki@nami.ru, [2] kirillov_kirill@inbox.ru

Abstract. This paper contains the analysis of materials and propositions for expert assessment of technical regulations in the automotive industry in case of design modifications by mounting the perspective technical equipment: computer vision, navigation systems, auxiliary equipment, on-board and stationary systems for modeling traffic conditions and traffic control. The international integration in the economy allows to establish the unified principles and mandatory safety requirements for standardization, metrology, accreditation, conformity assessment, monitoring (control), as well as for eliminating the technical barriers in commerce. The system of technical regulation in Russian automotive industry is built according to the international harmonized requirements for ensuring the road traffic safety. An illustrative registration procedure algorithm to introduce design modifications at the stage of operation in the Russian Federation is shown by infographics. An independent professional multi-stage check of the design safety is adapted to the modern realities and allows supporting the vehicle owners' technical initiative. The application of risk-oriented approach, unified forms and the methodology is another step in the improvement of the regulatory and technical basis in the field of the automotive industry. The paper gives commentaries to the effective national legislation, the summary dwells on the prospects of further scientific research. The article also presents information about the preparation of the national regulatory legal acts which are to establish relations in motor vehicles operation controlling. The analysis of the risks related to the vehicles design safety and security of road traffic is provided in details.

1. Introduction

The world level of road transport safety significance (TSS) is determined by sad statistics of road traffic incidents and accidents. According to the researches made by the safety institutions of the European Union countries, Japan and the USA, together with the results of studies conducted by various automobile companies, about 30% of all fatal accidents are caused by the loss of controllability and stability during braking [1]. Nowadays the Russian park of wheeled vehicles is approaching 60 million units. Annually there occur more than 2 million road accidents in the Russian Federation [2]. At that, the official tragic statistics covers only those that result in deaths and injuries. In Russia the main reason for more than 3% of accidents is a technical malfunction [3]. According to the data of Russian Federal Center for Forensic Expertise of the Russian Federation Ministry of Justice, the structural faulty elements that cause accidents could be presented statistically as follows: braking control system - 48.4%; steering system - 28%; chassis and transmission - 8.7%; wheels and tires - 9.4%; lighting devices and electrical equipment - 1.6; other - 3.7% [4]. Thus, the condition of braking and steering systems has the greatest impact on the road safety system throughout the entire park of operating vehicles. Therefore, the greatest potential damage associated with the death and injury of people as a result of an accident can be associated with the wrong interference in these systems.

From the terminology point of view, this interference is named a design modification or re-equipment (tuning), which is carried out both before the use of the component or in its operation. So, according to the current technical regulations, making changes to the installation or re-equipment of components which had not been provided by the design of the vehicle before its release into operation and affect road safety is an exception from the rules [5].

Changes to the design of vehicles in the Russian Federation, their "re-equipment" over the past 10-15 years have become widespread. Every year, the technical supervision departments of traffic police of the Russia Ministry of Internal Affairs issue 200 thousand safety certificates of vehicles conformity to safety requirements. It is estimated that the number of vehicles in which design changes have been made is at least twice higher, and there is a steady upward trend in these indicators.

The main reasons for a necessary "modification" are:

- insufficient number and diversity of vehicles manufactured by industry for intended purposes;
- an unfavorable economic situation in most of the enterprises operating motor vehicles, which does not allow them to order expensive specialized rolling stock and forces them to "modify" available vehicles;.
- rather a long service life of the vehicles in operation, while the production of a considerable number of them has stopped;
- the stoppage of production and the supply of automotive components for a number of types of vehicles, primarily of a foreign make.

In practice, the modification of the design (re-equipment) is the adaptation of a wheeled vehicle to:

- specific conditions of transport work (operation);
- specific types of passenger and/or cargo transportation (specialization);
- individual needs and tastes, in order to optimize (improve) the technical parameters, characteristics and economic indicators to satisfy the end-user's wishes for using a unique design ("stylization" or "customization");
- extention of the existing vehicle life while ensuring road safety, in spite of the absence of components not produced any longer [6].

The changes in the electronic control units of the vehicle, which lead to a change in the control algorithm provided by the program code, are equated to the introduction of modifications in the design [7].

Thus, the modification (re-equipment) makes a material and / or informational impact on the vehicle design safety and in case of ADAS (Advanced Driver Assistance Systems) installation, it has an impact on the active safety, that is the systems of braking and steering. In this case, as interference in these systems leads to the most unfavourable consequences, it should be officially authorized.

The exception is the case of installation of technical (additional) equipment that passed the conformity assessment as part of the type. Otherwise, for example, if changes are made to the design before its being released, the manufacturer or an official representative is required to conduct certification tests according to UN Regulation No. 10 with regard to electromagnetic compatibility as part of a complete vehicle and re-register the modified type. Afterwards a certificate of conformity for spare parts is issued according to a new complete set or a negative certification decision for a specific component is made.

2. Main part

Generally, experimental scientific developments (innovations) of Universities are carried out on vehicles in operation, registered in the traffic police, while tests are made at closed test sites. In such cases, there are difficulties with the delivery of testing equipment, as it is an unauthorized design modification which, according to the traffic rules, is subject to administrative punishment and results in a ban on the vehicle operation until the modification is eliminated.

In this article, we consider the longest stage in the life cycle of automotive products - their operation.

The modified systems must have legal and technical confirmation, that is, when installing technical means of perspective transport vehicles the structure safety requirements must be met and checked, namely: technical vision, navigation systems, additional equipment, on-board and stationary systems for modeling traffic conditions and traffic management, multimedia and computer equipment, communication equipment and their fasteners, security alarms, means of protection against unauthorized use of a vehicle, emergency call devices, other equipment of ADAS systems, the safety requirements for the design, such as unhindered access, the elimination of accidental or incorrect operation or failure, data protection, attachment reliability, identification and diagnostic capability, visibility, injury safety, fire safety, electromagnetic compatibility, electrical safety, cyber security. The requirements allow to estimate correctness, serviceability and interface operation of additional systems by monitoring them with minimal cost and time, and it should be done during the check of technical condition at the point of technical inspection.

The automotive industry designs and manufactures a huge number of components, including auto electronics. At the moment, the objects of technical regulation subject to mandatory conformity assessment include the following vehicle components: the tachograph, speed limiting device, electronic control unit, emergency call device, satellite navigation equipment and some others. "Vehicle components" are the components of the vehicle's construction, delivered to the manufacturing assembly plants as replacement parts for vehicles in operation.

Therefore, according to the certification rules, conformity certificates must be issued for these components.

In addition to the components certification conformity provided by technical regulation in Russia there is a Decree of the Government of the Russian Federation of September 26, 2016 N 969 "About approval of requirements to the functional properties of technical means of transport security provision" and "Rules of mandatory certification of technical means of transport security provision", according to which a wide range of technical systems and means of notification, access control, inspection, intelligent video surveillance, video recording, audio recording, reception and transmission of information, notification, collection and processing of information used to ensure transport security is subject to certification in Federal agencies, for example: FSB — in relation to systems and tools of screening, intelligent video surveillance of the Ministry of Internal Affairs of Russia in respect of systems and alarms, access control, audio and video recordings; the Ministry of the Russian Federation for civil defense, emergency situations and elimination of consequences of natural disasters in relation to the means of notification; Rossvyaz — in relation to communications, transmission and reception of information; the Ministry of Transport of Russia — in relation to information collection and processing systems.

The installation approval procedure for vehicles with a high degree of operating autonomy is carried out in accordance with the Decree No. 413 of 06.04.2019 of the Government of the Russian Federation "On approval of Rules changes in the design of operating wheeled vehicles and the follow-up to verify compliance with the requirements of technical regulations of the Customs Union "On safety of wheeled vehicles". Developed previously interstate standard GOST 33995-2016 "Motor vehicles. The procedure for assessing conformity assessment after the retrofitting" and approved by Rosstandart of September 21, 2017 No. 1194-st with administrative aspect has a voluntary character [8] (Figure 1). At the moment, the Ministry of internal Affairs of Russia is developing administrative regulations to provide public services for the registration of permits and certificates. The current regulations include individual transport vehicles making changes to the design, since a unified approach is used to assess the installation of additional equipment on both conventional vehicles and ITV.

The list of modification types in the design is not exhaustive and is determined by the level of scientific and technological development of society and its expanding needs.

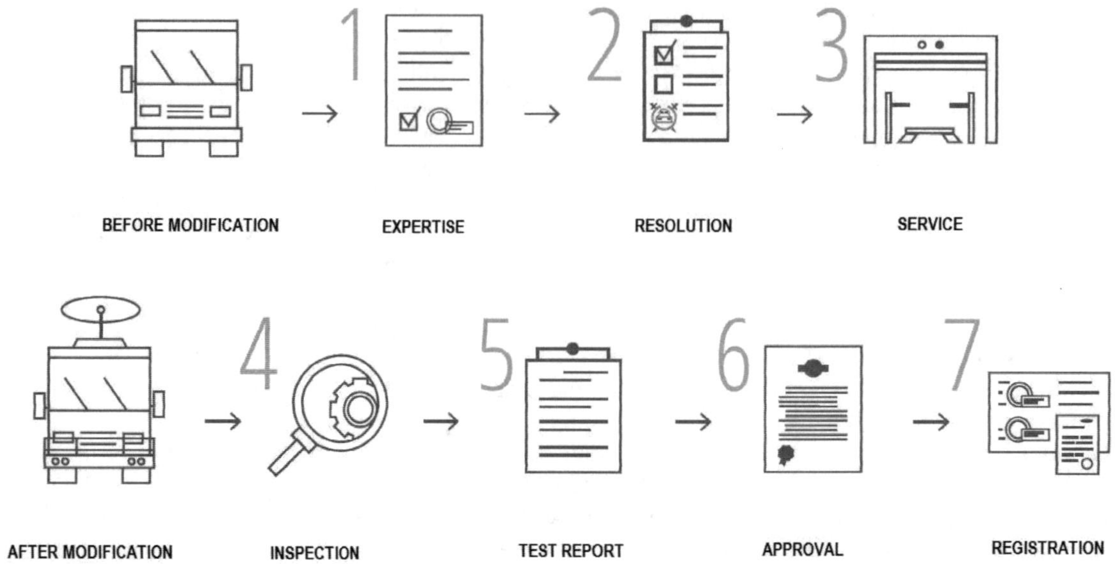

Figure 1. - Algorithm for the procedure for registration of design changes at the operational stage in the Russian Federation.

There come issues of providing methodological support and development of standards. A comprehensive program for developing standards for automotive products of the future until 2025 involves the development of a number of current ITS standards (about 100 standards) based on ISO, SAE, UN Regulations, GOST R and PNST, as well as of new ones, for example: "Motor vehicles. Safety requirements for the technical condition and methods of vehicles inspection with a high degree of control automation" [9].

So, the unauthorized changes to the design of the vehicle should be detected when its technical condition is checked at the point of regular technical inspection. An example is "chip tuning", which is gaining popularity, but its regulatory and technical confirmation is not properly implemented. A preliminary technical examination should be made to assess the state of delivered vehicles by means of diagnostic tools and identify the state of the electronics before making changes, and then determine the modification procedure according to the developed criteria.

Market leaders for ADAS systems are manufacturers such as Bosch, Continental, Valeo, Magna, Denso, Delphi, ZF, Avtoliv, Takata and others. A non-profit organization of European car manufacturers is testing ITS applications, V2X communication technologies, and the consortium includes well-known car manufacturers and component suppliers. Collective creativity has several advantages, but there are also disadvantages associated with joint responsibility.

Particular attention should be paid to cyber security, as the rest of the above requirements are typical of traditional wheeled vehicles, the test methods of which were established earlier.

As part of the development and testing of cyber security of technical means, including intelligent systems, standards and instructions are applied: practical recommendations of SAE, international standards ISO, DEF STAN (Defense Standards), NIST (National Institute of standards and technology (USA)) and others. The list provided is not exhaustive. Other standards and guidelines may be applied. In our point of view, it is advisable that, with respect to specific technologies or processes, corporations should review all applicable standards or guidelines that may be relevant to the process, as well as any new standards developed in this area to improve road safety. Thus, devices should be controlled considering both existing and new standards. Of course, the topic of modification (re-equippment) and road safety is typical not only of our country, but is also widespread abroad. International approval of the wheeled vehicles usage is of utmost importance in view of ensuring maximum safety for all road users. For the convenience of vehicles owners in countries from Australia to Europe, practical information about the engineering assessment of re-equipment of motor vehicles has been prepared including technically possible modifications of vehicles, and, if possible, subsequent registration procedures by making notes in identifying documents based on current national legislation [10].

Accredited persons conduct individual inspections of a specific complete vehicle, including the necessary studies with the help of certified measuring instruments and certified software, by identifying the components used and technical inspection. As a result of examinations, all changes made to the vehicle are listed, at that only approved components and positive results of the technical inspection are used.

On the basis of documents received from experts and containing information on assessing the results of individual technical creativity, as well as of re-equipment in accordance with the requirements of road safety, car owners apply to the authorities of the area to carry out registration actions in the documents identifying vehicles. For this reason, it is necessary to ensure proper control in this area.

All safety requirements must be properly tested before being permitted to participate in road traffic on public roads.

For example, according to the American media, during the certification tests, the US Federal Aviation Administration (FAA) granted broad authority to assess the safety of Boeing itself, as a consequence, two planes crashed, which raised questions about the objectivity of the tests. A similar practice was used in certification of other liners. In case of small batches certification of wheeled vehicles, the current technical regulation allowed the use of the internal factory tests results and the 7D declaration scheme was applied. Thus, we can conclude that having a permit for internal testing is not enough and requires independent examination and testing in an accredited testing laboratory for the standard ISO/IEC 17025.

As part of a legal experiment in Russia, Yandex conducts tests of autonomous vehicles and implements a project for assessing the first vehicle with autonomous control system based on Toyota/Hyundai passenger cars on public roads. In this case, there is a need for regulatory and technical confirmation of changes to the design, according to the above algorithm. Therefore, the Yandex drone is passing the necessary tests at the test site in the NAMI Test Center.

An example of equipping a vehicle with technical means is the creation of a universal multiservice platform (UMP) for buses, electric buses and trolleybuses. The purpose of the creation is the use of innovative technologies to monitor passenger flow using modern navigation and multimedia equipment in a perspective passenger vehicles with various power plants. This system involves the unification of exterior and interior modules, which allows to reduce the time and financial costs for the production, maintenance and repair of machines, to reduce their curb weight and, thus, achieve improved environmental performance. When retrofitting parks of buses in operation, a regulatory and technical confirmation procedure will be required.

It is also worth noting that the metrological component of technical means assessment of perspective transport is also important. For example: technical means of controlling drivers' modes of movement, work and rest (tachographs) are a measuring instrument and are registered as measuring instruments.

As in the TR CU 018/2011 the requirements for the location navigation error and speed of the vehicle are determined formally as the measurements provided by the Russian Federation legislation of technical regulation belong to the state regulation sphere of measurements uniformity in accordance with the rules prescribed in the Federal law from 26 June 2008 № 102-FZ, part 4 of article 1,, "On ensuring the uniformity of measurements"

In accordance with part 1 of art. 9 of the Law No. 102-FZ in the state regulation sphere of ensuring the unity of measurements, the measuring instruments have been approved and verified in accordance with the requirements of this Federal law. The compliance with the mandatory requirements established by the legislation of the Russian Federation has been ensured and established concerning the unity of measurements, including mandatory metrological requirements for measurements, measuring instruments to be used, and established by the legislation of the Russian Federation on technical regulation of mandatory requirements.

The risk analysis of the introduced metrology testing component into the accuracy determination of coordinates and speed of movement with the purpose of approving and verifying the measuring instrument type (the primary and repeatedly used in operation) shows that the introduction of these requirements is associated with risks of different nature (legal, organizational, technical, economic, reputational).

Periodic verification of technical means installed on the vehicle in operation will make it necessary (which is most appropriate from the point of view of providing "one window" services) to combine this procedure with periodic technical inspection of the vehicle at the relevant technical points. The approach, in turn, will require the additional accreditation of each point of technical inspection to verify the unity of measurements. Namely, it will require the providing of each point with necessary testing equipment and technical means and methods of verification without dismantling the vehicle, as well as competent technical specialists. There is also a more complex option of independent preliminary search of metrological services and timely verification or calibration.

With regard to the world experience in the application of the problem under consideration, the following can be noted. For example, the UN Regulation No. 144 developed on the initiative of Russia in respect of emergency call systems (devices) establishes requirements for the accuracy of determining coordinates using Global Navigation Satellite System signals, but they are not considered by the Rules as measuring instruments and they are not subject to "metrological" forms of conformity assessment - tests for the type approval of measuring instrument, as well as their verification

At the same time, in the course of timely maintenance and repair of technical means, a set of operations is implied:

- to confirm or update the data set stored in the memory: vehicle identification parameters and special characteristics of individual devices (sensor parameters, vehicle technical specifications, and other characteristics of the vehicle, etc.);
- to adjust (if necessary) the current time values and odometer indications.

A “third” package of amendments to TR CU 018/2011 is being prepared for vehicles in operation, which clarifies a number of the above terms, partially simplifies the procedure for re-equipment and introduces a number of requirements for typical cases of making changes to a vehicle design, and also specifies requirements for checking the technical condition in operation.

In addition to TR CU 018/2011, two more regulations are applied in the EAEU: TR CU 004/2011 Technical Regulations of the Customs Union “On the Safety of Low-Voltage Equipment” and TR CU 020/2011 Technical Regulations of the Customs Union “Electromagnetic Compatibility of Technical Means,” but we need to note that TR CU 004/2011 is not applied to electrical equipment designed for use on land transport, while electromagnetic compatibility is tested under UNECE Regulations No. 10 and in accordance with the Treaty on the Eurasian Economic Union of May 29, 2014, when mandatory requirements for products may be imposed exclusively by the Union technical regulations.

According to the bill, part of the supervisory reform, the changes made by the "regulatory guillotine" do not extend to road safety control. At the same time, in the context of control and supervisory activities, the Government of the Russian Federation made a list in which the road safety priority was stressed and which included 45 working groups together with the head department of the Ministry of Internal Affairs and the Ministry of Transport and Rostransnadzor head department. The list also covered: road transport: Rostransnadzor (Ministry of Transport, Ministry of Internal Affairs, Rosavtodor) and conformity assessment: Rosaccreditation (Ministry of Economic Development, Ministry of Industry and Trade, Rosstandart). The principle of operation groups was based on the implementation of the "Road map" to bring to life "regulatory guillotine" and was built in three stages:

1. determination of the objects that should be generally excluded from the list of supervised;
2. analysis of risks that will be controlled by the new regulatory system;
3. formulation of the risk management method.

From our point of view, the risk-based approach should influence not only the frequency of inspections, but also the composition of mandatory requirements, licensing activities and administrative responsibility for non-compliance and elimination of duplication of inspections by various departments.

For automatically controlled wheeled vehicles there exist the following traditional risk operations associated with the vehicle design : - the occurrence of an accident and the severity of the consequences;

IOP Conf. Series: Materials Science and Engineering **819** (2020) 012038 doi:10.1088/1757-899X/819/1/012038

- fire; - environmental damage to the environment; - theft; - electric shock; - interference with the operation of electronic equipment; - technical failure of elements (components) affecting road safety; - poor-quality work (services) on maintenance and repair, and also on modification of a design and additional: - geo-location errors, data loss, improper interaction with the driver (remote operator), unauthorized interference with the control system and technical failure [11].

The greatest aftermaths of an accident are buses, especially when transporting children, cars carrying dangerous goods, cars with gas equipment. In operation these vehicles are subject to more frequent checks. Table 1 presents a risk-based approach matrix for some types of retrofitting and identifies the resulting risk. For the conditional quantitative analysis we will accept five levels: - extremely high - "5"; - high - "4", - moderate - "3", - low - "2", - insignificant - "1" (Table 1).

Table 1. Analysis of risks associated with the modification.

Type of vehicle modification	Probability of an accident due to system failure during conversion (probability of occurrence) (V)	Indicator of severity of social potential negative consequences (public response) (T)	Number of appeals with this type of conversion (N)	Total risk (potential danger) (R)
Braking system	5	1-2	2	10-20
Steering	5	1-2	1	5-10
Into bus	1	5	2	10
ADR	2	5	3	30
Gas equipment	4	3	5	60
Uninstallation of seats	1	1-2	3	3-6
Body kit 4x4	2	2	3	12
Replacing the engine	1	1	3	3

Thus, the following formula is used in the processing of group expert evaluation of objects during direct evaluation:

$$R = \sum_{i=1}^{n} (V_i \times T_i \times N_i) , \tag{1}$$

where «n» is the number of possible types of structural changes. If there is only one type of design changes, the range of the total risk is from 1 to 125. Based on this approach, you can make a rating and schematically present it in a three-dimensional form as a surface with a gradient fill, where the color changes from green to red in terms of danger (Figure 2). The total risk is proposed to be divided into two groups: up to 30 and over 30, for increased attention when assessing the possibility of making changes to the design.

A risk-based approach in assessing the impact of changes in vehicle design on road safety during the control and supervision activities of the traffic police may be the subject of further research in case of collecting reliable statistics. For a detailed assessment it will be necessary to use the multi-measuring experiment of the State traffic inspectorate and to collect statistics on the following parameters: the number of yearly issued approvals, the vehicle types, the number of refitted vehicles in operation, the detailed analysis of a forensic examination results if re-equipment provoked or exacerbated the consequences of an accident.

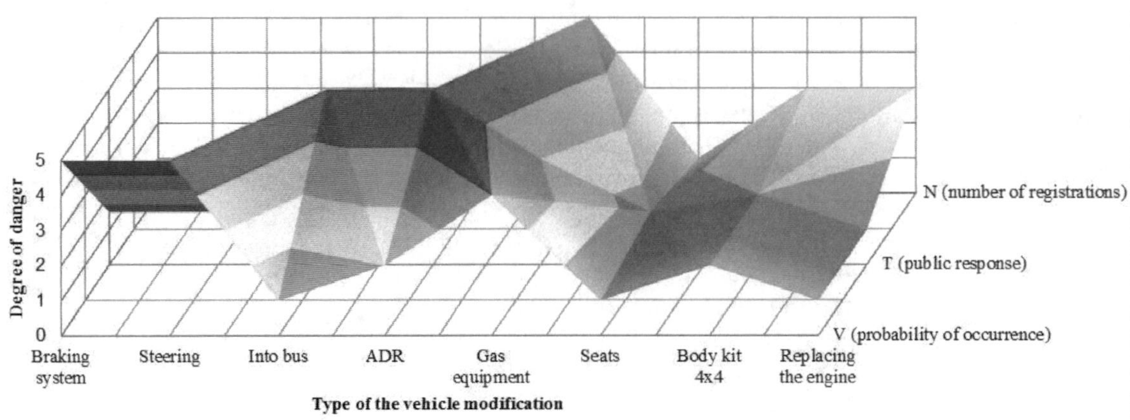

Figure 2. - Risk assessment for certain types of modification.

Only a systematic analysis will allow us to move from the total control to selective control and risk management, in which the intensity and volume of the checks by controlling organizations depends on the potential risk of harm level to the life and health of citizens. The work will help to compile the annual risky rating of organizations (technical inspection point, car service station, accredited persons) and determine the frequency of planned monitoring activities.

The input data for studying danger impact issue of re-equipment can be various classifications of both the main structural elements and additional equipment and their related properties.

By structural elements: - brake system; - steering; - lighting equipment; - glazing and mirrors; - wheels; - engine; - other; - weight and overall parameters; - reliability of additional elements;

By properties: - braking properties; - sustainability; - manageability; injury safety; - traction and speed properties; - visibility; - fire safety; - visibility in the dark; - evacuation ability; - environmental friendliness; - stability of the operation of ITS elements; - registration of vehicle motion parameters; - cyber-security; - software updates and security aspects of integrated electronic systems.

Naturally, there is a risk associated with the quality of work (services) in the system: - component manufacturer; - servicing; - diagnostics at a technical inspection point; - expertise by an accredited person. It is worth noting that the position of the manufacturer and servicing is more or less clear, the situation with checking the technical inspection is remains risky since at present, in many cases the control of the technical condition is not factually carried out.

In accordance with Part 1 of Article 27 of Federal Law dated December 28, 2013 No. 412-FZ "On Accreditation in the National Accreditation System", the federal state control over the activities of accredited persons is carried out using a risk-based approach, while in accordance with paragraph 5 of Part 2 of Article 5 Federal Law No. 412-FZ accreditation is based on the principle of voluntariness. At the moment, one of the main risks is ensuring impartiality. Thus, even the expert organizations that have long existed on the certification market, do not exclude the possibility of personnel actions leading to risks though they are minimized due to multi-level control. Using a unified approach to the examination of the vehicle design amendments allows ensuring the objectivity and independence of the procedure.

3. Conclusions

1. In most cases when retrofitting by installing additional equipment or replacing the existing units there is an improvement of technical parameters and characteristics, at that the risk depends on operating conditions and specialization. In case of using certified components, providing quality services when replacing and installing components as well as assembling additional equipment followed by the proper inspection, the modification, as a rule, is safe, which is confirmed by the final certificate.
2. Complicated work is to be done to improve regulatory and technical documents, taking into account a risk-based and verified approach in accordance with current and future standards together with the rejection of inspections duplication.

3. Based on existing experience, internal testing does not always guarantee the required safety and therefore independent expertise is required.
4. The application of a unified approach will improve the level and quality of inspections, and will make the basis for the collection of reliable statistics.
5. The costs of control and supervision activities may be higher than the prevented potential damage, therefore, a preliminary assessment of the socio-economic effect, that is the cost of administrative resource and obtaining useful result, is required.
6. One of the possible directions for the development of expert activity is the notification of expert organizations (inspection bodies).

References

[1] Bakhmutov S V, Endachev D V, Mezentsev N P 2018 Development of intelligent driver assistance systems (ADAS) in the Russian Federation *Transport systems*, **4** 17–21 Information on https://www.transport-systems.ru/images/2018/2018_04_003.pdf
[2] Information on the state of road safety in the Russian Federation for 2018. Information on https://stat-gibdd.ru/
[3] Kirillov K A 2016 Regulatory prospects and some monitoring results of the vehicle design and its technical condition in operation *Trudy NAMI* **265** 66–75
[4] Kapustin A V 2016 Technical inspection as an element of vehicle safety Report at the plenary session of the 95th international scientific and technical conference AAI at the Dmitrov autopoligon Information on: http://autoengineer.org/meropriyatiya/materials/konferenciya-aai-95.html
[5] Information system "Techexpert: 6 generation" Intranet. Information on http://lab2.cntd.ru/
[6] Kirillov K A 2016 Regulation of requirements for the transport vehicle safety in the Eurasian economic Union (EAEU) in case of design modifications. *Trudy NAMI* **266** 63–72
[7] Kisulenko B V, Gusakov N V, Bocharov AV, Shchepkin A I, Mironov A A, Anikeev S A 2015 Technical regulation in the automotive industry. Conformity assessment procedures. Moscow, 256 p. ISBN 978-5-9906820-6-1
[8] Girutsky O I, Kirillov K A 2017 The new standard on the administrative aspect of registration of changes in the design of the vehicle. *Izvestiya MGTU "MAMI"* **34** 83–92
[9] Comprehensive program for the development of standards for automotive products in the future until 2025.: Information on: https://www.gost.ru/portal/gost/
[10] Kirillov K A 2016 International experience in conformity assessment concerning safety requirements of vehicle design, including in operation. *Trudy NAMI* **267** 61–70
[11] Kisulenko B V 2017 Normative legal regulation of vehicles with a high degree of control automation: strategy and tactics of implementation in Russia. *Report at the international automotive scientific forum MANF-2017.* Information on: http://iasf.nami.ru/about/archive/2017/programm.php

Real-time diagnostic complex of automated car active safety system unit

V G Dygalo[1] and A V Keller[2]

[1] Department of Operation and maintenance of automobiles, Volgograd State Technical University, VSTU Volgograd, Russia
[2] FSUE "NAMI", 2 Avtomotornaya St., Moscow, 125438, Russia

Abstract. The article is concerned with the created variety of methods of drivegear active safety car system test. The analysis of factors of electronic control stability system test is displayed, as well as main disadvantages and test factors, characteristics and general terms. The system description is indicated, which gives opportunity to test active safety system. In the same way, it is possible to solve the problem of timely risk detection for key components, assemblies, and vehicle systems. And if in the traditional Driver-Car system an experienced driver can "feel" a possible malfunction, then an unmanned car has nothing but its own systems. In this case, it becomes obvious that on-board equipment is needed that monitors the state of the braking system and ABS in real time. As described above, the existing on-board diagnostic equipment does not provide the necessary information. The introduction of an additional diagnostic complex will allow tracking and detecting system failures. This system can be used with a "digital twin"

1. Introduction

Modern automated brake systems (ABS, ESP) belong to the group of complex electric drive systems (Fig. 1) [16]. This system of electric drive is an integral part of the brake system of a modern car. However, the efficient use of any electric drive system in operation is possible only if the methods for monitoring their technical condition are successfully implemented [1,2]. Otherwise (especially with regard to the use of automatic systems), it may result in a much more dangerous situation (caused by the incorrect operation of the actuators). In this case, by deflecting the working processes of a vehicle's brake system it is possible to detect the malfunction of the drive system at the early stages.

2. Main part

To control the state of automated brake system (ABS, ESP) manufacturers use a self-monitoring system, which is aimed at checking the integrity of electrical circuits and signal strength [10,11]. Once the modulator or sensor supply circuit breaks, it immediately turns off the automated brake system (ABS, ESP) and alerts the driver. Obviously, this system does not discover all the possible failures of its elements. The following can be added to that taken into account: changing the section of the channels of the modulator due to their clogging, jamming of the modulator valves, delay in the actuation of the valves, angular compliance of the ABS stator sensor, increased contact resistance, faults in the control unit automated brake system (ABS, ESP) blocks.

IOP Conf. Series: Materials Science and Engineering **819** (2020) 012039 doi:10.1088/1757-899X/819/1/012039

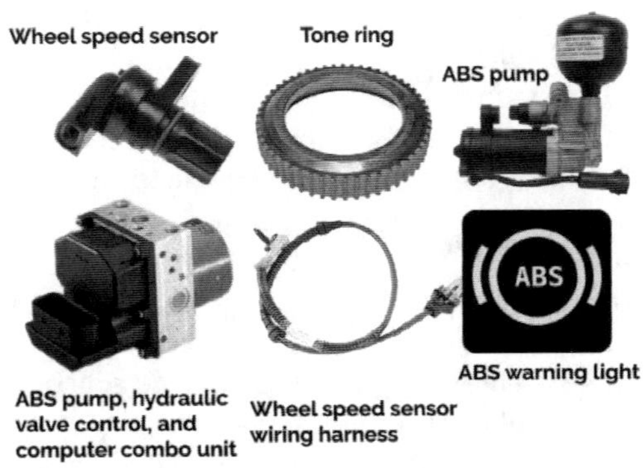

Figure 1. Components ABS actuator.

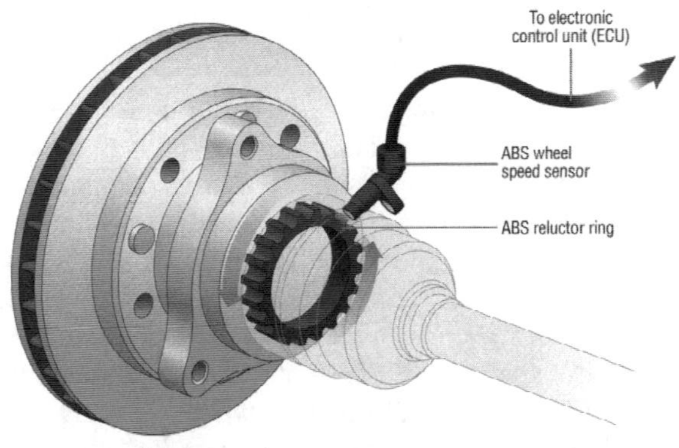

Figure 2. ABS Speed Sensor.

The system does not account for the state of the vehicle's service braking system either. Irreversible changes occurring with the brake system drive, such as wear of the brake pad linings, wear of the brake disc / drum, aging of the brake fluid (caused by the accumulation of rubber micro particles from wear of the piston cuffs of workers and the main brake cylinders, rust particles, moisture absorption from the air) etc., have inevitable negative effect on the efficiency of the vehicle braking system. In this case, the correct operation of the ABS is maintained, but the braking efficiency drops all the way to the system failure. It is important that with such malfunctions, the automated brake system (ABS, ESP) self-diagnosis system does not provide a failure signal.

When analyzing the self-diagnostics of the automated brake system (ABS, ESP) unit, it can be concluded that only a small list of system defects is currently available. The fault codes of the vehicle automated brake system (ABS, ESP) block can be divided into several groups:

1. Electrical networks control;
2. Monitoring communication with the units via the CAN bus;
3. Verification of incoming signal data and control of the control unit;
4. Brake system monitoring.

The most important groups shall be further considered in detail.

The failure codes of the first group are the control of electrical circuits occuring as the control unit of the automated brake system (ABS, ESP) detects voltage changes of the signals from each wheel speed sensor. If a circuit breaks or a short circuit is detected, the automated brake system (ABS, ESP) control unit saves the diagnostic code. It may be caused by damaged wiring harness or connectors, noise interference, wheel speed sensor malfunction, ABS-ECU malfunction.

Codes of failures C104B-C105B are of the same type and are responsible for the hydraulic block ABS valves. The cause of the failure codes is the ABS-ECU fault. These diagnostic codes are issued if the solenoid valve does not supply power, even if the ABS electronic unit has turned on the transistor or after the ABS computer turns off the transistor, the solenoid valve still receives power, and if a solenoid valve is detected [3].

The remaining codes of the electric circuits control group are also directed at the detection of defects in electrical circuits of the anti-lock system elements and connecting wires, as well as control of too high or low voltage in the vehicle's on-board network.

When monitoring communication with the units on the CAN bus, the codes are responsible for the communication between the units. Checks are in the procedure for searching for electrical faults on CAN bus wires, measuring the resistance of end-of-line resistors and estimating the traffic load of computer networks over medium voltage.

When the vehicle is moving, the ABS control unit monitors the signals from each wheel speed sensor. If the signal from these sensors shows a random change in the signal or the value of the signal,

IOP Conf. Series: Materials Science and Engineering **819** (2020) 012039 doi:10.1088/1757-899X/819/1/012039

the corresponding diagnostic code is recorded very high all the time and the control lamp on the instrument panel lights up.

During the operation of the system, the ABS-ECU continuously monitors its performance parameters. If any deviations are found, the code is output. The diagnosis is reduced to the CAN bus check. If there are no faults, the ABS unit must be replaced. From the practice of diagnostics, it is known that this code occurs when the ABS control unit is not correctly reprogrammed or interrupted. This brings up a conclusion that in this case the unit monitors the checksums of its software.

The fault may be caused by damage to the wiring harness or connectors, external noise interference, wheel speed sensor malfunction, ABS-ECU malfunction, too large a distance between the wheel speed sensor and the master dial, sticking of foreign materials to the wheel speed sensor and / or drive dial, malfunction wheel bearing, improper installation of the wheel speed sensor, deformation of the drive disk, violation of the magnetization pattern of the master disk, dyeing of the teeth of the master disk. A probable cause of the DTC is the brake fluid pressure remaining high or low for a long time. Since the automated braking system does not have the necessary set of pressure sensors, it is calculated based on the readings of wheel speed sensors. Accordingly, wheel speed sensors will require a diagnostic procedure as well. Checking for this group of codes is not much different from checking for other faults in the speed sensors. There is no checking of the braking system or hydraulic system state at any stage.

Code C1395 is responsible for the supply of brake fluid to the hydraulic system of the ABS. The cause of the fault code is an improper hydraulic system unit and a fault in the ABS control unit. The verification procedure for this code includes only two items - the CAN bus check (as mandatory for all codes) and the reset of the DTC. If the error re-appears, replace the ABS assembly.

The analysis of the list of failure codes of the on-board diagnostics system shows that the possible causes of the malfunction, such as noise interference, are not checked in any way. That means the electronic control unit can detect an interference, but there is no verification procedure. This leads to the fact that the search for the faulty element will fail, or indicate a working node, clearly following the diagnostic procedure. The braking system or hydraulic system health are not checked at any stage. This causes erroneous replacements and repairs.

The system overlooks a number of malfunctions, such as the adherence of ferromagnetic dust to the sensor or the master rotor, the clogging of the brake pipes, the change in the cross-section of the channels, the hanging of the ABS hydraulic valves, permanent interference (for example, due to installed optional equipment), erroneous data and commands in the CAN network, internal failures of the ABS control unit, damage to the printed circuit board of the ABS control unit, etc. Complex faults, meaning the joint impact of two or more faults, are not covered either. Such cases are the most complex, because their joint impact can mislead the self-diagnosis system.

The analysis of the existing diagnostic methods shows that their overwhelming majority are aimed at monitoring wheel speed sensors and a computer CAN bus. Even though the system can track deviations in pressure, lack of brake fluid, monitor the reliability of incoming parameters and the integrity of electrical circuits, the diagnosis is quite primitive and reduces to checking electrical circuits.

It is also worth considering that the electric brake system is an integral part of the brake system of the car. In this case, by deflecting the working processes of the vehicle's brake system, it is possible to determine the malfunction of the drive system at the early stages.

The presence of any deviations in the operation of the system, requires testing the executive mechanisms of the automotive brake system of the car. If left unattended, the slightest changes in the operation of the system can cause a system failure in the future.

In this case, there is an obvious need for the on-board equipment that monitors the state of the braking system and ABS in real time. As described above, the existing on-board diagnostic equipment does not provide the necessary information. The introduction of an additional diagnostic complex will allow tracking and detecting system failures [5,12,13].

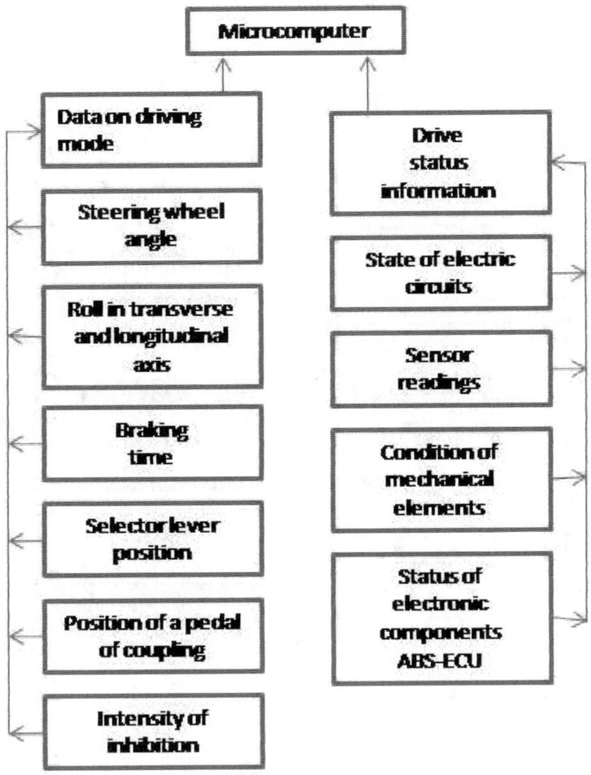

Figure 3. Electronic Components Status ABS-ECU Diagnostic Complex System

To perform all the necessary functions, the following blocks should be implemented:

- The array of diagnostic sensors;
- Data analysis unit.

The array of sensors allows capturing the parameters of the vehicle motion in braking mode, including the initial speed, deceleration, rotation of the car relative to the vertical axis, rotation of the vehicle relative to the transverse and longitudinal axis, braking distance, wheel lock. Obviously, the vehicle braking mode depends on the physical parameters of the driver and the mode of braking and the control of the trajectory that is being implemented, including the brake pedal force, the braking intensity, the steering wheel position during braking (angle of rotation), the braking time, the gearshift lever position (for the manual gearshift) position of the clutch pedal. The braking mode also depends on the environmental parameters, such as temperature, road surface type, weather conditions, time of day.

An adaptable plug-in diagnostic unit receives data on the work processes of the brake system components of the car and ABS in real time.

The data obtained by diagnostic sensors are analyzed by a microcomputer (diagnostic unit). In order for the diagnostic system to be scalable, the role of the diagnostic unit is performed by the Rospberry Pi 3 microcomputer.The sensors are connected via the integrated PLD-40 interface (GPIO, UART, I2C, SPI). This interface allows connecting various status sensors. The Raspberry Pi 3 operating system has the necessary software to process and analyze the signals coming to the PLD-40, as well as to arrange the signal analysis algorithm in the original software product created by the authors of the article with a custom graphical shell.

To test and debug the diagnostic complex, as well as study the regularities of work processes in the drive of the automotive brake system of the car, a full-size model of the automotive brake system of the car is used.

IOP Conf. Series: Materials Science and Engineering **819** (2020) 012039 doi:10.1088/1757-899X/819/1/012039

The model works to simulate the working processes of the drive of the car's automated brake system [6,7,8].

Figure 4. Assembled diagnostic unit.

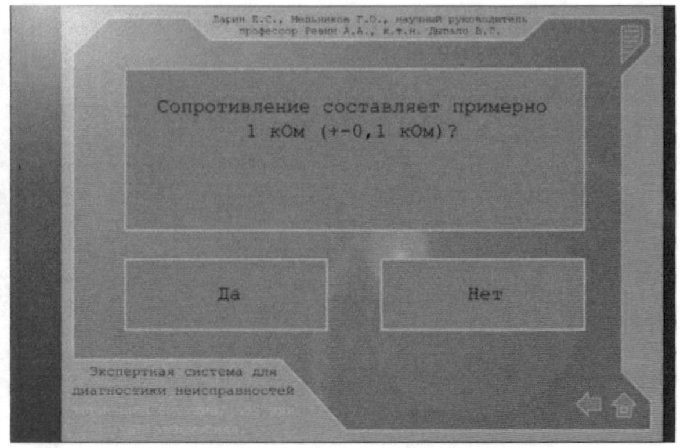

Figure 5. Graphical interface.

The unit is connected to the computer via a cable for relay control signals transmission, and with the ABS device using electric wires soldered to the solenoid terminals and the electric motor. To operate the solenoid, the computer sends a control signal to the relay, which activates the internal diode and allows supplying power to the controlled solenoid [9,11].

Direct control of solenoids and emulation of signals of wheel speed sensors allow simulating any modes of operation of the brake system.

Using a full-size model rather than a car makes it possible to simulate various malfunctions of the automated braking system and to study the corresponding deviations in the work processes without any risk to the driver / operator safety [4,14,15].

Figure 6. Connection to the ABS solenoids.

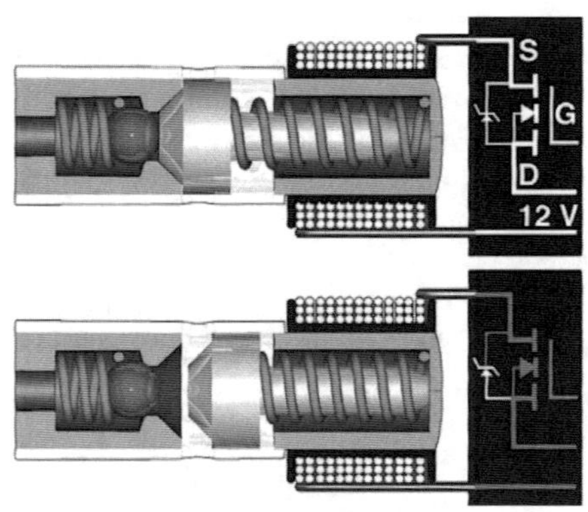

Figure 7. Solenoid ABS.

Thus, for example, consider the effect of an increase in the delay time of the operation of the modulator valves on the process of braking the wheel. The studies were carried out in the range of the time lag of 0.03 - 0.09 s. They showed that the manifestation of this malfunction is more pronounced when braking on a surface with average values of the coefficient of adhesion, for example, wet asphalt, from high initial speeds in the loaded condition of the vehicle. This circumstance is caused by the fact that under the given conditions, during the braking time, the greatest number of cycles of regulation of the ABS operation takes place, which makes it possible to more clearly trace the nature of the manifestation of the above-mentioned malfunction.

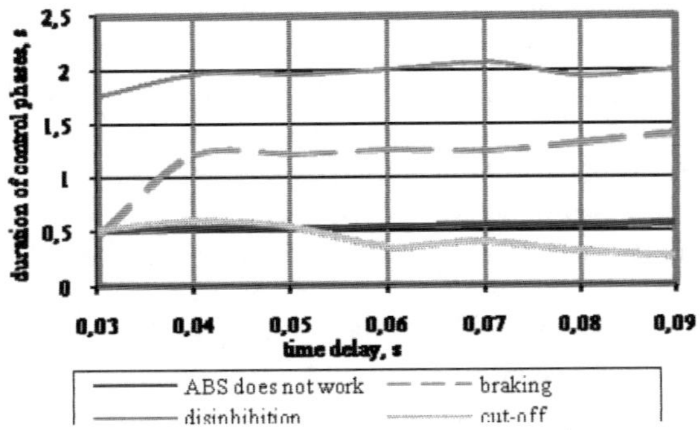

Figure 8. The change in the duration of the phases of the ABS operation when braking on wet asphalt concrete as a result of an increase in the delay time of the operation of the modulator valve.

As follows from the comparison of the physical parameters of the dynamic processes of braking the vehicle wheel on wet asphalt concrete with an increase in the delay time for the operation of the modulator valve to 0.05 s and 0.09 s, respectively, with the initial braking rate V0 = 19.4 m / s (70 km / h), with an increase in the delay time of the operation of the modulator valve, the time spent by the wheel in the state of the jib increases, and is unstable, especially in the middle part of the braking process, during a period of 1.5 to 3 s.

This phenomenon has a proven negative effect on the stability of the vehicle. In addition, with an insignificant increase in the delay time of the valve operation, the ABS transition is observed in some cycles for two-phase regulation, since a further increase in the delay time of the valve operation leads to instability of the "cut-off" phase. At the same time, when assessing the duration of the main phases of ABS operation, it was noted that a decrease in the efficiency of regulation is also a consequence of a significant increase (from 0.47 s to 1.42 s) in the inhibition phase and a less significant but clearly pronounced increase in time before ABS enters into operation.

To develop recommendations on the choice of diagnostic signs for increasing the delay time for the operation of the modulator valve, the authors consider the change in the average realized values of the main parameters when braking on wet asphalt (Fig. 9). As the graphs show, in all the considered range of the increase in the delay time of the operation of the modulator valve, the average realized coefficient of adhesion of the wheel with the road surface decreases (to 4.9%).

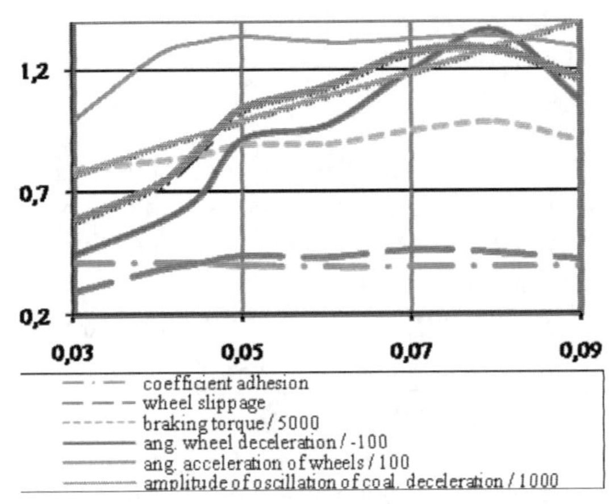

Figure 9. Modification of the average realistic values of the main parameters during braking on wet asphalt concrete as a result of an increase in the delay time of the operation of the modulator valve.

This confirms the negative impact of the analyzed fault on the performance of an ABS. It should be noted that the values of the remaining average parameters in the entire considered range exceed their initial value corresponding to the modulator's condition. However, as a diagnostic indication of an increase in the delay time of the operation of the modulator valve, an increase in the amplitude of the angular deceleration of the wheel should be used, since its variation over the entire range occurs practically along a linear relationship, which will positively influence the accuracy of the diagnosis. In addition, this parameter meets the requirements of both informativeness and sensitivity, since its increment reaches 44.5%.

Depending on the given initial conditions of braking, with the help of a microcomputer, it is possible to distinguish between braking modes (emergency braking, service braking, partial braking, total braking, braking by the engine, braking with a spare braking system) and memorizing the operating parameters of the braking system and the car's ABS.

These data are sorted and stored in the ROM of the microcomputer, taking into account the simulated braking mode. An example of data on the pressure in the brake drive, fixed on four wheels, is shown in the figure below.

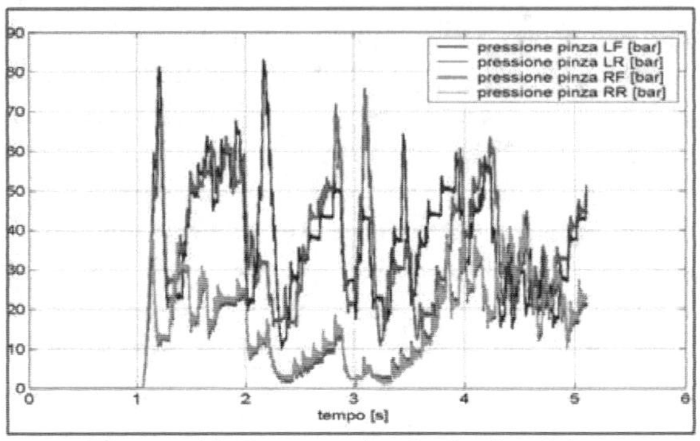

Figure 10. Pressure in the brake drive.

The data obtained during the braking process are recorded, and compared with data obtained under similar conditions and recorded in the ROM earlier. If there is a significant deviation, the microcomputer gives the operator information that the electric drive system is faulty. At the same time, based on the received data, the system makes a conclusion about possible malfunctions.

In the event that the microcomputer detects a malfunction, the collected data on the performance of the brake system and ABS can be useful in repairing and will make it possible to initially track the deviation of the operating parameters and prevent the failure of any element.

3. Conclusions

The diagnostic complex, based on the slightest deviations of the working processes of the automotive brake system of the car, allows determining the malfunction at the initial stage.

An important difference from the existing diagnostic systems is the possibility of establishing the exact fault in the element of the automotive brake system of the car, as well as determining the cause of the fault. The efficiency of the diagnostic process is increased, which leads to a reduction in the possibility of a repeated failure.

The use of the diagnostic system in conjunction with the chassis stand model, allows adapting of the system, if changes are made to the design of the automotive brake system of the car by the manufacturer.

Equipping with such a system will allow real-time monitoring of the technical condition, in particular of an automated brake system. The data obtained are compared with the reference "digital double" of the car. That is, the program itself analyzes whether all units and components behave normally, based on the year of release, mileage and working conditions of the vehicle. If there are deviations from the "digital" norm, the "smart" system gives a warning signal that prevention is needed. For example, replacing brake pads.

Note that this norm (that is, a digital model of car wear) is taken on the basis of tests, both full-scale and using virtual physical modeling technologies. After that, the mathematical laws of the digital double are specified. Thanks to the new system, the estimated service life of the car can be extended up to 20-25 percent. And the costs of maintenance and repair can be reduced by almost half - by 40-45 percent.

Such a system for monitoring the technical condition is certainly applicable to all types of vehicles. But it will bring the greatest economic effect for enterprises, both in freight traffic and in passenger traffic. Ideally, this should work like this: an analysis of the technical condition of a particular machine in real time is transmitted to a single control system. This allows the company to quickly respond to identified risks, adjust the operating mode of cars, check the availability of necessary components in the warehouse, and pre-order them if necessary. Then it is possible to find a window in the transportation cycle, form an application for targeted and timely service. At the same time, repair specialists will not need a complete assembly-disassembly in order to find a specific failure: the vulnerable node will be known in advance.

References

[1] GOST R 41.13-99 (UNECE Regulation No. 13). *Uniform provisions concerning the approval of motor vehicles of categories M, N and O with regard to braking*. - Introduced for the first time; Enter. 01.07.00. - M.: Standartinform, 2006 - 113 p

[2] GOST R 51709-2001. *Motor vehicles. Safety requirements for the technical state and methods of verification* - Introduced for the first time; Enter 01.01.02. - M: Standartinform, 2008 - p 39

[3] *Mitsubishi Lancer X Service Manual & Body Repair Manual*, MY 2011-2015 [Electronic resource]. - 2015 - Access mode: http://mmc-manuals.ru/manuals/lancer_x/online/Service_Manual_2008-2012/2015/index_M1.htm

[4] Dygalo V, Lyaschenko M, Boyko G and Dygalo L 2018 Application of the Virtual Physical Modeling Technology for the development of elements of the autonomous (unmanned) vehicles' systems *IOP Conference Series: Materials Science and Engineering* **315** p 11

[5] Revin A, Dygalo V, Lyaschenko M, Boyko G and Dygalo L 2018 Methods of monitoring the technical condition of the braking system of an autonomous vehicle during operation *IOP Conference Series: Materials Science and Engineering* **315** p 6

[6] Revin A and Dygalo V 2005 *Virtual tests in the design cycle of automated brake systems Science for Production* **1** pp 43-47

[7] Dygalo V and Revin A 2006 *Virtual physical technology of laboratory testing of active safety systems of vehicles* p 316

[8] Dygalo V and Revin A 2007 Virtual physical modeling technology in the design cycle of automated brake systems *Bulletin of the Volgograd State Technical University* **2** pp 13-15

[9] Revin A and Dygalo V 2001 *Research of brake dynamics of the car by methods of complex modeling technology* p 122

[10] Maier M and Müller K 1995 ABS 5.3: TheNew and Compact ABS5 Unit for Passenger Cars *SAE Technical Paper Series* 950757

[11] Emig R, Goebels H and Schramm H 1990 Antilock Braking Systems (ABS) for Commercial Vehicles - Status 1990 and Future Prospects *SAE Technical Paper Series* 901177

[12] Keller A and Aliukov S 2015 Efficient Power Distribution in an All-wheel Drive Truck *Lecture Notes in Engineering and Computer Science: Proceedings of The World Congress on Engineering* pp 1201-1206
[13] Keller A and Aliukov S 2016 Efficient Power Distribution in an All-Wheel Ground Vehicles *SAE Technical Paper* 2016-01-1105
[14] Aliukov S, Keller A, and Alyukov A 2015 On the Question of Mathematical Model of an Overrunning Clutch *Lecture Notes in Engineering and Computer Science: Proceedings of The World Congress on Engineering* pp 1319-1324
[15] Keller A , Aliukov S and Anchukov V 2017 Mathematical model of the truck for investigation of stability and control of movement *Lecture Notes in Engineering and Computer Science* pp 711-716
[16] Keller A , Shelepov A and Istomin D 2016 Estimation of Screw-ball Differential on Vehicle Dynamics *Procedia Engineering* **150** pp 1131-1136
[17] Robert Bosch GmbH 1995 Automotive Braking Systems p 336

IOP Conf. Series: Materials Science and Engineering **819** (2020) 012040 doi:10.1088/1757-899X/819/1/012040

HIL models formation principle in the design of automated vehicle braking system

V G Dygalo[1], A V Keller[2] and A M Zavatskiy[2]
[1] Department of Operation and maintenance of automobiles, Volgograd State Technical University, VSTU Volgograd, Russia
[2] FSUE "NAMI", 2 Avtomotornaya St., Moscow, 125438, Russia

Abstract. The general principles of the HIL models formation in the design of an ABS-featuring vehicle braking system are considered. The proposed modular principle of compiling a mathematical model of subsystems allows improving the model by the iterative method, which makes it possible to add or remove blocks, change their mathematical content, while maintaining the composition of formal relationships for a complete description of the object. The iterative process continues until a model has been obtained to solve the problem, which can be considered the most fully reflecting the physical and temporal processes taking place in real objects. Virtual physical simulation data help to adjust model parameters, apply numerical optimization, explore alternative scenarios for controller bandwidth, test development sensitivity, and study other factors. The application of this method also makes it possible to compare different implementations of control systems on different boards, analyze the development and evaluate the trade-offs. Thus, developers will be able to verify that the embedded system works exactly as planned, before its final implementation. This technology makes it possible to combine field systems with a mathematical description of the dynamics of ongoing processes and phenomena. At the same time, the Virtual Physical Modeling Technology, unlike the commercial Hardware-in-the-loop systems, is not tied to the hardware of any particular manufacturer.

1. Introduction

When modeling it is necessary to closely reflect the real picture of the flow of working processes in the "Car - Driver - Road" (C-D-R) system, thus solving the problems of forming the main operational properties of the car in braking mode, i.e. stability, efficiency and controllability, at the design stage [1,2]. In this system, both mechanical movements of masses (sprung and unsprung) and physical processes accompanying the operation of various vehicle systems affect the vehicle motion characteristics (interaction of the wheel with the road, hydrodynamic processes in the brake system, rotation of the steered wheels when controlled movement, etc.).

2. Main part

In a systematic study of the general case of car behavior in braking mode, it is necessary to adopt a unified methodology for describing the interaction of various parts of the studied C-D-R system, which is reflected in Figure 1.

IOP Conf. Series: Materials Science and Engineering **819** (2020) 012040 doi:10.1088/1757-899X/819/1/012040

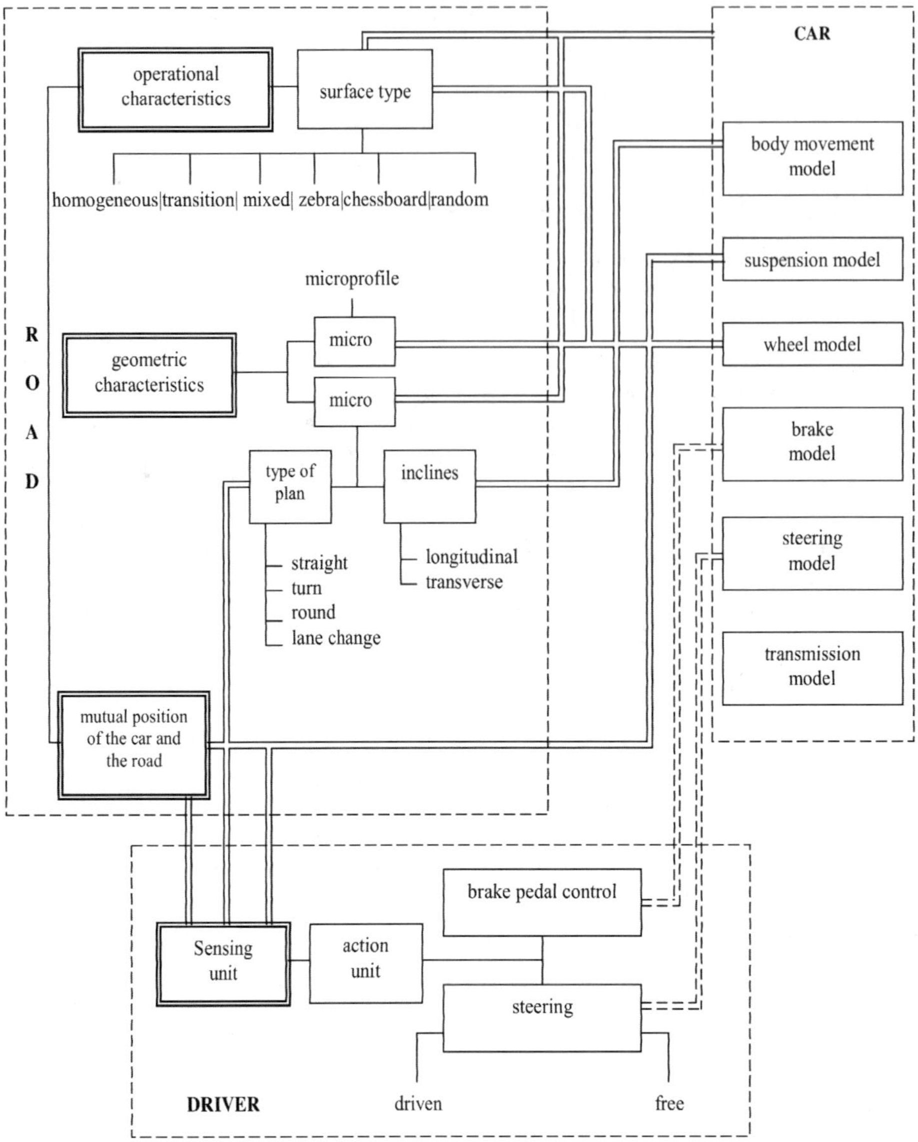

Figure 1. Diagram of the relationship between the models of the subsystems "Car - Driver - Road" and their main elements.

The most complex part of the C-D-R system is the subsystem "Car". Each module of this subsystem makes certain transformations of quantities and, therefore, is characterized by input and output parameters. In this regard, a structural diagram of the relationship between the models of modules in the "Car" subsystem is presented in Figure 2.

As illustrated by the above diagram, all modules can be arranged with some priority. The “Wheel” and “Body” modules have the highest priority, since without them it is generally impossible to simulate the movement of a car. The second priority is the module "Brake System", which, in accordance with the goal set above, is necessary to model vehicle braking.

The third priority are modules aimed at refining the model or solving any subproblems [3,4] (modules "Suspension", "Steering", "Transmission").

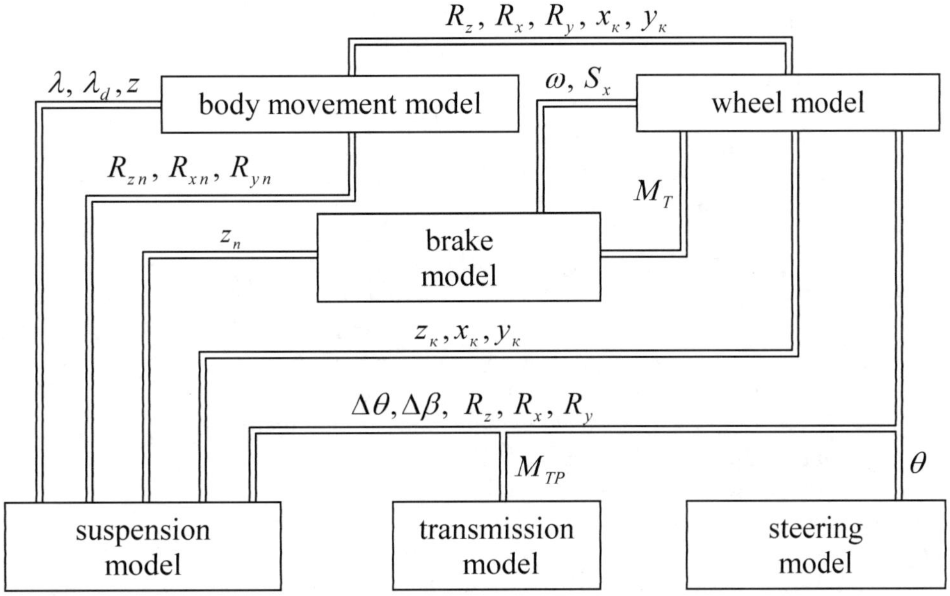

Figure 2. Scheme of interaction of module models in the subsystem "Car" in the general case of braking.

Such an arrangement of modules in the “Car” subsystem, when the modules with the highest priority are processed first, allows detecting the impossibility of solving a problem, insufficient communications with other modules, or a lack of initial data in the early stages of modeling. In addition, such a partition of the “Car” subsystem minimizes the number of interactions between the modules, which best affects one of the research tasks — the speed of solving a model on a computer.

The proposed modular principle of compiling a mathematical model of the subsystem "Car" makes it possible to improve the model by the iterative method, which allows adding or removing blocks, changing their mathematical content, while maintaining the composition of formal relationships for a complete description of the object. The iterative process continues until a model has been obtained to solve the problem, which can be considered the most fully reflecting the physical and temporal processes taking place in real objects. The effective operation of the mathematical model is possible only if reliable initial parameters are available. Otherwise, the complication of the model can lead to the opposite result, i.e. a decrease in its reliability. In our case, when modeling the braking dynamics of a car with ABS, such data include the parameters of the car itself (geometric, weight and inertia), wheels, brake system, steering, suspension, transmission, driver and road. When designing an ABS from the initial design work to start it in series, developers primarily have to solve the following types of main tasks:

- ✓ form the algorithm of the braked wheel and the selection of the characteristics of the modulator;
- ✓ choose the control structure;
- ✓ optimise the control structure with account to the actions of the driver.

The studies of E.V. Hertz, G.V. Kreinin, N.F. Metlyuka, V.P. Autushko, A.A. Revina et al. established that describing the flow of fluid or gas in the brake line and the ABS modulator requires knowing a large number of parameters, which at the design stage are often impossible to obtain with the required accuracy. At the same time, the dynamics of the processes of interaction of the elastic

IOP Conf. Series: Materials Science and Engineering **819** (2020) 012040 doi:10.1088/1757-899X/819/1/012040

wheel with the road surface is well described mathematically. The most effective solution for this is the use of Virtual Physical Modeling (VPhM), when part of the object of study is presented in the form of a real unit (often a prototype), which allows the designer to carry out operational verification of the decisions made even at the stage of debugging and debugging the system, tied to the hardware of a particular manufacturer.

Therefore, when solving the problems of optimizing the working processes of automated brake systems and studying the vehicle's dynamism in the braking mode, it is generally advisable to represent the brake circuit lines together with ABS modulators and working cylinders (including hose flexibility and stiffness of the block-drum system), subject to the preservation of the geometry of the highways and the location of the studied units. Then the car chassis, brake mechanism, steering, suspension, a pair of "tire - road", etc. is represented in the form of a mathematical model on a control computer complex. However, when using complex modeling technology, one of the most important requirements - to ensure real-time operation - must be met. At the first stage, when solving relatively simple problems, it is possible to select control computers based on the use of analog computers that have the property of working in real time. The studies conducted at Volgograd State Technical University [1] showed that the results of bench studies fully reflect the physical picture of the braking process and give good convergence in the final values of the parameters, which confirmed the possibility of their further use in studies of a complex modeling setup.

However, the complication of the mathematical description of various elements of the car necessitated the use of modern PCs, while it was necessary to ensure its most important property, i.e. real-time operation. Without the implementation of the latter, the use of complex modeling technology is impossible. When using a digital computer, this requirement is achieved when the following indispensable condition is fulfilled: the calculation time of the process should be less than the time of its actual course.

The practice of using the Virtual Physical Modeling Technology (VPhM) (or Hardware-in-the-loop) at Volgograd State Technical University showed that using system design and modeling, it is possible to build stands that combine new and existing mechanical, hydraulic and electrical actuators and sensors [5,6,7], while not tied to the hardware of any particular manufacturer.

Virtual Physical Modeling (or Hardware-in-the-loop) is especially effective at this stage of design, when a real object has not yet been built.

Figure 3 shows an example of an implemented stand for solving the problems of formation and optimization of the operation algorithm of the Electro-hydraulic brake system (EHB) [11,12,14].

The next task to be solved by the designers is the selection of a control structure.

The selection and optimization of the control structure can be implemented on the model of the class "chassis". At the same time, developers can use Virtual Physical Modeling (or Hardware-in-the-loop) to expand the capabilities of testing the chassis and electronic automotive systems that meet the stringent requirements of certification for health and safety. Physical simulation allows studying the entire system, which combines several different subsystems, and analyzing its performance in the conditions and environments too dangerous or expensive to conduct their field testing [15-18].

Virtual Physical Modeling is especially effective at this stage of design, since the requirements of security and speed demand to test the system before attracting people to its implementation, while it is necessary to minimize the downtime of a real object.

Depending on the objectives of the study, the mathematical models of the chassis can take a different configuration. Volgograd State Technical University created the mathematical models of the chassis of the following types of automatic telephone exchanges: biaxial, triaxial, truck and trailer trains. Practice has shown the need to solve the problem of choosing a control structure in addition to traditional ones, taking into account the following physical processes: the gyroscopic moments of the steered wheels during their self-rotation and the fading effect in the braking mechanism under the influence of the ABS, hysteresis and inertia of the braking mechanism when debugging the algorithm.

Figure 3. Test bench for EHB (for one wheel contour).

The stand shown in Figure 4, implemented in the laboratory of the Turin Polytechnic Institute as a result of scientific cooperation between research groups, clearly illustrates the possibility of solving problems of the chassis class.

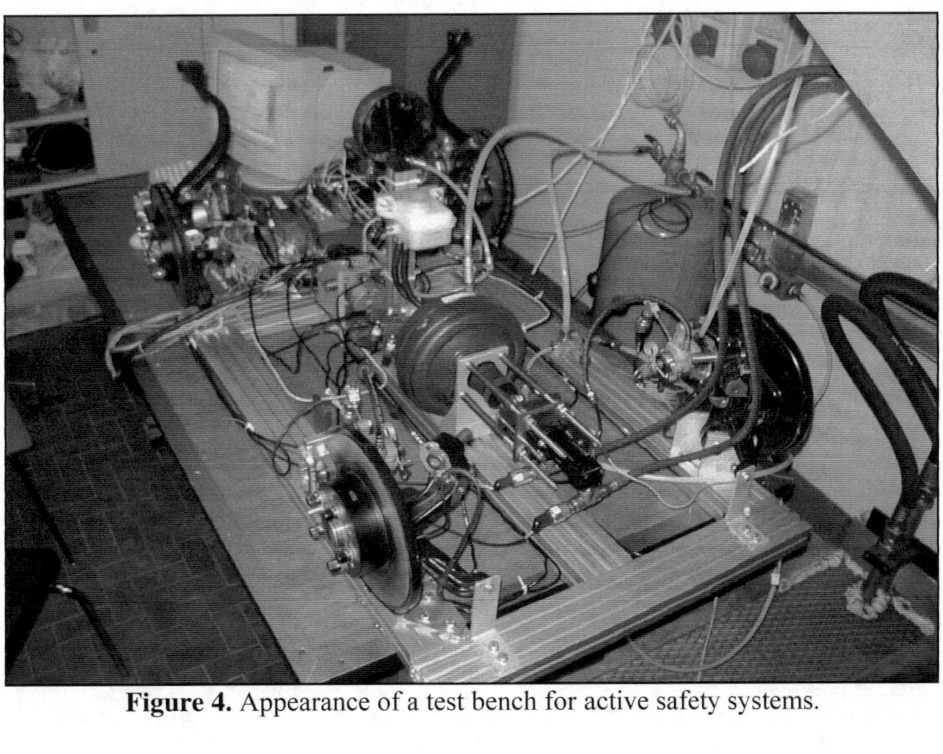

Figure 4. Appearance of a test bench for active safety systems.

IOP Conf. Series: Materials Science and Engineering **819** (2020) 012040 doi:10.1088/1757-899X/819/1/012040

During the tests, various technical solutions for connecting the blocks were used. So, to verify the operation of the stand, the electronic filling of the Bosch ABS and Bosch ESP blocks was used, and to develop our own algorithms, direct connection to the valve body valves was used [7, 14]. This made it possible to reduce the volume of road tests of active safety systems that combine several subsystems.

However, the optimization of the control structure is impossible without taking into account the driver's actions to correct the direction of movement. Thus, the third task is the optimization of the control structure with account to the "man-machine" system interaction. Virtual Physical Modeling is especially effective at this design stage, as the safety requirements to the system testing are very strict. This technology (VPM) provides for testing the system performance in extreme modes, which are complex and dangerous to re-create, before its actual implementation by the staff. Figure 5 shows an example of the implementation of the stand for solving "chassis + driver" problems.

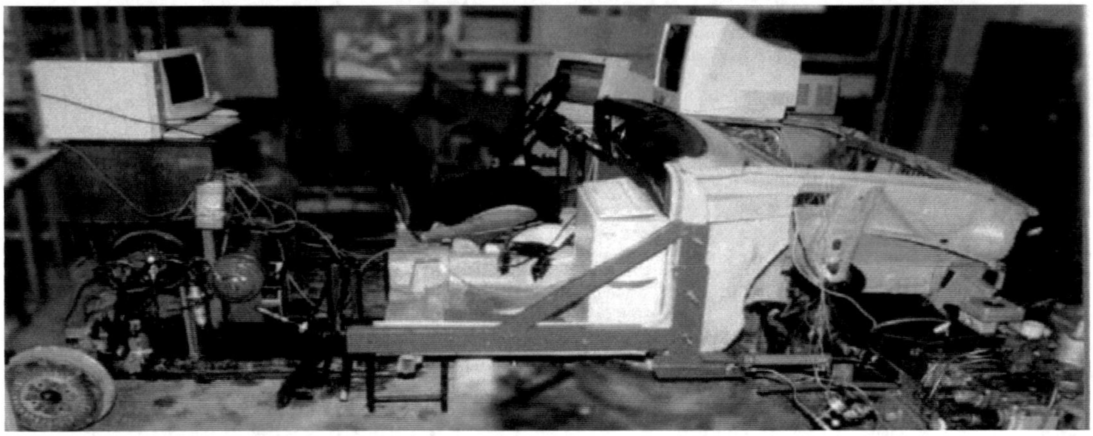

Figure 5. A stand for solving task of the "chassis + driver" class.

The problem can be solved by expanding the capabilities of bench equipment by:

- Providing visualization of the movement process perceived by the real driver;
- The implementation of channels of tactile information transmitted to the driver.

Rapid prototyping provides the developers of control and data processing systems with a quick and cost-efficient way to test a project in the early stages of development and analysis in order to choose the best solution. Depending on the tasks to be solved, a selection of implemented functional blocks is made (Figure 6).

3. Conclusions

Implementation of the Virtual Physical Modeling Technology makes it possible to automate many tasks, providing for:

- the performance of hardware testing of new active security systems and algorithms in real time;
- creation of alternative project options as soon as possible;
- spending more time optimizing the system.

VPM allows confirming the developed system functional capacity in-situ at the very early stages of its design. Verification is performed on equipment operating in real time. At the same time, the project can be quickly refined and modified until the required result is achieved. Virtual physical simulation data help to adjust model parameters, apply numerical optimization, explore alternative scenarios for controller bandwidth, test development sensitivity, and study other factors.

The application of this method also makes it possible to compare different implementations of control systems on different boards, analyze the development and evaluate the trade-offs. Thus, developers will be able to verify that the embedded system works exactly as planned, before its final implementation.

Creating a system for Virtual Physical Modeling, a developer can choose the right equipment from hundreds of items by various manufacturers, in contrast to existing commercial solutions of Hardware-in-the-loop systems. The developer is absolutely free to choose I / O boards, processor power, type of execution, and other features of the iterative development environment.

These stands, combining several subsystems, make it possible to perform the following tasks:

- ✓ to develop and test the chassis and safety management systems without expensive road tests of vehicles;
- ✓ develop and test algorithms using models at earlier stages of development;
- ✓ implement rapid prototyping of controller algorithms;
- ✓ implement hardware and software modeling of electrical and mechanical components.

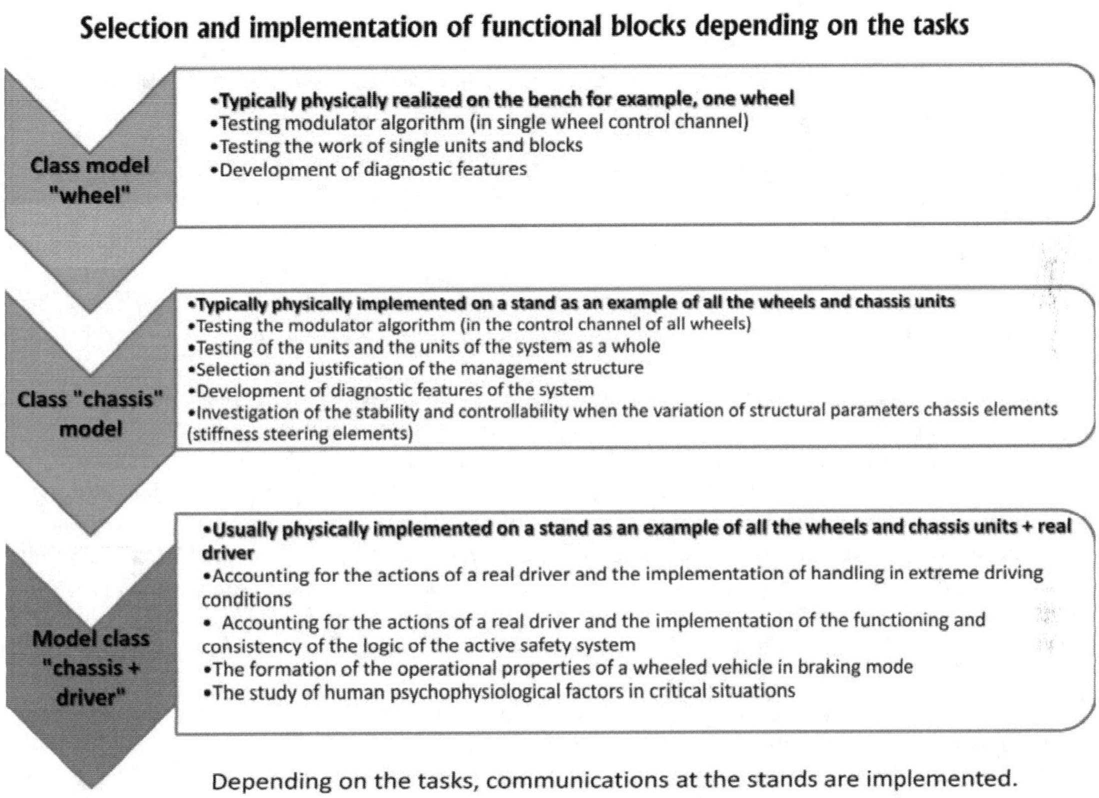

Figure 6. Selection and implementation of functional blocks depending on the tasks.

Verification and confirmation technologies used in the development process provide for the identification of errors in the early stages of the design process. Most errors occur at the stage of formation of the primary specification, though, they appear only during testing. Using models for virtual testing in the early stages of design, specialists can reduce development time by up to 50%.

Verification, confirmation and testing operations can be performed at all stages of the design process based on the model.

Virtual Physical Modeling Technology (VPhM) makes it possible to solve the above problems that the designers and testers of new systems [6, 7, 8, 9, 10, 13] face. With the modern design and test methods already in place, so-called V-cycles can be used in the design of active safety systems.

At the stage of designing and calculating the system, and subsequently a specific module, software packages of mathematical modeling (SIL) make it possible to facilitate the work as much as possible. At the stage of verification, control and testing, the widespread use of the Virtual Physical Modeling Technology (VPhM) is possible. This technology allows combining field systems with a mathematical description of the dynamics of ongoing processes and phenomena. In this case, the control of the physical object, as well as the calculation of the model, is conducted in real time. If, for example, an error was found at the stage of system verification, then it is possible to conduct rapid prototyping and apply the VPhM by adjusting the system specification, which does not cause significant delays in the design and the problem can be solved locally.

Using prototypes for early verification of system development gives users the opportunity to test, optimize, and test the developed algorithms on hardware before starting production. Early verification through Virtual Physical Modeling can dramatically reduce development time, while not being tied to the hardware of any particular manufacturer

References

[1] Keller A, Aliukov S and Anchukov V 2017 Mathematical model of the truck for investigation of stability and control of movement *Lecture Notes in Engineering and Computer Science* pp 711-716

[2] Dubrovskiy A, Aliukov S, Keller A, Dubrovskiy S and Alyukov A 2016 Adaptive Suspension of Vehicles with Wide Range of Control *SAE Technical Papers* 2016-01-8032

[3] Keller A and Aliukov S 2015 Effectiveness of Methods of Power Distribution in Transmissions of All-Wheel-Drive Trucks *SAE Technical Paper* 2015-01-2732

[4] Dygalo V and Revin A 2012 *Technologies for testing active safety systems of vehicles* p 387

[5] Dygalo V and Revin A 2006 *Virtual physical technology of laboratory testing of active safety systems of vehicles* p 316

[6] Dygalo V, Lyaschenko M, Boyko G, and Dygalo L 2018 Application of the Virtual Physical Modeling Technology for the development of elements of the autonomous (unmanned) vehicles' systems *IOP Conference Series: Materials Science and Engineering* **315** p 11

[7] Dygalo V and Revin A 2014 The principles of synthesis of virtual physical models designed to develop a brake system of a car with ABS *Automotive Industry* **8** pp 17-19

[8] Tumasov A, Groshev A, Kostin S, Saunin M, Trusov Yu and Dygalo V 2011 Investigation of the active safety properties of vehicles by simulation method *Journal of automotive engineers* **2** pp 34-37

[9] Revin A and Dygalo V 2002 *Integrated modeling in the design cycle of cars and their systems Automotive industry* **11** pp 29-30

[10] Dygalo V, Kotov V and Revin A 2012 Adequacy assessment when modeling the braking dynamics of a car with ABS *Automotive Industry* **12** pp 16-18

[11] Dygalo V, Revin A, Sorniotti A and Vellardocchia M 2006 Stand for complex laboratory tests of EHB *Automotive industry* **3** pp 34-36

[12] Dygalo V and Revin A 2012 Virtual-physical modeling technology in the V-cycle when designing active safety systems *Izv. Volgograd State Technical University Series "Ground transportation systems"* **2** pp 35-38

[13] Dygalo V 2010 Development of an on-off valve control algorithm for an electro-hydraulic brake system using virtual physical simulation technology *Izv. Volgograd State Technical University Series "Ground transportation systems* **10** pp 37-40

[14] Dygalo V 2012 Development of a device for simulating signals from wheel sensors of an active safety system *Izv. Volgograd State Technical University. Series "Ground transportation systems"* **2** pp 32-35

[15] Keller A and Aliukov S 2015 Efficient Power Distribution in an All-wheel Drive Truck *Lecture Notes in Engineering and Computer Science: Proceedings of The World Congress on Engineering* pp 1201-1206

IOP Conf. Series: Materials Science and Engineering **819** (2020) 012040 doi:10.1088/1757-899X/819/1/012040

[16] Keller A and Aliukov S 2016 Efficient Power Distribution in an All-Wheel Ground Vehicles *SAE Technical Paper* 2016-01-1105

[17] Aliukov S, Keller A, and Alyukov A 2015 On the Question of Mathematical Model of an Overrunning Clutch *Lecture Notes in Engineering and Computer Science: Proceedings of The World Congress on Engineering* pp 1319-1324

[18] Keller A, Aliukov S and Anchukov V 2017 Evaluation of Stability and Control of Movement of 6x6 Truck for Different Operating Modes *SAE Technical Paper* 2017-01-1575

IOP Conf. Series: Materials Science and Engineering **819** (2020) 012041 doi:10.1088/1757-899X/819/1/012041

Vibration analysis of the F-segment car exhaust system

V S Popov[1], S S Gavryushin[2], G G Nadareishvili[1] and S I Yudin [1]

[1] FSUE "NAMI", 2, Avtomotornaya Str., Moscow 125438, Russian Federation,
[2] Bauman Moscow state technical University, Moscow

E-mail: vladislav.popov@nami.ru

Abstract. The paper considers the problem of improving the comfort and environmental friendliness of a modern f-segment car with an internal combustion engine by reducing the noise and vibration levels of the exhaust system. Gas-dynamic and mechanical excitations from the car engine and road surface cause vibrations in the exhaust system, which are transmitted through the body and the bracket structure to the body causing discomfort of the driver and passengers. To reduce the negative impact, when designing the exhaust system, it is necessary to ensure such a location of the exhaust system suspension elements that will meet the maximum operation requirements for the vibration comfort. An important task is to build a numerical experiment for the entire assembly structure to determine the nature of the similar systems loading. The solution of the above tasks is aimed at the insurance of specific strength and durability of both the entire system as a whole and the most important support elastic elements. This calculated study is of practical importance not only for reducing the vibration load of car exhaust systems and the car body, but also for improving the vibro-acoustic comfort of passengers.

1. Introduction

In cities, the problem of increased noise and vibration of vehicles is becoming more and more actual. Vibrations are the cause of discomfort for residents, drivers and passengers of vehicles. The most dangerous is considered to be the impact on a person of general and combined vibration in the low-frequency range of 4-30 Hz.

All modern vehicles must meet the requirements of increased comfort and environmental friendliness. It is achieved mostly by increasing the protection against noise and vibrations of the power plant, namely, by improving the exhaust systems of any purpose cars. The solution of the problem lies in the study of natural oscillation frequencies of this system and differentiation of their values from the resonant frequencies values of the car structure elements. The main features of these systems and occurring processes in them are the aggressiveness of the exhaust gas environment, high temperature, gas flow rate, etc. [1].

2. Main part

The car's exhaust system consists of pipelines, the main purpose of which is to transfer engine exhaust gases into the atmosphere and prevent them from entering the car cabin. Baffles and chambers are designed for partial treatment of exhaust gases and reduction of noise and vibration which are caused by the gas flow. The exhaust system consists of a muffler, resonator, and connecting pipes. An example of the exhaust system under study is shown in figure 1.

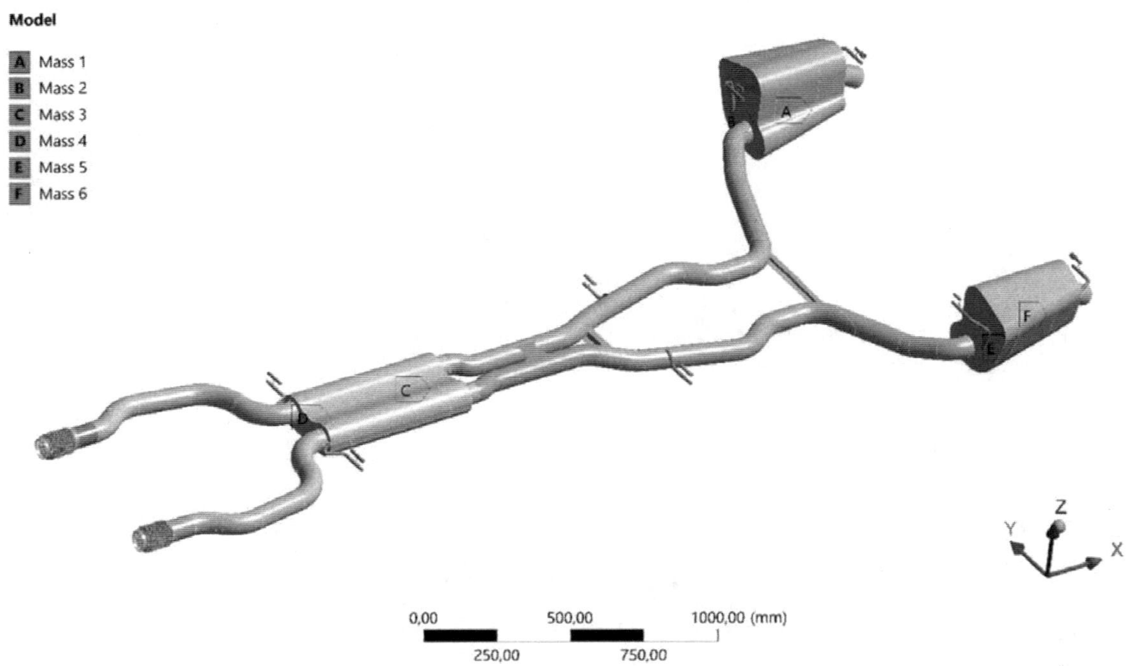

Figure 1. General view of the exhaust system.

Studying the dynamics of the exhaust system design will reduce vibration and noise. The exhaust system should not be rigidly attached to the car body, because this will transmit vibrations to the cab of the vehicle. Therefore, the exhaust system must be installed through elastic elements. One of the parameters that can be effective in reducing the noise and vibration of the exhaust system is the choice of the elastic elements stiffness.

A high-performance exhaust system becomes unusable because of its insufficient durability caused by excessive vibration levels. This excessive level of vibration resulting from the engine exciting forces and road surfaces is transferred to the suspension brackets, reducing their service life. Therefore, the participation of NVH (noise, vibration, harshness) engineers when selecting the elastic elements location becomes necessary to ensure their minimal loads and increase the durability of the elements.

For the calculation analysis, a finite element model divided into 389 thousand shell elements has been created. The elastic suspension elements of the exhaust system were modeled as springs with corresponding stiffness. The exhaust system is made of stainless steel with a Young's modulus E = 2,3 E+11 PA, a Poisson's ratio μ = 0,3 and a density p= 7800 kg / m3.

The calculation was performed with preloading of the exhaust system by gravitational forces and thermal boundary conditions of the 2nd kind. Dissipative basalt packing was modeled by concentrated masses at the corresponding points of the structure [2].The sagging of the exhaust system was 3 mm. Setting the boundary conditions for the calculation is shown in figure 2.

IOP Conf. Series: Materials Science and Engineering **819** (2020) 012041 doi:10.1088/1757-899X/819/1/012041

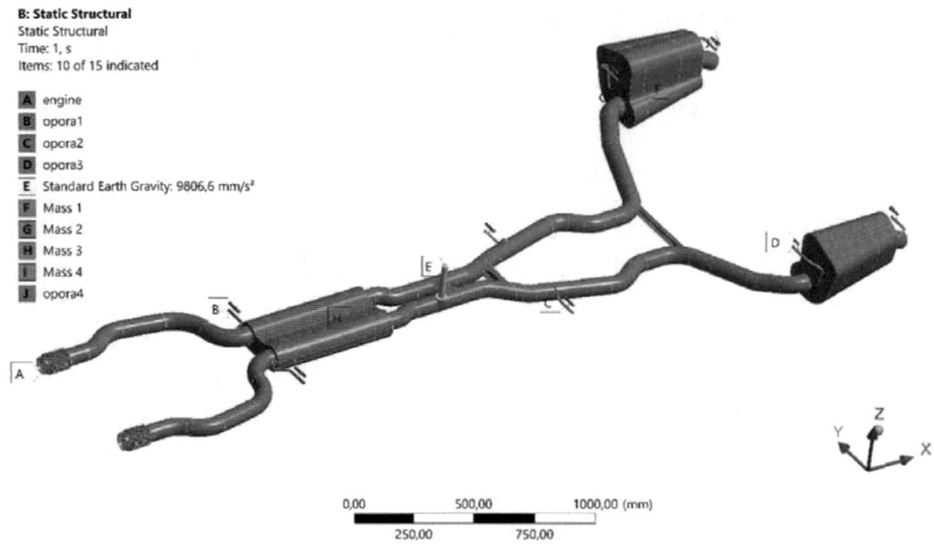

Figure 2. Boundary condition.

Modal analysis is performed using the ANSYS software package to determine the frequencies and mode shapes of the system. The results of calculations of the first ten natural frequencies of the exhaust system are shown in table 1.

Table 1.

Mode	1	2	3	4	5	6	7	8	9	10
Frequency [Hz]	3,8	7,9	8,4	10,1	14,3	15.8	17,8	18,4	26,6	38,8

For example, figure 3 shows the mode shapes of the exhaust system at a frequency of 18.4 Hz.

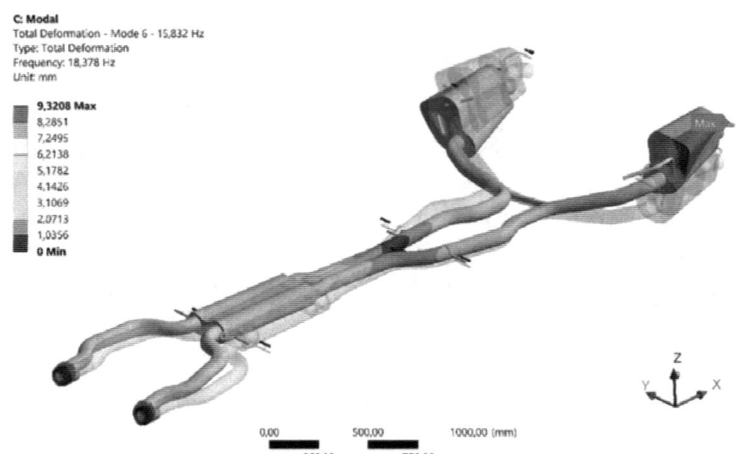

Figure 3. Sixth form of oscillation of the exhaust system.

The quality of the exhaust system suspension points can be estimated by the vibrations amplitude of the exhaust system at resonance frequencies [3]. The harmonic response of the system with an amplitude of 1.8 g was estimated. Figure 4 shows a dependency graph of the vertical axis movements amplitude on the oscillations frequency for the four points of suspension of the exhaust system.

The results of the analysis showed that the amplitudes of oscillations at different points of suspension differ significantly. The oscillation amplitude at the second suspension point is two time greater than at the third and fourth points. When optimizing the design, the output parameters will be a reduction in the vibration amplitudes of the suspension points, followed by the further comparison of the amplitudes.

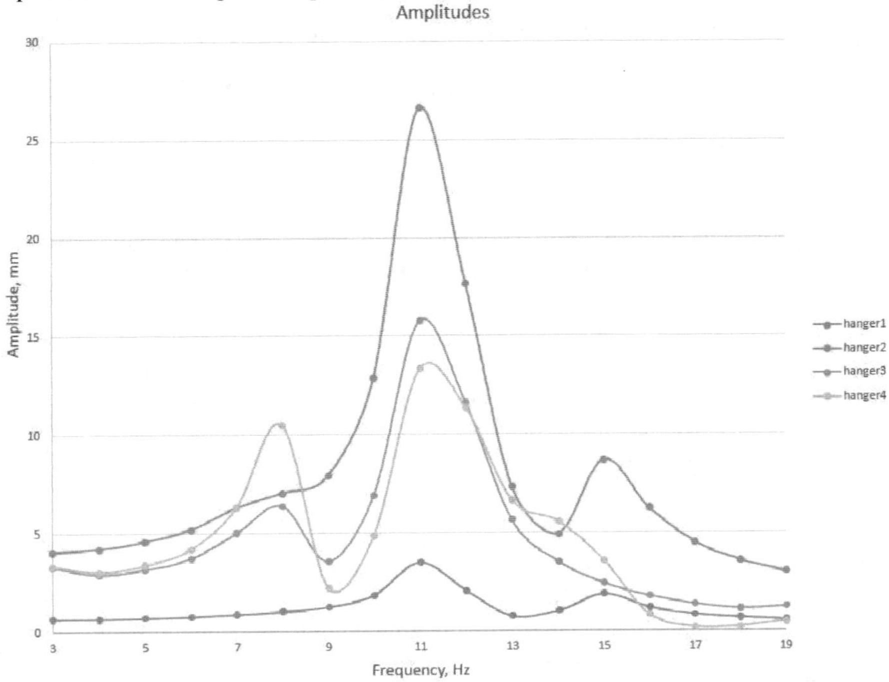

Figure 4. Dependency graph of amplitude of vertical axis movements on the frequency for four points of the exhaust system suspension.

3. Conclusions

Thus, Finite Element Modeling allows to determine location of the vibration nodes at different frequencies. The results obtained for the considered exhaust system showed that the location of the suspension points can be optimized by calculations.

References

[1] Popov V. S., Osipov N. L., Pirozhkov V. A. Calculation and experimental study of exhaust systems of the internal combustion engine // Mechanical Engineering and Innovation XXIX Conference of Young Scientists and Students (MEICYSS-2017) 2018. Moscow: 2017. Pp. 427-429

[2] Shojaeifard M, Ebrahimi-Nejad R.S, Kamarkhani S. Optimization of Exhaust System Hangers for Reduction of Vehicle Cabin Vibrations . IJAE. 2017; 7 (1) :2314-2324 p

[3] Noorazizi M.S., Aminudin B.A. and Zetty M.I. Systematic Fea Study of Vehicle Exhaust System Hanger Location Using Addofd Method. Applied Mechanics and Materials, 2013, 392 p., 161-164 p

IOP Conf. Series: Materials Science and Engineering **819** (2020) 012042 doi:10.1088/1757-899X/819/1/012042

Classification of tablesemantically binary relevant information for drivers in highly automated vehicles

V V Savchenko and V V Litarovich

The State Scientific Institution "The Joint Institute of Mechanical Engineering of the National Academy of Sciences of Belarus", Minsk, Republic of Belarus

E-mail: uus@tut.by

Abstract. The search for ways to solve the problem of transfer control to the driver in highly automated vehicles, involves the development of new methods and adaptation of known under the solution of the problem. The development of the method of monitoring the perception of semantically binary relevant information by the driver implies the identification of sources of significant information and classification of semantically relevant information for drivers coming from ADAS, ITS and the dashboard. Determination the driver's time to response to the relevant information and the subsequent analysis will allow in real time, during the execution of the algorithms of the activity on the management of the highly automated vehicle, without using the additional equipment, to monitor and update the particular driver's database containing quantitative values that characterize a number of its professionally important qualities, in automatic mode, with the use of cloud servers. The obtained results are focused on the solution of the problem control transfer to the driver in highly automated vehicles, when the vehicle's onboard systems cannot support further "unmanned" mode of control and the vehicle is not within a specific operational design domain.

1. Introduction

In accordance with the UNECE Resolution (ECE/TRANC/WP.1/165) on the introduction into practice of highly automated and fully automated vehicles in traffic conditions, "highly automated vehicle" means a vehicle equipped with an automated driving system. This automated driving system operates "within a specific operational design domain" for some or all trips without the need for human intervention as a fallback to road safety; "operational design domain" means the ambient and geographical conditions, time of day, and road, infrastructure, weather, and other conditions for which the automated driving system is specifically designed to operate.

In the work [1], to solve the problem of the control transfer to the driver in highly automated vehicles when the vehicle's onboard systems cannot support "unmanned" mode of control further, being at the level of exploratory research today [2], a method is introduced to monitor the perception of semantically binary relevant information by the driver in highly automated vehicles taking into account the circulating information flows in the vehicle's onboard systems, the driver's response to semantically binary relevant information, which is based on the earlier works [3-6], the dynamics of a particular driver's professionally important qualities in the context of solving the problem of transferring control to the driver in highly automated vehicles, when the vehicle's onboard systems cannot support "unmanned" control mode further, i.e. do not correspond to the operational design domain. The current main problems of functioning safety of highly automated and unmanned vehicles are considered. The structure of the top-level system for transferring control to the driver in highly automated vehicles is presented (figure 1).

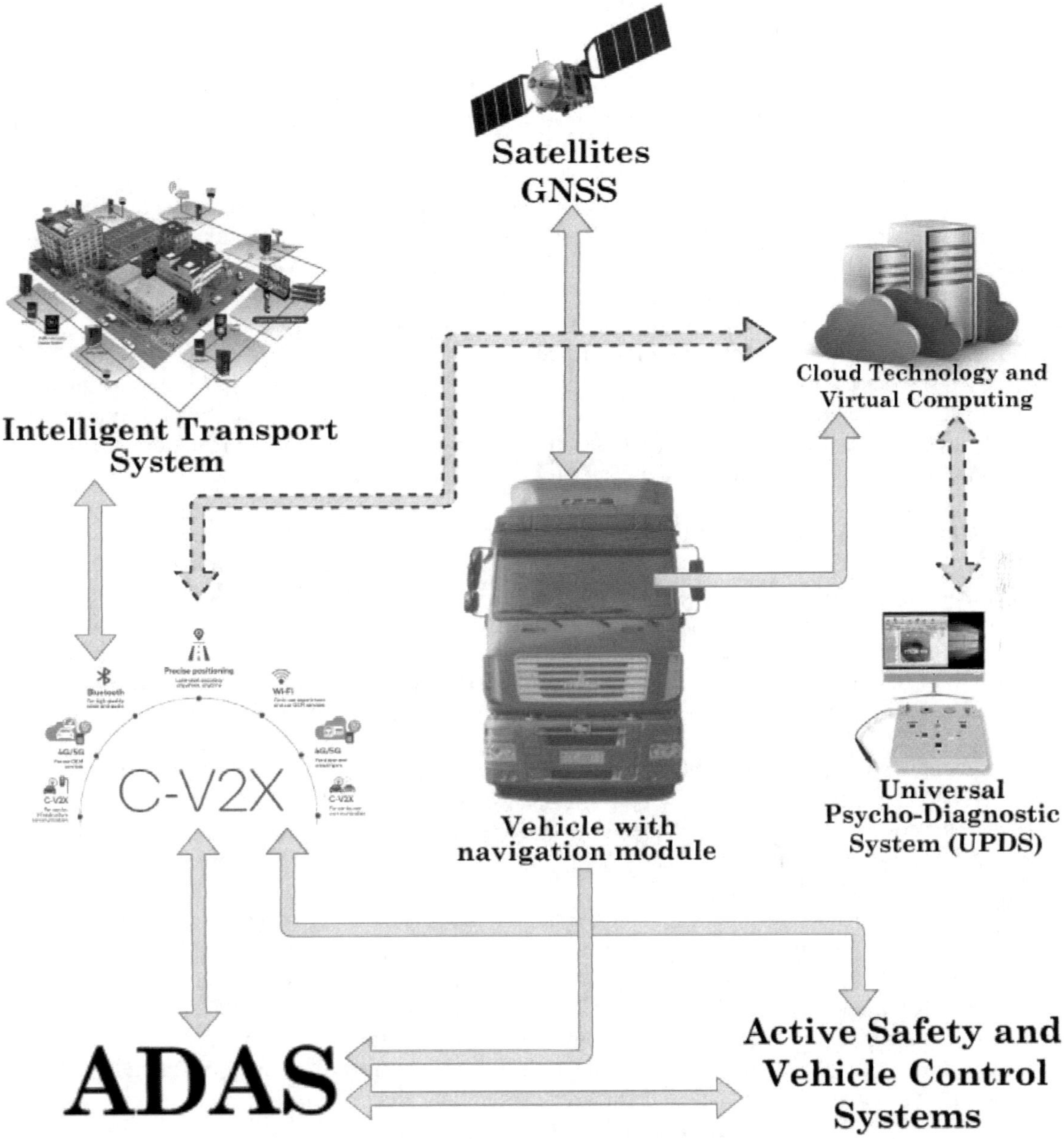

Figure 1. Structure of the top-level system for transferring control to the driver in highly automated vehicles.

For implementation of perception monitoring method of the semantically binary relevant information by the driver in the highly automated vehicles, it is necessary to identify the sources and to classify such information.

2. Semantically binary relevant information sources for drivers

Relevant information falls into three categories [6]:

- Warning information is informative, contains information about the general situation (including emergency) and recommendations for action, preserving the ability of the driver to choose the final decision.
- Prescribing information is commanding and requires or permits execution of strictly defined actions.
- Prohibiting information is of an emergency nature, imposes strict restrictions on the performance or prohibition of certain actions, indicates a component unavailability or malfunction.

In highly automated vehicles semantically binary information for the driver is: warning of ADAS about changing situational circumstances and/or a particular driver's wrong actions in the current situation (crossing the road marking lines without turning on the direction indicator, attempt to change lines when another car is in the dead area of visibility, the sudden emergence of a pedestrian or cyclist, overspeeding, etc.), other relevant information about the vehicle systems operation presented on information boards (panels), in some cases duplicated by an audio signal.

It is also cross–modal relevant information for drivers "circulating" in a communication platform C-V2X for the "connected" vehicles, for example, "transmitted" via V2I protocols (interaction vehicle-to-infrastructure), for example, traffic light sign change, adverse weather conditions warning, etc., according to protocols V2V (interaction vehicle-to-vehicle), for example, an accident along the route of the vehicle, warning of the overtaking danger, "problems" with the road carpet, etc.

3. Semantically binary relevant information classification for drivers

In general, the semantically binary relevant information is an informational message to the driver that can have a different modality, being significant in specific traffic conditions and directing the driver to the immediate implementation of appropriate algorithms activities ("here and now") or posing requirements to perform important specific actions in the short term. In ADAS, the driver can be informed in three ways: visual, audio and tactile. To classify semantically binary relevant information for drivers coming from ADAS, the following systems were analyzed: FCW – Forward Collision Warning; UFCW – Urban Forward Collision Warning; PCW – Pedestrian Collision Warning; LDW – Line Departure Warning; HMW – Headway Monitoring and Warning; SLI – Speed Limit Indicator; TSR – Traffic Sign Recognition; TLR – Traffic Light Recognition; BSM – Blind Spot Monitoring; DM – Driver Monitoring (monitoring the driver's attention); RCTA – Rear Cross Traffic Alert; LKAS – Line Keeping Assist System; LCA – Lane Change Assist; PD – Pedestrian Detection; APA – Advanced Parking Assist; NV – Night Vision.

Visual information provision for the driver of the vehicle is carried out using special icons displayed on the vehicle's digital display or a specially installed system; information reflected on the vehicle's digital display; highlighting of indicators that are located on the side mirrors or on the rear-view mirror; information display on the vehicle's windshield in the driver's field of visibility. Audio information provision of the driver is carried out using embedded audio systems (often with a signal tone and frequency change), tactile information provision of the driver is performed using a vibration mechanism installed in the steering wheel or under the driver's seat of the vehicle.

Table 1 presents the above mentioned systems and the signal modality to inform the driver of the vehicle.

Table 1 shows that all systems provide visual information to the driver of the vehicle. Audio information is not provided in the SLI and TSR, and in TLR, LCA and NV, the availability of informing depends on the manufacturers of the systems, all the other systems provide audio information. Tactile information is not provided in TSR, TLR, DM, RCTA, PD, APA and NV, in DSM and LCA, availability of informing depends on the manufacturer of the system, all the rest provide tactile information.

Semantically binary relevant messages are messages that come from all presented systems, to one degree or another, as well as always messages that have audio and/or tactile information to the driver of the vehicle. Since all systems except NV (there is no data on the system), have a CAN-bus, it is possible to register the information emergence time to analyze it subsequently, for example, to determine the driver's time to respond to the relevant information, which makes it possible in real-time during the execution of algorithms for controlling of highly automated vehicle without the use of additional equipment, to monitor and update the specific driver's database containing quantitative values characterizing a number of his professionally important qualities in automatic mode.

Table 1. Modality of signals for informing the driver of the vehicle.

ADAS	Informing the driver of the vehicle		
	visual	audio	tactile
FCW	+	+	+
UFCW	+	+	+
PCW	+	+	+
LDW	+	+	+
HMW	+	+	+
SLI	+	–	+
TSR	+	–	–
TLR	+	+/–	–
BSM	+	+	+/–
DM	+	+	–
RCTA	+	+	–
LKAS	+	+	+
LCA	+	+/–	+/–
PD	+	+	–
APA	+	+	–
NV	+	+/–	–

4. Systems that are part of ITS and inform the driver of the vehicle

ITS, providing information to the driver: RWIS – Road Weather Information System; ATIS – Advanced Traveler Information Systems; ATMS – Advanced Traffic Management System, collects statistical data on vehicle traffic, aimed at unloading and redistribution of traffic flows based on the collected data analysis; the ATMS consists of several subsystems of narrow focus, which monitor the traffic of certain areas of the road network and provide information to the driver of the vehicle. ATMS subsystems:

- BTM – Bridge/Tunnel Management. The system controls road traffic on bridges and tunnels and includes:
- – HWWS – High Wind Warning System. High Wind Warning Systems monitor wind speed and direction and inform the driver when wind conditions pose a potential hazard. High wind warnings are typically part of traffic management systems on bridges. Typically, wind warnings are divided into different levels according to pre-set thresholds for various parameters, including wind speed, direction, and stability. Depending on the conditions, response plans are executed, ranging from spreading warning messages for approaching traffic to complete closure of the bridge in severe conditions. As the impact of high winds varies depending on the vehicle's size, warning messages often target specific vehicle types, primarily alerting the most exposed vehicles and advising drivers to reduce vehicle speed to minimize potential hazards. Depending on the conditions, the response plan may include restricting access only of small vehicles to the bridge and eventually close the bridge completely. Depending on the subsystems involved in the traffic management system on the bridge, warning messages may be distributed through the media, HAR and/or DMS [7];
- – OOD – Overheight/Overweight Detection of the vehicle. Overheight detection systems include the installation of devices on the track prior to the object emergence with a height restriction. The aim is to detect the excess, and alert the driver so that he can choose a different route. These devices can range from a simple mechanical device, such as chains suspended over the road

surface, to more sophisticated electronic overheight detection systems. The devices are used in conjunction with road signs (both static and dynamic) to inform the driver of a vehicle exceeding the permissible clearance and to redirect him to the appropriate route. With regard to the detection of overweight, all commercial vehicles must be weighed at commercial vehicle inspection stations located at specific points throughout the road network or using specialized equipment Weigh-in-Motion (WIM) [7]. WIM can be used on selected sites located in front of stations to verify and confirm that the weight and dimensions of the commercial vehicle are within acceptable limits. WIM is an effective sorting tool that helps inspectors identify problem vehicles while at the same time allowing the relevant vehicles to pass through the inspection site unhindered.

- SWZ – Smart Work Zones. Applications that inform the driver in construction work areas. Technologies and systems that are used to inform [7]:
  - RI/MS – Roadway Information/Monitoring Systems. The system is intended for monitoring and informing in repair areas. One or more vehicle detectors located in the working areas are used. They are often used together with portable variable message signs (PVMS) to inform drivers about the reduction of speed down the route, creating an autonomous system that automatically generates messages for PVMS;
  - MC – Merge Control. Several ITS products have been developed for merge control, which provide a smooth traffic flow leading to the work area by creating a dynamically impassable zone in front of the construction site, which reduces conflicts, reduces collisions and reduces "road rage". Trailer-mounted portable signs usually consist of flashing lights and a "DO NOT PASS WHEN FLASHING" sign. The system purpose is to merge the car flow at an early stage to prevent "backups" of car flows that may occur when merging at the last minute.

Table 2 presents the above mentioned systems and the modality of signals for informing the driver of the vehicle.

Table 2. Modality of signals for informing the driver of the vehicle.

ITS		Informing the driver of the vehicle	
		visual	audio
RWIS		+	+
ATIS		+	+
ATMS		+	+
BTM	HWWS	+	+
	OOD	+	–
SWZ		+	+
SWZ	RI/MS	+	–
	MC	+	–

Table 2 shows that all systems of this class provide visual information to the driver of the vehicle, audio information provision is carried out in all systems except the OOD, RI/MS and MC subsystems.

Semantically binary relevant information would constitute as information that is critical to the driver at a given time and/or in the given situational circumstances:

- RWIS – bad weather icons that are reflected on the vehicle display, and discrete audio signals.
- ATIS – information is displayed on vehicle information systems (warning of adverse conditions such as sharp turns, wet road surface, etc.).
- ATMS – information is provided by the vehicle's navigation systems and RDMS.
- HWWS – information is provided using DMS.
- OOD – information is provided by dynamic and static road signs and specialized height control devices.
- SWZ – information is provided using PVMS, Queue Warning, and WZIA.

- RI/MS – information is provided using PVMS.
- MC – information is provided by signs that include flashing lights and an explanatory inscription to them.

5. Informing the driver using the potential of the communication platform C-V2X

Informing the driver is provided mainly in two ways – visual and audio, and only in some cases – tactile impact. Informing the driver using C-V2X, main subsystems:

- V2V – vehicle-to-vehicle interaction. It is designed to provide communication between the vehicles, the transmitted information using V2V contains the following data: speed, direction, braking status, vehicle location, etc. The range of V2V messages reaches more than 300 meters, which makes it possible to detect dangerous situations that cannot be observed due to traffic density, terrain or weather conditions. Informing the driver is carried out by special icons on the onboard computer of the vehicle, audio signals, and the tactile information provision to the driver is being developed, presumably on the steering wheel [8].
- V2I – vehicle-to-infrastructure interaction. V2I allows vehicles to exchange information with components that support the highway system. Such components include RFID readers, cameras, traffic lights, lane signs, street lights, signs and parking meters. V2I communication is wireless and bidirectional: data from infrastructure components can be delivered to the vehicle via a special network and vice versa. In ITS, V2I sensors can collect infrastructure data and provide travelers with real-time information on issues such as road conditions, traffic jams, accidents, construction areas and parking availability [9]. Similarly, traffic control systems can use data from infrastructure and vehicles to set variable speed limits and adjust the phase and timing of traffic lights (SPaT – Signal Phase and Timing) to increase fuel economy and speed of traffic flow. Information provision of the driver of the vehicle is carried out using special icons displayed on the vehicle's onboard computer or monitor, audio signals, tactile effects (vibration on the pedal or steering wheel) [10].
- V2P – vehicle-to-pedestrian interaction. V2P makes it possible to exchange location information between a vehicle and a pedestrian. Information provision of the driver of the vehicle is carried out using the display of pedestrian icons and action icons, for example, the system developed by Honda based on V2P can recognize some pedestrian actions (talking on the phone, texting, listening to music with headphones) when a dangerous approach, the inscription "WARNING" is displayed on the vehicle's onboard computer or on the HUD (Head-Up Display), and emergency braking is performed. The icon display is duplicated by short audio signals [11].

Table 3 presents the above mentioned systems and the modality of signals for informing the driver of the vehicle.

Table 3. Modality of signals for informing the driver of the vehicle.

C-V2X	Informing the driver of the vehicle		
	visual	audio	tactile
V2V	+	+	+/–
V2I	+	+	+
V2P	+	+	–

Table 3 shows that all technologies provide visual and audio information to the driver of the vehicle. Provision of tactile information is performed in V2I, and it is under development for V2V. All types of alerts of vehicle driver are semantically binary relevant information.

6. Informing the driver of the vehicle from the dashboard

Informing the driver of the vehicle from the dashboard (indicator panel) can be carried out in two ways – visual and audio. The information is provided to the driver using the indicator panel.

Informing the driver is carried out by the following devices: speedometer, tachometer, odometer, fuel level indicator, engine coolant temperature gage, econometer, indicator lights, alarms reflected on the LCD display.

Table 4 presents the above mentioned devices and the modality of signals for informing the driver of the vehicle.

Table 4. Modality of signals for informing the driver of the vehicle.

Dashboard	Informing the driver of the vehicle	
	visual	audio
Speedometer	+	–
Tachometer	+	–
Odometer	+	–
Fuel level indicator	+	–
Engine coolant temperature gage	+	–
Econometer	+	–
Indicator lights	+	+/–
LCD display	+	–

The information coming from all instrument panel components is semantically binary relevant information to one degree or another. As a rule, indicator lights marked in red are duplicated by an audio signal.

Table 4 shows that visual information provision of the driver is carried out by all presented devices. Audio signal duplication is provided for the indicator lights that carry out the alarm, for the fault of the brake system, the oil pressure in the engine lubrication system.

7. Conclusion

To develop the perception monitoring method of semantically binary relevant information by the driver in highly automated vehicles and the subsequent problem solution of control transfer to the driver in highly automated vehicles, when onboard vehicle systems cannot support further "unmanned" control mode, the classification of semantically binary relevant information for the drivers coming from ADAS, ITS and dashboard is presented. In general, semantically binary relevant information is information message to the driver, it can be of different modalities, significant in specific traffic conditions and prescribing the driver to immediately perform the relevant algorithms of activity ("here and now") or imposing requirements for the implementation of important specific actions in the short term. Determining the driver's time to respond to the relevant information and its subsequent analysis allows in real time, during the execution of algorithms for the management of highly automated vehicle, without using the additional equipment, to monitor and update the particular driver's database containing quantitative values characterizing a number of his professionally important qualities in automatic mode, for which cloud services are used.

References

[1] Savchenko V V 2019 Problem of transferring control to the driver in highly automated vehicles: method of monitoring the perception of semantically binary relevant information by the driver *Mechanics of machines, mechanisms and materials* **2(47)** pp 14–9

[2] Effects of adaptive cruise control and highly automated driving on workload and situation awareness: a review of the empirical evidence Information on https://www.vdi-wissensforum.de/news/effects-of-adaptive-cruise-control-and-highly-automated-driving/

[3] Savchenko V V 1991 Methods and means of improving the efficiency functioning of ergatic system operators (abstract of the thesis SPb) p 16

[4] Savchenko V V 1992 *Methods and means of improving the efficiency of ergatic system operators (control of the level of waking, control of the perception of relevant information by the*

operator, algorithms of increasing the efficiency of operators of precision production (Minsk: Institute of Technical Cybernetics of the Academy of Sciences of Belarus)
[5] Man'shin G G, Savchenko V V and Shunevich N N 1994 *Operator functional state control device* (Patent RU 2020871)
[6] Savchenko V V 2005 Methods and means of improving the efficiency functioning of operators of human-machine transport systems *Vesti National Academy of Sciences of Belarus* **2** pp 29–37
[7] *Advanced Traffic Management Systems. Ontario Traffic Manual* 2007 **B 19** p 148
[8] Vehicle-to-Vehicle Communication Information on https://www.nhtsa.gov/technology-innovation/vehicle-vehicle-communication
[9] Vehicle-to-Infrastructure (V2I or v2i) Information on https://whatis.techtarget.com/definition/vehicle-to-infrastructure-V2I-or-V2X
[10] *Definition of necessary vehicle and infrastructure systems for Automated Driving* 2011 (European Commission) p 111
[11] Honda V2P Overview Information on https://www.qualcomm.com/videos/honda-v2p-overview

AUTHOR INDEX